Bovine Viral Diarrhea Virus and Related Pestiviruses

Bovine Viral Diarrhea Virus and Related Pestiviruses

Editor

Helle Bielefeldt-Ohmann

MDPI • Basel • Beijing • Wuhan • Barcelona • Belgrade • Manchester • Tokyo • Cluj • Tianjin

Editor
Helle Bielefeldt-Ohmann
The University of Queensland
Australia

Editorial Office
MDPI
St. Alban-Anlage 66
4052 Basel, Switzerland

This is a reprint of articles from the Special Issue published online in the open access journal *Viruses* (ISSN 1999-4915) (available at: https://www.mdpi.com/journal/viruses/special_issues/BVDV_Pestiviruses).

For citation purposes, cite each article independently as indicated on the article page online and as indicated below:

LastName, A.A.; LastName, B.B.; LastName, C.C. Article Title. *Journal Name* **Year**, *Article Number*, Page Range.

ISBN 978-3-03943-577-7 (Hbk)
ISBN 978-3-03943-578-4 (PDF)

Contents

About the Editor

Helle Bielefeldt-Ohmann, DVM, Ph.D, Senior Research Fellow. Veterinary pathologist and virologist with special interests in infectious diseases pathobiology, viral persistence, flaviviruses, transplacental virus infections, co-infections, viral zoonoses, arbo-viruses, and influenza viruses.

viruses

MDPI

Editorial

Special Issue: Bovine Viral Diarrhea Virus and Related Pestiviruses

Helle Bielefeldt-Ohmann [1,2,3]

1 Australian Infectious Diseases Research Centre, The University of Queensland, St. Lucia, QLD 4072, Australia; h.bielefeldtohmann1@uq.edu.au
2 School of Veterinary Science, The University of Queensland, Gatton Campus, QLD 4343, Australia
3 School of Chemistry and Molecular Biosciences, The University of Queensland, St. Lucia, QLD 4072, Australia

Received: 13 October 2020; Accepted: 16 October 2020; Published: 19 October 2020

The genus *Pestivirus*, encompassing small positive-strand RNA viruses in the family *Flaviviridae*, comprises four viruses of very significant economic impact to the cattle, swine and sheep industries worldwide: bovine viral diarrhoea virus (BVDV) type 1 and type 2, classical swine fever virus (CSFV) and border disease virus (BDV). Both BVDV- and CSFV-related disease syndromes have been recognised for over 70 years and major progress has been made in elucidating the pathogenesis of these important infections of ruminants and pigs. While much research effort rightfully has gone into epidemiology, diagnostics and prevention—and continues to do so—BVDV and CSFV have also served as excellent models for understanding mechanisms in RNA-virus biology [1,2], transplacental virus infection [3], and host responses to persistent virus infection [4,5]. More recently, a number of novel viruses have been detected in wild and domestic animals by isolation and/or virome studies, and which appear to be related to the pestiviruses and qualify as new pestivirus species. Much still remains to be learned about these latter viruses, including host spectrum, virus–host interactions, epidemiology and clinical spectrum.

This Special Issue of *Viruses* encompasses a range of reports on various aspects of pestiviruses, including an overview of the bovine pestiviruses with an update on the current—albeit still not widely used—nomenclature of the pestiviruses [6]. The epidemiology of pestiviruses remains a subject of considerable interest, as it impinges on biosecurity, susceptibility to secondary infections, management strategies, including vaccination versus eradication, and the importance of potential wildlife reservoirs [7–11].

A first and fundamental step in a viral infection, is attachment to a cellular membrane molecule, a viral receptor. In this issue, two articles focus on the purported receptor for BVDV, bovine CD46 [12,13], describing new tools for the study of the interaction between BVDV and CD46 and suggesting, perhaps not surprisingly considering what we know about receptors for other viruses [14], that CD46 may be but one of two or more molecules on the target cell surface membrane necessary for viral uptake [12].

The interaction of pestiviruses with the host immune system has been a focus of research for decades, notably for BVDV. In this issue the effect of congenital (transplacental) BVDV infection and persistence on the development of innate and adaptive immune functions is explored in an experimental infection model [15]. Monocytes and macrophages are known target cells in both acute-transient [16] and persistent [17,18] BVDV infections. Given the central role of these cells in both the innate and adaptive immune responses, one might expect profound and potentially adverse effects, and this is explored in one article in this issue [19].

One of the characteristics of the pestiviruses, at least the ruminant pestiviruses, is the lack of strict species specificity [6,20]. This can present challenges with regard to serology-based diagnostics, biosecurity and eradication programmes [6,20]. It may also disguise the appearance and diagnosis of new and emerging pestivirus infections, and it is therefore paramount that we gain a better understanding of these new entities and refine diagnostic capabilities—both at the clinical and the virological level. In this issue, several articles focus on three of these new pestiviruses and the diseases they may cause in pigs and ruminants: LINDA virus [21], atypical porcine pestivirus (APPV or Pestivirus K), which by now has been detected in many parts of the world [22–25], and Bungowannah virus (Pestivirus F) [26–28].

It is the hope that this issue, which has brought together contributions from multiple disciplines—virology, immunology, veterinary clinical medicine, epidemiology and pathology—will stimulate further exploration of this fascinating group of viruses in the future. There remains many questions to be addressed even in those very same areas dealt with in this issue, including the conditions, at both the cellular and organismic level, that are conducive to events leading to biotype-switch, viral persistence, emergence of new pestivirus diseases as well as some of the discrepancies between in vitro and in vivo results in regard to immune responses—to mention just a few. Development of new approaches to the investigation of "old" and new pestiviruses are already under way [12,13,29] and others are likely to become available in the future.

Funding: This research received no external funding.

Conflicts of Interest: The authors declares no conflict of interest.

References

1. Tautz, N.; Tews, B.A.; Meyers, G. The molecular biology of pestiviruses. *Adv. Virus Res.* **2015**, *93*, 47–160. [CrossRef] [PubMed]
2. Becher, P.; Tautz, N. RNA recombination in pestiviruses: Cellular RNA sequences in viral genomes highlight the role of host factors for viral persistence and lethal disease. *RNA Biol.* **2011**, *8*, 216–224. [CrossRef] [PubMed]
3. Hansen, T.R.; Smirnova, N.P.; Webb, B.T.; Bielefeldt-Ohmann, H.; Sacco, R.E.; Van Campen, H. Innate and adaptive immune responses to in utero infection with bovine viral diarrhea virus. *Anim. Health Res. Rev.* **2015**, *16*, 15–26. [CrossRef]
4. Schweizer, M.; Peterhans, E. Pestiviruses. *Annu. Rev. Anim. Biosci.* **2014**, *2*, 141–163. [CrossRef]
5. Georges, H.M.; Knapek, K.J.; Bielefeldt-Ohmann, H.; Van Campen, H.; Hansen, T.R. Attenuated lymphocyte activation leads to the development of immunetolerance in bovine fetuses persistently infected with bovine viral diarrhea virus. *Biol. Reprod.* **2020**, *103*, 560–571. [CrossRef] [PubMed]
6. Riitho, V.; Strong, R.; Larska, M.; Graham, S.; Steinbach, F. Bovine pestivirus heterogeneity and its potential impact on vaccination and diagnosis. *Viruses* **2020**, *12*, 1134. [CrossRef] [PubMed]
7. Basqueira, N.; Ramos, J.; Torres, F.; Okuda, L.; Hurley, D.; Chase, C.; Gomes, A.; Gomes, V. An assessment of secondary clinical disease, milk production and quality, and the impact on reproduction in Holstein heifers and cows from a single large commercial herd persistently infected with bovine viral diarrhea virus type 2. *Viruses* **2020**, *12*, 760. [CrossRef]
8. McGowan, M.; McCosker, K.; Fordyce, G.; Kirkland, P. Epidemiology and management of BVDV in rangeland beef breeding herds in Northern Australia. *Viruses* **2020**, *12*, 1063. [CrossRef]
9. das Neves, C.; Johansson Wensman, J.; Nymo, I.; Skjerve, E.; Alenius, S.; Tryland, M. Pestivirus infections in semi-domesticated Eurasian tundra reindeer (*Rangifer tarandus tarandus*): A retrospective cross-sectional serological study in Finnmark County, Norway. *Viruses* **2020**, *12*, 29. [CrossRef]
10. Huaman, J.; Pacioni, C.; Forsyth, D.; Pople, A.; Hampton, J.; Carvalho, T.; Helbig, K. Serosurveillance and molecular investigation of wild deer in Australia reveals seroprevalence of *Pestivirus* infection. *Viruses* **2020**, *12*, 752. [CrossRef]
11. Read, A.; Gestier, S.; Parrish, K.; Finlaison, D.; Gu, X.; O'Connor, T.; Kirkland, P. Prolonged detection of bovine viral diarrhoea virus infection in the semen of bulls. *Viruses* **2020**, *12*, 674. [CrossRef]
12. Riedel, C.; Chen, H.; Reichart, U.; Lamp, B.; Laketa, V.; Rümenapf, T. Real time analysis of bovine viral diarrhea virus (BVDV) infection and its dependence on bovine CD46. *Viruses* **2020**, *12*, 116. [CrossRef] [PubMed]

13. Szillat, K.; Koethe, S.; Wernike, K.; Höper, D.; Beer, M. A CRISPR/Cas9 generated bovine CD46-knockout cell line—A tool to elucidate the adaptability of bovine viral diarrhea viruses (BVDV). *Viruses* **2020**, *12*, 859. [CrossRef] [PubMed]
14. Maginnis, M.S. Virus-Receptor Interactions: The key to cellular invasion. *J. Mol. Biol.* **2018**, *430*, 2590–2611. [CrossRef]
15. Knapek, K.; Georges, H.; Van Campen, H.; Bishop, J.; Bielefeldt-Ohmann, H.; Smirnova, N.; Hansen, T. Fetal lymphoid organ immune responses to transient and persistent infection with bovine viral diarrhea virus. *Viruses* **2020**, *12*, 816. [CrossRef]
16. Liebler-Tenorio, E.M.; Ridpath, J.F.; Neill, J.D. Distribution of viral antigen and development of lesions after experimental infection of calves with a BVDV 2 strain of low virulence. *J. Vet. Diagn. Investig.* **2003**, *15*, 221–232. [CrossRef] [PubMed]
17. Bielefeldt-Ohmann, H.; Ronsholt, L.; Bloch, B. Demonstration of bovine viral diarrhoea virus in peripheral blood mononuclear cells of persistently infected, clinically normal cattle. *J. Gen. Virol.* **1987**, *68*, 1971–1982. [CrossRef]
18. Bielefeldt-Ohmann, H. In situ characterization of mononuclear leukocytes in skin and digestive tract of persistently bovine viral diarrhea virus-infected clinically healthy calves and calves with mucosal disease. *Vet. Pathol.* **1988**, *25*, 304–309. [CrossRef]
19. Abdelsalam, K.; Rajput, M.; Elmowalid, G.; Sobraske, J.; Thakur, N.; Abdallah, H.; Ali, A.; Chase, C.C.L. The effect of bovine viral diarrhea virus (BVDV) strains and the corresponding infected-macrophages' supernatant on macrophage inflammatory function and lymphocyte apoptosis. *Viruses* **2020**, *12*, 701. [CrossRef]
20. de Oliveira, L.; Mechler-Dreibi, M.; Almeida, H.; Gatto, I. Bovine viral diarrhea virus: Recent findings about its occurrence in pigs. *Viruses* **2020**, *12*, 600. [CrossRef]
21. Kiesler, A.; Seitz, K.; Schwarz, L.; Buczolich, K.; Petznek, H.; Sassu, E.; Dürlinger, S.; Högler, S.; Klang, A.; Riedel, C.; et al. Clinical and serological evaluation of LINDA virus infections in post-weaning piglets. *Viruses* **2019**, *11*, 975. [CrossRef] [PubMed]
22. Dall Agnol, A.; Alfieri, A.; Alfieri, A. *Pestivirus K* (atypical porcine pestivirus): Update on the virus, viral infection, and the association with congenital tremor in newborn piglets. *Viruses* **2020**, *12*, 903. [CrossRef] [PubMed]
23. Sozzi, E.; Salogni, C.; Lelli, D.; Barbieri, I.; Moreno, A.; Alborali, G.; Lavazza, A. Molecular survey and phylogenetic analysis of atypical porcine *pestivirus* (APPV) identified in swine and wild boar from Northern Italy. *Viruses* **2019**, *11*, 1142. [CrossRef]
24. Choe, S.; Park, G.; Cha, R.; Hyun, B.; Park, B.; An, D. Prevalence and genetic diversity of atypical porcine pestivirus (APPV) detected in South Korean wild boars. *Viruses* **2020**, *12*, 680. [CrossRef] [PubMed]
25. Folgueiras-González, A.; van den Braak, R.; Simmelink, B.; Deijs, M.; van der Hoek, L.; de Groof, A. Atypical porcine pestivirus circulation and molecular evolution within an affected swine herd. *Viruses* **2020**, *12*, 1080. [CrossRef]
26. Finlaison, D.; Kirkland, P. The outcome of porcine foetal infection with Bungowannah virus is dependent on the stage of gestation at which infection occurs. Part 1: Serology and virology. *Viruses* **2020**, *12*, 691. [CrossRef]
27. Finlaison, D.; Kirkland, P. The outcome of porcine foetal infection with Bungowannah virus is dependent on the stage of gestation at which infection occurs. Part 2: Clinical signs and gross pathology. *Viruses* **2020**, *12*, 873. [CrossRef]
28. Read, A.; Finlaison, D.; Kirkland, P. Infection of ruminants, including pregnant cattle, with Bungowannah virus. *Viruses* **2020**, *12*, 690. [CrossRef]
29. Dalmann, A.; Wernike, K.; Snijder, E.; Oreshkova, N.; Reimann, I.; Beer, M. Single-round infectious particle production by DNA-launched infectious clones of Bungowannah virus. *Viruses* **2020**, *12*, 847. [CrossRef]

viruses

MDPI

Review

Bovine Pestivirus Heterogeneity and Its Potential Impact on Vaccination and Diagnosis

Victor Riitho [1,†], Rebecca Strong [1], Magdalena Larska [2], Simon P. Graham [3,4] and Falko Steinbach [1,4,*]

1 Virology Department, Animal and Plant Health Agency, APHA-Weybridge, Woodham Lane, New Haw, Addlestone KT15 3NB, UK; vriitho@gmail.com (V.R.); Rebecca.Strong@apha.gov.uk (R.S.)
2 Department of Virology, National Veterinary Research Institute, Al. Partyzantów 57, 24-100 Puławy, Poland; maglar7@wp.pl
3 The Pirbright Institute, Ash Road, Pirbright GU24 0NF, UK; simon.graham@pirbright.ac.uk
4 School of Veterinary Medicine, University of Surrey, Guilford GU2 7XH, UK
* Correspondence: falko.steinbach@apha.gov.uk
† Current Address: Centre of Genomics and Child Health, The Blizard Institute, Queen Mary University of London, London E1 2AT, UK.

Received: 4 September 2020; Accepted: 3 October 2020; Published: 6 October 2020

Abstract: Bovine Pestiviruses A and B, formerly known as bovine viral diarrhoea viruses (BVDV)-1 and 2, respectively, are important pathogens of cattle worldwide, responsible for significant economic losses. Bovine viral diarrhoea control programmes are in effect in several high-income countries but less so in low- and middle-income countries where bovine pestiviruses are not considered in disease control programmes. However, bovine pestiviruses are genetically and antigenically diverse, which affects the efficiency of the control programmes. The emergence of atypical ruminant pestiviruses (Pestivirus H or BVDV-3) from various parts of the world and the detection of Pestivirus D (border disease virus) in cattle highlights the challenge that pestiviruses continue to pose to control measures including the development of vaccines with improved cross-protective potential and enhanced diagnostics. This review examines the effect of bovine pestivirus diversity and emergence of atypical pestiviruses in disease control by vaccination and diagnosis.

Keywords: bovine pestiviruses; bovine viral diarrhoea; vaccination; control; diagnosis; antigenic cross-reactivity

1. Bovine Pestiviruses

The Pestivirus genus within the family Flaviviridae of single stranded positive sense RNA viruses comprises eleven recognized species, Pestivirus A-K [1]. The previously recognised species included bovine viral diarrhoea virus 1 (BVDV-1, now known as Pestivirus A), BVDV-2 (Pestivirus B), classical swine fever virus (CSFV, Pestivirus C) and border disease virus (BDV, Pestivirus D). In addition, a further 7 other species were designated as Pestivirus E-K: Pestivirus E (pronghorn antelope virus), Pestivirus F (porcine Pestivirus), Pestivirus G (giraffe Pestivirus), Pestivirus H (Hobi-like Pestivirus, atypical ruminant Pestivirus, also known as BVDV-3), Pestivirus I (Aydin-like Pestivirus, sheep Pestivirus), Pestivirus J (rat Pestivirus) and Pestivirus K (atypical porcine Pestivirus). The reclassification of species names is relatively new, and still not widely used, and thus for ease of comparing the literature, both nomenclatures will be used here throughout.

The classification of pestiviruses is based on genetic and antigenic relatedness as well as the host of origin [2]. Genetic likeness of pestiviruses has been shown to be consistent with antigenic relatedness as defined by binding assays with monoclonal antibodies (mAbs) or serum cross-neutralisation relative to the type virus of a particular species [3]. Pestiviruses differ in their host tropism with Pestivirus A,

B and H mainly found in Bovidae or material thereof, hence the original naming as BVDV-1, -2 and -3. Phylogenetic analysis has identified 21 Pestivirus A subtypes (BVDV-1a-u) and 4 Pestivirus B subtypes (BVDV-2a-d) [4]. Pestiviruses H (BVDV-3), formerly referred to as atypical bovine pestiviruses, are a species with similar variability, but no defined subtypes as of yet [5]. Bovine pestiviruses can also infect other domestic livestock species such as sheep, goats and pigs [6]. Conversely, publications have demonstrated the infection of cattle with Pestivirus D (border disease virus) normally associated with the infection of small ruminants, including the ability to establish a persistent infection in bulls [7–9]. There is a need to understand the role of heterologous hosts in the transmission, spill over and emergence of bovine pestiviruses [10].

Pestivirus H (BVDV-3) represents a group of atypical ruminant pestiviruses that were first detected in commercial foetal bovine serum (FBS), originating from South America [11], Southeast Asia [12] or with unknown origin [13]. Other viruses have also been isolated from aborted bovine foetuses [14] and from buffalo in Brazil [15]. It is not clear whether cattle or other bovids are the natural reservoir/host of atypical ruminant pestiviruses. More recently, a Pestivirus H strain has been associated with a severe respiratory disease outbreak and abortions in multiparous cows in Italy [16]. Accordingly, there is evidence suggesting that Pestivirus H is spreading in cattle in South America [14], Southeast Asia [17] and Europe [16]. The extent to which they are present in the cattle population worldwide needs to be further assessed since the genetic and antigenic diversity between bovine pestiviruses poses a significant challenge in BVD diagnosis and vaccination [18].

2. Impact and Control of Bovine Pestiviruses

Bovine pestiviruses are an important group of pathogens that cause significant economic losses to the cattle industry worldwide [19]. BVD is well recognised as an economic factor of cattle production in the western world but less so in the developing world including emerging economies, such as Brazil, where it is not yet considered in disease control programmes. A meta-analysis of bovine pestivirus prevalence in 325 studies across 73 countries showed global prevalence with significantly higher prevalence in countries without BVDV control programmes [20]. The prevalence and impact of Pestivirus H is yet to be fully considered in such studies.

As a result of their economic impact, significant efforts are being made to prevent and control bovine pestiviruses in many developed countries, particularly in Europe. BVD control programmes have been classified as either systematic, involving a monitored, goal-oriented reduction in incidence and prevalence across a regional or national cattle industry, non-systematic, where measures are implemented on a herd basis as a bottom-up approach without wider systematic monitoring [21]. Three key elements for the systematic control of bovine pestiviruses have been described: the identification and elimination of congenitally persistently infected (PI) immunotolerant animals [22]; increased surveillance to monitor the progress of interventions and detect new infections [23] and measures to prevent the infection of pestivirus naïve animals and (re-) introduction of the virus into BVD-free herds. All of this might be achieved by strict biosecurity protocols, including quarantine for incoming animals, but in most cases will require the assistance of vaccines to make the control measures more sustainable, particularly in farms where biosecurity is difficult to maintain [24].

3. Immunity to Bovine Pestiviruses

Antibodies against pestiviruses are acquired from maternal colostrum or following an active immune response due to infection or vaccination and the importance of neutralising antibodies has been well documented [25,26]. Neutralising antibody responses have been described to target envelope glycoproteins E1 and E2, with E2 being immunodominant, playing a major role in the attachment of the virus on to a target/host cell that the neutralising antibodies inhibit [27,28]. Studies of the immune response to Pestivirus A (BVDV-1) have suggested a role for both antibody and T cell responses in protection [29–32]. The further characterisation of antibody and T cell targets that elicit protective immune responses to pestiviruses therefore remains an important prerequisite in the design of next-generation vaccines.

There is good evidence for the role of T cell responses in BVD immunity. Calves vaccinated in the presence of maternal antibodies, whilst unable to mount an effective antibody response, do generate memory T cells sufficient to protect against subsequent viral challenge [33]. Both $CD4^+$ and $CD8^+$ T cell responses have been shown to be evoked by Pestivirus A (BVDV-1) [34], although the antibody depletion of $CD4^+$ T cells, but not $CD8^+$ or γ/δ T cells, has been shown to increase the duration of virus shedding [35]. These $CD4^+$ T cell responses have been shown to be directed principally against E2 and NS3 but also to other proteins such as the N^{pro}, C and E^{rns} proteins [36–38]. An assessment of longitudinal responses to all the different BVDV proteins in the course of natural infection has not been conducted. However, E2 and NS3 have been shown to be the immunodominant proteins by assessment of ex vivo T cell responses to peptide pools representing the whole Pestivirus A (BVDV-1) proteome following experimental infection [39].

4. Cross-Protection between Bovine Pestiviruses

Given the range of Pestivirus species that may infect cattle and cause BVD, it is evident that broadly cross-protective vaccines are required for the control of pestiviruses in cattle, buffalo and other bovid species. While good cross-protection has been observed within the wide range of Pestivirus A (BVDV-1) genotypes, the need to adapt vaccines to include atypical bovine pestiviruses (Pestiviruses H/BVDV-3) has been considered in the development of vaccines [40,41].

Clinical cross-protection against Pestivirus B (BVDV-2) challenge has been reported using Pestivirus A based MLV vaccines [42]. However, the inability to fully prevent foetal infection, postnatal infection and virus shedding has, in part at least, been attributed to antigenic diversity [18,43]. Accordingly, Pestivirus B has been included in some newer vaccine preparations [44,45]. The comparison of T cell and antibody responses observed after infection with a Pestivirus A and/or a Pestivirus H (BVDV-3) virus showed limited immune reactivity to Pestivirus H in Pestivirus A infected animals but good reactivity to Pestivirus A in Pestivirus H inoculated animals [46]. Highly conserved targets such as non-structural protein NS3 generated recall T cell responses in both groups, while viral glycoprotein E2 responses were virus species specific.

Considering the proteomes of bovine pestiviruses these results are not surprising, but they allow for the rationale design of future vaccines that should convey a broad cross-protection. It has been well described that Pestivirus A can be separated into genotypes a-u [4,47], but all of these belong to the same serotype with antibodies widely cross-reacting. Indeed, the amino acid (aa) identity of Pestivirus A genotypes is >85% for the polyprotein, compared to the identity between Pestiviruses A, B and H of 71–76% (data not shown).

The differences within and between pestiviruses are the highest in the E2 glycoprotein that mediates the viral attachment and is under selective pressure from neutralising antibodies. Here, the identity among Pestivirus A E2 proteins is 68–78% and between Pestivirus A to B is 61–66%, while between Pestivirus A and H it is only 57–63%. Accordingly, the aforementioned lack of T-cell cross-reaction between E2 of different pestiviruses and a limited ability to cross-neutralise is not surprising. More so, when we focus on the E2 aa 1-271 that are known to contain the host-cell interaction mediating domains DA, DB, and DC [48], the aa identity among Pestivirus A is further reduced to 66–76% and between the three pestivirus species A, B and H to 50–62% (Table 1). These figures give an approximation of the challenge for vaccine design, without taking discontinuous, conformational epitopes into account while conversely some of the changes might not affect the neutralising epitopes. More research is needed to precisely identify the molecular events and binding partners involved in virus attachment and fusion, as well as defining neutralising epitopes on the pestivirus envelope E2 protein.

Table 1. E2 amino acid homology of selected Pestivirus A, B and H strains. Full genome sequences of selected strains were downloaded from GenBank; the respective accession numbers are provided. Additional information as per strain name and/or country of origin are provided in the left column. Sequences were translated into proteins where necessary and aligned with Clustal W and the sequences trimmed to contain the aa 1-271 of the E2 protein using the MacVector software package. The matrix depicts identities of amino acids (aa) above and homologies below the diagonal. The identities between the same virus species: Pestivirus A (BVDV-1), B (BVDV-2) or H (BVDV-3) are highlighted in green. The identities of Pestivirus B or H compared to Pestivirus A are highlighted in yellow.

	MH379638.1	M96687	KX987157	KR866116	LC089876	JN400273	JQ799141	AF502399	KJ000672	MH231148	KY683847	FJ040215	JQ612704
MH379638.1 BVDV-1a GB Ho916	100	75.6	73.3	73.7	71.9	69.6	66.3	58.5	56.7	60.4	54.1	52.6	53
M96687 BVDV-1b OSLOSS	84.5	100	69	72.3	72.3	69.7	64.9	57.2	56.1	59.4	52.8	52.4	52.4
KX987157 BVDV-1f SLO	82.6	79.3	100	72.2	75.2	73.7	71.1	59.6	58.1	61.5	56.7	54.4	56.7
KR866116 BVDV-1m CN	84.1	81.2	85.2	100	70.7	74.4	66.7	55.9	57.4	59.6	57.4	56.3	58.1
LC089876 BVDV-1n JP	79.6	81.2	83	81.5	100	67.4	68.5	58.5	57	58.5	55.9	55.2	54.4
JN400273 BVDV-1q SD0803 pig	82.2	81.9	84.1	83.7	78.9	100	65.9	58.1	59.3	60.7	55.2	53.3	55.9
JQ799141 BVDV-1u Yak CN	78.5	77.1	81.1	80.4	79.6	81.9	100	58.1	60.4	63	52.6	52.2	52.6
AF502399 BVDV-2a NY-93	71.9	70.5	73.3	72.2	68.1	74.4	75.2	100	80.6	80.6	52.2	54.1	51.1
KJ000672 BVDV-2b SD1301	70.4	67.5	71.1	71.5	66.3	73.7	74.1	88.4	100	81	50.7	53	50.4
MH231148 BVDV-2c 12-149150	72.2	70.5	73.3	73.3	68.1	73.7	77	90.7	87.7	100	53.7	54.4	53
KY683847 BVDV-3 BRA Bos Ind	67	67.2	71.9	71.9	71.1	70.4	67.8	65.9	65.2	66.3	100	90	94.1
FJ040215 BVDV-3 Thai KK	68.5	67.9	71.9	70.7	73	70	68.5	67.8	67	67.4	94.8	100	86.2
JQ612704 BVDV-3 IT	66.3	66.4	71.1	72.2	69.6	71.1	68.1	65.6	65.9	65.6	97	92.6	100

Conversely, NS3, the other main immunogenic protein identified, is highly conserved among Pestivirus A (>94%) and between Pestiviruses A, B and H (>89%, Table 2). Accordingly, broad cross-reactivity for the T-cells directed against NS3 can be assumed. This cross-reactivity would support both cytotoxic $CD8^+$ T cell and $CD4^+$ helper T cell responses and thus indirectly support B cell responses. Accordingly, it is tempting to speculate that the observed clinical cross-protection at least observed by Pestivirus A vaccines, or after infection of cattle with other pestiviruses is indeed T cell driven.

Table 2. NS3 amino acid homology of selected Pestivirus A, B and H strains. Approach and methods used are similar to Table 1. For this analysis, the full length NS3 protein sequences were used.

	MH379638.1	M96687	KX987157	KR866116	LC089876	JN400273	JQ799141	AF502399	KJ000672	MH231148	KY683847	FJ040215	JQ612704
MH379638.1 BVDV-1a GB Ho916	100	97.5	97.4	97.4	97.4	97.4	94	91.9	91.1	92.1	92.4	91.8	92.2
M96687 BVDV-1b OSLOSS	99.4	100	97.4	97.8	97.8	97.8	94.3	92.1	90.6	91.9	92.5	91.9	92.4
KX987157 BVDV-1f SLO	99.4	99.4	100	98.1	97.4	98	94.4	92.1	90.8	91.9	92.5	91.9	92.4
KR866116 BVDV-1m CN	99.4	99.4	99.4	100	97.8	98.2	94	92.7	91.4	92.5	92.4	91.5	92.2
LC089876 BVDV-1n JP	99.4	99.4	99.4	99.4	100	97.7	94	91.8	90.3	91.7	92.4	91.9	91.9
JN400273 BVDV-1q SD0803 pig	99.1	99.4	99	99.1	99.1	100	94.6	91.7	90.5	91.5	92.4	91.7	92.2
JQ799141 BVDV-1u Yak CN	98.2	98.2	98.1	98	98	98	100	90.6	89.5	89.9	91.5	91.4	91.5
AF502399 BVDV-2a NY-93	98.2	98	98.4	98.2	98	97.7	97.1	100	97.5	98.1	91.7	90.9	91.5
KJ000672 BVDV-2b SD1301	97.5	97.5	98	97.8	97.5	97.2	96.3	99	100	97.7	90.3	89.9	90.8
MH231148 BVDV-2c 12-149150	98.1	97.8	98.2	98.1	97.8	97.5	96.9	99.6	99.1	100	91.1	90.3	91.2
KY683847 BVDV-3 BRA Bos Ind	97.2	97.2	96.9	97.4	97.2	97.2	96.6	96.6	95.9	96.5	100	98.5	99.6
FJ040215 BVDV-3 Thai KK	97.2	97.2	96.9	97.2	97.2	96.9	96.6	96.3	95.9	96.2	99.3	100	98.4
JQ612704 BVDV-3 IT	97.2	97.2	96.9	97.4	96.9	97.2	96.6	96.6	96.2	96.5	99.7	99.3	100

It seems feasible to consider a vaccine design that takes only a limited amount of conserved T cell reactive antigens such as NS3 and a mix of E2 (possibly only domains DA-DC) to induce neutralising antibodies into account. This, however, highlights crucial limitations of both the immune system and some of the current state-of-the-art approaches to design modern vaccines: the immune system cannot make a rational decision; it will induce B (antibody) and T cell responses to all antigens presented in a vaccine. Accordingly, traditional inactivated vaccines would likely distract the immune response into too many unnecessary and unhelpful directions. Conversely, some of the promising modern platforms that can mimic MLV will struggle with this design too. Both replication deficient and recombinant pestiviruses use the exchange of one protein (or parts of it). Such a vaccine cannot be designed carrying multiple E2 variants. Similarly, the delivery of multiple proteins is a challenge for viral vector vaccines—unless variants of these are combined in one product. DNA, RNA or subunit vaccines would have an advantage here but are not (yet) ready in their design and proof of concept for production that they can be used with same efficacy as MLV or surrogates thereof.

5. Vaccination against Bovine Pestiviruses

According to VeVax, an online licensed veterinary vaccines database [49], there are more than 120 registered BVD vaccine products currently in use around the world, mostly in North and South America. These are conventional modified live virus (MLV) or inactivated/killed virus vaccines, formulated as either Pestivirus A and/or B (BVDV-1 and/or -2) preparations or multivalent vaccines including other pathogens implicated in the bovine respiratory disease complex, such as members of the Pasteurellaceae family (including *Mannheimia haemolytica*, *Pasturella multicoda* and *Haemophilus somni*), bovine herpesvirus-1, parainfluenza type 3 virus and bovine respiratory syncytial virus [50].

Vaccination against bovine pestiviruses is an additional control measure aimed at the protection of post-natal calves against infection after maternally-derived antibody wanes and of heifers to prevent foetal infections which may result in reproductive failure, foetal losses and birth of PI calves, which continuously spread the virus [51]. The adoption of BVD vaccination in Europe is voluntary with uptake ranging from 20–75% [52]. BVD eradication without vaccination has been successfully carried out by large-scale eradication schemes in Scandinavian countries where 90–99% of herds are considered free of Pestiviruses A and B (BVDV-1 and -2) [53]. These schemes are, however, expensive and intensive and take a long time to implement. The addition of vaccination remains a cost-effective measure for disease control [54]. Furthermore, in regions with high cattle densities and BVDV prevalence, and therefore an increased probability of virus reintroduction into naïve herds, vaccination can easily be incorporated into systematic control strategies [55].

5.1. *Modified Live Virus Vaccines*

MLV vaccines are more efficacious since they induce high titres of virus neutralising antibodies and provide a longer duration of protection from clinical disease than inactivated vaccines that often require booster immunizations to achieve sustained protection [56]. However, there is a risk that MLVs may revert to a virulent form or recombine with field viruses and cause disease and vaccinated animals have been reported to develop transient viremia and to shed vaccine virus [57,58]. MLV vaccines have also been shown to confer foetal protection after Pestivirus A (BVDV-1) and, in a few instances, Pestivirus B (BVDV-2) challenge following vaccination with Pestivirus A based MLV vaccines [59–61]. In pregnant animals, however, live vaccines pose the risk of the vertical transmission of vaccine virus that can occasionally result in foetal complications or birth of PI calves [62]. MLV vaccination has also been implicated in post-vaccination mucosal disease, when PI animals do not mount an immune response to BVDV, due to their intra-uterine infection, and a fatal condition develops, characterised by severe lesions of the oral and intestinal mucosa [63]. As a result of these safety concerns, MLVs are not licensed in all countries.

5.2. *Inactivated Vaccines*

Whilst inactivated vaccines are generally safer and therefore preferred for the vaccination of breeding cattle, bovine neonatal pancytopenia (BNP) has highlighted that there too problems may arise. BNP was a syndrome associated with an inactivated BVD vaccine that affected calves in their first month of life, characterised by pancytopenia, severe bleeding and high lethality that had originally been described in Europe. The vaccine design was rational: prepare the virus in a bovine cell line to reduce allergic or other reactions [64]. What had not been considered sufficiently was the ability of such a preparation to induce some form of auto-immune reaction as it seems to have occurred. The vaccine in question contained a significant amount of bovine (cell line derived) non-viral antigens [65] and the transfer (ingestion) of colostrum from vaccinated affected dams (i.e., those that had given birth to BNP calves) was sufficient to induce disease in several (albeit not all) calves [66]. The underlying pathogenesis of BNP is far from being fully resolved as blood transfusion to affected calves was not sufficient to overcome this problem. The nature of alloreactive antibodies associated with BNP [67] remains debatable, with bovine MHC class I molecules suggested to be the alloantigens responsible [10,68,69]. It is worth noting that BNP affected only some dams in some herds, thus an unresolved genetic component also seems to have played a role. The disease was a highly unusual event that could have been avoided through improved vaccine preparations. In principle, however, this highlights the potential for a safety problem in the use of crude inactivated BVDV preparations that contain significant amounts of host cell derived material.

5.3. *Differentiation of Infection from Vaccination (DIVA)*

Importantly, neither MLV nor inactivated vaccines allow for DIVA [70], which reduces their suitability for use in BVD eradication efforts when vaccination and control could be monitored by Ab ELISA tests. Inactivated vaccines were previously thought to facilitate DIVA because of a lack of production of non-structural proteins such as NS3 and hence the diminished responses to these target antigens [71]. This has, however, in a number of studies proved not be the case [72,73], due to the presence of non-structural proteins in the crude virus preparations used in inactivated vaccines.

5.4. *Sub-Unit and Next-Generation Vaccines*

Recent BVD vaccine developments aim to address the shortcomings of the existing vaccines. An ideal vaccine should prevent disease, prevent vertical and horizontal virus transmission, be safe in pregnant animals, unable to revert to virulence, have broad efficacy to account for virus diversity and permit DIVA [40]. Various approaches towards the development of the next generation of BVD vaccines have been made and some already evaluated in cattle, but most still face challenges [74–77]. One such approach, DNA vaccination, has been trialled with mixed success and is generally deemed to be effective but comes with challenges such as the optimal route of delivery. The use of recombinant subunit proteins had earlier been deemed insufficient to provide effective protection, not least when focusing on E2 [39,78–81]. The combination of antigens (e.g., E2 and NS3) in formulations with molecular adjuvants to stimulate antigen-presenting cells [58,59] has been more successful, but requires further refinement. The combination of approaches such as a DNA prime, protein boost regimes [82–84] is overall more promising than either of the two alone, but comes with the inherent drawback of requiring at least two vaccinations. Accordingly, these types of vaccines are a challenge to the manufacturer, the veterinarian and the farmer. Vaccinations of animals in remote areas of under-developed countries are a challenge already, but additionally, farmers in developed countries and industrialised settings prefer single, possibly multivalent, or fewer multiple vaccines that require minimal inoculations. Defective viral vectored replicons [85–89], synthetic attenuated infectious cDNA clones [90,91], as well as chimeric pestivirus marker vaccines [92], all aim to build upon the success of MLV vaccines, but face regulatory challenges in some countries as they are genetically engineered microorganisms and those built on recombinant pestiviruses might (with regards to recombination) come with similar limitations as MLVs.

In summary, BVD vaccine development efforts are the subject of ongoing research, following the trends of modern vaccine innovation, and have contributed to pave the way for such, including the use of RNA vaccination [70]. To further improve the design of safe and efficacious vaccines that cross-protect against several related pestiviruses and are sufficiently cheap to manufacture and apply, an improved understanding of the immunology in bovine pestivirus infections is required.

6. Diagnosis of Bovine Pestiviruses

Virus isolation has historically been considered the gold standard for the detection of pestiviruses in blood and milk samples. Antigen detection by immunohistochemistry, antigen-capture ELISA and nucleic acid detection by polymerase chain reaction (PCR) based tests are now broadly applied [93,94]. The detection of bovine pestivirus-specific antibody in milk or blood samples by ELISA or by virus neutralisation test is also possible, although these tests are not able to differentiate infected from vaccinated animals (DIVA), which is a significant limitation for BVD control programmes [70]. Antibody and antigen tests, when used together, help to distinguish acutely infected from persistently infected (PI) animals. The diagnoses of both virus and immune responses can be hampered by the genetic and antigenic variability of pestiviruses. More robust approaches are required to assess antigenic relationships between the various Pestivirus species and subtypes [95]. Most tests were established against Pestiviruses A and B (BVDV-1 and -2) and have not been fully validated to detect Pestivirus H (BVDV-3) or Pestivirus D (BDV). Whilst the comparative performance of different Pestivirus A and B assays to detect antibodies against atypical Pestiviruses has been assessed [96], the existing tests are not sensitive enough and no specific tests are available. This makes the comprehensive detection of all potential bovine pestiviruses a diagnostic challenge that can be addressed by focusing on conserved parts of the genome/proteome [11,17,97].

7. Conclusions

Bovine pestivirus diversity, heterologous hosts and the emergence of novel ruminant pestiviruses all pose a significant challenge to control and vaccination [10]. Vaccine efficacy studies should include challenges with heterotypic bovine pestivirus species and vaccines will have to be designed with at least Pestivirus A, B and H (BVDV-1, -2 and -3) in mind. A better understanding of this diversity and consideration of vaccine correlates of protection, cross-protective potential and efficacy against various bovine Pestivirus species will improve vaccine design and thus support global BVD control to reduce the disease burden and economic impact.

Author Contributions: Conceptualization: V.R. and S.P.G.; writing—original draft preparation: V.R.; writing—review and editing: R.S., M.L., S.P.G. and F.S.; supervision: S.P.G. and F.S. All authors have read and agreed to the published version of the manuscript.

Funding: Work at APHA has previously been supported by project grants from Defra on BVDV. Current support to RS and FS at APHA is through core funding (ED1000). SPG is supported by a UKRI Biotechnology and Biological Sciences Research Council (BBSRC) Institute Strategic Programme Grant to The Pirbright Institute (BBS/E/I/00007031).

Conflicts of Interest: The authors declare no conflict of interest.

References

1. King, A.M.Q.; Lefkowitz, E.J.; Mushegian, A.R.; Adams, M.J.; Dutilh, B.E.; Gorbalenya, A.E.; Harrach, B.; Harrison, R.L.; Junglen, S.; Knowles, N.J.; et al. Changes to taxonomy and the International Code of Virus Classification and Nomenclature ratified by the International Committee on Taxonomy of Viruses (2018). *Arch. Virol.* **2018**, *163*, 2601–2631. [CrossRef]
2. Simmonds, P.; Becher, P.; Bukh, J.; Gould, E.A.; Meyers, G.; Monath, T.; Muerhoff, S.; Pletnev, A.; Rico-Hesse, R.; Smith, D.B.; et al. ICTV Virus Taxonomy Profile: Flaviviridae. *J. Gen. Virol.* **2017**, *98*, 2–3. [CrossRef]
3. Becher, P.; Ramirez, R.A.; Orlich, M.; Cedillo-Rosales, S.; König, M.; Schweizer, M.; Stalder, H.; Schirrmeier, H.; Thiel, H.-J. Genetic and antigenic characterization of novel pestivirus genotypes: Implications for classification. *Virology* **2003**, *311*, 96–104. [CrossRef]
4. Yeşilbağ, K.; Alpay, G.; Becher, P. Variability and Global Distribution of Subgenotypes of Bovine Viral Diarrhea Virus. *Viruses* **2017**, *9*, 128. [CrossRef]
5. Bauermann, F.V.; Ridpath, J.F.; Weiblen, R.; Flores, E.F. HoBi-like viruses: An emerging group of pestiviruses. *J. Veter. Diagn. Investig.* **2013**, *25*, 6–15. [CrossRef]

6. Tao, J.; Liao, J.; Wang, Y.; Zhang, X.; Wang, J.; Zhu, G. Bovine viral diarrhea virus (BVDV) infections in pigs. *Veter. Microbiol.* **2013**, *165*, 185–189. [CrossRef]
7. Braun, U.; Hilbe, M.; Peterhans, E.; Schweizer, M. Border disease in cattle. *Veter. J.* **2019**, *246*, 12–20. [CrossRef]
8. Strong, R.; La Rocca, S.A.; Ibata, G.; Sandvik, T. Antigenic and genetic characterisation of border disease viruses isolated from UK cattle. *Veter. Microbiol.* **2010**, *141*, 208–215. [CrossRef]
9. McFadden, A.M.; Tisdall, D.J.; Hill, F.I.; Otterson, P.; Pulford, D.J.; Peake, J.; Finnegan, C.J.; La Rocca, S.A.; Kok-Mun, T.; Weir, A.M. The first case of a bull persistently infected with Border disease virus in N. Z. *N. Z. Veter. J.* **2012**, *60*, 290–296. [CrossRef]
10. Walz, P.H.; Chamorro, M.F.; Falkenberg, S.M.; Passler, T.; Van Der Meer, F.; Woolums, A.R. Bovine viral diarrhea virus: An updated American College of Veterinary Internal Medicine consensus statement with focus on virus biology, hosts, immunosuppression, and vaccination. *J. Veter. Intern. Med.* **2020**. [CrossRef]
11. Schirrmeier, H.; Strebelow, G.; Depner, K.; Hoffmann, B.; Beer, M. Genetic and antigenic characterization of an atypical pestivirus isolate, a putative member of a novel pestivirus species. *J. Gen. Virol.* **2004**, *85*, 3647–3652. [CrossRef]
12. Liu, L.; Xia, H.; Baule, C.; Belák, S. Maximum likelihood and Bayesian analyses of a combined nucleotide sequence dataset for genetic characterization of a novel pestivirus, SVA/cont-08. *Arch. Virol.* **2009**, *154*, 1111–1116. [CrossRef]
13. Peletto, S.; Zuccon, F.; Pitti, M.; Gobbi, E.; De Marco, L.; Caramelli, M.; Masoero, L.; Acutis, P.L. Detection and phylogenetic analysis of an atypical pestivirus, strain IZSPLV_To. *Res. Veter. Sci.* **2012**, *92*, 147–150. [CrossRef]
14. Cortez, A.; Heinemann, M.B.; De Castro, A.M.M.G.; Soares, R.M.; Pinto, A.M.V.; Alfieri, A.F.; Flores, E.F.; Leite, R.C.; Richtzenhain, L.J. Genetic characterization of Brazilian bovine viral diarrhea virus isolates by partial nucleotide sequencing of the 5'-UTR region. *Pesqui. Veterinária Bras.* **2006**, *26*, 211–216. [CrossRef]
15. Stalder, H.; Meier, P.; Pfaffen, G.; Canal, C.W.; Rufenacht, J.; Schaller, P.; Bachofen, C.; Marti, S.; Vogt, H.; Peterhans, E. Genetic heterogeneity of pestiviruses of ruminants in Switzerland. *Prev. Veter. Med.* **2005**, *72*, 37–41. [CrossRef]
16. DeCaro, N.; Lucente, M.S.; Mari, V.; Cirone, F.; Cordioli, P.; Camero, M.; Sciarretta, R.; Losurdo, M.; Lorusso, E.; Buonavoglia, C. Atypical Pestivirus and Severe Respiratory Disease in Calves, Europe. *Emerg. Infect. Dis.* **2011**, *17*, 1549–1552. [CrossRef]
17. Ståhl, K.; Kampa, J.; Alenius, S.; Wadman, A.P.; Baule, C.; Aiumlamai, S.; Belák, S. Natural infection of cattle with an atypical 'HoBi'-like pestivirus – Implications for BVD control and for the safety of biological products. *Veter. Res.* **2007**, *38*, 517–523. [CrossRef]
18. Bolin, S.R.; Grooms, D.L. Origination and consequences of bovine viral diarrhea virus diversity. *Veter. Clin. N. Am. Food Anim. Pract.* **2004**, *20*, 51–68. [CrossRef]
19. Pinior, B.; Garcia, S.; Minviel, J.J.; Raboisson, D. Epidemiological factors and mitigation measures influencing production losses in cattle due to bovine viral diarrhoea virus infection: A meta-analysis. *Transbound. Emerg. Dis.* **2019**, *66*, 2426–2439. [CrossRef]
20. Scharnböck, B.; Roch, F.-F.; Richter, V.; Funke, C.; Firth, C.L.; Obritzhauser, W.; Baumgartner, W.; Käsbohrer, A.; Pinior, B. A meta-analysis of bovine viral diarrhoea virus (BVDV) prevalences in the global cattle population. *Sci. Rep.* **2018**, *8*, 1–15. [CrossRef]
21. Lindberg, A.; Houe, H. Characteristics in the epidemiology of bovine viral diarrhea virus (BVDV) of relevance to control. *Prev. Veter. Med.* **2005**, *72*, 55–73. [CrossRef] [PubMed]
22. Schweizer, M.; Peterhans, E. Pestiviruses. *Annu. Rev. Anim. Biosci.* **2014**, *2*, 141–163. [CrossRef] [PubMed]
23. Lindberg, A.; Brownlie, J.; Gunn, G.J.; Houe, H.; Moennig, V.; Saatkamp, H.W.; Sandvik, T.; Valle, P. The control of bovine viral diarrhoea virus in Europe: Today and in the future. *Rev. Sci. Tech.* **2006**, *25*, 961–979. [CrossRef] [PubMed]
24. Negron, M.; Raizman, E.A.; Pogranichniy, R.; Hilton, W.M.; Levy, M. Survey on management practices related to the prevention and control of bovine viral diarrhea virus on dairy farms in Indiana, United States. *Prev. Veter. Med.* **2011**, *99*, 130–135. [CrossRef] [PubMed]
25. Bolin, S.R.; Ridpath, J.F. Assessment of protection from systemic infection or disease afforded by low to intermediate titers of passively acquired neutralizing antibody against bovine viral diarrhea virus in calves. *Am. J. Veter. Res.* **1995**, *56*, 755–759.

26. Howard, C.J.; Clarke, M.C.; Brownlie, J. Protection against respiratory infection with bovine virus diarrhoea virus by passively acquired antibody. *Veter. Microbiol.* **1989**, *19*, 195–203. [CrossRef]
27. Bolin, S.R. Immunogens of bovine viral diarrhea virus. *Veter. Microbiol.* **1993**, *37*, 263–271. [CrossRef]
28. Mignon, B.; Waxweiler, S.; Thiry, E.; Boulanger, D.; Dubuisson, J.; Pastoret, P.-P. Epidemiological evaluation of a monoclonal ELISA detecting bovine viral diarrhoea pestivirus antigens in field blood samples of persistently infected cattle. *J. Virol. Methods* **1992**, *40*, 85–93. [CrossRef]
29. Chase, C.C.; Elmowalid, G.; Yousif, A.A. The immune response to bovine viral diarrhea virus: A constantlychanging picture. *Veter. Clin. N. Am. Food Anim. Pract.* **2004**, *20*, 95–114. [CrossRef]
30. Kapil, S.; Walz, P.; Wilkerson, M.; Minocha, H. Immunity and Immunosuppression. In *Bovine Viral Diarrhea Virus*; Blackwell Publishing Ltd.: Hoboken, NJ, USA, 2008; pp. 157–170. [CrossRef]
31. Howard, C.J. Immunological responses to bovine virus diarrhoea virus infections. *Rev. Sci. Tech.* **1990**, *9*, 95–103. [CrossRef]
32. Potgieter, L.N. Immunology of Bovine Viral Diarrhea Virus. *Veter. Clin. N. Am. Food Anim. Pract.* **1995**, *11*, 501–520. [CrossRef]
33. Endsley, J.J.; Roth, J.A.; Ridpath, J.; Neill, J. Maternal antibody blocks humoral but not T cell responses to BVDV. *Boiologicals* **2003**, *31*, 123–125. [CrossRef]
34. Rhodes, S.G.; Cocksedge, J.M.; A Collins, R.; I Morrison, W. Differential cytokine responses of CD4+ and CD8+ T cells in response to bovine viral diarrhoea virus in cattle. *J. Gen. Virol.* **1999**, *80*, 1673–1679. [CrossRef] [PubMed]
35. Howard, C.; Clarke, M.; Sopp, P.; Brownlie, J. Immunity to bovine virus diarrhoea virus in calves: The role of different T-cell subpopulations analysed by specific depletion in vivo with monoclonal antibodies. *Veter. Immunol. Immunopathol.* **1992**, *32*, 303–314. [CrossRef]
36. Lambot, M.; Letesson, J.-J.; Douart, A.; Pastoret, P.P.; Joris, E. Characterization of the immune response of cattle against non-cytopathic and cytopathic biotypes of bovine viral diarrhoea virus. *J. Gen. Virol.* **1997**, *78 Pt 5*, 1041–1047. [CrossRef]
37. Collen, T.; Morrison, W.I. CD4+ T-cell responses to bovine viral diarrhoea virus in cattle. *Virus Res.* **2000**, *67*, 67–80. [CrossRef]
38. Collen, T.; Carr, V.; Parsons, K.; Charleston, B.; Morrison, W.I. Analysis of the repertoire of cattle CD4(+) T cells reactive with bovine viral diarrhoea virus. *Veter. Immunol. Immunopathol.* **2002**, *87*, 235–238. [CrossRef]
39. Riitho, V.; Walters, A.A.; Somavarapu, S.; Lamp, B.; Rümenapf, T.; Krey, T.; Rey, F.A.; Oviedo-Orta, E.; Stewart, G.R.; Locker, N.; et al. Design and evaluation of the immunogenicity and efficacy of a biomimetic particulate formulation of viral antigens. *Sci. Rep.* **2017**, *7*, 13743. [CrossRef]
40. Fulton, R.W. Vaccines. In *Bovine Viral Diarrhea Virus*; Blackwell Publishing Ltd.: Hoboken, NJ, USA, 2008; pp. 209–222. [CrossRef]
41. Fulton, R.W. Impact of species and subgenotypes of bovine viral diarrhea virus on control by vaccination. *Anim. Health Res. Rev.* **2015**, *16*, 40–54. [CrossRef]
42. Kelling, C.L.; Hunsaker, B.D.; Steffen, D.J.; Topliff, C.L.; Eskridge, K.M. Characterization of protection against systemic infection and disease from experimental bovine viral diarrhea virus type 2 infection by use of a modified-live noncytopathic type 1 vaccine in calves. *Am. J. Veter. Res.* **2007**, *68*, 788–796. [CrossRef]
43. Newcomer, B.W.; Chamorro, M.F.; Walz, P.H. Vaccination of cattle against bovine viral diarrhea virus. *Veter. Microbiol.* **2017**, *206*, 78–83. [CrossRef] [PubMed]
44. Ridpath, J.F. Practical significance of heterogeneity among BVDV strains: Impact of biotype and genotype on U.S. control programs. *Prev. Veter. Med.* **2005**, *72*, 17–30. [CrossRef] [PubMed]
45. Fulton, R.W.; Cook, B.J.; Payton, M.E.; Burge, L.J.; Step, D.L. Immune response to bovine viral diarrhea virus (BVDV) vaccines detecting antibodies to BVDV subtypes 1a, 1b, 2a, and 2c. *Vaccine* **2020**, *38*, 4032–4037. [CrossRef] [PubMed]
46. Riitho, V.; Larska, M.; Strong, R.; La Rocca, S.A.; Locker, N.; Alenius, S.; Steinbach, F.; Liu, L.; Uttenthal, A.; Graham, S.P. Comparative analysis of adaptive immune responses following experimental infections of cattle with bovine viral diarrhoea virus-1 and an Asiatic atypical ruminant pestivirus. *Vaccine* **2018**, *36*, 4494–4500. [CrossRef]
47. Mirosław, P.; Polak, M. Increased genetic variation of bovine viral diarrhea virus in dairy cattle in Poland. *BMC Veter. Res.* **2019**, *15*, 1–12. [CrossRef]

48. El Omari, K.; Iourin, O.; Harlos, K.; Grimes, J.M.; Stuart, D.I. Structure of a Pestivirus Envelope Glycoprotein E2 Clarifies Its Role in Cell Entry. *Cell Rep.* **2013**, *3*, 30–35. [CrossRef]
49. He, Y.; Racz, R.; Sayers, S.; Lin, Y.; Todd, T.; Hur, J.; Li, X.; Patel, M.; Zhao, B.; Chung, M.; et al. Updates on the web-based VIOLIN vaccine database and analysis system. *Nucleic Acids Res.* **2014**, *42*, D1124–D1132. [CrossRef]
50. Allen, J.W.; Viel, L.; Bateman, K.G.; Nagy, E.; Røsendal, S.; Shewen, P.E. Serological titers to bovine herpesvirus 1, bovine viral diarrhea virus, parainfluenza 3 virus, bovine respiratory syncytial virus and Pasteurella haemolytica in feedlot calves with respiratory disease: Associations with bacteriological and pulmonary cytological variables. *Can. J. Veter. Res.* **1992**, *56*, 281–288.
51. Fulton, R.W.; Ridpath, J.; Confer, A.W.; Saliki, J.T.; Burge, L.J.; Payton, M.E. Bovine viral diarrhoea virus antigenic diversity: Impact on disease and vaccination programmes. *Boiologicals* **2003**, *31*, 89–95. [CrossRef]
52. Moennig, V.; Brownlie, J. Position paper: Vaccines and vaccination strategies. *EU Themat. Netw. BVDV Control* **2006**.
53. Bitsch, V.; Rønsholt, L. Control of Bovine Viral Diarrhea Virus Infection Without Vaccines. *Veter. Clin. N. Am. Food Anim. Pract.* **1995**, *11*, 627–640. [CrossRef]
54. Van Oirschot, J.T.; Bruschke, C.J.; Van Rijn, P.A. Vaccination of cattle against bovine viral diarrhoea. *Veter. Microbiol.* **1999**, *64*, 169–183. [CrossRef]
55. Moennig, V.; Eicken, K.; Flebbe, U.; Frey, H.-R.; Grummer, B.; Haas, L.; Greiser-Wilke, I.; Liess, B. Implementation of two-step vaccination in the control of bovine viral diarrhoea (BVD). *Prev. Veter. Med.* **2005**, *72*, 109–114. [CrossRef] [PubMed]
56. Theurer, M.E.; Larson, R.L.; White, B.J. Systematic review and meta-analysis of the effectiveness of commercially available vaccines against bovine herpesvirus, bovine viral diarrhea virus, bovine respiratory syncytial virus, and parainfluenza type 3 virus for mitigation of bovine respiratory disease complex in cattle. *J. Am. Veter. Med. Assoc.* **2015**, *246*, 126–142. [CrossRef]
57. Fulton, R.W.; Saliki, J.T.; Burge, L.J.; Payton, M.E. Humoral Immune Response and Assessment of Vaccine Virus Shedding in Calves Receiving Modified Live Virus Vaccines Containing Bovine Herpesvirus-1 and Bovine Viral Diarrhoea Virus 1a. *J. Veter. Med. Ser. B* **2003**, *50*, 31–37. [CrossRef] [PubMed]
58. Grooms, D.L.; Brock, K.V.; Ward, L.A. Detection of cytopathic bovine viral diarrhea virus in the ovaries of cattle following immunization with a modified live bovine viral diarrhea virus vaccine. *J. Veter. Diagn. Investig.* **1998**, *10*, 130–134. [CrossRef]
59. Xue, W.; Mattick, D.; Smith, L.; Maxwell, J. Fetal protection against bovine viral diarrhea virus types 1 and 2 after the use of a modified-live virus vaccine. *Can. J. Veter. Res.* **2009**, *73*, 292–297.
60. Fairbanks, K.K.; Rinehart, C.L.; Ohnesorge, W.C.; Loughin, M.M.; Chase, C.C. Evaluation of fetal protection against experimental infection with type 1 and type 2 bovine viral diarrhea virus after vaccination of the dam with a bivalent modified-live virus vaccine. *J. Am. Veter. Med. Assoc.* **2004**, *225*, 1898–1904. [CrossRef]
61. Kovács, F.; Magyar, T.; Rinehart, C.; Elbers, K.; Schlesinger, K.; Ohnesorge, W.C. The live attenuated bovine viral diarrhea virus components of a multi-valent vaccine confer protection against fetal infection. *Veter. Microbiol.* **2003**, *96*, 117–131. [CrossRef]
62. Ficken, M.D.; Ellsworth, M.A.; Tucker, C.M.; Cortese, V.S. Effects of modified-live bovine viral diarrhea virus vaccines containing either type 1 or types 1 and 2 BVDV on heifers and their offspring after challenge with noncytopathic type 2 BVDV during gestation. *J. Am. Veter. Med. Assoc.* **2006**, *228*, 1559–1564. [CrossRef]
63. Becher, P.; Orlich, M.; Thiel, H.-J. RNA Recombination between Persisting Pestivirus and a Vaccine Strain: Generation of Cytopathogenic Virus and Induction of Lethal Disease. *J. Virol.* **2001**, *75*, 6256–6264. [CrossRef] [PubMed]
64. Benedictus, L.; Bell, C.R. The risks of using allogeneic cell lines for vaccine production: The example of Bovine Neonatal Pancytopenia. *Expert Rev. Vaccines* **2016**, *16*, 65–71. [CrossRef] [PubMed]
65. Euler, K.N.; Hauck, S.M.; Ueffing, M.; Deeg, C.A. Bovine neonatal pancytopenia—Comparative proteomic characterization of two BVD vaccines and the producer cell surface proteome (MDBK). *BMC Veter. Res.* **2013**, *9*, 18. [CrossRef] [PubMed]
66. Friedrich, A.; Büttner, M.; Rademacher, G.; Klee, W.; Weber, B.K.; Müller, M.; Carlin, A.; Assad, A.; Hafner-Marx, A.; Sauter-Louis, C.M. Ingestion of colostrum from specific cows induces Bovine Neonatal Pancytopenia (BNP) in some calves. *BMC Veter. Res.* **2011**, *7*, 10. [CrossRef]

67. Pardon, B.; Stuyven, E.; Stuyvaert, S.; Hostens, M.; Dewulf, J.; Goddeeris, B.M.; Cox, E.; Deprez, P. Sera from dams of calves with bovine neonatal pancytopenia contain alloimmune antibodies directed against calf leukocytes. *Veter. Immunol. Immunopathol.* **2011**, *141*, 293–300. [CrossRef]
68. Bastian, M.; Holsteg, M.; Hanke-Robinson, H.; Duchow, K.; Cussler, K. Bovine Neonatal Pancytopenia: Is this alloimmune syndrome caused by vaccine-induced alloreactive antibodies? *Vaccine* **2011**, *29*, 5267–5275. [CrossRef]
69. Foucras, G.; Corbière, F.; Tasca, C.; Pichereaux, C.; Caubet, C.; Trumel, C.; Lacroux, C.; Franchi, C.; Burlet-Schiltz, O.; Schelcher, F.; et al. Alloantibodies against MHC Class I: A Novel Mechanism of Neonatal Pancytopenia Linked to Vaccination. *J. Immunol.* **2011**, *187*, 6564–6570. [CrossRef]
70. Van Oirschot, J.T. Diva vaccines that reduce virus transmission. *J. Biotechnol.* **1999**, *73*, 195–205. [CrossRef]
71. Makoschey, B.; Sonnemans, D.; Bielsa, J.M.; Franken, P.; Mars, M.; Santos, L.; Alvarez, M. Evaluation of the induction of NS3 specific BVDV antibodies using a commercial inactivated BVDV vaccine in immunization and challenge trials. *Vaccine* **2007**, *25*, 6140–6145. [CrossRef]
72. Álvarez, M.; Donate, J.; Makoschey, B. Antibody responses against non-structural protein 3 of bovine viral diarrhoea virus in milk and serum samples from animals immunised with an inactivated vaccine. *Veter. J.* **2012**, *191*, 371–376. [CrossRef]
73. Raue, R.; Harmeyer, S.S.; Nanjiani, I.A. Antibody responses to inactivated vaccines and natural infection in cattle using bovine viral diarrhoea virus ELISA kits: Assessment of potential to differentiate infected and vaccinated animals. *Veter. J.* **2011**, *187*, 330–334. [CrossRef] [PubMed]
74. Nobiron, I.; Thompson, I.; Brownlie, J.; Collins, M.E. DNA vaccination against bovine viral diarrhoea virus induces humoral and cellular responses in cattle with evidence for protection against viral challenge. *Vaccine* **2003**, *21*, 2082–2092. [CrossRef]
75. Young, N.J.; Thomas, C.J.; Thompson, I.; Collins, M.E.; Brownlie, J. Immune responses to non-structural protein 3 (NS3) of bovine viral diarrhoea virus (BVDV) in NS3 DNA vaccinated and naturally infected cattle. *Prev. Veter. Med.* **2005**, *72*, 115–120. [CrossRef] [PubMed]
76. Liang, R.; van den Hurk, J.V.; Zheng, C.; Yu, H.; Pontarollo, R.A.; Babiuk, L.A.; van Drunen Littel-van den Hurk, S. Immunization with plasmid DNA encoding a truncated, secreted form of the bovine viral diarrhea virus E2 protein elicits strong humoral and cellular immune responses. *Vaccine* **2005**, *23*, 5252–5262. [CrossRef] [PubMed]
77. Harpin, S.; Hurley, D.J.; Mbikay, M.; Talbot, B.; Elazhary, Y. Vaccination of cattle with a DNA plasmid encoding the bovine viral diarrhoea virus major glycoprotein E2. *J. Gen. Virol.* **1999**, *80 Pt 12*, 3137–3144. [CrossRef]
78. Thomas, C.J.; Young, N.J.; Heaney, J.; Collins, M.E.; Brownlie, J. Evaluation of efficacy of mammalian and baculovirus expressed E2 subunit vaccine candidates to bovine viral diarrhoea virus. *Vaccine* **2009**, *27*, 2387–2393. [CrossRef]
79. Bolin, S.R.; Ridpath, J.F. Glycoprotein E2 of bovine viral diarrhea virus expressed in insect cells provides calves limited protection from systemic infection and disease. *Arch. Virol.* **1996**, *141*, 1463–1477. [CrossRef]
80. Zoth, S.C.; Leunda, M.R.; Odeón, A.; Taboga, O. Recombinant E2 glycoprotein of bovine viral diarrhea virus induces a solid humoral neutralizing immune response but fails to confer total protection in cattle. *Braz. J. Med. Biol. Res.* **2007**, *40*, 813–818. [CrossRef]
81. Wang, S.; Yang, G.; Nie, J.; Yang, R.; Du, M.; Su, J.; Wang, J.; Wang, J.; Zhu, Y. Recombinant Erns-E2 protein vaccine formulated with MF59 and CPG-ODN promotes T cell immunity against bovine viral diarrhea virus infection. *Vaccine* **2020**, *38*, 3881–3891. [CrossRef]
82. Liang, R.; van den Hurk, J.V.; Landi, A.; Lawman, Z.; Deregt, D.; Townsend, H.; Babiuk, L.A.; van Drunen Littel-van den Hurk, S. DNA prime–protein boost strategies protect cattle from bovine viral diarrhea virus type 2 challenge. *J. Gen. Virol.* **2008**, *89*, 453–466. [CrossRef]
83. Liang, R.; van den Hurk, J.V.; Babiuk, L.A.; van Drunen Littel-van den Hurk, S. Priming with DNA encoding E2 and boosting with E2 protein formulated with CpG oligodeoxynucleotides induces strong immune responses and protection from Bovine viral diarrhea virus in cattle. *J. Gen. Virol.* **2006**, *87*, 2971–2982. [CrossRef] [PubMed]
84. Cai, D.; Song, Q.; Duan, C.; Wang, S.; Wang, J.; Zhu, Y. Enhanced immune responses to E2 protein and DNA formulated with ISA 61 VG administered as a DNA prime-protein boost regimen against bovine viral diarrhea virus. *Vaccine* **2018**, *36*, 5591–5599. [CrossRef] [PubMed]

85. Elahi, S.M.; Shen, S.H.; Talbot, B.G.; Massie, B.; Harpin, S.; Elazhary, Y. Recombinant adenoviruses expressing the E2 protein of bovine viral diarrhea virus induce humoral and cellular immune responses. *FEMS Microbiol. Lett.* **1999**, *177*, 159–166. [CrossRef] [PubMed]
86. Elahi, S.M.; Bergeron, J.; Nagy, E.; Talbot, B.G.; Harpin, S.; Shen, S.H.; Elazhary, Y. Induction of humoral and cellular immune responses in mice by a recombinant fowlpox virus expressing the E2 protein of bovine viral diarrhea virus. *FEMS Microbiol. Lett.* **1999**, *171*, 107–114. [CrossRef]
87. Toth, R.L.; Nettleton, P.F.; McCrae, M.A. Expression of the E2 envelope glycoprotein of bovine viral diarrhoea virus (BVDV) elicits virus-type specific neutralising antibodies. *Veter. Microbiol.* **1999**, *65*, 87–101. [CrossRef]
88. Baxi, M.K.; Deregt, D.; Robertson, J.; Babiuk, L.A.; Schlapp, T.; Tikoo, S.K. Recombinant Bovine Adenovirus Type 3 Expressing Bovine Viral Diarrhea Virus Glycoprotein E2 Induces an Immune Response in Cotton Rats. *Virology* **2000**, *278*, 234–243. [CrossRef]
89. Elahi, S.M.; Shen, S.-H.; Talbot, B.G.; Massie, B.; Harpin, S.; Elazhary, Y. Induction of Humoral and Cellular Immune Responses against the Nucleocapsid of Bovine Viral Diarrhea Virus by an Adenovirus Vector with an Inducible Promoter. *Virology* **1999**, *261*, 1–7. [CrossRef]
90. Meyer, C.; Von Freyburg, M.; Elbers, K.; Meyers, G. Recovery of Virulent and RNase-Negative Attenuated Type 2 Bovine Viral Diarrhea Viruses from Infectious cDNA Clones. *J. Virol.* **2002**, *76*, 8494–8503. [CrossRef]
91. Meyers, G.; Tautz, N.; Becher, P.; Thiel, H.J.; Kümmerer, B.M. Recovery of cytopathogenic and noncytopathogenic bovine viral diarrhea viruses from cDNA constructs. *J. Virol.* **1996**, *70*, 8606–8613. [CrossRef]
92. Luo, Y.; Yuan, Y.; Ankenbauer, R.G.; Nelson, L.D.; Witte, S.B.; Jackson, J.A.; Welch, S.-K.W. Construction of chimeric bovine viral diarrhea viruses containing glycoprotein Erns of heterologous pestiviruses and evaluation of the chimeras as potential marker vaccines against BVDV. *Vaccine* **2012**, *30*, 3843–3848. [CrossRef]
93. Saliki, J.T.; Dubovi, E.J. Laboratory diagnosis of bovine viral diarrhea virus infections. *Veter. Clin. N. Am. Food Anim. Pract.* **2004**, *20*, 69–83. [CrossRef] [PubMed]
94. Sandvik, T. Selection and use of laboratory diagnostic assays in BVD control programmes. *Prev. Veter. Med.* **2005**, *72*, 3–16. [CrossRef] [PubMed]
95. Mosena, A.C.S.; Falkenberg, S.M.; Ma, H.; Casas, E.; Dassanayake, R.P.; Walz, P.H.; Canal, C.W.; Neill, J.D. Multivariate analysis as a method to evaluate antigenic relationships between BVDV vaccine and field strains. *Vaccine* **2020**, *38*, 5764–5772. [CrossRef] [PubMed]
96. Larska, M.; Polak, M.P.; Liu, L.; Alenius, S.; Uttenthal, A. Comparison of the performance of five different immunoassays to detect specific antibodies against emerging atypical bovine pestivirus. *J. Virol. Methods* **2013**, *187*, 103–109. [CrossRef]
97. Bauermann, F.V.; Flores, E.F.; Ridpath, J.F. Antigenic relationships between Bovine viral diarrhea virus 1 and 2 and HoBi virus: Possible impacts on diagnosis and control. *J. Veter. Diagn. Investig.* **2012**, *24*, 253–261. [CrossRef]

viruses

MDPI

Article

An Assessment of Secondary Clinical Disease, Milk Production and Quality, and the Impact on Reproduction in Holstein Heifers and Cows from a Single Large Commercial Herd Persistently Infected with Bovine Viral Diarrhea Virus Type 2

Natália Sobreira Basqueira [1], Jean Silva Ramos [1], Fabricio Dias Torres [1], Liria Hiromi Okuda [2], David John Hurley [3], Christopher C. L. Chase [4], Anny Raissa Carolini Gomes [5] and Viviani Gomes [1,*]

1 Department of Internal Medicine, College of Veterinary Medicine and Animal Science, University of São Paulo, São Paulo 05508-270, Brazil; na_sobreira@usp.br (N.S.B.); ramos.jan@outlook.com (J.S.R.); torres@axysanalises.com.br (F.D.T.)
2 Biological Institute, 1252 Conselheiro Rodrigues Alves Ave, Vila Mariana, São Paulo 04014-900, Brazil; okuda@biologico.sp.gov.br
3 Food Animal Health and Management Program, College of Veterinary Medicine, University of Georgia, Athens, GA 30602, USA; djhurley@uga.edu
4 Department of Veterinary and Biomedical Sciences, South Dakota State University, Brookings, SD 57007, USA; christopher.Chase@SDSTATE.EDU
5 Department of Veterinary Medicine, Federal University of Paraná, Curitiba 80040-380, Brazil; anny.gomes7@gmail.com
* Correspondence: viviani.gomes@usp.br; Tel.: +55-11-3091-1331

Received: 31 May 2020; Accepted: 8 July 2020; Published: 15 July 2020

Abstract: The aim of this study was to evaluate secondary clinical disease, milk production efficiency and reproductive performance of heifers and cows persistently infected (PI) with bovine viral diarrhea virus type 2 (BVDV type 2). PI animals (n = 25) were identified using an antigen capture ELISA of ear notch samples. They were distributed into three age groups: $\leq$ 12 (n = 8), 13 to 24 (n = 6) and 25 to 34 (n = 11) months old. A control group of BVDV antigen ELISA negative female cattle that were age matched to the PI animals was utilized from the same herd. The PI group had a 1.29 higher odds ratio for diarrhea than controls (p = 0.001, IC95% = 1.032–1.623) and 1.615 greater chance of developing bovine respiratory disease (BRD) (p = 0.012, IC95% = 1.155–2.259). The age at first insemination (p = 0.012) and number of insemination attempts required to establish the first pregnancy (p = 0.016) were both higher for PI than controls. Milk production was higher for control cows than PI cows during most of the sampling periods. Somatic cell counts (SCC) were higher in PI cows than the controls at all sampling points across lactation ($p \leq 0.042$). PI cattle had a higher incidence of disease, produced less milk, a higher SCC, and poorer reproductive performance than control cattle in this study.

Keywords: diarrhea; bovine respiratory disease; milk production; somatic cells count (SCC); reproductive performance; BVDV persistent infection

1. Introduction

Bovine viral diarrhea virus (BVDV) is a ubiquitous infectious agent that affects the productivity and reproduction efficiency of both dairy and beef cattle. BVDV is responsible for significant monetary losses to producers that have been estimated to range between $0.50 to $687.80 US dollars per infected animal per year [1]. Losses have been reported to be associated with morbidity, mortality, premature voluntary culling, reduced slaughter value, stillbirths, abortion and other reproductive losses, the cost of veterinary services and treatments, the cost of replacement stock, the costs of additional labor and reduction in milk production [1–3].

The documented clinical impacts of BVDV infection provide the rationale for continual BVDV testing in the herd [4]. The clinical disease associated with BVDV infected cattle can present as a myriad of effects. These range from mild and unapparent infection to severe disease, leading to rapid death. The disease observed depends on the properties of the infecting strain, and the immune and physiological state of the host. Host factors such as gestational age at the time of exposure, general immunocompetence, immunotolerance to BVDV viral antigens due to prior exposure or vaccination, and the level of physical, social and environmental stress impact the severity and outcome of the disease [5].

The typical clinical signs and symptoms associated with BVDV infection have been reported in both acutely infected and persistently infected (PI) cattle [6,7]. These signs and symptom can routinely be induced during experimental challenge with any of several commonly used strains [8–10]. However, challenge with BVDV under experimental conditions often does not reproduce the full variety of manifestations of the disease seen in the field. Specifically, there is a lack of authoritative information about the typical clinical findings throughout the lifespan of PI cattle [11]. This is, in part, due to the short lifespan of PI cattle, which is often less than one year [12].

We have observed that there are few studies that have been published detailing the associated clinical picture of BVDV-associated secondary diseases in PI cattle. The diseases most frequently detected in PI cattle infected with the major BVDV 1 subtypes (a, b, c) were diarrhea (41%), bronchopneumonia (20%), and their combination (9%). In addition, neurological signs were observed at necropsy in 10% of the BVDV infected PI cattle. Further, abortion, laminitis and other diseases of the hoof, weakness, and anemia were only occasionally observed in PI animals [13].

Kane et al. (2015) described an outbreak of BVDV type 2a that results in abortion and PI births, despite neonatal deaths. Many of the PI cattle died as yearlings. Seventeen of 36 died with lesions consistent with mucosal disease, whereas six died without gross lesions, and two were euthanized because of chronic ill thrift. The 11 PI animals appeared healthy and were sold for slaughter. The virus population of BVDV type 2a in PI animals from this outbreak differs in size and diversity, and it could be used as a quantifiable phenotype correlated with clinical presentation of BVDV such as growth rate, congenital defects, viral shed and cytokine expression [14].

BVDV PI animals are often difficult to breed successfully. Further, herds with PI cattle can have reproductive problems. BVDV has been identified in the ovaries and uterus of PI animals [15]. The virus appears to inhibit conception through physiological alterations that induce morphologic changes in the ovaries leading to an increased interval between calving. In addition, BVDV inhibits normal embryo cleavage and reduces the number of viable embryos produced during in vitro fertilization [15–17]. BVDV can also induce congenital defects during fetal development, abortions (in 6% to 10% of fetuses), and stillbirth at any point in gestation [18]. Thus, PI cattle may have a significant role in poor reproductive function in herds.

A dairy herd containing PI infected animals generally has a higher incidence of mastitis (both subclinical and clinical), and frequently has a higher somatic cell count (SCC) than comparably managed BVDV free herds [19–22]. The effects of BVDV on milk production include: an average loss of 368 kg of milk, 9.35 kg of protein and 10.2 kg of fat per cow per lactation [23].

The majority of studies estimating the losses caused by BVDV infection have been reported for animals with acute infection. There is little reliable information about the impact of PI cattle with respect to secondary clinical disease, efficacy of reproduction, or its impact on milk production and quality. Thus, the aim of this study was to collect data about associated secondary clinical diseases, document reproductive efficacy, and document milk production and quality for Holstein PI heifers and cows that had naturally acquired infection with BVDV type 2 viruses.

Here, we present data collected from one large commercial herd with a well-documented history of BVDV infections over a long period. Prior to the PI screening in this herd, BVDV-associated diseases and production losses were inferred from farm records. Historically, this farm had a high incidence and recurrence of bronchopneumonia in young calves, characterized by an interstitial bronchopneumonia or fibrinosuppurative bronchopneumonia. In August 2013, the etiological cause of bovine respiratory disease (BRD) was investigated for ten young heifers. Two of these animals were positive in transtracheal wash for BVDV as detected by RT-PCR [24]. Between 2013 and 2015, BVDV was confirmed for some cases of fetal infection by the detection of serum BVDV antibody in calves before colostrum intake. In addition, neonate calves were found to be BVDV positive by RT-PCR [25,26]. In this herd, our team investigated the evidence of specific antibodies against the BVDV p80 protein in heifers from 1 to 390 days of age ($n = 585$). We detected low levels of p80 antibody in serum over the first 120 days in 10–20% of the calves. There was an abrupt rise in the number of p80 positive to 70% of the calves being serum positive after 120 days of age. This was associated with a higher mortality of calves in the post-weaning period due to anaplasmosis and babesiosis [27]. In 2015, mucosal disease was observed in a heifer during the post-weaning period. This herd also had a decline in reproductive performance. The associated reduction in the production of calves was not initially associated with BVDV infection. However, evidence linking BVDV to calf disease lead to a global screening for PI cattle in the herd in 2016.

Based on this history, the study presented here was conducted to determine the impact of naturally occurring BVDV PI animals on production parameters.

2. Materials and Methods

2.1. Herd Management

This study was performed on the largest commercial dairy farm in Brazil. It is located in São Paulo State (22°21′25″ S and 47°23′03″ W). The research was approved by the Committee for Ethics in Animal Use of the School of Veterinary Medicine and Animal Science from University of São Paulo in 1 May 2017 (Protocol number 5131190216). Written informed consent was obtained from the farm responsible.

This dairy herd was composed of approximately 3700 Holstein cattle. There were 1750 cows in lactation, with an average daily milk production of 38.5 L per cow. Calves were housed in individual suspended cages (1.0 × 0.9 m) from birth until they were abruptly weaned at about day 75 of age. They were transferred to group housing of 9, 18 or 36 heifers as they got older. The pens contained a sand bed, common drinking water trough, and a bunker with free choice calf starter diet (Guabi, São Paulo, Brazil). The calves had free access to grass (Tifton) pasture inside the pen.

Beginning at 10 months of age, heifers were kept in a confinement pasture with about 100 animals per unit. These heifers were enrolled into the standard reproductive protocols for the farm at about 13 months of age when they reached about 1.20 m tall at the withers. The majority of the heifers had entered the reproductive protocol by 13 to 14 months of age and their average weight was 380 kg. In the reproduction protocol, each heifer received a dose of prostaglandin (MSD Animal Health, Montes Claros, Minas Gerais) weekly. The heifers were inseminated when they demonstrated they were in heat. Pregnancy diagnosis using ultrasound (Mindray, Shenzhen, China) was performed every 21 days, and heifers previously inseminated, but not pregnant at this time were submitted to the fixed-time artificial insemination protocol, or fixed-time embryo transfer. A majority of the heifers (92–93%) were pregnant by 17 months of age. They were kept in a confinement pasture with about 100 animals per pasture with free choice TMR ration formulated and produced in the farm, water and access to grass pasture until the seventh month of gestation.

In the eighth month of pregnancy, heifers were moved to the maternity pen. This pen had a cross-ventilation system and held 200 heifers or dry cows. After calving, lactating cows were housed in a force cross-ventilated barn, divided into six pens, with about 290 animals in each pen.

During early calfhood, passive BVDV control was assumed to be provided by the transfer of maternal BVDV antibody in colostrum (see BVDV vaccine protocol below). Calves were fed a total volume of 5 L of fresh maternal colostrum (containing ≥50 g/L of IgG, documented using a colostrometer (MS Schippers, Campinas, Brazil), divided into two feedings. Both feedings were delivered within the first 18 h after birth.

2.2. Herd Vaccination History

The young heifers were vaccinated against BVDV (using a killed commercial vaccine—Cattle Master® GOLD FP 5/L5, Zoetis, Parsippany-Troy Hills, NJ, USA), beginning at 60 days of age with a priming dose. This was followed by a booster dose of the same vaccine 30 days after the first vaccination. From this point forward, all animals in this herd (heifers ≥2 months of age and cows) were vaccinated biannually with the same vaccine in April and October. This herd previously used a commercial vaccine without BVDV type 2 for several decades until 2012 (Cattle Master 4 + L5®, Zoetis). Subsequent vaccination utilized an inactivated BVDV 1 (5960) and BVDV 2 (53637) containing vaccine with the addition of a thermosensitive BoHV-1 (Cooper) and BPIV-3 (RLB 103), a modified-live BRSV (375), and five species of killed *Leptospira* spp. This vaccine was diluted in an immune stimulating complexes (ISCOM) adjuvant (Cattle Master® GOLD FP 5/L5, Zoetis).

2.3. PI Screening and Choice of Experimental Groups

Ear skin samples (ear notches measuring 1 cm × 0.5 cm), were obtained from the dorsal pinna margin of each young heifer or calf (n = 2247) using a stainless-steel ear notching clamp (type V pig, Walmur, Porto Alegre, Rio Grande do Sul). The sample obtained was stored in a sterile microtube (Eppendorf, São Paulo, Brazil), then frozen at −20 °C until it was processed. Each sample was assessed individually for BVDV using an antigen capture, Erns antigen specific ELISA test (IDEXX BVDV Ag/Serum Plus Test, IDEXX, Westbrook, ME, USA). After 30 days, animals positive at initial screening were retested using the same protocol with new biopsy samples. This was followed by testing of all live dams and grandams of the animals identified as PI among the heifers and calves. After the removal of all PI animals identified in this herd, all newborn calves were tested monthly by ear notch sampling. The duration of PI screening lasted about 13 months from September of 2015 until October of 2016.

A total of 26 PI cattle, including 19 calves and heifers, 4 dams and 3 neonates (born after removal of PI animals from the investigated herd) were found. Twenty-five of the 26 heifers and cows were included in this study. Animals were distributed by age: ≤12 months (n = 8), 13 to 24 months (n = 6) and 25 to 36 months (n = 11). The control group was composed of animals free of persistent BVDV (n = 25). They were selected as aged matched pairs for the PI cows and heifers.

2.4. Virus Characterization

Blood samples were collected via jugular venipuncture using an 18 gauge × 2.5 cm single sample needles into two 8 L glass tubes (Vacutainer ACD Solution A REF 364606, BD Diagnostics, San Jose, CA, USA) from all animals in the study. The blood collected in ACD solution A were processed for buffy coat isolation as previously described by Harpin et al. [28] and stored at −80 °C.

For BVDV characterization, the RNA was extracted from the buffy coat using TRIZOL LS reagent (Invitrogen, Thermo Fisher Scientific Inc., Waltham, MA, USA), according to manufacturer's instructions and stored at −80 °C until molecular testing was performed. For RT-PCR, a one-step RT-PCR kit (Access Quick RT-PCR System, Promega Corporation, Madison, WI, USA) was utilized and a set of primers from the 5′ UTR region (5′ TAG CCA GCT CCT TAG TAG GAC 3′ and 5′ ACT CCA TGT GCC ATG TAC AGC 3′) were selected. This allowed the detection of all BVDV strains (types I, II and Hobi-like) [23]. Using a total volume of 25 μL, the reaction was performed using 12.5 μL AcessQuick Master mix 1× (Tris-HCl, KCl, dNTP, $MgCl_2$ and Taq Polymerase), 1.25 μL of sense primer and 1.25 μL of anti-sense primer (1 mol/μL), 0.5 μL of reverse transcriptase enzyme AMV, 4.5 μL of nuclease-free water and 5 μL of isolated RNA. The amplification cycle was 45 °C for 45 min for disassociation and reverse transcriptase activation at 94 °C for 5 min. This was followed by 40 cycles of 94 °C for 30 s, 57 °C for 30 s and 72 °C for 30 s. An incubation at 72 °C for 5 min allowed the final extension. The assay was held at 4 °C following amplification. Amplicons were visualized on 1.5% agarose gels by horizontal electrophoresis. The gels were stained with Gel Red at a dilution of 1:150 (Biotium, Fremont, CA, USA). The products were visualized and measured using a transilluminator under ultraviolet light (302 nm). The samples were considered positive, as compared to the BVDV positive control, if the product was 29 bp. The standard 100 bp ladder (Thermo Fisher Scientific Inc. Waltham, MA, USA) was used to establish the product size.

From 19 BVDV positive samples by RT-PCR, five were submitted to sequencing to verify which type was involved in this herd. Three PIs aged from 1 to 12 months of age, 1 PI from 14 to 24 months of age, and 1 PI from 25 to 32 months of age were selected so that there was at least one PI representative of each age group. The PCR products were purified using the Wizard DNA and PCR Clean-up kit (Promega™, Madison, WI, USA) following manufacturer's instructions. The reaction consisted of the BigDye 3.1 Xterminator kit® (Applied Biosystems™, Foster City, CA, USA) and universal BVDV primers (3.2 pmol/μL) from the 5′ UTR region. The total reaction volume was 10 μL. The sequencing cycle consisted of 1 cycle of 95 °C for 1 min, 35 cycles of 95 °C for 30 s, 50 °C for 15 s and 60 °C for 4 min. This amplification was followed by holding the product at 4 °C to preserve the DNA. The sequencing reaction plate was precipitated using the BigDye Xterminator purification kit according to manufacturer's instructions. Finally, the plate was applied to a 350 L Sequencer (Applied Biosystems™). The quality of sequences generated was analyzed using Sequence Analyzer software (Applied Biosystems™). The BVDV sequence data was edited using the BioEdit program, and after obtaining a consensus sequence fit, it was assessed using Blast and aligned with published BVDV sequences using the ClustalW program. The criteria utilized for building the phylogeny tree were based on a better than 98% identity with the core BVDV sequence. The MEGA 7.0 program was used to determine which algorithm method [29] could be applied to these samples for relatedness analysis, and a bootstrap of 1000 replicates was used to build a relational tree.

2.5. Data from the Farm Record

The farm records were utilized to document the reproductive performance, milk production and milk quality. Retrospective data correspond to the longevity of PI animals, since they were diagnosed as persistently infected at different ages and were immediately slaughtered after diagnosis. The digital record of each animal from the Dairy Comp Program® (Valley Agricultural Software, Tulare, CA, USA) was used to generate the milk production records.

The reproduction records were collected from the farm register. The parameters collected were: age at first insemination, number of artificial insemination (AI) attempts required to achieve conception, the frequency of abortion, and the age of each heifer at the first calving. Fourteen of 16 PI heifers (12 to 36 months of age) were presented for artificial insemination. The success in reproduction (based on the parameters collected), was compared to control heifers paired for age and size ($n = 14$).

Eight PI heifers calved. These PI heifers were assessed relative to their BVDV negative pair heifer ($n = 8$). Data were also collected comparatively for the quantity and quality of the milk produced from the monthly milk report of the farm based on program records. All data collected were assessed relative to windows for days in milk: M1 = 15–39 days; M2 = 44–73 days; M3 = 74–104 days; M4 = 107–132 days; M5 = 139–166 days and M6 = 167–195 days. Not all data was recorded for each cow at each interval, because all PI cows were slaughtered immediately after being confirmed positive for BVDV by antigen ELISA results, removing them from the study in different phrases of lactation.

We were able to collect data for immediate post-partum diseases in eight PI and eleven paired cows with respect to the treatment given. The data for the treatment of cows for metritis, mammary edema, retained placenta, ketosis, clinical mastitis and bovine respiratory disease were collected.

2.6. Clinical Scores

Clinical scores were collected for the heifers (both PI and paired controls) at the initiation of the study. Over the course of the study, all pairs remaining at each assessment point were collected for health status. The health status of each heifer (or cow) was evaluated using a scoring system developed at the University of Wisconsin (https://www.vetmed.wisc.edu/fapm/svm-dairy-apps/calf-health-scorer-chs/). At each assessment point, a clinical examination was performed for each pair of heifers or cows, beginning on the day of the second BVD ear notch collection. The fecal score was recorded as 0 (normal consistency), 1 (pasty, semi-formed), 2 (pasty with large amount of water), or 3 (watery, fluid feces with fecal content in the perineum and on tail). The BRD Score was compiled from the rectal temperature, evidence of cough, character of nasal secretion, presence of ocular secretion, observation of abnormal head and ear positioning. These were classified on a scale from 0 to 3 for each observation, representing the intensity of each clinical manifestation. Animals having a fecal score of 2 or 3, or a summed BRD score of ≥4.0 were considered to have clinical disease.

2.7. Statistical Analysis

Statistical analysis was performed using the Statistical Package for the Social Science (SPSS) version 19.0 (IBM Corporation, Armonk, NY, USA).

The association between experimental groups (PI and paired control) for the prevalence of diarrhea, BRD, post-partum disease and the rate of successful calving was evaluated using either Chi-square or Fisher's exact test. Fisher's exact test was chosen when the groups had less than 5 pairs represented. Binary logistic regression was performed to estimate the odds ratios and confidence intervals of 95% between outcomes in the paired heifers (cows).

The normality of continuous variables was tested using the Shapiro–Wilk test. If the data were not normally distributed, the variables were transformed using one of the following: quadratic root, log10 or inverse transformation. Student's t-test for normally distributed independent samples, and the Mann–Whitney U test for samples that were not normally distributed were performed to compare between groups for parametric and non-parametric analysis, respectively. The time-series analyses of milk production and milk composition were evaluated by using a one-way ANOVA for repeated measures coupled with a Bonferroni post-hoc test to compare data across time within each experimental group.

3. Results

3.1. Identification of PI Animals and the Outcomes for PI Heifers and Cows

PI screening was conducted on samples from 2247 heifers and calves from young animals born prior to the start of the testing program, and on the cows in the herd that had male or stillborn calves. Thirty-four of 2247 (1.5%) were BVDV infected. BVDV positive animals included: 19 aged 1 to 12; seven aged 13 to 24, and eight aged 25 to 36 months.

Results of the second round of testing indicated that 19 of the remaining 30 animals (four animals died or were euthanized before retesting) were confirmed as PI. Following this round of testing, all live dams ($n = 11$) and grandams ($n = 4$) of the PI heifers and calves were tested. Four dams (25 to 36 months of age) were found to be PI. After the removal of the PI cows from the herd, all newborn calves were tested on a monthly basis as they were born. Three additional PI calves (of 103 new calves) born over the next two months were identified with no additional PI calves identified during the next nine months on this farm.

3.2. Virus Characterization

Nineteen of 25 (76%) buffy coat samples from confirmed PI cows tested were positive for BVDV. All positive samples in the RT-PCR appeared to be closely related BVDV type 2 viruses. Five of these positive samples were subjected to sequencing to confirm the genetic relatedness based on their having suitable quality for sequencing and representing the diversity of the age groups established in the study. The BVDV virus 5′ UTR sequences were analyzed and grouped in a maximum likelihood tree. The phylogenetic tree is shown in Figure 1, and it was observed that the five samples were all identified as BVDV II strains with 98% of confidence. This was based on published BVDV II strain sequences. No BVDV virus was recovered from any of the control animals using the RT-PCR test.

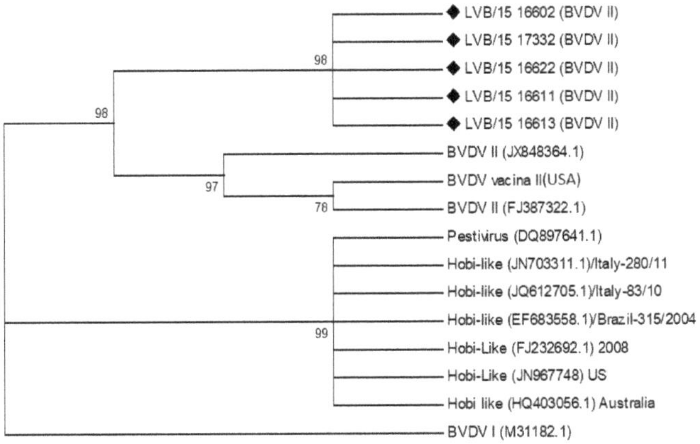

Figure 1. Phylogram displaying the relationships of the nucleotide sequences obtained in the current study as aligned against bovine viral diarrhea virus (BVDV) and other pestivirus sequences reported in GenBank. The phylogenetic tree was constructed using a maximum likelihood algorithm under the Kimura 2 evolutionary model with 1000 bootstrap replicates testing.

3.3. *Impact of Being PI on Reproductive Performance*

Reproductive performance was evaluated in sexually mature females (13 to 36 months of age). Artificial insemination of control pair heifers was initiated earlier (14.3 ± 1.19 months) than for PI heifers (15.58 ± 1.42 months) ($p = 0.012$). In addition, the number of artificial insemination attempts required to achieve the first successful pregnancy was greater for the PI heifers (mean of 4.0) than the paired control heifers (mean of 1.0) ($p = 0.016$). Only half of the 16 PI heifers calved during this study. All 16 paired control heifers produced a calf during the course of the study. There was an association observed between successful calving if a heifer was PI in the study ($p = 0.003$).

Only eight PI within a total of 14 heifers calved following artificial insemination. Failure to calf by PI heifers was generally due to abortion of her calf or being sent to slaughter before she could calve. In contrast, all paired control heifers ($n = 14$) calved following the first insemination. Control heifers were 1.75 more times more likely to deliver a calf than PI heifers (IC 95%, 1.112–2.755).

3.4. *The Quantity of Milk Produced and Quality and Composition of Milk*

A summary of the quantity of milk produced and of milk quality measurements are shown in Table 1. The difference between groups means (PI vs. control) for quantity of milk produced were: 4.71, 10.46, 11.38, 12.98, 21.23 and 16.63 L, respectively for the periods designated M1 to M6. Higher quantities of milk production were observed in control cows relative to PI cows during the sampling periods M2 ($p = 0.019$), M4 ($p = 0.013$), M5 ($p = 0.002$) and M6 ($p = 0.004$).

Table 1. Mean and standard deviation (±SD) of the quantity of milk produced and the composition of the milk produced by PI and paired control cows (NI) over the course of lactation. SCC: somatic cell count.

Variables	Groups	M1 (15–39 DIM)		M2 (44–73 DIM)		M3 (74–104 DIM)		M4 (107–132 DIM)		M5 (139–166 DIM)		M6 (167–195 DIM)		*p*-Value
		Mean ± SD	*N*	Mean ± SD	*N*	Mean ± SD	*N*	Mean ± SD	*N*	Mean ± SD	*N*	Mean ± SD	*N*	
Yield (Liters)	PI	26.29 ± 11.77	7	24.29 ± 10.61	7	27.5 ± 5.28	6	28.4 ± 4.56	5	23.4 ± 12.82	5	26 ± 2.16	4	0.217
	NI	31 ± 13.06	7	34.75 ± 14.71	8	38.88 ± 15.42	8	41.38 ± 8.57	8	44.63 ± 7.29	8	42.63 ± 7.82	8	0.948
	p-value	0.492	-	0.143	-	0.112	-	0.010	-	0.003	-	0.002	-	-
SCC ($\times 10^5$)	PI	203.43 ± 246.21	7	308.86 ± 380.88	7	530 ± 476.28	6	1046.2 ± 970.37	5	919.2 ± 979.85	5	1045.25 ± 1156.22	4	0.555
	NI	50.57 ± 50.21	7	23.88 ± 16.99	8	41.38 ± 72.17	8	372.38 ± 978.72	8	21.75 ± 18.61	8	24 ± 14.4	8	0.445
	p-value	0.042	-	0.001	-	0.001	-	0.004	-	0.001	-	0.006	-	-
Lactose (%)	PI	4.66 ± 0.53	7	4.86 ± 0.3	7	4.75 ± 0.48	6	4.6 ± 0.49	5	4.72 ± 0.37	5	4.7 ± 0.44	4	0.954
	NI	4.63 ± 0.51	7	4.84 ± 0.16	8	4.79 ± 0.22	8	4.85 ± 0.18	8	4.79 ± 0.2	8	4.84 ± 0.2	8	0.941
	p-value	0.920	-	0.438	-	0.734	-	0.380	-	0.904	-	0.901	-	-
Fat (%)	PI	3.4 ± 0.81	7	3.77 ± 0.8	7	3.3 ± 0.36	6	3.34 ± 0.63	5	3.86 ± 0.46	5	3.38 ± 0.69	4	0.673
	NI	3.11 ± 0.38	7	3.45 ± 0.46	8	3.49 ± 0.57	8	3.18 ± 0.52	8	3.38 ± 0.27	8	3.23 ± 0.49	8	0.574
	p-value	0.475	-	0.712	-	0.512	-	0.710	-	0.054	-	0.961	-	-
Protein (%)	PI	2.91 ± 0.21a	7	3.07 ± 0.18ab	7	3.13 ± 0.16ab	6	3.22 ± 0.13ab	5	3.24 ± 0.05ab	5	3.35 ± 0.13b	4	0.001
	Control	2.83 ± 0.26a	7	2.88 ± 0.21a	8	2.96 ± 0.18ab	8	3.06 ± 0.16ab	8	3.16 ± 0.16b	8	3.26 ± 0.21b	8	0.003
	p-value	0.586	-	0.082	-	0.094	-	0.078	-	0.403	-	0.175	-	-

N = sample size. Difference between group means was assessed using Student's *t* test, and changes over time using a one-way ANOVA. *p*-value was considered significant if $p < 0.05$, and tendency declared when $p > 0.05$–1.00.

The SCC was higher in PI than paired control cows at each timepoint analyzed during the course of lactation. The differential in the quantity of milk produced between the PI and paired control cows increased over the course of lactation ($p \leq 0.001$). In general, PI milk was more concentrated than milk from the paired controls. PI had higher fat content (in %) at the M5 sampling ($p = 0.054$), and showed a tendency for a higher percent of protein at the samplings M2 ($p = 0.082$), M3 ($p = 0.094$) and M4 ($p = 0.078$). The level of lactose was not different between the PI and paired control cows at any sampling point. No differences were observed within either of the experimental groups over the course of the experiment (M1–M6).

3.5. The Assessment of Clinical Signs and Disease

There was an association between the prevalence of diarrhea and the incidence of BRD that was different between the PI and control groups (Table 2). An analysis of the dataset, without respect to age, demonstrated an association between a higher diarrhea score ($p = 0.012$) and higher BRD score ($p = 0.001$) linked to the PI group of animals. When age was added as a factor in analysis, a tendency was detected ($p = 0.074$) toward a higher frequency of diarrhea in PI (33%) than control (0%) cattle of 25 to 36 months of age. There was no difference observed in diarrhea frequency among calves <12 months of age ($p = 0.500$), or heifers 12 to 24 months of age ($p = 0.455$) between groups. The PI heifers had a 1.29 greater odds ratio for development of diarrhea (fecal score ≥ 2.0) than the control heifers ($p = 0.001$, IC 95% = 1.032–1.623). This was assessed using binary logistic regression of the global PI population, without respect to age.

Table 2. The percentage (number of cases/number of subjects *100) of diarrhea and bovine respiratory disease (BRD) in persistently infected (PI) and control Holstein heifers and cows.

Clinical Scores	Groups	Results	Global Population	Age (Months)		
				<12	13–24	25–36
Diarrhea (Score ≥2.0)	PI	Positive	22.7 (5/22)	12.5 (1/8)	20.0 (1/5)	33.3 (3/9)
		Negative	77.3 (17/22)	87.5 (7/8)	80.0 (4/5)	66.7 (6/9)
	Control	Positive	0 (0/25)	0	0	0
		Negative	100 (25/25)	100 (8/8)	100 (6/6)	100 (11/11)
	p-value		0.012	0.500	0.455	0.074
BRD (Sum of Score ≥4.0)	PI	Positive	38.1 (8/21)	37.5 (3/8)	20.0 (1/5)	50 (4/8)
		Negative	61.9 (13/21)	62.5 (5/8)	80.0 (4/5)	50 (4/8)
	Control	Positive	0 (0/25)	0 (0/8)	0 (0/6)	0 (0/11)
		Negative	100 (25/25)	100 (8/8)	100 (6/6)	100 (11/11)
	p-value		0.001	0.055	0.251	0.008

Global population means all PI and non-infected BVDV, without respect to age. Difference between groups was considered significant if $p \leq 0.05$, and a tendency declared if $p > 0.05 \leq 0.10$, using the Chi-square or Fisher exact test as necessary.

The frequency of animals with BRD (sum of scores ≥ 4) was higher for PI heifers (38.5%) than the paired control heifers (0%) based on analysis of the global data ($p = 0.001$). Similar results were observed for cows 25 to 36 months of age and young heifers (<12 month). For these cows, PI animals had a higher frequency of BRD (50.0%) than control cows (0%). No difference in BRD was observed for heifers 12 to 24 months of age ($p = 0.251$) between experimental groups in this study. The PI heifers had a 1.615 greater chance of developing BRD (sum of scores > 4.0) than the control heifers ($p = 0.012$, IC 95% = 1.155–2.259) when assessed using binary logistic regression analysis applied to the global population.

The relative occurrence of mammary edema ($p = 0.322$), retained placenta ($p = 0.183$), ketosis ($p = 0.421$), clinical mastitis ($p = 0.297$) and BRD ($p = 0.297$) in PI cows was determined by assessing values from eight lactating PI cows and 11 control cows (originally paired with PI heifers). We observe a higher frequency of the occurrence of metritis in PI (37.5%, 3/8) than the control cows (0%, 0/11) ($p = 0.058$). The PI cows had a 1.6 greater chance of developing metritis than the control cows (IC 95% = 0.935–2.737).

When all types of post-partum disease were combined, the total occurrence was higher in the PI cows (100%, 8/8) than in control cows (36.4%, 4/11) ($p = 0.007$). The PI heifers had a three-times greater chance of developing post-partum disease than paired control heifers (IC 95% = 1.348–6.678).

4. Discussion

This research was focused on understanding the impact of persistent infection with BVDV type 2 on secondary clinical disease, the quantity of milk produced and its quality, and on factors affecting reproductive efficacy in Holstein heifers and cows. The study comprehensively tested the herd and discovered a significant number of PI heifers, calves and even established cows (25–36 months of age). To our knowledge, this paper represents the first study to examine the impacts of BVDV type 2 persistent infection relevant to dairy cattle production and reproduction in a large dairy herd. Previously, Kane et al. (2015) reported information about clinical presentation in PI caused by a single strain of BVDV2, however it was in a Angus and Angus-cross beef herd.

This study collected all of the data from PI cattle with naturally acquired infections. Previous papers in the literature have almost exclusively presented the effects of induced, or naturally acquired, acute infection, or induced PI resulting from BVDV challenge. The experimental challenge cannot induce disease that has identical complexity as naturally acquired infections. It is likely that challenge models do not fully replicate the interaction between virus, host and the environment that occur in a true production setting.

The investigated herd was not closed. Animals from southern Brazil were added to the herd continually. To accelerate the genetic improvement in the herd, heifers and cows from the farm were sent to an embryo harvesting center in Paraná state in the south of Brazil. The embryos resulting from their utilization of this reproductive technology were implanted in cows from the farm of origin. The embryo transfer facility also had cows and heifers from other farms that did not utilize formal BVDV control programs. BVDV testing and vaccination for these animals from other farms was not implemented until the cows and heifers arrived at the embryo facility. After delivery of the calves, half of the new heifers resulting from this program were returned to the farm where the cows originated. Among the PI cows in this study, at least one animal infected with BVDV type 2 came from this embryo harvesting facility.

During the last two decades (1998–2018), over 300 bovine pestiviruses have been partially or fully sequenced in Brazil. These include viruses identified in a number of geographical regions, representing different backgrounds of the cattle, and from cattle presenting with diverse clinical pictures. Phylogenetic analysis of these viruses demonstrated a predominance of BVDV 1 (54.4%) in Brazil. The BVDV type 1 subgenotypes identified were: 1a (33.9%) and 1b (16.3%) most frequently, and subgenotypes-1d, -1e, and -1i at very low frequencies. The overall BVDV type 2 frequency was 25.7%, but it varied considerably by the region of the country. It reached as high as 48% of the BVDV identified in one state. HobiPeV accounted for 19.9% of the viruses identified. HobiPeV had the highest frequency in northeast Brazil [30].

Understanding regional genetic diversity of ruminant pestiviruses is important for establishing appropriate vaccine protocols. It makes sense that vaccines should include viral genotypes that are present in each region [31]. In Brazil, most commercial vaccines against BVDV contain only BVDV type 1 strains. This is in the face of epidemiological evidence of a broad distribution of the BVDV type 2 viruses among several Brazilian states. The farm where this research was conducted utilized a vaccine containing BVDV type 1a for several decades prior to our study. The vaccine protocol was changed in 2012. This was due to the introduction of the first commercial BVDV vaccine containing both BVDV type 1 and type 2 genotypes of virus.

The detection of 17 persistently infected cattle of 13 to 36 months of age was a surprise. Taylor et al. reported that only 4/51 PI animals survived longer than one year in a beef herd that was studied.

Noncytopathic (NCP) BVDV establish lifelong PI in fetal calves following infection between 40 and 120 days of gestation, prior to functional immune system development. Cytopathic BVDV strains arise from NCP strain via mutation in the NS23 gene. Superinfection of PI animals with a closely related CP virus will generate mucosal disease [32]. Darweesh et al. reported a mortality rate of 23/41 (56%) in PI cattle also infected with BVDV type 2a that subsequently developed mucosal disease. The adult PI reported in this research did not develop fatal mucosal disease [33].

The prevalence of diarrhea disease and BRD in PI cattle infected with BVDV type 2 was established by scoring fecal and respiratory signs into disease scores in this study. PI cattle had a higher prevalence of both diarrhea (22.7%) and BRD (38.1%) than the paired control animals (0%). This finding was most clear in adult animals (25–36 months of age). It is our contention that while these animals did not develop mucosal disease they did develop subacute chronic intestinal and/or pulmonary inflammatory problems that resulted in the development of secondary disease.

Bachofen et al. published a retrospective and prospective study that analyzed 86 clinical reports of PI animals (median: 12 months of age) between January 1995 to 2005. The animals were under the care of the ruminant clinic at the Department of Farm Animals, University of Zurich. Within this population, 26% (about 30 cattle) were 24 months of age, and four animals were more than three years old. Most of the cases had a history of recurrent, or untreated, diarrhea (41%), pneumonia (20%) or both together (9%). These finding are similar to the findings reported from our study.

Unfortunately, the short time between the completion of the second PI testing cycle and the PI slaughter by the farm did not allow us to conduct a complete physical examination of all PI animals. Therefore, our clinical findings were limited to an examination of the intestinal and respiratory tract in the slaughtered cattle. However, our physical examination of lactating cows allowed us to detect enlarged lymph nodes and enlarged hemolymphatic nodules throughout the body under the skin. We were also able to observe any asymmetry of the mammary gland (quarters), and identify vulvovaginitis, periodontitis and circular alopecia lesions around the eyeball. Further, all lactating PI cows had at least one post-partum disease. These included hypocalcemia, mammary edema, mastitis, retained placenta, metritis, ketosis or BRD. It is our position that this is the first report of the spectrum of post-partum disease that occurs within PI cows.

BRD is frequently reported in cattle when acutely or persistently infected with BVDV. The nasopharynx and respiratory tract are the main routes of entry for BVDV virus. BVDV antigen has routinely been detected in both the upper and lower respiratory tract of animals with BRD [34,35]. Moreover, the genetic material of BVDV has been previously detected using trans-tracheal lavage from both slaughterhouse and necropsied animals, some showing no symptoms [36,37].

This assessment of reproductive records indicated that PI heifers were older than their paired controls at their first artificial insemination attempt. This was due to a delay in the development of sexual maturity in the PI heifers. The earliest inseminations were done at 12 to 13 months of age. The heifers weighed about 350 kg and were about 1.20 m tall at the withers. Heifers should have attained at least 40% to 50% of their adult body weight (~300 kg) by the time of the first service. Stokstad and Loken reported that the PI calves born to heifers that were challenged with an experimental infection during pregnancy were found to be smaller, less active and to grow more slowly than calves from uninfected heifer dams [38].

The number of inseminations required to achieve the first pregnancy was greater for PI heifers than the paired control heifers. BVDV has been localized in ovarian tissue for prolong periods of the following acute infection with cytophatic and noncytophatic strains [18,39]. Altamiranda et al. identified NCP BVDV throughout the ovarian tissue of PI cattle. This virus appeared to be associated with alterations in the structure of the follicular regions of the ovaries. These alterations are believed to directly impact embryo development, leading to reduced rates of ovum cleavage and impaired embryo development.

Grooms, Ward and Brock evaluated the morphological differences in ovaries of six PI cows, compared to six cows with no documented BVDV, using classical histological methods. PI cows appeared to have ovarian hypoplasia and significant morphological changes in the number of tertiary, graafian, atretic, corpus luteum hemorrhagic and albicans body follicles.

Fray et al. suggested that BVDV compromises ovarian function through three mechanisms: (1) BVDV may affect the gonadotrophic function of the pituitary; (2) BVDV suppresses the plasma estrogen level affecting ovulation and estrus; (3) BVDV induces generalized leukopenia, affecting the leukocyte population of the ovaries and impairing follicular dynamics [40].

The reproductive impact in PI cows continues after calving. PI cows have reduced milk production volume and produce more concentrated milk. It is important to note that the decrease in milk production and the increase in SCC are gradual processes that persist throughout the entire lactation period. We found no previously published data about individual PI cows with respect to milk production or quality. However, the reduced milk production, high SCC, and higher occurrence of subclinical and clinical mastitis have been reported in herds with very high BVDV antibody titers. This was observed both before and immediately after the institution of significant BVDV eradication and control programs in Europe [19–23].

To summarize, in a single large dairy herd, we identified 25 PI cattle infected with BVDV type 2. The impact of the virus on production measures (both reproduction and milk production) appears to be significant and consistent. It appears that PI heifers are only half as successful in achieving a first calf as paired controls, and that milk production and quality are significantly poorer in PI cows. Further, PI cattle had diarrhea disease at a higher frequency and in older cows than the paired controls, and PI cattle had a greater frequency of BRD.

5. Conclusions

PI cattle suffered more diarrhea and BRD than their paired controls. The PI heifers had a delayed initial breeding success, and only half of PI heifers delivered a calf following their initial series of AI service. In addition, the PI cows had reduced milk production and poorer milk quality, with an increased SCC throughout lactation.

Author Contributions: V.G., N.S.B., F.D.T., D.J.H. and C.C.L.C. designed the experiments. V.G., N.S.B., L.H.O. and F.D.T. performed the experiments. V.G., J.S.R., N.S.B. and A.R.C.G. analyzed the data. V.G., N.S.B., D.J.H., A.R.C.G. and C.C.L.C. wrote the manuscript. V.G., D.J.H., A.R.C.G. and C.C.L.C. reviewed the manuscript. All authors have read and agreed to the published version of the manuscript.

Funding: This research received no external funding.

Conflicts of Interest: The authors declare no conflict of interest.

References

1. Richter, V.; Lebl, K.; Baumgartner, W.; Obritzhauser, W.; Käsbohrer, A.; Pinior, B.A. Systematic Worldwide Review of the Direct Monetary Losses in Cattle due to Bovine Viral Diarrhoea Virus Infection. *Vet. J.* **2017**, *220*, 80–87. [CrossRef] [PubMed]
2. Chi, J.; Weersink, A.; VanLeeuwen, J.A.; Keefe, G.P. The Economics of Controlling Infectious Diseases on Dairy Farms. *Can. J. Agric. Econ.* **2002**, *50*, 237–256. [CrossRef]
3. Yarnall, M.J.; Thrusfield, M.V. Engaging Veterinarians and Farmers in Eradicating Bovine Viral Diarrhoea: A Systematic Review of Economic Impact. *Vet. Rec.* **2017**, *181*, 347. [CrossRef] [PubMed]
4. Grooms, D.L.; Givens, M.D.; Sanderson, M.W.; White, B.J.; Grotelueschen, D.M.; Smith, D.R. Integrated BVD control plans for beef operations. *Bov. Pract.* **2009**, *43*, 106–116.
5. Baker, J.C. The Clinical Manifestations of Bovine Viral Diarrhea Infection. *Vet. Clin. N. Am. Food. Anim. Pr.* **1995**, *11*, 425–445. [CrossRef]
6. Givens, M.D.; Marley, M.S.D. Infectious Causes of Embryonic and Fetal Mortality. *Theriogenology* **2008**, *70*, 270–285. [CrossRef]
7. Walz, P.H.; Grooms, D.L.; Passler, T.; Ridpath, J.F.; Tremblay, R.; Step, D.L.; Callan, R.J.; Givens, M.D. Control of Bovine Viral Diarrhea Virus in Ruminants. *J. Vet. Intern. Med.* **2010**, *24*, 476–486. [CrossRef]
8. Zimmerman, A.D.; Buterbaugh, R.E.; Schnackel, J.A.; Chase, C.C. Efficacy of a Modified-live Virus Vaccine Administered to Calves with Maternal Antibodies and Challenged Seven Months Later with a Virulent Bovine Viral Diarrhea type 2 virus. *Bov. Pract.* **2009**, *43*, 35–43.

9. Zimmerman, A.D.; Klein, A.L.; Buterbaugh, R.E.; Hartman, B.; Rinehart, C.L.; Chase, C.C. Vaccination with a Multivalent Modified-live Virus Vaccine Administered One Year Prior to Challenge with Bovine Viral Diarrhea Virus Type 1b and 2a in Pregnant Heifers. *Bov. Pract.* **2013**, *47*, 22–33.
10. Falkenberg, S.M.; Dassanayake, R.P.; Neill, J.D.; Ridpath, J.F. Evaluation of Bovine Viral Diarrhea Virus Transmission Potential to Naïve Calves by Direct and Indirect Exposure Routes. *Vet. Microbiol.* **2018**, *217*, 144–148. [CrossRef]
11. Kane, S.E.; Holler, L.D.; Braun, L.J.; Neill, J.D.; Young, D.B.; Ridpath, J.F.; Chase, C.C.L. Bovine viral diarrhea virus outbreak in a beef cow herd in South Dakota. *J. Am. Vet. Med. Assoc.* **2015**, *246*, 1358–1362. [CrossRef] [PubMed]
12. Smirnova, N.P.; Bielefeldt-Ohmann, H.; Van Campen, H.; Austin, K.J.; Han, H.; Montgomery, D.L.; Shoemaker, M.L.; van Olphen, A.L.; Hansen, T.R. Acute non-cytopathic bovine viral diarrhea virus infection induces pronounced type I Interferon Response in Pregnant Cows and Fetuses. *Virus Res.* **2008**, *132*, 49–58. [CrossRef] [PubMed]
13. Bachofen, C.; Braun, U.; Hilbe, M.; Ehrensperger, F.; Stalder, H.; Peterhans, E. Clinical Appearance and Pathology of Cattle Persistently Infected with Bovine Viral Diarrhoea Virus of Different Genetic Subgroups. *Vet. Microbiol.* **2010**, *141*, 258–267. [CrossRef]
14. Ridpath, J.F.; Bayles, D.O.; Neill, J.D.; Falkenberg, S.M.; Bauermann, F.V.; Holler, L.; Braun, L.J.; Young, D.B.; Kane, S.E.; Chase, C.C.L. Comparison of the breadth and complexity of bovine viral diarrhea (BVDV) populations circulating in 34 persistently infected cattle generated in one outbreak. *Virology* **2015**, *485*, 297–304. [CrossRef] [PubMed]
15. Grooms, D.L.; Ward, L.A.; Brook, K.V. Morphologic Changes and Immunohistochemical Detection of Viral Antigen in Ovaries from Cattle Persistently Infected with Bovine Viral Diarrhea Virus. *Am. J. Vet. Res.* **1996**, *57*, 830–833. [PubMed]
16. Pinto, V.D.S.C.; Alves, M.F.; Martins, M.D.S.N.; Basso, A.C.; Tannura, J.H.; Pontes, J.H.F.; Romaldini, A.H.C.N. Effects of Oocytes Exposure to Bovine Diarrhea Viruses BVDV-1, BVDV-2 and Hobi-like Virus on In Vitro Produced Bovine Embryo Development and Viral Infection. *Theriogenology* **2017**, *97*, 67–72. [CrossRef]
17. Altamiranda, E.A.G.; Kaiser, G.G.; Mucci, N.C.; Verna, A.E.; Campero, C.M.; Odeón, A.C. Effect of Bovine Viral Diarrhea Virus on the Ovarian Functionality and in Vitro Reproductive Performance of Persistently Infected Heifers. *Vet. Microbiol.* **2013**, *165*, 326–332. [CrossRef] [PubMed]
18. Grooms, D.L. Reproductive Consequences of Infection with Bovine Viral Diarrhea Virus. *Vet. Clin. North. Am. Food. Anim. Pract.* **2004**, *20*, 5–19. [CrossRef]
19. Waage, S. Influence of New Infection with Bovine Virus Diarrhoea Virus on Udder Health in Norwegian Dairy Cows. *Prev. Vet. Med.* **2000**, *43*, 123–135. [CrossRef]
20. Berends, I.M.G.A.; Swart, W.A.J.M.; Frankena, K.; Muskens, J.; Lam, T.J.G.M.; Van Schaik, G. The Effect of Becoming BVDV-free on Fertility and Udder Health in Dutch Dairy Herds. *Prev. Vet. Med.* **2008**, *84*, 48–60. [CrossRef]
21. Laureyns, J.; Piepers, S.; Ribbens, S.; Sarrazin, S.; De Vliegher, S.; Van Crombrugge, J.M.; Dewulf, J. Association Vetween Herd Exposure to BVDV-Infection and Bulk Milk Somatic Cell Count of Flemish Dairy Farms. *Prev. Vet. Med.* **2013**, *109*, 148–151. [CrossRef] [PubMed]
22. Tschopp, A.; Deiss, R.; Rotzer, M.; Wanda, S.; Thomann, B.; Schüpbach-Regula, G.; Meylan, M. A Matched Case-control Study Comparing Udder Health, Production and Fertility Parameters in Dairy Farms Before and After the Eradication of Bovine Virus Diarrhoea in Switzerland. *Prev. Vet. Med.* **2017**, *144*, 29–39. [CrossRef] [PubMed]
23. Tiwari, A.; Vanleeuwen, J.A.; Dohoo, I.R.; Keefe, G.P.; Haddad, J.P.; Tremblay, R.; Scott, H.M.; Whiting, T. Production Effects of Pathogens Causing Bovine Leukosis, Bovine Viral Diarrhea, Paratuberculosis, and Neosporosis. *J. Dairy Sci.* **2007**, *90*, 659–669. [CrossRef]
24. Basqueira, N.S.; Martin, C.C.; França, J.; Okuda, L.H.; Pituco, M.E.; Batista, C.F.; Maria, A.; Paiva, M.; Gomes, V. Bovine Respiratory Disease (BRD) Complex as a Signal for Bovine Viral Diarrhea Virus (BVDV) Presence in the Herd. *Acta Sci. Vet.* **2017**, *55*, 1–6. [CrossRef]
25. Silva, B.T.; Baccili, C.C.; Henklein, A.; Oliveira, P.L.; Oliveira, S.M.F.N.; Sobreira, N.M.; Gomes, V. Transferência de Imunidade Passiva (TIP) e Dinâmica de Anticorpos Específicos em Bezerros Naturalmente Expostos para as Viroses Respiratórias. *Arq. Bras. Med. Vet. Zootec.* **2018**, *70*, 1414–1422. [CrossRef]

26. Martin, C.C.; Baccili, C.C.; Silva, B.T.; Novo, S.M.F.; Sobreira, N.M.; Pituco, E.M.; Gomes, V. Detection of Bovine Viral Diarrhea Virus Infection in Newborn Calves Before Colostrum Intake. *Semina Ciências Agrárias* **2016**, *37*, 1379–1388. [CrossRef]
27. Baccili, C.C.; Sobreira, N.M.; Silva, B.T.; Pituco, E.M.; Gomes, V. Interface Between Maternal Antibodies and Natural Challenge for Bovine Viral Diarrhea Virus (BVDV) in Holstein Heifers. *Acta Sci. Vet.* **2016**, *44*, 1–6. [CrossRef]
28. Harpin, S.; Hurley, D.J.; Mbikay, M.; Talbot, B.; Elazhary, Y. Vaccination of Cattle with a DNA Plasmid Encoding the Bovine Viral Diarrhoea Virus Major Glycoprotein E2. *J. Gen. Virol.* **1999**, *80*, 3137–3144. [CrossRef] [PubMed]
29. Kumar, S.; Hedges, S.B. Advances in Time Estimation Methods for Molecular Data. *Mol. Biol. Evol.* **2016**, *33*, 863–869. [CrossRef]
30. Flores, E.F.; Cargnelutti, J.F.; Monteiro, F.L.; Bauermann, F.V.; Ridpath, J.F.; Weiblen, R. A genetic profile of bovine pestiviruses circulating in Brazil (1998–2018). *Anim. Health Res. Rev.* **2018**, *19*, 134–141. [CrossRef]
31. Mahony, T.J.; McCarthy, F.M.; Gravel, J.L.; Corney, B.; Young, P.L.; Vilcek, S. Genetic Analysis of Bovine viral Diarrhoea Viruses from Australia. *Vet. Microbiol.* **2005**, *106*, 1–6. [CrossRef]
32. Taylor, L.F.; Janzen, E.D.; Ellis, J.A.; Van Den Hurk, J.V.; Ward, P. Performance, Survival, Necropsy, and Virological Findings from Calves Persistently Infected with the Bovine Viral Diarrhea Virus Originating from a Single Saskatchewan Beef Herd. *Can. Vet. J.* **1997**, *38*, 29–37. [PubMed]
33. Darweesh, M.F.; Rajput, M.K.; Braun, L.J.; Ridpath, J.F.; Neill, J.D.; Chase, C.C. Characterization of the Cytopathic BVDV Strains Isolated from 13 Mucosal Disease Cases Arising in a Cattle Herd. *Virus Res.* **2015**, *195*, 141–147. [CrossRef] [PubMed]
34. Shin, T.; Acland, H. Tissue Distribution of Bovine Viral Diarrhea Virus Antigens in Persistently Infected Cattle. *J. Vet. Sci.* **2001**, *2*, 81–84. [CrossRef] [PubMed]
35. Liebler-Tenorio, E.M.; Ridpath, J.E.; Neill, J.D. Distribution of Viral Antigen and Tissue Lesions in Persistent and Acute Infection with the Homologous Strain of Noncytopathic Bovine Viral Diarrhea Virus. *J. Vet. Diagn. Investig.* **2004**, *16*, 388–396. [CrossRef]
36. Weinstock, D.; Bhudevi, B.; Anthony, E.C. Single-tube Single- enzyme Reverse Transcriptase PCR Assay for Detection of Bovine Viral Diarrhea Virus in Pooled Bovine Serum. *J. Clin. Microbiol.* **2001**, *39*, 343–346. [CrossRef]
37. Fulton, R.W.; Purdy, C.W.; Confer, A.W.; Saliki, J.T.; Loan, R.W.; Briggs, R.E.; Burge, L.J. Bovine Viral Diarrhea Viral Infections in Feeder Calves with Respiratory Disease: Interactions with Pasteurella spp., Parainfluenza-3 Virus, and Bovine Respiratory Syncytial Virus. *Can. J. Vet. Res.* **2000**, *64*, 151–159.
38. Stokstad, M.; Loken, T. Pestivirus in Cattle: Experimentally Induced Persistent Infection in Calves. *J. Vet. Med. B Infect. Dis. Vet. Public Health* **2002**, *49*, 494–501. [CrossRef]
39. Grooms, D.L.; Brock, K.V.; Pate, J.L.; Day, M.L. Changes in ovarian follicles following acute infection with bovine viral diarrhea virus. *Theriogenology* **1998**, *49*, 595–605. [CrossRef]
40. Fray, M.D.; Paton, D.; Alenius, S. The Effects of Bovine Viral Diarrhoea Virus on Cattle Reproduction in Relation to Disease Control. *Anim. Reprod. Sci.* **2000**, *60*, 615–627. [CrossRef]

viruses

MDPI

Review

Epidemiology and Management of BVDV in Rangeland Beef Breeding Herds in Northern Australia

Michael McGowan [1,*], Kieren McCosker [2], Geoff Fordyce [3] and Peter Kirkland [4]

1 School of Veterinary Science, The University of Queensland, Gatton, QLD 4343, Australia
2 Department of Primary Industry and Resources, Katherine, NT 0851, Australia; kieren.mccosker@nt.gov.au
3 Queensland Alliance for Agriculture and Food Innovation, Centre for Animal Science, The University of Queensland, St Lucia, QLD 4072, Australia; g.fordyce@uq.edu.au
4 Elizabeth Macarthur Agricultural Institute, PMB 4008, Narellan, NSW 2567, Australia; peter.kirkland@dpi.nsw.gov.au
* Correspondence: m.mcgowan@uq.edu.au

Received: 7 August 2020; Accepted: 13 September 2020; Published: 23 September 2020

Abstract: Approximately 60% of Australia's beef cattle are located in the vast rangelands of northern Australia. Despite the often low stocking densities and extensive management practices of the observed herd, animal prevalence of BVDV infection and typical rates of transmission are similar to those observed in intensively managed herds in southern Australia and elsewhere in the world. A recent large three- to four-year study of factors affecting the reproductive performance of breeding herds in this region found that where there was evidence of widespread and/or recent BVDV infection, the percentage of lactating cows that became pregnant within four months of calving was reduced by 23%, and calf wastage was increased by 9%. BVDV is now considered the second most important endemic disease affecting beef cattle in northern Australia, costing the industry an estimated AUD 50.9 million annually. Although an effective killed vaccine was released in Australia in 2003, the adoption of routine whole herd vaccination by commercial beef farmers has been slow. However, routine testing to identify persistently infected replacement breeding bulls and heifers has been more widely adopted.

Keywords: epidemiology; BVDV; rangeland beef herds; northern Australia

1. Introduction

Approximately 60% of Australia's beef cattle are located in northern Australia, which includes the state of Queensland, the Northern Territory, and the northern part of the state of Western Australia [1]. This is a subtropical-tropical region with a characteristic wet and dry season dominated by a summer rainfall pattern. Cattle predominantly graze either native rangeland pastures or those containing introduced tropical grasses and legumes, which all vary considerably in dry matter digestibility and crude protein content according to the season. Approximately 85% of beef cattle in this region contain at least some *Bos indicus* genetics, enabling them to better cope with high environmental temperatures, low quality grazing pastures, and internal and external parasitism, in particular cattle tick (*Rhipicephalus microplus*) and buffalo fly (*Haematobia irritans exigua*) infestations. Median property and paddock size [1] vary considerably across this region (60 to 1250 km^2 and 419 to 2611 ha, respectively) and stocking rates are typically moderate to low; one adult equivalent (AE) in 5–30 ha, but in some areas, this is as low as 1 AE to 150 ha. Cattle are extensively managed, typically only being mustered (brought in from the paddock and handled through the cattle yards and crush or chute) twice a year, between April to June and August to October, when a number of husbandry and management practices are performed, such as branding and weaning, pregnancy diagnosis, drafting cattle into new management groups, and identification and removal of cull cows and bulls. Notably, this is generally the only time when disease control measures, such as the vaccination of cattle, are conducted. Approximately two-thirds of cow herds in the northern dry tropical rangelands are continuously mated [1], whereas in the more intensively managed southern areas of this region, herds are control mated, typically for periods of four to seven months. Peak calving occurs from October through to January. A consequence of the long mating periods in this region is that at any time there are always some cattle being mated through to six months of gestation, and thus at risk of bovine viral diarrhoea virus (BVDV) induced disease.

In Australia, BVDV was first reported to have been isolated from cases of acute and chronic mucosal disease of cattle in 1964 [2]. Several years later, the findings of the first serological survey of Australian cattle were published [3], and reported a 91% crude prevalence of seropositive cattle sampled from herds located north of the Tropic of Capricorn compared to a prevalence of 54% for those located south of the Tropic. The proportion of seropositive herds was similar, 96% and 84%, respectively. Overall, the estimated prevalence of seropositive herds was 89%, confirming that the national herd was already endemically infected. In 2007 [4], phylogenetic analysis of Australian isolates of BVDV, primarily from persistently infected (PI) cattle collected over a 25-year study, found that 96.3% were BVDV-1c strains, with the remainder being either BVDV-1a (3.1%) or 1b strains (0.3%). No type-2 isolates or Hobi-like (type 3) were identified. Notably, subgenotype 1c has only been reported in Japan, Chile, Argentina, Spain, and South Africa [4]; the first cattle imported into Australia came from South Africa.

It has been estimated [5] that approximately 1% of cattle in the Australian national herd are PI. This is supported by the results of antigen capture ELISA testing of young cattle (9–24 months of age) for export (n = 24,035) and health certification (n = 13,800) conducted at the Elizabeth MacArthur Agricultural Institute between approximately 2004 and 2010; 1.1% and 1.53% of cattle were confirmed PI, respectively (P. Kirkland pers. comm.). The prevalence of PI cattle amongst breeding-age bulls and heifers in Queensland, where approximately half of Australia's cattle reside, was 0.35%. This review focuses on updating the knowledge and understanding of the epidemiology and management of BVDV in northern Australia.

2. Prevalence

In the mid-1990s, as part of a structured animal health surveillance programme conducted by the Queensland government, a sample of heifers and cows from 213 Queensland beef herds was bled, and the BVDV seroprevalence was determined using a virus neutralisation test [6]. The overall individual animal seroprevalence was 45%. Despite significant differences between regions with respect to environment and herd management, there was no significant difference between two consecutive years in the percentage of heifer groups and herds, which were estimated to be entirely seronegative (28–41% and 9–13%, respectively). However, herds with 500 or more cattle had a significantly higher likelihood of containing one or more seropositive cattle. In small herds, spontaneous elimination of infection is likely more common because following an outbreak of infection, most cattle become naturally immune and PI cattle have a significantly shortened life expectancy. In larger herds with multiple management groups, typically there are some groups with a high proportion of naïve cattle, with the mixing of cattle from different groups resulting in the ongoing birth of PI cattle. For example, the seroprevalence of seven breeding groups of heifers and cows on a farm in south-east Queensland ranged from 0–80%, with four groups having a very low seroprevalence and one group having a high seroprevalence (Marbach pers. comm.).

In a four-year (2007–2011) longitudinal study [1] of factors affecting the reproductive performance of 73 commercial beef herds across northern Australia, sera from a cross-sectional sample of cattle in each enrolled breeding group were tested using a BVDV Agar Gel Immunodiffusion test (AGID) [7]. The advantage of using this serological test is that both the prevalence of seropositive cattle and the prevalence of recently infected cattle (those with an AGID test result of three or greater) can be estimated, and test results are not affected by vaccination.

The overall seroprevalence was similar for heifers and cows and between years, varying between 50% and 55%. The median management group seroprevalence was also similar across regions and years. Approximately one in four and one in five heifer and cow groups, respectively, were mostly naïve (<20% seroprevalence), and 3 in 10 and 4 in 10 heifer and cow groups, respectively, were mostly naturally immune (>80% seroprevalence; Table 1). Although the stocking rate is generally low, the typical behaviour of cattle in this hot region of Australia, including the daily close congregation of animals around watering points and nutritional supplementation sites, is likely to encourage the transmission of BVDV.

Table 1. Distribution of cow and heifer mobs by observed BVDV seroprevalence category [1].

Animal Class	Year	No. of Management Groups	Group Seroprevalence Category *		
			Low	Moderate	High
Heifers	2009	42	31.0%	35.7%	33.3%
	2011	25	24.0%	48.0%	28.0%
Cows #	2009	62	21.0%	38.7%	40.3%
	2011	60	15.0%	50.0%	35.0%

Some cow groups contained both heifers and cows; * Seroprevalence category defined as: Low, <20%; Moderate, 20–80%; and High, >80% seropositive.

The frequency of management groups likely to have experienced an outbreak of BVDV infection (i.e., >30% of AGID test results were ≥3; Table 2) varied between years. In 2009, approximately 3 in 10 and 2 in 10 heifer and cow groups, respectively (Table 2), were likely to have experienced an outbreak of BVDV infection. However, in 2011 only 1 in 10 and 1 in 20 heifer and cow groups, respectively (Table 2), were likely to have experienced an outbreak of BVDV infection. Outbreaks of BVDV infection in groups of heifers are relatively common due to the recognised generally lower age-specific seroprevalence [6]. A serological study of groups of mating age heifers and first-lactation cows in 12 herds in the Northern Territory found that in nearly half of the herds most of the cattle were likely to be susceptible to infection [8]. Further, these estimates of the prevalence of outbreaks of BVDV infection are similar to those derived from collation of laboratory investigations conducted in Queensland and the Northern Territory [9]. Differences between years in the prevalence of outbreaks of infection are likely to reflect the cyclical changes in proportion of naïve and PI cattle in management groups, strongly influenced by herd culling and heifer replacement practices.

Overall, there is no evidence that the prevalence of infection in beef cattle in northern Australia, either at the herd or animal level, has changed significantly between 1967 and 2009 to 2011.

Table 2. Distribution of management groups of unvaccinated heifers and cows by prevalence of recent BVDV infection [1].

Animal Class	Year	No. of Management Groups	Frequency of Recent Infection *		
			Low	Moderate	High
Heifers	2009	36	41.7%	30.6%	27.8%
	2011	22	63.6%	27.3%	9.1%
Cows #	2009	57	42.1%	42.1%	15.8%
	2011	54	68.5%	27.8%	3.7%

Some cow mobs contained both heifers and cows; * Mob prevalence of recent BVDV infection defined as Low: <10%; Moderate: 10–30% and High: >30% AGID test result ≥3.

3. Transmission

Reported rates of transmission of BVDV vary considerably depending particularly on the opportunities for close contact between susceptible and PI cattle. When susceptible cattle were placed in close proximity to a PI animal (e.g., kept in a small yard with common water and hay feeder),

60% became infected [10,11] within 24 hours. Under stocking rates of approximately one animal per hectare, transmission rates of 30–60% per month (1–2% per day) are often observed from the temperate beef grazing regions of Australia [10]. However, the birth of a cluster of PI calves can result in approximately 1–4% of susceptible cows becoming infected per day [10].

There have been few reports on transmission of BVDV in the extensively managed beef herds of northern Australia. A study of BVDV transmission on two commercial beef properties in central Queensland [12] found that between two days prior to artificial insemination (AI) through to day 51 after AI, 70% and 32% of the seronegative cows and heifers, respectively, seroconverted, but between day 51 and 210, only 17% and 3% of the seronegative cows and heifers, respectively, seroconverted. Thus, during the period in which there was frequent handling to synchronise oestrous and conduct AI and later pregnancy diagnosis (up to day 51), 1% and 0.6% of susceptible cows and heifers became infected per day, respectively; though after this, when cattle were not handled, only 0.1% and 0.02% of susceptible cows and heifers became infected per day, respectively. However, the marked decrease in the rate of transmission after the period of frequent handling cannot be simply explained by the lack of handling alone, but is more likely to have been due to the removal (at the time of pregnancy diagnosis), relocation (planned or inadvertent) or death of some or all of the PI cattle in these herds.

In a north Queensland study [9] of rates of transmission over a two- to six-month period during the dry season in groups (n = 199 to 506) of recently weaned calves (3 to 7 months of age), managed in large paddocks (~4000 ha) at stocking rates of one AE per 2–8 ha, 0.5% to 1% of susceptible cattle became infected per day. These rates of transmission are similar to those observed in intensively managed beef herds. Typically in these rangeland herds, calves weaned from different breeding groups are accumulated in large holding yards over one or more weeks, and after processing (castration, dehorning, etc.) graze together in a single paddock, all of which increases the likelihood of transmission of BVDV. A serological study of heifers in herds in the Northern Territory [8] found that although only approximately half the cattle had seroconverted by the time of weaning, by the age of first mating (two to three years of age), virtually all had seroconverted. However, in one herd, there was no evidence of seroconversion at the time of weaning, and by the time of mating, only about 20% had seroconverted. Therefore, although the typical management of young cattle in these herds is likely to encourage the transmission of BVDV, by chance alone in a given year, PI animal(s) may not be present in a paddock of weaned calves, or the PI may associate with only some of the group (particularly in very large paddocks with multiple watering points). This results in a high proportion of cattle being susceptible to infection at the time of their first mating at around two years of age.

In a serological study [13] of BVDV infection in central Queensland, beef herds showed no significant difference between heifers and young cows in the estimated annual incidence of infection (22% to 29%). A larger study of beef herds across Queensland conducted in 1994–1995 [7] reported similar, albeit lower, estimates of annual incidence of seroconversion (12% to 24%), with again no observed age-related effect.

4. Clinical Presentation

In the first 20 years after its initial detection in Australia, BVDV was considered to primarily cause acute or chronic mucosal disease, with the latter being a much more common presentation. However, fundamental research conducted at the Central Veterinary Laboratory, Glenfield, in the mid-1980s [11] demonstrated that infection of immunologically naïve female cattle around the time of mating resulted in a significantly lower conception rate and an increased rate of late embryonic and early fetal loss without any evidence of clinical disease in the females. Further work by this research group [14] demonstrated that infection of females between 25 to 40 days of gestation resulted in almost all surviving fetuses being born persistently infected.

Although in northern Australia most mixing of cattle from different paddocks and properties typically occurs around the bi-annual mustering events, problems with maintaining fence security (floods, bushfires, wildlife, and bulls) means that introductions of new cattle to a breeding group can occur at any time. Based on the findings [1] presented in Table 2, about 2 in 10 and 1 in 10 management groups of unvaccinated heifers and cows, respectively, are at risk of experiencing an outbreak of BVDV infection during mating across northern Australia. The most common clinical presentations observed as a consequence of outbreaks of infection in this region are lower than expected pregnancy rates and/or weaning rates, and a higher than expected incidence of weak non-viable calves born and/or ill-thrifty calves at time of weaning. Cases of chronic mucosal disease in yearling and older cattle presenting primarily as marked loss of body condition, and cases of BVDV-induced neurological disease in calves, are only sporadically observed due to paddocks being large and the terrain and vegetation markedly hinder observation of individual cattle.

In these extensively managed herds, the recommended approach to determining whether BVDV infection is likely to have contributed to the observed lower than expected reproductive performance requires, firstly, accounting for the likely impact of the major factors known to affect performance [1], and, secondly, cross-sectional blood sampling of cattle from affected management groups for AGID serology [15]. Multivariable modelling [1] of the impact of the magnitude of BVDV seroprevalence predicted that in management groups with a high seroprevalence (>80% seropositive) at the time of pregnancy diagnosis, the mean percentage of lactating cows that became pregnant within four months of calving was 23% lower ($p < 0.05$; Table 3) than in management groups with a low seroprevalence (<20% seropositive) [1]. In the majority of high seroprevalence management groups, there was evidence from AGID serology of some (>10% of samples tested) recent BVDV infection. Modelling also showed that in management groups that showed evidence of widespread recent BVDV infection (>30% of samples had an AGID test result of ≥3) either at the time of pregnancy diagnosis or weaning, the predicted calf wastage (losses between confirmed pregnancy and weaning) was approximately twice as great (Table 4) as that in groups with evidence of only a low-to-moderate prevalence of recent infection.

Table 3. Predicted percentage of lactating cows that became pregnant within four months of calving (P4M) by BVDV seroprevalence category [1].

BVDV Seroprevalence *	Mean % P4M	95% Confidence Interval	
		Lower	Upper
Low	57.3 ^{A}	43.8	70.9
Moderate	43.2 ^{AB}	26.2	60.1
High	34.3 ^{B}	17.0	51.6

* Seroprevalence category defined as Low: <20%; Moderate: 20–80% and High: >80% seropositive. Means not sharing a common superscript are significantly different ($p < 0.05$).

Table 4. Predicted mean percentage of calf wastage by category of recent BVDV infection [1].

Prevalence * of Recent BVDV	Mean % Calf Wastage	95% Confidence Interval	
		Lower	Upper
Low	11.45 ^{A}	6.51	16.39
Moderate	12.08 ^{A}	7.00	17.16%
High	20.84 ^{B}	12.49	29.19

* Mob prevalence of recent BVDV infection defined as Low: <10%; Moderate: 10–30% and High: >30% AGID test result ≥3. Means not sharing a common superscript are significantly different ($p < 0.05$).

5. Economic Impact

In Meat Livestock Australia's most recent (2015) assessment of the economic cost of endemic disease in the cattle and sheep industries [16], BVDV was considered the second most important disease problem in both southern and northern Australia. Using Australian findings from both experimental and field studies [11,12,14,17,18] of the impact of BVDV infection during mating and gestation, the estimated prevalence of PI cattle [9] and the distribution of prevalence of susceptible cattle in herds in northern Australia [1], it was estimated that a population-level weaning rate would be conservatively reduced by 1–4.5% as a result of between 3% to 7% of heifers/cows being infected each year. Overall the estimated annual economic cost of BVDV infection in the beef industry in northern Australia was AUD 50.9 million.

6. Management Strategies

Australian cattle veterinarians and farmers only began to recognise the potential impact of BVDV infection on beef herd performance in the late 1980s, with those involved in the artificial breeding industry being most interested in strategies to prevent outbreaks of infection in groups of cattle undergoing artificial insemination or embryo collection and transfer. However, in contrast to other countries, Australia has only had access to a BVDV vaccine since 2003. Therefore, initial strategies such as blood sample testing of cattle suspected of being PI (e.g., ill-thrifty cattle) and then deliberately exposing selected unmated/non-pregnant females to confirmed PI cattle, or inoculation of these cattle with the live virus either derived from a PI animal [19] or from laboratory culture were used.

Due to recognised differences in strains of BVDV found in Australian cattle and national biosecurity concerns, an Australian killed vaccine was developed by Pfizer Animal Health (Zoetis) using isolates provided by Dr. Peter Kirkland's laboratory (EMAI). A field evaluation [20] conducted in 2007–2009 demonstrated that in groups of heifers experiencing significant outbreaks of BVDV during mating, the efficacy of the Pestigard®vaccine in preventing infection in progeny of vaccinates was 80% (95% CI, 71–86). Despite the demonstrated efficacy of the vaccine, adoption of routine vaccination of breeding cattle has been relatively slow. A survey [1] of 73 commercial breeding herds located across northern Australia found that in only 6% of herds all breeding cattle were vaccinated, and in 3% of herds, only heifers were vaccinated. Cost of vaccinations and requirement to initially vaccinate cattle twice have been cited as reasons for lack of adoption. The latter issue has been greatly assisted by research, which demonstrated that the interval between the first and second vaccination could be extended to six months, which more readily enables vaccination to be incorporated into routine herd management. Further, with the greater recognition of the impact on beef breeding businesses of an outbreak of BVDV in bull-breeding herds, those using artificial breeding and those using short mating periods are increasingly adopting either vaccination of their heifers only or whole herd vaccination. In some cases, the decision on whether to vaccinate is based on the findings of AGID testing of a cross-section of cattle in each breeding group.

Perhaps the most common BVDV control measure being used by cattle producers in northern Australia is requiring that all purchased replacement bulls be certified as being not PI. Seedstock producers increasingly test to assess the PI status of their bulls at the time of conducting breeding soundness examinations prior to sale. The availability of a reliable antigen capture by ELISA with high sensitivity and specificity, combined with the convenience of testing either hair or ear-notch samples, has encouraged the adoption of testing to confirm that replacement breeding animals are not PI.

To support the ongoing efforts of veterinarians and cattle producers to control BVDV in Australia, a technical advisory group has developed comprehensive guidelines defining the approach to investigating and managing BVDV in beef and dairy herds [15]. Further, although there has been some discussion about the possibility of eradicating BVDV from Australian cattle, due to the high prevalence of infection at a herd and animal level, strategic prevention of infection via testing and quarantining of introduced cattle, and vaccination of groups/herds of cattle at risk of an outbreak of infection will be the main approach to BVDV control in Australia for the foreseeable future.

Funding: This review received no external funding.

Acknowledgments: Funding for major research projects cited in this review was provided by Meat Livestock Australia. The authors wish to thank the collaborating beef cattle producers and their veterinarians across Northern Australia for their wonderful co-operation in completing these projects.

Conflicts of Interest: The authors declare no conflict of interest.

References

1. McGowan, M.R.; McCosker, K.; Fordyce, G.; Smith, D.; O'Rourke, P.K.; Perkins, N.; Barnes, T.; Marquet, L.; Morton, J.; Newsome, T.; et al. *North Australian Beef Fertility Project: Cash Cow*; Final Report, Project B.NBP.0382; Meat and Livestock Australia: Sydney, Australia, 2014.
2. French, E.L.; Snowdon, W.A. Mucosal Disease in Australian Cattle. *Aust. Vet. J.* **1964**, *40*, 99–105. [CrossRef]
3. St George, T.D.; Snowdon, W.A.; Parsonson, I.M.; French, E.L. A serological survey of mucosal disease and infectious bovine rhinotracheitis in cattle in Australia and New Guinea. *Aust. Vet. J.* **1967**, *43*, 549–556. [CrossRef] [PubMed]
4. Ridpath, J.F.; Fulton, R.W.; Kirkland, P.D.; Neill, J.D. Prevalence and antigenic differences observed between Bovine viral diarrhea virus subgenotypes isolated from cattle in Australia and feedlots in the southwestern United States. *J. Vet. Diagn. Investig.* **2010**, *22*, 84–191. [CrossRef] [PubMed]
5. Littlejohns, I.R.; Horner, G.W. Incidence, Epidemiology and control of bovine pestivirus infections and disease in Australia and New Zealand. *Rev. Sci. Tech.* **1990**, *9*, 195–205. [CrossRef] [PubMed]
6. Taylor, L.F.; Black, P.F.; Pitt, D.J.; MacKenzie, A.R.; Johnson, S.J.; Rodwell, B.J. A seroepidemiological study of bovine pestivirus in Queensland beef and dairy herds conducted in 1994/95. *Aust. Vet. J.* **2006**, *84*, 163–168. [CrossRef] [PubMed]
7. Kirkland, P.D.; MacKintosh, S.G. *Ruminant Pestivirus Infections*; Australia and New Zealand Standard Diagnostic Procedures: Camden, Australia, 2006.
8. Schatz, T.J.; Melville, L.F.; Davis, S.S. Pestivirus (BVDV) prevalence on Northern Territory cattle properties. *Proc. Aust. Soc. Anim. Prod.* **2008**, *27*, 38.
9. Kirkland, P.D.; Fordyce, G.; Holroyd Taylor, J.; McGowan, M.R. *Impact of Infectious Diseases on Beef Cattle Reproduction: Investigations of Pestivirus and Neospora in Beef Herds in Eastern Australia*; Final Report, Project AHW.042; Meat and Livestock Australia: Sydney, Australia, 2009.
10. Littlejohns, I.R. Complications to the study of the relationship between bovine lymphocyte antigens and mucosal disease. In *Characterization of the Bovine Immune System and the Genes Regulating Expression of Immunity with Particular Reference to their Role in Disease Resistance*; Davis, W.C., Shelton, J.N., Weems, C.W., Eds.; Washington State University: Pullman, DC, USA, 1985; pp. 179–190.
11. McGowan, M.R.; Kirkland, P.D.; Richards, S.G.; Littlejohns, I.R. Increased reproductive losses in cattle infected with bovine pestivirus around the time of insemination. *Vet. Rec.* **1993**, *133*, 39–43. [CrossRef]
12. McGowan, M.R.; Kirkland, P.D.; Rodwell, B.J.; Kerr, D.R.; Carroll, C.L. A field investigation of the effects on reproductive performance of Bovine Viral Diarrhoea (BVD) virus infection around the time of insemination. *Theriogenology* **1993**, *39*, 443–449. [PubMed]
13. McGowan, M.R.; Baldock, F.C.; Kirkland, P.D.; Ward, M.P.; Holroyd, R.G. A Preliminary Estimate of the Economic Impact of Bovine Pestivirus Infection in Beef Cattle Herds in Central Queensland. In Proceedings of the Annual Conference of Australian Association of Cattle Veterinarians, Townsville, Australia, 14–18 May 1992; pp. 129–132.
14. Kirkland, P.D.; McGowan, M.R.; MacKintosh, S.G. Factors influencing the development of persistent infection of cattle with pestivirus. In *Proceedings of the 2nd Symposium on Pestiviruses*; Edwards, S., Ed.; Foundation Marcel Merieux: Lyon, France, 1993; pp. 117–121.
15. McGowan, M.; Kirkland, P.; Howard, R.; Morton, J.; Younis, P.; Bergman, E.; Cusack, P. *Guidelines for the Investigation and Control of BVDV in Beef and Dairy Herds and Feedlots*, 3rd ed.; Pfizer Animal Health: Sydney, Australia, 2011.
16. Lane, J.; Jubb, T.; Shepherd, R.; Webb-Ware, J.; Fordyce, G. *Priority List of Endemic Diseases for the Red Meat Industries*; Final Report, Project B.AHE.0010; Meat and Livestock Australia: Sydney, Australia, 2015.
17. Kirkland, P.D.; Hart, K.G.; Moyle, A.M.; Rogan, E. The Impact of pestivirus on an artificial breeding program for cattle. *Aust. Vet. J.* **1990**, *67*, 261–263. [CrossRef] [PubMed]
18. Kirkland, P.D.; MacKintosh, S.G.; Moyle, A.M. The outcome of widespread use of semen from a bull persistently infected with pestivirus. *Vet. Rec.* **1994**, *135*, 527–529. [CrossRef] [PubMed]

19. Cook, L.G.; Littlejohns, I.R.; Jessep, T. Induced sero-conversion in heifers with a field strain of bovine pestivirus—A comparison of methods and doses. *Aust. Vet. J.* **1990**, *67*, 393–395. [CrossRef] [PubMed]
20. Morton, J.M.; Phillips, N.J.; Taylor, L.F.; McGowan, M.R. Bovine viral diarrhea virus in beef heifers in commercial herds in Australia: Mob-level seroprevalences and incidences of seroconversion, and vaccine efficacy. *Aust. Vet. J.* **2013**, *91*, 517–524. [CrossRef] [PubMed]

viruses MDPI

Article

Pestivirus Infections in Semi-Domesticated Eurasian Tundra Reindeer (*Rangifer tarandus tarandus*): A Retrospective Cross-Sectional Serological Study in Finnmark County, Norway

Carlos G. das Neves [1,2,*], Jonas Johansson Wensman [3], Ingebjørg Helena Nymo [1,4], Eystein Skjerve [5], Stefan Alenius [3] and Morten Tryland [4,*]

1 Norwegian Veterinary Institute, N-0106 Oslo/N-9010 Tromsø, Norway; ingebjorg.nymo@vetinst.no
2 Faculty of Health Sciences, UiT The Arctic University of Norway, N-9037 Tromsø, Norway
3 Department of Clinical Sciences, Swedish University of Agricultural Sciences, P.O. Box 7054, SE-75007 Uppsala, Sweden; jonas.wensman@slu.se (J.J.W.); stefan.alenius@slu.se (S.A.)
4 Department of Arctic and Marine Biology, UiT The Arctic University of Norway, N-9037 Tromsø, Norway
5 Centre for Epidemiology and Biostatistics, Faculty of Veterinary Sciences, Norwegian University of Life Sciences, N-0033 Oslo, Norway; eystein.skjerve@nmbu.no
* Correspondence: carlos.dasneves@vetinst.no (C.G.d.N.); morten.tryland@uit.no (M.T.); Tel.: +47-96231702 (C.G.d.N.); +47-77625215 (M.T.)

Received: 8 December 2019; Accepted: 24 December 2019; Published: 26 December 2019

Abstract: Members of the Pestivirus genus (family *Flaviviridae*) cause severe and economically important diseases in livestock. Serological studies have revealed the presence of pestiviruses in different cervid species, including wild and semi-domesticated Eurasian tundra reindeer. In this retrospective study, serum samples collected between 2006 and 2008 from 3339 semi-domesticated Eurasian reindeer from Finnmark County, Norway, were tested for anti-pestivirus antibodies using an enzyme linked immunosorbent assay (ELISA) and a subset of these by virus neutralization test (VNT). A seroprevalence of 12.5% was found, varying from 0% to 45% among different herding districts, and 20% in western Finnmark, as compared to 1.7% in eastern Finnmark. Seroprevalence increased with age. Pestivirus-specific RNA was not detected in any of the 225 serum samples tested by real-time RT-PCR. Based on VNT results, using a panel of one bovine viral diarrhea virus (BVDV) strain and two border disease virus (BDV) strains, the virus is most likely a reindeer-specific pestivirus closely related to BDV. A characterization of the causative virus and its pathogenic impact on reindeer populations, as well as its potential to infect other domestic and wild ruminants, should be further investigated.

Keywords: bvdv; epidemiology; reindeer; border disease virus; Norway

1. Introduction

Members of the *Pestivirus* genus, belonging to the family *Flaviviridae*, cause severe and economically important diseases in livestock [1] such as: bovine viral diarrhea virus (BVDV), causing bovine viral diarrhea (BVD) and mucosal disease (MD) in cattle; classical swine fever virus (CSFV)/hog cholera virus, causing classical swine fever in pigs; and border disease virus (BDV), causing border disease (BD) in sheep.

Several serological studies have revealed the presence of pestiviruses in a variety of free-ranging and captive wild cervid species [2,3]. The observed sero-prevalence has varied significantly between studies, geographic regions, cervid species, and proximity to other ruminant species known to harbor e.g., BVDV or BDV [4,5].

A few serological studies have been carried out in different reindeer subspecies and populations across the Arctic region, usually targeting BVDV but allowing for some cross-reactivity with other pestiviruses. In Sweden, several studies in semi-domesticated Eurasian tundra reindeer (*R. t. tarandus*) have revealed a prevalence ranging from 0 to 35% [5–7]. In Norway a study in several wild cervid species identified a prevalence of 4.2% in wild reindeer in southern Norway [8], while a study of 48 carcasses of emaciated semi-domesticated reindeer from Finnmark County, Norway, revealed a prevalence of 33% (BVDV virus neutralization test; VNT) [9]. A study of Svalbard reindeer (*R. t. platyrhynchus*) from the Arctic Archipelago of Svalbard in the 1990s revealed no seropositive animals [10].

Pestiviruses have never been isolated from wild or semi-domesticated reindeer. However, one isolate (Reindeer-1) was obtained from a reindeer that died with signs of severe diarrhea and anorexia at Duisburg Zoo in Germany in 1996 [11]. Phylogenetic studies have revealed that the strain was most closely related to BDV type 2 (BDV-2) strains isolated from German sheep in 1999 and 2000 [11,12].

In an experimental pestivirus infection (BVDV-1) of reindeer [13], the animals displayed clinical signs, such as serous and mucopurulent nasal discharge, bloody diarrhea, laminitis, and coronitis, thus indicating that this species is susceptible to BVDV.

Studies in the early 80s revealed that nearly 30% of the dairy herds in Norway had antibodies to BVDV [14]. An eradication program for cattle was initiated in 1992 with large-scale serological screenings focusing on PI animals and the enforcement of restrictions and culling. In November 2006, Norway was considered free of BVDV [15,16]. BD was identified in both sheep and goats in Norway in the early 1980s, but the disease was associated with BVDV strains rather than BDV strains, which have to our knowledge never been identified in Norway [14,17–20]. Even though the eradication program focused on cattle alone, infections from small ruminants have also not been reported since the 1990s. Studies in goats in Norway in the 1980s revealed low prevalence for pestivirus (3.6%) [21] and only a single PI goat kid was diagnosed [19].

Finnmark County hosts approximately 69% of the semi-domesticated reindeer in Norway, with an estimated population in 2017 of 147,500 reindeer (down from 170,000 reindeer in 2006 and 188,000 in 2008 at the time of sampling). Reindeer are kept under semi-nomadic husbandry conditions, free ranging but herded within well-defined husbandry districts, with seasonal migrations between summer pastures in coastal areas and winter pastures on the inland mountain plateau. Animals are usually gathered at least twice a year for tagging calves, anti-parasitic treatment, sorting, and slaughter.

Reindeer mortality in Finnmark County reached 37% among calves during the years of 2005/2006. Although predators accounted for the majority of mortality, 11% of calf mortalities were of unknown etiology [22]. Fecundity of semi-domesticated reindeer is difficult to evaluate as they usually give birth unattended, while scavengers quickly remove aborted materials. It is therefore difficult to assess the role of abortion or weak-born calves to calf mortality [23].

The goal of this study was to conduct a retrospective cross-sectional serological screening in selected reindeer husbandry districts, to address prevalence and to identify risk factors for becoming infected with pestivirus. It was also anticipated that a study of this nature could help identify which pestivirus species and/or strains are circulating among reindeer in Norway and if reindeer can serve as reservoir hosts for viruses known to cause disease in livestock.

2. Materials and Methods

2.1. Sampling

In this retrospective study we used blood samples collected from reindeer (n = 3339) from 15 summer herding districts at four different slaughterhouses from 2004 to 2008 during the winter slaughtering periods in Finnmark County, Norway. Blood was centrifuged at 3500 rpm for 15 min and sera collected and stored at −20 °C until testing.

2.2. Animal Data

Information on gender, age, and carcass weight was collected from each herding district. Animals were grouped into two age classes: calves (≤1 year) and adults (>1 year).

For each reindeer herding district included in the study, information on total available reindeer pasture area, numbers of adults and calves, and mortality rates were obtained from the 2008 Reindeer Husbandry Authority Report [24]. Animal density, i.e., the number of animals per square kilometer of available reindeer summer pasture (n/km^2), was calculated for each herding district. Districts with animal densities lower than the mean reindeer density for Finnmark (4.6/km^2; range, 1.02–15.93/km^2) were classified as low-density districts, whereas those with a density higher or equal to the mean were classified as high-density districts.

2.3. Enzyme-Linked Immunosorbent Assay (ELISA)

Antibodies to pestivirus were detected using a commercial blocking ELISA (SERELISA® BVD p80 Ab Mono Blocking, Synbiotics, Lyon, France) that detects specific antibodies to a protein highly conserved in sequence between all strains of BVDV and BDV (p80/125 non-structural protein) [25,26]. Competition percentages and cut-off values were calculated according to the manufacturer's instructions for testing small ruminant samples (i.e., sheep and goats). Samples were classified as positive if the competition percentage was greater than 40% and doubtful if it was between 20 and 40%.

To evaluate the kit's performance with reindeer serum samples, we included in addition to the positive and negative bovine control sera supplied by the manufactures, fourteen reindeer sera previously classified as having pestivirus antibodies by VNT [9]. This ELISA kit has previously been used to detect antibodies against pestivirus in red deer (*Cervus elaphus*) [8], and other ruminant pestivirus ELISA kits with the same protein as antigen have been used to test other wild ruminant populations [5].

2.4. Virus Neutralization Test (VNT)

VNT was carried out to investigate to which ruminant virus the reindeer had been exposed. A panel of 30 samples, representative of six different geographical locations (three in eastern and three in western Finnmark), both age groups and both genders were selected. From each herding district, five samples were selected based on the ELISA results: two positive, one doubtful and two negatives. A panel of three different pestivirus strains (BVDV-1 strain NADL; BDV-1 strain 137/4, and BDV-2 strain Reindeer-1) were used to compare the capacity for neutralization of the selected reindeer test sera. Prior to incubation with the virus strains, the sera were heat-inactivated at 56 °C for 30 min. The VNT was performed as described previously by Kautto and co-workers [5]. Samples were considered positive if neutralization was observed in at least one of two wells (replicates) at a dilution ≥ 1:4. BVDV negative serum was used as a negative control. Anti-BVDV polyclonal sera (VLA Weybridge) and ovine antisera raised against BDV-1 strain X818 and BDV-2 strain Reindeer-1 [12] were used as positive controls.

The neutralizing titers were calculated according to the Spearman–Kärber method [27] as the serum dilution necessary to neutralize the virus in 50% of the cell culture wells (effective dose 50%; ED_{50}).

2.5. Real Time Reverse Transcriptase Polymerase Chain Reaction (Real-Time RT-PCR)

A total of 225 reindeer sera from all sampled districts, 15 samples per district, including animals with strong ELISA positive sera, as well as weak positive, doubtful, and negative sera were screened for the presence of pestivirus RNA by real-time reverse transcriptase polymerase chain reaction (real time RT-PCR). Whenever possible the following proportions, 2+1+12, between positive/weak positive, doubtful and negative animals were used when selecting samples per district, resulting in a total of 185 seronegative, 14 doubtful, and 26 seropositive samples being tested by real-time RT-PCR. Total RNA was extracted using QIAamp® Viral RNA Mini Kit (Qiagen, Venlo, The Netherlands) according to the manufacturer's instructions and eluted in a final volume of 40 μL. Reverse transcription was carried out with the iScript™ cDNA synthesis kit (random hexamer primers) from BioRad (Hercules, CA, USA) using the protocol supplied by the manufacturer [5]. For the following real-time PCR, we used the pan-pestivirus primers OPES13A: 5′-GCTAGCCATGCCCTTAGTAGGA -3′ and OPES14A: 5′- ATCAACTCCATGTGCCATTTACAGC -3′ at recommended primer concentrations [28] and the iQ™ SYBR® Green Supermix from BioRad (CA, USA), in a total volume of 20 μL of PCR mix and 5 μL of template (cDNA) [5]. The PCR cycling conditions applied were as follows: primary denaturation at 95 °C for 10 min, and 40 two-step amplification cycles at 95 °C for 15 s and at 60 °C for 1 min. Amplification products were verified for each run using a melting curve analysis of the gradient of temperature between 55 °C and 95 °C, after the final cycle. In addition, the sizes of the amplicons were verified by visualization in 1.5% agarose gel.

2.6. Statistical Analyses

Statistical analysis was carried out using Stata/SE 14 for Windows (Stata Corp., College Station, TX-USA). A non-parametric kernel density estimation of the probability density function for seroprevalence was performed to analyze the distribution of serological status between samples classified as positive, negative, or doubtful and to determine whether different cut-off calculations would have significant effects on results.

Weight was classified into 10 quantiles according to age and seroprevalence in order to assess the relationship between age and carcass weight and to verify that there were no major discrepancies.

Prevalence estimates were established using the survey commands of STATA; where district was the primary sampling unit for the model and data were stratified according to geographical area, carcass weight, and age class. Estimates were corrected using the following sample weighting procedure: 1/(N sampled reindeer/N total reindeer per district).

A multivariable logistic regression model was established using a backwards procedure adding initially all biological and ecological variables (gender, age, weight, geographical area, animal density, and year of sampling) and subsequently removing those with a *p*-value of the likelihood-ratio test >0.05. District was used as a cluster variable in the analyses and sample weighting was performed as for the prevalence estimates. The Hosmer and Lemeshow test for goodness of fit was carried out [29]. A classification table for sensitivity and specificity and a receiving operating characteristic (ROC) curve was calculated to assess the predictive qualities of the model.

ELISA and VNT titer results were categorized as follows: ELISA as positive if ≥ 40%, otherwise negative, and VNT as positive if titer ≥ 1:4, otherwise negative. The agreement between ELISA and VNT results was measured using the Cohen's kappa (κ) test. A threshold of $p = 0.05$ was used when appropriate.

3. Results

3.1. Overall Results

The ELISA classified 418 of the 3339 reindeer samples as positive (12.5%) with an additional 89 samples classified as doubtful. The distribution of the percentage competition values of the ELISA results, using kernel density estimation, is shown in Figure 1. Positive and negative results formed two clearly distinguishable clusters and the positive results were concentrated above a competition percentage of 70%.

The overall seroprevalence by herding district level is shown in Table 1 and Figure 2A (calves) and Figure 2B (adults). Seroprevalence varied from 0% in district 13 to 44.8% in district 34. Table 1 further shows animal densities and the mean carcass weights according to district, which taken together are good indicators of asymmetries in sample composition. Seroprevalence (ELISA) was 1.7% in eastern Finnmark and 20.0% in western Finnmark.

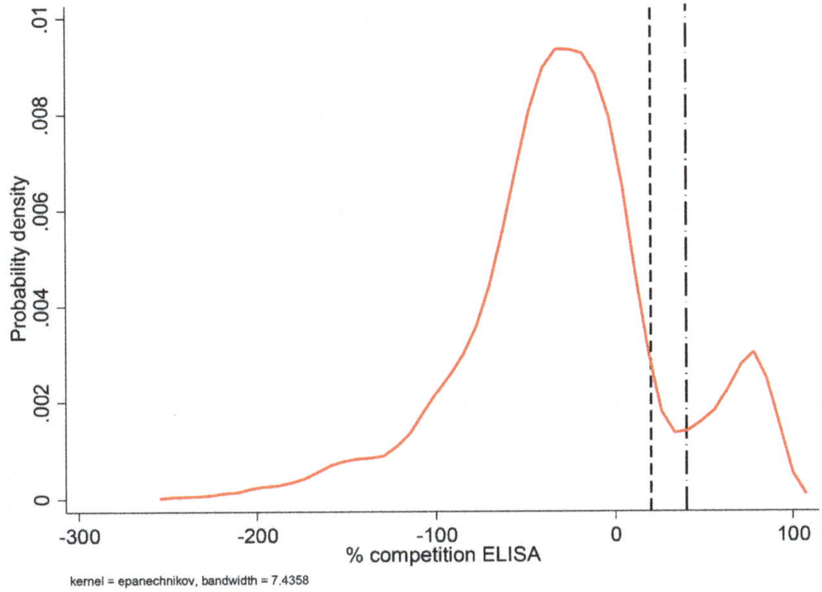

Figure 1. Distribution of percentage competition values for 3339 reindeer sera tested for antibodies against ruminant pestivirus using an ELISA. The density curve represents a kernel estimation of the probability density. The ELISA cut-off values are indicated in the graph by vertical lines; the dashed line (left) indicates a negative cut-off value of 20% and the dash-dot line (right) represents a positive cut-off value of 40%. Percentage competition values between 20% and 40% were considered doubtful.

Table 1. Distribution of ruminant pestivirus seroprevalence according to reindeer herding district in Finnmark County, Norway.

Area	District	Density	Weight	Age	Seroprevalence		
		(n/km²)	Mean	(C/A) [a]	N +/total *(Doubt)* [b]	(%) [c]	95% CI [d]
East	5A	1.0	20.3	90/33	1/185 *(1)*	0.5–1.1	[0–2.6]
	5B	4.2	— [e]	20/53	1/73 *(1)*	1.4–2.7	[0–6.5]
	6	2.1	19.5	110/5	4/119 *(1)*	3.4–4.2	[0.1–7.8]
	7	1.4	24.7	45/23	12/73 *(2)*	16.4–19.2	[7.9–28.3]
	13	4.4	21.8	175/73	0/256 (2)	0–0.8	[0–1.9]
	14A	4.3	31.8	10/56	1/66 *(0)*	1.5	[0–4.5]
	16	7.9	24.2	150/426	4/588 *(7)*	0.7–1.9	[0–3.0]
E subtotal		**2.9**	**23.5**	**600/699**	**23/1360 *(14)***	**1.7–2.7**	**[1.0–3.6]**
West	19	4.7		245/38	19/283 *(5)*	6.7–8.5	[3.8–11.7]
	27	14.9	20.9	211/241	74/452 *(13)*	16.4–19.2	[13.0–22.9]
	29	6.0	35.9	1/42	18/43 *(1)*	41.9–44.2	[26.9–59.2]
	33	11.2	23.9	10/145	71/234 *(15)*	30.3–36.8	[24.4–42.9]
	34	11.5	23.7	63/163	162/362 *(25)*	44.8–51.7	[39.6–56.8]
	35	5.5	19.2	194/28	4/224 *(5)*	1.8–4.0	[0–6.6]
	36	3.9	25.9	49/131	5/202 *(0)*	2.5	[0.3–4.6]
	40	15.9	24.5	2/177	42/179 *(11)*	23.5–29.6	[17.2–36.3]
W subtotal		**6.6**	**22.9**	**775/965**	**395/1979 (75)**	**20.0–23.7**	**[18.2–25.6]**
Total		**4.1**	**23.2**	**1375/1634**	**418/3339 *(89)***	**12.5–15.2**	**[11.4–16.4]**

[a] Age distribution between calves (≤1 year) and adults (>1 year). Information on age was only available for 3009 animals. [b] Bracketed values represent the number of animals classified as doubtful after retesting. [c] Seroprevalence is presented as an interval between % calculated from ELISA positive animals only and % including all doubtful results as positives. [d] Confidence intervals calculated for the maximum interval when animals classified as doubtful were included as if they were positive. [e] No information regarding weight was available for animals sampled in district 5B.

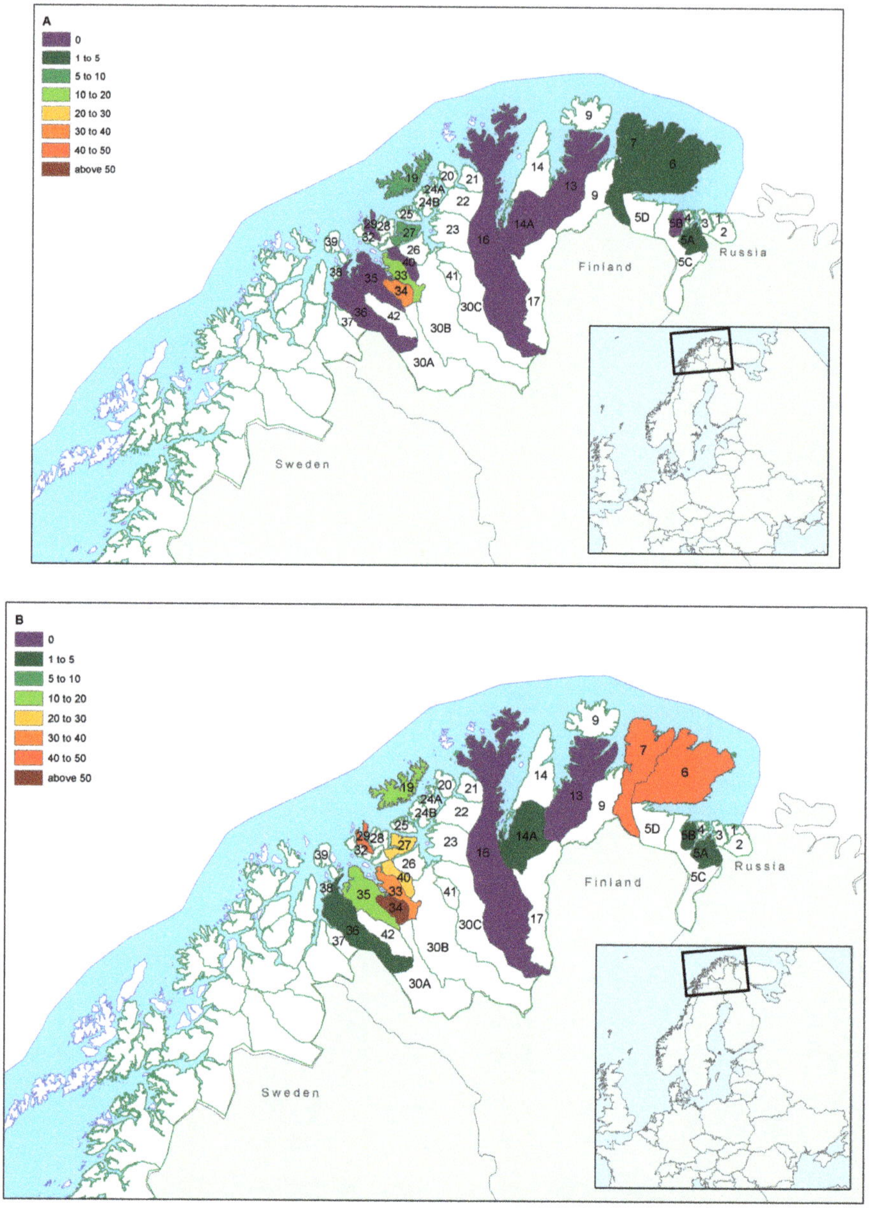

Figure 2. Seroprevalence of pestivirus infection according to herding district and age group (Map (**A**): calves; map (**B**): adults) in Finnmark County, Norway. Tables 1 and 2 provide detailed information on age distribution within districts that should be considered when extrapolating the seroprevalence to the general population in each district. The border between western and eastern Finnmark is between districts 16 and 14A.

Table 2. Distribution of seroprevalence in 10 carcass weight quantiles and classification of age.

	Carcass Weight		Number of Animals			Seroprevalence	
Quantile	Mean (kg)	[Min–Max] (kg)	Calves	Adults	N	%	95% CI
1	13.9	[9.2–15.7]	352	5	22	6.2	[3.7–8.7]
2	16.8	[15.7–17.7]	255	18	13	4.8	[2.2–7.3]
3	18.7	[17.7–19.5]	232	34	13	4.9	[2.3–7.5]
4	20.7	[19.5–21.2]	195	95	18	6.2	[3.4–9.0]
5	22.1	[21.2–22.9]	142	182	36	11.1	[7.7–14.5]
6	23.8	[22.9–24.8]	108	236	47	13.7	[10.0–17.3]
7	25.9	[24.8–27.0]	65	281	55	16.0	[12.1–20.0]
8	27.9	[27.0–29.0]	15	269	59	20.8	[16.0–25.5]
9	30.5	[29.0–32.3]	11	247	41	15.9	[11.4–20.4]
10	38.9	[32.3–69.7]	1	265	45	16.9	[12.4–21.4]

Data from 3008 animals were used since data on weight and age were not available for all animals. Animals were classified as calves (≤1 year old) or adults (>1 year old).

Extrapolation based on the inverse of the number of animals sampled per district divided by the total number of animals in the respective district showed that the samples included in this serosurvey were representative of 111,224 animals, accounting for 66.0% of the total reindeer population in Finnmark County. The overall seroprevalence was 12.5 to 15.2% before extrapolation and 8.2 to 10.2% after extrapolation. The seroprevalence ranged from 1.9 to 2.8% in eastern Finnmark and from 18.1 to 21.2% in western Finnmark.

3.2. *Age Classification and Weight Classes*

Seroprevalence increased from the lowest weight class (quantile 1; seroprevalence = 3.7%) to the highest weight class (quantile 10; seroprevalence = 21.4%). Table 2 shows the relationships between carcass weight and seroprevalence when carcass weight was stratified into 10 quantiles.

3.3. *Factors Affecting Seroprevalence*

Carcass weight was positively correlated with seroprevalence in calves and adults up to 35 kg, irrespective of gender or geographical location (Table 2). No interactions were detected between carcass weight and any of the other variables.

The logistic regression model (Table 3) showed that geographic location, age and gender affected seroprevalence to differing extents. Age (odds ratio = 5.57, CI [2668–11,629]) clearly affected seroprevalence with adult animals having a much higher chance of being seropositive than calves. Geographic location had a minimal effect (odds ratio = 0.084), with animals in western Finnmark having a slightly higher chance of being positive. Concerning gender there seemed to be a slightly higher chance of males being positive than females (odds ratio = 0.472). Differences in seroprevalence between variations in animal density by district or year of sampling were not statistically significant ($p > 0.05$).

Table 3. Logistic regression model for risk factors associated with becoming infected with pestivirus.

Pestivirus Seroprevalence	Odds Ratio	Robust SE	z	P > |z|	95% CI
Region (west to east)	0.084	0.062	−3.36	0.001	[0.020–0.356]
Age (calves to adults)	5.570	2.092	4.57	<0.0001	[2.668–11.629]
Gender (male to female)	0.472	0.167	−2.12	0.034	[0.236–0.944]
Constant	0.093	0.038	−5.82	<0.0001	[0.042–0.207]
No. of observations	3009				

The standard error was adjusted for 15 clusters over districts.
p values < 0.05 were regarded as statistically significant.
Non-significant interactions were removed from the model

None of the additional tests indicated that the model did not fit the data. The Hosmer and Lemeshow goodness of fit test had a chi-square value of 0.46 and a p value of 0.977 (in this instance, the p value is significant if it is greater than 0.05), implying that the model's estimates fitted the data at an acceptable level.

A classification table was compiled to determine the accuracy of the logistic regression model. For a cut-off value of 50% probability ($p = 0.5$), the model correctly predicted the classification of 88.4% of the samples. The area under the ROC curve was 0.81.

3.4. Virus Neutralization Test (VNT)

VNT results are summarized in Table 4, in which the ED_{50} values are presented as the neutralizing titer that corresponded with the respective $\log_2 ED_{50}$. VNT showed that samples negative according to the ELISA were unable to neutralize any of the ruminant pestiviruses, thus being in concordance with the ELISA results for these samples. Doubtful samples had on average low titers for BDV1 and BDV2 and none against BVDV1. All samples that were positive according to the ELISA had higher neutralization titers for BDV1 (average 1:126) and BDV2 (average 1:131) than BVDV1 (average 1:19). Titers against BDV1 and BDV2 were usually very close to each other (within 1 log and non-statistically relevant), with the exception of district 16 where neutralizing titers against BDV1 were not observed.

Table 4. Virus neutralization test of reindeer sera samples for ruminant pestiviruses.

		ELISA	VNT ($\bar{x}$ Titer)		
Region	District	$\bar{x}$ % Compet.	BVDV NADL	BDV1 137/4	BDV2 Reindeer-1
East	5	64.8	6	68	130
	7	78.4	12	192	96
	16	76.2	6	6	192
West	19	84.2	66	264	132
	34	85.5	6	160	144
	40	79.5	18	128	160
	TOTAL (positives)	**78.1**	**19**	**136**	**142**
	Doubtful (all districts)	33.5	7	33	24
	Negative (all districts)	−68.0	<4	<4	<4

A panel of 30 samples representative of different geographical locations. Both age groups and both genders were selected. From each district, five samples were selected based on the ELISA results: two seropositive, one doubtful, and two seronegative. The neutralizing titers were calculated according to the Spearman–Kärber method as the serum dilution necessary to neutralize the virus in 50% of the wells (effective dose 50%; ED50). Average titers were calculated per district.

Titers against BVDV1 were consistently much lower in every district than those observed for BDV1 and BDV2. No serum sample neutralized BVDV1 at a higher dilution than it neutralized the BDV1 and BDV2.

Statistical analysis of the agreement between ELISA and VNT showed a good correlation between the two tests, where samples with increasing competition ability (ELISA) also had increasing VNT titers. The best correlation between the ELISA and the VNT was found when BDV2 was used in the VNT ($\kappa = 0.580$ for ELISA and BVDV; $\kappa = 0.723$ for ELISA and BDV1; $\kappa = 0.862$ for ELISA and BDV2).

3.5. *Realtime RT-PCR*

None of the reindeer serum samples selected for real time RT-PCR were positive. Pestivirus positive controls were positive, with an amplicon of the expected size of around 295 base pairs confirmed by electrophoresis.

4. Discussion

The presence of seropositive animals in several districts across eastern and western Finnmark confirmed that a pestivirus is enzootic among semi-domesticated reindeer in Norway. VNT results seem to indicate that the virus is more closely related to BDV than to BVDV1.

Serum samples classified as positive and negative in the ELISA comprised two clearly separated clusters, whereas the probability of a sample being doubtful (i.e., 20–40% competition) was low (Figure 1). Furthermore, all samples classified as negative in the ELISA and subsequently tested in the VNT failed to neutralize the actual virus tested. Thus, we believe that the ELISA cut-off values chosen with this serological kit can be applied when testing reindeer samples. The viral envelope non-structural protein p80 is highly conserved between the different BVDV strains and represents one of the most important markers of cp BVDV [25,26].

The seroprevalence in this study ranged from 0 to 51%, with a mean value of 12 to 15%, and increased with age. This is in line with previous studies, although mean values of prevalence vary considerably between studies. Studies in Sweden identified a prevalence of 35% (5], while previous studies in Norway found a prevalence between 4% [8] and 33% [9]. Differences in analytical methods, geographic regions, and selection criteria for test animals from the herds are some factors that might help explain these differences. It is, however, beyond doubt that pestiviruses are circulating continuously in most tested reindeer populations.

Figure 2B shows two clusters of relatively high prevalence in adults. The cluster around districts 33, 34, and 40 shows a pattern already described for alphaherpesvirus infections [30], in a study that tested the same animals for cervid herpesvirus 2. While these are areas with a higher number of livestock, as compared to districts in eastern Finnmark, the indications that both alphaherpes- and pestiviruses circulating in the reindeer populations are reindeer-specific, suggest that the presence of domestic animals cannot explain the high prevalence in reindeer. Whether this could be the result of husbandry practices, reduced area for the animals to graze (temporary high-density episodes) or presence of other predisposing factors that may increase infection with these two viruses, remains unknown and should be a target for further investigations.

Carcass weight was an important risk factor for seropositivity. The mean weight value for each quantile (Table 2) can be used to correctly interpret seroprevalence at district level (Table 1). Our method of expressing the results as a cross interpretation between weight and seroprevalence avoids misclassification of age. This could have affected the data from 275 animals in this study and resulted in misinterpretation of the results (as shown by some discrepancies in the weight quantiles shown in Table 2, i.e., calves in quantiles 5 to 10 and adults in quantiles 1 to 4).

The strong correlation between seroprevalence and weight/age identified in this study is supported by previous studies [5,8] and supports previous findings of life-long immunity to pestivirus infection. Most likely, the source of infection are PI animals.

Geographic origin (district) and gender seemed to have only a small impact on the risk of infection. We find no compelling argument to explain the slightly higher likelihood of males becoming infected as compared to females, since there is no special difference in the management of males and females that could account for this difference. We thus argue that this association might be the result of a possible bias in the sampling between genders. In fact, several studies of pestivirus infections in wildlife have tried to assess the effect of gender, often obtaining weak or non-significant associations or sometimes-contradictory associations [31].

VNT results indicate that the virus circulating in reindeer is more closely related to BDV1 and BDV2 than BVDV1. Since BVDV was eradicated from domestic animals in Norway in 2006 [15], and BDV has never been reported in Norway [17], the VNT results might therefore indicate the circulation of an unknown virus more closely related to BDV. The relatively low titers against the BDV1 and BDV2 could indicate reduced affinity (i.e., heterologous viruses), but may also suggest a reduction of circulating antibodies over time. The lack of new outbreaks of BVDV1 in cattle in Norway also supports the conclusion that it is unlikely that reindeer would be harboring this virus (as confirmed by low titers in our VNT). Reindeer have sporadic contact with both cattle and small ruminants and one could presume if BVDV1 were circulating in reindeer, it would have been detected during routine screenings of livestock.

Real-time RT-PCR failed to identify any sample positive for viral RNA. Detection of viral RNA is only possible either from PI animals or from transiently infected animals sampled during the period of viremia, which in cervids may last less than 5 days [32]. The lack of detection of PI animals by RT-PCR is not surprising as these animals usually represent a very small fraction of any given host population [33,34]. A study in Sweden including more than 276 reindeer serum samples from districts with high pestivirus seroprevalence also failed to detect viral RNA by real-time RT-PCR [5].

The high seroprevalence in adult animals likely represents continuous episodes of transmission. This requires the presence of PI animals, generated by transplacental infection from a transiently infected mother. Since the fetus must be infected during early pregnancy to become PI, and since this period coincides with the peak of winter when domestic animals are housed, the chance of domestic animals contributing to the maintenance of the infection cycle in reindeer is small or non-existent [5]. This is further supported by the absence of pestivirus infections in domestic ruminants in Norway.

Infections known to be involved in abortion or reduced survival of newborns, such as herpes or pestivirus infections, should be further studied to gain a better understanding of factors affecting mortality and reproductive success. The identification of abortion related to pestiviruses or the mortality of young animals due to mucosal disease—both often described in cattle associated with BVDV [35]—is however difficult, given that aborted calves and dead animals are quickly removed by scavengers.

Another important factor that remains unknown at this point is the potential pathogenicity of a reindeer-specific pestivirus to other ruminant species (wild or domesticated). PI sheep with BDV1 have been shown to transmit the virus to seronegative calves and adult cattle [36,37]. Given the relatively high seroprevalence among reindeer and the possibility of contact with cattle, goats or sheep, it might be that this transmission potential is reduced, or that the virus in question produces mild or no clinical signs in other species.

In conclusion, the present results confirm the circulation of a hitherto unknown pestivirus in the semi-domesticated reindeer population of northern Norway that seems to be serologically closely related to the BDV genotype group 2. Characterization of the causative virus, its infection biology and pathogenic impact in the reindeer populations, as well as its potential to infect other domestic and wild ruminants, should be further investigated.

Author Contributions: Data curation, C.G.d.N. and E.S.; Formal analysis, C.G.d.N.; Funding acquisition, M.T. and C.G.d.N.; Investigation, C.G.d.N., I.H.N., J.J.W. and M.T.; Methodology, C.G.d.N., J.J.W., I.H.N., S.A., and M.T.; Project administration, M.T.; Validation, C.G.d.N., E.S., J.J.W. and S.A.; Writing—original draft, C.G.d.N.; Writing—review and editing, C.G.d.N., J.J.W., I.H.N., E.S., S.A., and M.T. All authors have read and agreed to the published version of the manuscript.

Funding: This project was funded by the Norwegian Reindeer Development Fund (Reindriftens Utviklingsfond, RUF).

Acknowledgments: For irreplaceable help in the field, we acknowledge veterinary students Therese Berger, Anett K. Larsen, Trine Marhaug, Veronique Poulain, Matthieu Roger, and Gina Petrovich, and guest researcher Frederico Morandi. We thank the staff at the Karasjok, Kautokeino and Šuoššjávri slaughterhouses for their help and hospitality. We also acknowledge the help from Eva M. Breines and Ellinor Hareide in our laboratory in Tromsø, and Irene Dergel, Maj Hjort, and Lena Renström at National Veterinary Institute, Uppsala, Sweden, for laboratory assistance. Finally, we are very thankful to Sophie Ellen Scotter for language corrections.

Conflicts of Interest: The authors declare no conflict of interest.

References

1. Murphy, F.A.; Gibbs, E.P.; Horzinek, M.C.; Studdert, M.J. Laboratory Diagnosis of Viral Diseases. In *Veterinary Virology*, 3rd ed.; Murphy, F.A., Gibbs, E.P.J., Horzinek, M.C., Studdert, M.J., Eds.; Academic Press: London, UK, 1999; pp. 193–224.
2. Vilcek, S.; Nettleton, P.F. Pestiviruses in wild animals. *Vet. Microbiol.* **2006**, *116*, 1–12. [CrossRef] [PubMed]
3. Ridpath, J.F.; Neill, J.D. Challenges in Identifying and Determining the Impacts of Infection with Pestiviruses on the Herd Health of Free Ranging Cervid Populations. *Front. Microbiol.* **2016**, *7*, 921. [CrossRef] [PubMed]
4. Larska, M. Pestivirus infection in reindeer. *Front. Microbiol.* **2015**, *6*, 1187. [CrossRef] [PubMed]
5. Kautto, A.H.; Alenius, S.; Mossing, T.; Becher, P.; Belak, S.; Larska, M. Pestivirus and alphaherpesvirus infections in Swedish reindeer (*Rangifer tarandus tarandus* L.). *Vet. Microbiol.* **2012**, *156*, 64–71. [CrossRef] [PubMed]
6. Rehbinder, C.; Nordkvist, M. A suspected virus infection of the oral mucosa in Swedish reindeer (*Rangifer tarandus*). *Rangifer* **1985**, *5*, 22–31. [CrossRef]
7. Rehbinder, C.; Belak, K.; Nordkvist, M. A serological, retrospective study in reindeer on five different viruses. *Rangifer* **1992**, *12*, 191–195. [CrossRef]
8. Lillehaug, A.; Vikoren, T.; Larsen, I.L.; Akerstedt, J.; Tharaldsen, J.; Handeland, K. Antibodies to ruminant alpha-herpesviruses and pestiviruses in Norwegian cervids. *J. Wildl. Dis.* **2003**, *39*, 779–786. [CrossRef]
9. Tryland, M.; Mørk, T.; Ryeng, K.A.; Sørensen, K.K. Evidence of parapox-, alphaherpes- and pestivirus infections in carcasses of semi-domesticated reindeer (*Rangifer tarandus tarandus*) from Finnmark, Norway. *Rangifer* **2005**, *25*, 75–83. [CrossRef]
10. Stuen, S.; Krogsrud, J.; Hyllseth, B.; Tyler, N.J.C. Serosurvey of three virus infections in reindeer in northern Norway and Svalbard. *Rangifer* **1993**, *13*, 215–219. [CrossRef]
11. Becher, P.; Orlich, M.; Kosmidou, A.; Konig, M.; Baroth, M.; Thiel, H.J. Genetic diversity of pestiviruses: Identification of novel groups and implications for classification. *Virology* **1999**, *262*, 64–71. [CrossRef]
12. Avalos-Ramirez, R.; Orlich, M.; Thiel, H.J.; Becher, P. Evidence for the presence of two novel pestivirus species. *Virology* **2001**, *286*, 456–465. [CrossRef] [PubMed]
13. Morton, J.; Evermann, J.F.; Dieterich, R.A. Experimental infection of reindeer with bovine viral diarrhea virus. *Rangifer* **1990**, *10*, 75–77. [CrossRef]
14. Loken, T.; Krogsrud, J.; Larsen, I.L. Pestivirus infections in Norway. Serological investigations in cattle, sheep and pigs. *Acta Vet. Scand.* **1991**, *32*, 27–34. [PubMed]
15. Kampen, A.H.; Åkerstedt, J.; Gudmundsson, S.; Hopp, P.; Grøneng, G.; Nyberg, O. *The Surveillance and Control Programme for Bovine Virus Diarrhoea (BVD) in Norway*; Norwegian Veterinary Institute: Oslo, Norway, 2007; pp. 65–71.
16. Loken, T.; Nyberg, O. Eradication of BVDV in cattle: The Norwegian project. *Vet. Rec.* **2013**, *172*, 661. [CrossRef] [PubMed]
17. Loken, T. Pestivirus infections in ruminants in Norway. *Revue Sci. Tech.-Off. Int. Épizoot.* **1992**, *11*, 895–899. [CrossRef]
18. Loken, T.; Barlow, R.M. Border disease in Norway. *Acta Vet. Scand.* **1981**, *22*, 137–139.
19. Loken, T.; Bjerkas, I.; Hyllseth, B. Border disease in goats in Norway. *Res. Vet. Sci.* **1982**, *33*, 103–131. [CrossRef]
20. Loken, T.; Hyllseth, B.; Larsen, H.J. Border disease in Norway. Serological examination of affected sheep flocks. *Acta Vet. Scand.* **1982**, *23*, 46–52.

21. Loken, T. Pestivirus infections in Norway. Epidemiological studies in goats. *J. Comp. Pathol.* **1990**, *103*, 1–10. [CrossRef]
22. *Ressursregnskap for Reindriftsnaeringen*; Reindriftsforvaltningen/Norwegian Reindeer Husbandry Authority: Alta, Norway. Available online: https://docplayer.me/49637426-Ressursregnskap-for-reindriftsnaeringen-r-e-i-n-d-r-i-f-t-s-f-o-r-v-a-l-t-n-i-n-g-e-n-j-u-n-i.html (accessed on 26 December 2019).
23. Tveraa, T.; Fauchald, P.; Henaug, C.; Yoccoz, N.G. An examination of a compensatory relationship between food limitation and predation in semi-domesticated reindeer. *Oecologia* **2003**, *137*, 370–376. [CrossRef]
24. *Ressursregnskap for Reindriftsnaeringen*; Reindriftsforvaltningen/Norwegian Reindeer Husbandry Authority: Alta, Norway, 2008. Available online: https://www.landbruksdirektoratet.no/no/reindriften/for-siidaandeler/publikasjoner?index=0 (accessed on 26 December 2019).
25. Collett, M.S. Molecular genetics of pestiviruses. *Comp. Immunol. Microbiol. Infect. Dis.* **1992**, *15*, 145–154. [CrossRef]
26. Deregt, D.; Masri, S.A.; Cho, H.J.; Bielefeldt, O.H. Monoclonal antibodies to the p80/125 gp53 proteins of bovine viral diarrhea virus: Their potential use as diagnostic reagents. *Can. J. Vet. Res.* **1990**, *54*, 343–348. [PubMed]
27. Thrusfield, M. Serological epidemiology. In *Veterinary Epidemiology*; Thrusfield, M., Ed.; Butterworths: London, UK, 1986; pp. 175–186.
28. Elvander, M.; Baule, C.; Persson, M.; Egyed, L.; Ballagi-Pordany, A.; Belak, S.; Alenius, S. An experimental study of a concurrent primary infection with bovine respiratory syncytial virus (BRSV) and bovine viral diarrhoea virus (BVDV) in calves. *Acta Vet. Scand.* **1998**, *39*, 251–264. [PubMed]
29. Hosmer, D.W.; Lemeshow, S. *Applied Logistic Regression*; John Wiley &Sons, Inc.: New York, NY, USA, 1989.
30. das Neves, C.G.; Thiry, J.; Skjerve, E.; Yoccoz, N.G.; Rimstad, E.; Thiry, E.; Tryland, M. Alphaherpesvirus infections in semidomesticated reindeer: A cross-sectional serological study. *Vet. Microbiol.* **2009**, *139*, 262–269. [CrossRef] [PubMed]
31. Pioz, M.; Loison, A.; Gibert, P.; Dubray, D.; Menaut, P.; Le Tallec, B.; Artois, M.; Gilot-Fromont, E. Transmission of a pestivirus infection in a population of Pyrenean chamois. *Vet. Microbiol.* **2007**, *119*, 19–30. [CrossRef]
32. Ridpath, J.F.; Mark, C.S.; Chase, C.C.; Ridpath, A.C.; Neill, J.D. Febrile response and decrease in circulating lymphocytes following acute infection of white-tailed deer fawns with either a BVDV1 or a BVDV2 strain. *J. Wildl. Dis.* **2007**, *43*, 653–659. [CrossRef]
33. Hessman, B.E.; Fulton, R.W.; Sjeklocha, D.B.; Murphy, T.A.; Ridpath, J.F.; Payton, M.E. Evaluation of economic effects and the health and performance of the general cattle population after exposure to cattle persistently infected with bovine viral diarrhea virus in a starter feedlot. *Am. J. Vet. Res.* **2009**, *70*, 73–85. [CrossRef]
34. Loneragan, G.H.; Thomson, D.U.; Montgomery, D.L.; Mason, G.L.; Larson, R.L. Prevalence, outcome, and health consequences associated with persistent infection with bovine viral diarrhea virus in feedlot cattle. *J. Am. Vet. Med. Assoc.* **2005**, *226*, 595–601. [CrossRef]
35. Schweizer, M.; Peterhans, E. Pestiviruses. *Annu. Rev. Anim. Biosci.* **2014**, *2*, 141–163. [CrossRef]
36. Braun, U.; Bachofen, C.; Buchi, R.; Hassig, M.; Peterhans, E. Infection of cattle with Border disease virus by sheep on communal alpine pastures. *Schweiz. Arch. Tierheilkd.* **2013**, *155*, 123–128. [CrossRef]
37. Braun, U.; Reichle, S.F.; Reichert, C.; Hassig, M.; Stalder, H.P.; Bachofen, C.; Peterhans, E. Sheep persistently infected with Border disease readily transmit virus to calves seronegative to BVD virus. *Vet. Microbiol.* **2014**, *168*, 98–104. [CrossRef] [PubMed]

viruses

MDPI

Article

Serosurveillance and Molecular Investigation of Wild Deer in Australia Reveals Seroprevalence of *Pestivirus* Infection

Jose L. Huaman [1,2], Carlo Pacioni [3,4], David M. Forsyth [5], Anthony Pople [6], Jordan O. Hampton [4,7], Teresa G. Carvalho [2] and Karla J. Helbig [1,*]

1. Department of Physiology, Molecular Virology Laboratory, Anatomy and Microbiology, School of Life Sciences, La Trobe University, Melbourne 3086, Australia; j.huamantorres@latrobe.edu.au
2. Department of Physiology, Molecular Parasitology Laboratory, Anatomy and Microbiology, School of Life Sciences, La Trobe University, Melbourne 3086, Australia; t.carvalho@latrobe.edu.au
3. Department of Environment, Land, Water and Planning, Arthur Rylah Institute for Environmental Research, Heidelberg 3084, Australia; carlo.pacioni@delwp.vic.gov.au
4. School of Veterinary and Life Sciences, Murdoch University, South Street, Murdoch, WA 6150, Australia; jordan.hampton@murdoch.edu.au
5. NSW Department of Primary Industries, Vertebrate Pest Research Unit, Orange 2800, Australia; dave.forsyth@dpi.nsw.gov.au
6. Department of Agriculture and Fisheries, Invasive Plants & Animals Research, Biosecurity Queensland, Ecosciences Precinct, Brisbane 4102, Australia; tony.pople@daf.qld.gov.au
7. Ecotone Wildlife, P.O. Box 76, Inverloch, VIC 3996, Australia

* Correspondence: k.helbig@latrobe.edu.au; Tel.: +61-3-9479-6650

Received: 28 May 2020; Accepted: 8 July 2020; Published: 13 July 2020

Abstract: Since deer were introduced into Australia in the mid-1800s, their wild populations have increased in size and distribution, posing a potential risk to the livestock industry, through their role in pathogen transmission cycles. In comparison to livestock, there are limited data on viral infections in all wildlife, including deer. The aim of this study was to assess blood samples from wild Australian deer for serological evidence of exposure to relevant viral livestock diseases. Blood samples collected across eastern Australia were tested by ELISA to detect antigens and antibodies against *Pestivirus* and antibodies against bovine herpesvirus 1. A subset of samples was also assessed by RT-PCR for *Pestivirus*, Simbu serogroup, epizootic hemorrhagic disease virus and bovine ephemeral fever virus. Our findings demonstrated a very low seroprevalence (3%) for ruminant *Pestivirus*, and none of the other viruses tested were detected. These results suggest that wild deer may currently be an incidental spill-over host (rather than a reservoir host) for *Pestivirus*. However, deer could be a future source of viral infections for domestic animals in Australia. Further investigations are needed to monitor pathogen activity and quantify possible future infectious disease impacts of wild deer on the Australian livestock industry.

Keywords: Australia; deer; prevalence; *Pestivirus*; ruminants; serosurveillance; virology; wildlife disease

1. Introduction

Deer (family Cervidae) often attain high densities when introduced to new areas [1]. As cervids are ungulates and closely related to economically important livestock species including cattle, sheep and goats, it is unsurprising that they share many pathogens, including several of major agricultural relevance. Many viral pathogens of farm ruminants have been detected in wild cervids globally, the most important of which are bovine viral diarrhea virus (BVDV), bovine herpesvirus 1 (BoHV-1), epizootic hemorrhagic disease virus (EHDV) and bovine ephemeral fever virus (BEFV) [2–6].

In Australia, wild populations of six non-native deer species became established in the late 1800s and early 1900s. These populations have expanded in abundance and distribution [1,2]. Wild deer in Australia commonly share grazing areas with livestock, and their susceptibility to a wide range of viral infections of importance to the livestock industry has been demonstrated [2]. Hence, wild deer represent a significant potential source of pathogen transmission to livestock [1].

Australian agriculture is currently free from many viral diseases that impact livestock industries in other parts of the world, including foot-and-mouth disease (FMD) and louping ill virus. However, the issue of cervid-transmitted disease in Australia is important because incursions or outbreaks of emerging, exotic or endemic disease could cause serious production losses, resulting in substantial economic impacts for the livestock industry [1,2]. Transmission of disease by cervids could also prevent effective control, management or eradication of a livestock disease, resulting in prolonged epidemics [2].

There is currently limited information about the infection status of Australia's wild deer populations. A small number of studies assessed the distribution of BVDV in wild fallow (*Dama dama*) and red (*Cervus elaphus*) deer populations in the 1970s and 1980s [7,8], but these were conducted in small geographical areas. Moreover, serological studies of endemic livestock viruses including BEFV, EHDV and Simbu serogroup virus were performed in red deer from Queensland [9] and rusa deer (*Rusa timorensis*) from New South Wales [10]. Therefore, the role wild deer might play in the spread of diseases to livestock remains unclear. Addressing this knowledge gap is important for anticipating how viruses might be transmitted to other animals, and how diseases might be controlled. The objective of this study was to assess whether Australian wild deer populations are exposed to relevant viral livestock diseases by testing blood samples for antibodies and antigens or through detection of viral genetic material. Our results establish a baseline exposure level, and possible spreading patterns in Australia's wild deer.

2. Materials and Methods

2.1. Sampling

Blood samples were collected by recreational hunters and professional culling staff from deer populations across eastern Australia. Most of the sampling areas were located within a 1- to 10-kilometer radius from agricultural grazing areas (Table 1, Figure 1A). Samples were generally collected during field necropsies in cooler winter months (Table 1, Figure 1B), when most culling and hunting occurs in Australia. Blood was drawn from the jugular vein and/or the heart and thoracic cavity, and collected in tubes (Becton Dickinson, Franklin Lakes, NJ, USA) with and without anticoagulant (EDTA) to obtain plasma and serum, respectively. Collection tubes were immediately refrigerated and transported to the laboratory. At the laboratory, serum and plasma were separated by centrifugation (10 min at 2000 g), transferred to 1.5 mL screw capped vials and stored at −80 °C until assayed.

Table 1. Population characteristics and distribution of deer sampled and tested in this study.

States or Territory	Animals	Sampling Location	Species	Sex			Age Groups				Month of Sampling									No. Deer Tested by		
				M	F	N.r.	Ad	Yrl	Fw	N.r.	Feb	Mar	Apr	May	Jun	Jul	Aug	Oct	Nov	ELISA Ab	ELISA Ag	PCR
NSW	244	Liverpool Plains *	Fallow	74	52	0	74	47	5	0	0	0	0	0	39	0	87	0	0	126	126	42
		Eden *		12	21	5	21	9	3	5	0	0	0	0	0	0	18	15	5	38	0	18
		Wollongong *	Rusa	69	11	0	68	12	0	0	43	0	14	0	13	0	0	10	0	80	66	7
ACT	34	Canberra *	Fallow	14	20	0	26	8	0	0	0	0	0	0	34	0	0	0	0	34	0	10
VIC	44	Alpine National Park	Sambar	17	14	1	19	6	6	1	10	0	15	7	0	0	0	0	0	32	16	17
		Upper Yarra Flats		2	6	1	6	1	1	1	0	0	0	0	8	0	0	1	0	9	8	4
		Yellinbo		0	1	0	0	0	1	0	0	0	0	0	0	1	0	0	0	1	0	1
			Fallow	0	2	0	1	0	1	0	0	0	0	0	0	2	0	0	0	2	0	2
QLD	110	North east Queensland *	Chital	41	69	0	90	20	0	0	0	47	0	0	5	0	0	0	58	110	105	43
Total	432			229	196	7	305	103	17	7	53	47	29	7	99	3	105	26	63	432	321	144

* Sampling was conducted in areas close to livestock farms; NSW: New South Wales; ACT: Australian Capital Territory; VIC: Victoria; QLD: Queensland; F: female, M: male, Ad: adult, Yrl: yearling, Fw: fawn, N.r.: Not recorded; chital deer (*Axis axis*), rusa deer (*Rusa timorensis*), sambar deer (*Rusa unicolor*), fallow deer (*Dama dama*).

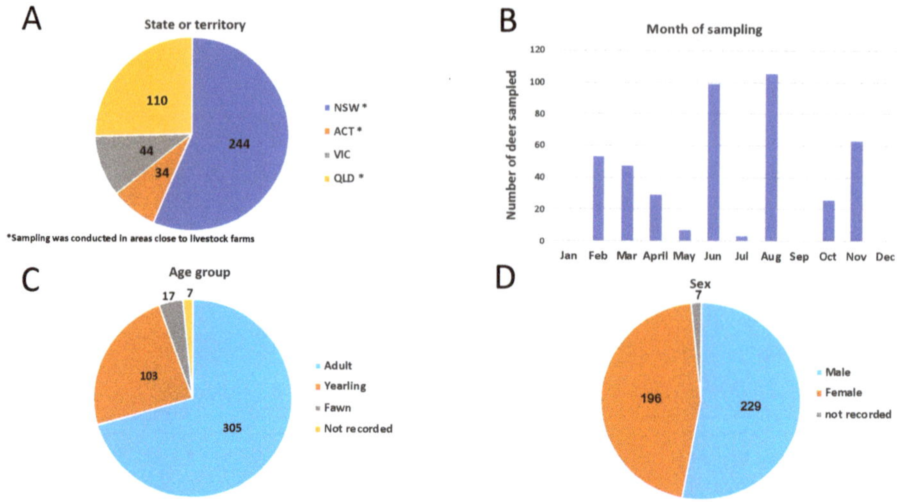

Figure 1. Population characteristics and distribution of deer sampled and tested in this study. Total numbers of deer sampled per region (**A**), per month (**B**), by age group (**C**) and by sex (**D**) are represented graphically.

2.2. Serological Methods

Serum and plasma were tested for *Pestivirus* antigen (Ag) and antibodies (Ab) against BoHV-1 and *Pestivirus* using the commercially available immunoenzymatic assay (ELISA) kits SERELISA® BVD p80 Ag Mono Indirect (Synbiotics SAS, Lyon, France), SERELISA BHV-1 Ab Mono Blocking ELISA (Synbiotics SAS, Lyon, France) and the PrioCHECK™ BVD & BD p80 Serum Kit (ThermoFisher Scientific, Rockford, USA) according to the manufacturer's instructions. For BoHV-1, samples with a ratio OD (optical density) sample/OD negative control ≤0.5 were classified as positive. This kit presents sensitivity (sn) of 96% and specificity (sp) 99%, according with the manufacturer. The *Pestivirus* kit detects antibodies targeted against the protein (p80/125), common to all BVDV and border disease virus (BDV) strains, with a manufacturer reported 97% sn and 98% sp for BVDV and 96% sn and 100% sp for BDV. Inhibition percentage (%INH) for each sample was calculated according to the manufacturer's kit insert.

Samples with a percentage inhibition (%INH) of <50 were classified as negative, those with 50 ≤ %INH < 80 as weak positive, and %INH ≥ 80 as strong positive. Results for the *Pestivirus* antigen detection kit are expressed as an index = 0.5 × OD sample−OD Positive control (P). Any sample having an index ≥(−0.15 × OD P) was considered positive, <(−0.3 × OD P) was considered negative and between (−0.15 × OD P) and (−0.3 × OD P) was considered doubtful according to the manufacturer's instructions.

Positive and negative controls were included in each run following the manufacturer's recommendations. Furthermore, all deer samples were initially tested in pools of three, with all serum samples in positive pools being additionally sampled individually and in duplicate. Optical density was measured using a plate reader (ClarioStar—BMG Labtech, Ortenberg, Germany) at 450 nm wavelength.

2.3. RNA Extraction and RT-PCR

Due to the large number of animals sampled, only a subset of 144 sera was selected across all sampled regions to be screened by PCR (Table 1) for four agriculturally relevant viruses (EHDV, BEFV, *Pestivirus* and Simbu serogroup). These included all the samples with ELISA-Ag positive and doubtful results. RNA was extracted from 140 µL of serum or cell culture supernatant (positive controls) using a QIAamp® Viral RNA Mini Kit (Qiagen, Valencia, CA, USA), according to the manufacturer's instructions. Viral RNA was reverse transcribed using a Tetro cDNA Synthesis Kit (Bioline, London, UK) using random hexamers according to the manufacturer's directions. RNA extracted from in vitro cultures for Akabane Virus, BEFV, EHDV and one bovine serum sample confirmed to be positive for BVDV, were used as positive controls. All culture material and BVDV positive sera were kindly donated by the Department of Jobs, Precincts and Regions, Victoria. PCR amplification was performed in a 25 µL reaction mixture containing 1× Green GoTaq Flexi buffer, 2 mM of $MgCl_2$, 10 mM of dNTPs, 0.2 µM of both forward and reverse primers (Table 2), 0.625 units of GoTaq G2 DNA polymerase (Promega, Madison, WI, USA) and 1 µL of total genomic DNA template. PCR primers were obtained from the literature for the four viruses included in this study (Table 2 and references therein). Amplification was carried out in a T100 thermal cycler (BioRad, Hercules, CA, USA), and amplification products visualized by gel electrophoresis, using 2% agarose gel, RedSafe™ (iNtRON Biotechnology, Gyeonggi-do, Korea), and a high-resolution imaging system—ChemiDoc™ MP Imaging System (Bio-Rad, Hercules, CA, USA).

2.4. Statistical Analysis

Samples were categorized as "positive" or "negative" based on the results of the ELISA-Ab. Seroprevalence was calculated from the proportion of seropositive results of those tested and is presented with 95% confidence interval (CI), calculated using the Wilson score interval (www.epitools.ausvet.com.au). Binary logistic regression models were used to evaluate the effect of the sex, age category and sampling site on the antibody status. Logistic regression was performed using R version 4.0.0 (R Development Core Team, Vienna, Austria). Due to the sparse nature of the data, we performed the logistic regression analysis only on fallow deer (in which most of the positive samples were detected). Lastly, we used the two-tailed Fisher's exact test to evaluate the hypothesis of non-random distribution of positive samples between fallow and rusa deer. $p < 0.05$ was considered statistically significant.

Table 2. List of oligonucleotides and PCR conditions used in this study.

Virus	Target Region	Primer Name	Sequence 5′–3′	Amplicon Length (bp)	PCR Condition	Reference
Pestivirus	5′UTR	324 326	ATGCCCWTAGTAGGACTAGCA WCAACTCCATGTGCCATGTAC	288	95 °C × 2 min 40 cycles (95 °C × 45 s, 52 °C × 45 s, 72 °C × 45 s) 72 °C × 5 min	[11]
Simbu Serogroup	Segment S	Uni-S-59F Uni-S-254R	GATGWCCWCAACGGAAT TGGGGAAAATGGTTATTAAC	215	95 °C × 2 min 40 cycles (95 °C × 45 s, 55 °C × 45 s, 72 °C × 45 s) 72 °C × 5 min	[12]
BEFV	Glucoprotein G	GF GR	ATGTTCAAGGTCCTCATAATTACC TAATGATCAAAGAACCTATCATCA	1871	95 °C × 2 min 40 cycles (95°C × 45 s, 52 °C × 45 s, 72 °C × 2 min) 72 °C × 5 min	[13]
EHDV	NS3	NS3F NS3R	CAGCGCYWTATWCGATATTG TCCGGAGATACCTCCATTAC	533	95 °C × 2 min 40 cycles (95 °C × 45 s, 55 °C × 45 s, 72 °C × 60 s) 72 °C × 5 min	[14]

3. Results

3.1. Deer Sampling and Distribution

During the sampling period, 432 wild deer were sampled encompassing four deer species (200 fallow deer, 110 chital deer (*Axis axis*), 80 rusa deer and 42 sambar deer (*Rusa unicolor*)) across eastern Australia (Table 1). Sampling was conducted from November 2017 to November 2019, with most samples (72%) collected between June and October (Table 1, Figure 1B). Slightly more females (n = 229) than males (n = 196) were sampled, while no information was available for seven animals (Table 1, Figure 1D). Individuals were classified in three age categories based on morphological characteristics, including body size, tooth wear and antler growth: fawn (<1 year), yearling (1 to <2 years) and adult (≥2 years). Most of the samples came from adult individuals (n = 305), followed by yearlings (n = 103) and fawns (n = 17). Information on age was not available for seven animals (Table 1, Figure 1C).

3.2. ELISA Testing

Sera and plasma samples from all 432 wild deer were screened by ELISA-Ab for *Pestivirus* and BoHV-1 (Table 1). All samples were negative for BoHV-1 antibodies. However, a total of 13 wild deer reacted positive for *Pestivirus* antibodies, resulting in an overall seroprevalence of 3.0%. Of the *Pestivirus* seropositive deer, 46.2% were sampled in June and 17.8% in August. Of the positive samples, 85% were obtained from adults. Additionally, all *Pestivirus* positive samples were collected from fallow and rusa deer, with a 5.5% and a 2.5% *Pestivirus*-seropositivity for each species, respectively (Figure 2).

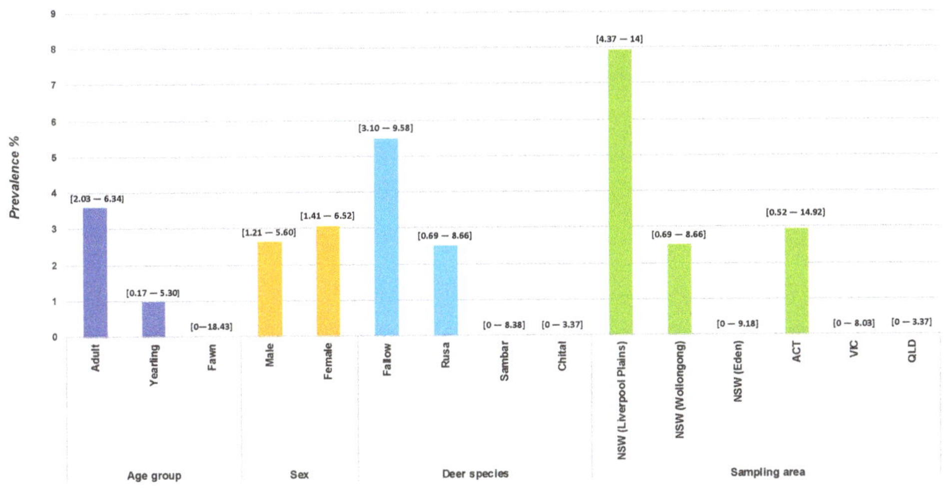

Figure 2. Prevalence of *Pestivirus* Antibodies: antibody prevalence was determined by ELISA and is represented as prevalence across age group, sex, species and sampling area. The 95% confidence interval is shown in brackets.

The animals positive for *Pestivirus* ELISA-Ab were sampled in Australian Capital Territory (ACT; 7.7%) and New South Wales (NSW; 92.3%). Two out of the three sampling areas from this last state showed seropositive deer with a local prevalence of 7.94% in central NSW (Liverpool Plains), and 2.5% in coastal NSW (Wollongong) (Figure 2). Five of the 13 wild deer reacted as strong positives with test values (%INH) ranging between 80–92%; the remaining eight wild deer were weak positives with percentages ranging between 51–79% (Table 3).

Table 3. Description of deer samples that tested positive for *Pestivirus* antibodies.

Sample	Species	Sampling Month	Sex	Age	Location	Anti-*Pestivirus* ELISA			
						Serum		Plasma	
						Result	%INH [a]	Result	%INH [a]
1	Fallow	June	F	Ad	New South Wales	WP	65	Neg	-
2	Fallow	June	F	Ad	New South Wales	SP	91.9	SP	91.4
3	Fallow	June	F	Ad	New South Wales	WP	73.7	WP	57.4
4	Fallow	June	F	Yrl	New South Wales	WP	73.6	WP	59.0
5	Fallow	June	F	Ad	New South Wales	WP	61.6	WP	67.0
6	Fallow	August	M	Ad	New South Wales	WP	71.6	WP	67.1
7	Fallow	August	M	Ad	New South Wales	SP	80.1	WP	79.4
8	Fallow	August	M	Ad	New South Wales	SP	84.4	SP	83.3
9	Fallow	August	M	Ad	New South Wales	SP	87.2	SP	80.1
10	Fallow	August	M	Ad	New South Wales	SP	87.2	WP	78.9
11	Rusa	February	ND	ND	New South Wales	WP	51.8	NS	-
12	Rusa	October	F	Ad	New South Wales	Neg	-	WP	60.1
13	Fallow	June	M	Ad	Australian Capital Territory	NS	-	WP	56.4

[a] %INH: percentage of inhibition obtained by ELISA; ND: no data, F: female, M: male, Ad: adult, Yrl: yearling; WP: weak positive, SP: strong positive, NS: no sample, Neg: negative.

Additionally, 321 out the 432 wild deer were also tested for *Pestivirus* antigens (see Table 1). Of the 321 deer tested, 278 (86.6%) showed a negative result, 24 (7.5%) revealed a doubtful result and 19 (5.9%) reacted positive. Additionally, only two of these positives overlapped with positive results shown for the ELISA-Ab screen.

Due to quasi-complete separation (i.e., most of the positives were in one category), we initially limited the predictors to the main effect of sex and age category (adult, yearling and fawn) or sex and location, and then tried to further simplify the model by considering sex only. In all models, the effect of the variables was not significant ($p > 0.05$). The proportion of seropositive *Pestivirus* results for fallow and rusa deer was similar ($p = 0.36$).

3.3. RT-PCR Screening

From the whole biobank of sera, 144 samples (see Table 1) were selected to be tested for *Pestivirus*, EDHV, BEFV and Simbu serogroup using previously validated RT-PCR primers sets [11–14]. These included all samples that had a doubtful or positive reaction for ELISA-Ag. Samples with hemolysis and insufficient volume were not included in the screening. No positive PCR amplification was obtained for any of the viruses screened, however, a positive result was obtained for all positive control samples in all runs.

4. Discussion

Australia's wild deer populations have increased in abundance and distribution during recent decades, and the close interaction between deer and livestock is a risk for pathogen transmission. However, little is known about the epidemiological role of wild deer in Australia. This study constitutes the largest number of deer, and deer species, sampled in Australia to date for viral pathogens, and complements a recent study performed on a similarly large number of animals across multiple geographic locations for the presence of parasitic infections [15]. Moreover, this study indicates exposure of deer to *Pestivirus* and is the first report of antibodies to ruminant *Pestiviruses* in rusa deer. This baseline information is of value for monitoring the status of endemic livestock pathogens in Australian deer, and for evaluating the risk of disease transmission between wild deer and livestock.

Although numerous pathogens have been detected in different cervids worldwide, including agriculturally relevant viruses [2], there is scarce knowledge about the viral infection status of Australian wild deer populations, with only four previous serosurveys being performed on red, fallow and rusa deer, but limited in their geographical coverage [7–10]. In contrast, the present study is based on the assessment of large sample sizes from four deer species collected throughout eastern Australia.

Our findings reveal the presence of antibodies in Australian wild deer species, against ruminant *Pestiviruses* (BVDV and BDV), which infect a variety of wild and domestic ungulate species and are associated with severe economic losses in livestock production worldwide [16]. In this study, positive serologic reactions were recorded in fallow and rusa deer at similar rates. Although the serosurveys have been proven to be a fundamental tool when it comes to disease surveillance, serological testing used in this study cannot distinguish between BVDV and BDV infections.

To our knowledge, this is the first report of *Pestivirus* antibodies in rusa deer. Additionally, in accordance with our findings, *Pestivirus* antibodies were identified previously in Australian fallow deer [7,8]. Seropositive fallow deer were first reported by Munday et al. (1972) [7] in Tasmania, with a prevalence of 14.5%. Ten years later, another serosurvey described a prevalence of 1.2% (1/86) in fallow deer sourced from New South Wales [8], in an area also included in our study. Moreover, McKenzie et al. [9] reported a prevalence of 4% in red deer sourced from 20 localities in eastern Australia. In comparison with the previous report for fallow deer from New South Wales [8], a higher seroprevalence is reported in the present study. The use of a different detection assay could have influenced the prevalence obtained. Compared with previous Australian serosurveys that used a virus neutralization test targeting only BVDV1, we used a more robust immunoassay that could detect antibodies against BVDV1, BVDV2 and BDV. The higher seroprevalence in this study might also reflect a real change in prevalence in the last 40 years, possibly due to an increase in deer density leading to greater interaction with livestock. Although samples from red deer were not included in our analyses, seroprevalence detected in fallow and rusa deer is comparable to that described in red deer by McKenzie et al. [9]. These findings may indicate that red, fallow and rusa deer are similarly exposed to *Pestivirus*.

Fallow deer, chital deer and, to a lesser extent, rusa deer, are gregarious, and hence, their transmission of pathogens would be expected to be higher compared to sambar deer, which are usually solitary [17]. No seroprevalence for *Pestivirus* was observed for sambar deer in this study, however, sampling numbers were low in comparison to other species. The sampling location for sambar deer in this study was composed of forested areas within mountain landscapes mostly distant from livestock grazing areas; however, the chital deer samples were collected from within pastoral areas where they are known to interact with cattle.

There is conflicting evidence in the literature surrounding prevalence of *Pestivirus* in cervids and known contact with cattle. Studies [18–20] have reported that close contact with cattle can induce high seroprevalence in wild cervids; however, a study by Frolich et al. [21] contrasted these findings and hypothesized an independent cycle as responsible for intrapopulation persistence. Additionally, the identification of persistently infected mule deer (*Odocoileus hemionus*) [22] and white-tailed deer (*Odocoileus virginianus*) [23–25] suggests that BVDV can sustain itself in deer populations without contact with cattle. The susceptibility of sambar and chital deer to BVDV has not been demonstrated, and we cannot completely discount the lack of susceptibility as a possible cause of these negative results. However, it is possible that mortality of persistently infected animals or differences in social behavior between chital and sambar deer with the other deer species, which could also explain these negative results.

Most of the seropositive *Pestivirus* outcomes were recorded in adult deer, which is not unexpected, as they constituted ~70% of the animals sampled in this study. Although antibody levels could be detected in the sampled deer, it is not possible to know when the animals were exposed to the virus. Seroconversion in cattle usually appears two weeks after infection, with the titer continuing to rise for 10 to 12 weeks; after this time, a plateau is reached [26,27]. Similarly, deer experimentally infected with BVDV exhibit a similar seroconversion course, developing antibodies at 8 to 15 days post infection [28,29]. However, antibodies against BVDV have been demonstrated to remain in serum for longer than three years [30].

Pestivirus antibody detection was performed in serum and plasma using a blocking ELISA kit with sensitivity and specificity comparable with other similar kits [31], also validated for non-bovine samples [32–34]. Concordance in results were found in eight overlapped specimens (serum and plasma), and two of the remaining overlapped specimens revealed positive outcomes only in one of the samples. Although this discrepancy is not enough to state which sample is better for *Pestivirus* antibody detection in deer, previous studies did not find variation between serum and plasma for viral antibody detection [35–37]. It is possible that other factors including the quality of the sample and antibody concentration could have affected the results in some samples.

Overall, the seroprevalence for *Pestivirus*es was 3%, considerably lower than the 52.6% observed in Australian cattle [16]. Similar findings were detected in European countries during serosurveys of *Pestivirus*es in deer to evaluate the epidemiological importance of deer in BVDV eradication programs: Belgium, 1.3% [38], Germany, 2% [39], Switzerland, 2.7% [3], and Italy, 4.5% [40]. The authors concluded that despite regular interactions with farmed ruminants, infection in deer was occasional with virus transmission from cattle to deer, and therefore, the possibility of deer being a source of infection for cattle was remote.

In contrast to other reports, we did not detect any infected deer by PCR, although the RT-PCR assay we used detects a broad range of *Pestivirus*es from pigs, cattle and sheep [11]. Thus, based on detection of a low seroprevalence for ruminant *Pestivirus*es in the deer population studied, the fact that we did not identify any persistently infected (virus positive, antibodies negative) deer, and given the high number of seropositive cattle in Australia, we consider it more probable that deer are an accidental spillover host rather than a reservoir host for ruminant *Pestiviruses*, and that persistently infected cattle could transmit these viruses to wild deer.

Pestivirus antigen detection by ELISA resulted in 5.9% and 7.5% of samples testing positive and doubtful, respectively. However, RT-PCR negative results were obtained in all the samples. The ELISA kit utilized is reported to detect *Pestivirus* antigens in ruminant samples, however, there is no published information about validation with cervid samples. A similar antigen detection methodology was previously used for roe deer (*Capreolus capreolus*) [41], with positive results obtained. However, those findings were not confirmed by RT-PCR. A possible explanation for the false positive antigen results could be variations in the primer target region not detectable by RT-PCR used. A second explanation is that there were unspecific cross-reactions, but further work is needed to determine the exact cause of the potential false positive antigen ELISA results obtained in this study.

We acknowledge that a limitation of this study is the type of tissue utilized for detection of BVDV genetic material. Although there are studies that used serum to detect BVDV by PCR [42,43], BVDV has trophism for epithelia of both the alimentary and integumentary systems [44], however, due to the collection strategies available to us during the sampling procedures, specimens that would be more reliable in detecting low levels of viral nucleic acid over a longer period of infection time, such as reproductive tissues [45,46] or spleen [47], were not available for testing.

BoHV-1 is the best characterized member of the subfamily *Alphaherpesvirus*, responsible for Infectious Bovine Rhinotracheitis (IBR), a cattle disease of major economic concern worldwide [48], and widespread in Australian cattle with a seroprevalence of 25–40% [49]. Susceptibility to BoHV-1 of wild cervids has previously been demonstrated [5,50], and additionally, several ruminant *Alphaherpesviruses* related to BoHV-1 have been isolated from cervids, including cervid herpesvirus 1 (CvHV-1) in red deer [51] and cervid herpesvirus 2 (CvHV-2) in reindeer (*Rangifer tarandus*) [52]. Our failure to detect antibodies against BoHV-1 is consistent with previous reports in Australia, which demonstrated an absence of BoHV-1 antibodies in fallow deer in Tasmania [7] and red deer in Queensland [9]. Furthermore, we can state that there is no evidence of cervid alphaherpesviruses in Australian wild deer, since serological cross reactivity by both virus neutralization and ELISA between ruminant alphaherpesviruses, including CvHV-1 and -2, is well documented [53–55].

The assessment of a subset of 144 serum samples by PCR also revealed no evidence of acute infection for the other viral livestock pathogens screened in this study. BEFV, EHDV and Akabane virus (a member of the Simbu serogroup) are endemic in Queensland and they have a seasonal spread in New South Wales [56–58]. Moreover, these viruses remain undetected in Victoria [57,58]. As they are vector-borne viruses, their occurrence is limited by the effect of cold weather, which restricts the distribution of their vectors. Previous studies performed in Australian deer reported serological evidence for BEFV, EHDV and Akabane virus in red deer from south-eastern Queensland [9]. Moreover, Moriarty et al. [10] found seropositive outcomes in a small sample of rusa deer in coastal central NSW for Akabane virus and EHDV. All the samples screened in the present study were negative for these viruses; however, the presence of vector species and previous evidence highlights the need for further serologic analysis to determine the role of deer as a spillover or reservoir host for these viruses, particularly in Queensland and New South Wales where BEFV, EHDV and Akabane virus are endemic. One limitation of this study was that all deer were sampled in the colder winter months, which would also lessen the activity of the vectors necessary for virus transmission.

5. Conclusions

Our findings provide an overview of the current *Pestivirus* infection status of wild deer in eastern Australia. The low prevalence of *Pestivirus* antibodies and negative findings for the viruses tested suggests that wild deer are an incidental spill-over host, and not a reservoir host. However, considering the substantial increase observed in fallow deer seroprevalence compared with a previous report [8], and the expected increase in distribution and abundance [1] (in the absence of substantial control), we cannot rule out the possibility that deer species sampled in this study could be a future source of infection for livestock.

Author Contributions: Conceptualization, C.P., T.G.C. and K.J.H.; methodology, J.L.H., C.P. and K.J.H.; formal analysis, J.L.H.; investigation, J.L.H., C.P. and K.J.H.; resources, C.P., D.M.F., A.P., J.O.H. and K.J.H.; data curation, J.L.H.; writing—original draft preparation, J.L.H. and K.J.H.; writing—review and editing, C.P., D.M.F., A.P., J.O.H., T.G.C. and K.J.H.; visualization, J.L.H. and K.J.H.; supervision, C.P., T.G.C. and K.J.H.; project administration, C.P., D.M.F., A.P., T.G.C. and K.J.H.; funding acquisition, C.P., D.M.F., A.P., T.G.C. and K.J.H. All authors have read and agreed to the published version of the manuscript.

Funding: This study was funded by the Centre for Invasive Species Solutions (PO1-L-002).

Acknowledgments: We would like to thank Richard Francis (ABZECO), Jake Haddad (VPAC), Kirk Stone (Strathbogie Wildlife), Andrew Bengsen, Troy Crittle and Quentin Hart (all New South Wales Department of Primary Industries), Bob McKinnon and Amy Sheridan (North West Local Land Services), Michael Brennan and Matt Amos (Biosecurity Queensland) and the staff from Parks Victoria for assisting with sample collection. In addition, we thank Kim O'Riley and Peter Mee for technical support. Finally, we thank Tao Zheng, PhD and Stacey Lynch, PhD from the Department of Jobs, Precincts and Regions, Victoria, for providing samples and cell culture supernatant used as positive controls.

Conflicts of Interest: The authors declare no conflict of interest.

References

1. Davis, N.E.; Bennett, A.; Forsyth, D.M.; Bowman, D.M.J.S.; Lefroy, E.C.; Wood, S.W.; Woolnough, A.P.; West, P.; Hampton, J.O.; Johnson, C.N. A systematic review of the impacts and management of introduced deer (family Cervidae) in Australia. *Wildl. Res.* **2016**, *43*, 515–532. [CrossRef]
2. Cripps, J.K.; Pacioni, C.; Scroggie, M.P.; Woolnough, A.P.; Ramsey, D.S.L. Introduced deer and their potential role in disease transmission to livestock in Australia. *Mammal Rev.* **2019**, *49*, 60–77. [CrossRef]
3. Casaubon, J.; Vogt, H.R.; Stalder, H.; Hug, C.; Ryser-Degiorgis, M.P. Bovine viral diarrhea virus in free-ranging wild ruminants in Switzerland: Low prevalence of infection despite regular interactions with domestic livestock. *BMC Vet. Res.* **2012**, *8*, 204. [CrossRef] [PubMed]
4. Duncan, C.; Backus, L.; Lynn, T.; Powers, B.; Salman, M. Passive, opportunistic wildlife disease surveillance in the Rocky Mountain Region, USA. *Transbound. Emerg. Dis.* **2008**, *55*, 308–314. [CrossRef]
5. Graham, D.A.; Gallagher, C.; Carden, R.F.; Lozano, J.M.; Moriarty, J.; O'Neill, R. A survey of free-ranging deer in Ireland for serological evidence of exposure to bovine viral diarrhoea virus, bovine herpes virus-1, bluetongue virus and Schmallenberg virus. *Irel. Vet. J.* **2017**, *70*, 13. [CrossRef]
6. Roug, A.; Swift, P.; Torres, S.; Jones, K.; Johnson, C.K. Serosurveillance for livestock pathogens in free-ranging mule deer (*Odocoileus hemionus*). *PLoS ONE* **2012**, *7*, e50600. [CrossRef]
7. Munday, B.L. A serological study of some infectious diseases of Tasmanian wildlife. *J. Wildl. Dis.* **1972**, *8*, 169–175. [CrossRef]
8. English, A.W. Serological survey of wild fallow deer (*Dama dama*) in New South Wales, Australia. *Vet. Rec.* **1982**, *110*, 153–154. [CrossRef]
9. McKenzie, R.A.; Green, P.E.; Thornton, A.M.; Chung, Y.S.; MacKenzie, A.R.; Cybinski, D.H.; St George, T.D. Diseases of deer in south eastern Queensland. *Aust. Vet. J.* **1985**, *62*, 424. [CrossRef]
10. Moriarty, A. Ecology and Enviromental Impact of Javan Rusa Deer (*Cervus timorensis russa*) in the Royal National Park. Ph.D. Thesis, University of Western Sydney, Sydney, NSW, Australia, 2004.
11. Vilcek, S.; Herring, A.J.; Herring, J.A.; Nettleton, P.F.; Lowings, J.P.; Paton, D.J. Pestiviruses isolated from pigs, cattle and sheep can be allocated into at least three genogroups using polymerase chain reaction and restriction endonuclease analysis. *Arch. Virol.* **1994**, *136*, 309–323. [CrossRef]
12. Golender, N.; Bumbarov, V.Y.; Erster, O.; Beer, M.; Khinich, Y.; Wernike, K. Development and validation of a universal S-segment-based real-time RT-PCR assay for the detection of Simbu serogroup viruses. *J. Virol. Methods* **2018**, *261*, 80–85. [CrossRef] [PubMed]
13. Kato, T.; Aizawa, M.; Takayoshi, K.; Kokuba, T.; Yanase, T.; Shirafuji, H.; Tsuda, T.; Yamakawa, M. Phylogenetic relationships of the G gene sequence of bovine ephemeral fever virus isolated in Japan, Taiwan and Australia. *Vet. Microbiol.* **2009**, *137*, 217–223. [CrossRef]

14. Ohashi, S.; Yoshida, K.; Yanase, T.; Kato, T.; Tsuda, T. Simultaneous detection of bovine arboviruses using single-tube multiplex reverse transcription-polymerase chain reaction. *J. Virol. Methods* **2004**, *120*, 79–85. [CrossRef] [PubMed]
15. Huaman, J.L.; Pacioni, C.; Forsyth, D.M.; Pople, A.; Hampton, J.O.; Helbig, K.J.; Carvalho, T. Screening of Blood Parasites in Australian Wild Deer. *Authorea* **2020**. [CrossRef]
16. Scharnbock, B.; Roch, F.F.; Richter, V.; Funke, C.; Firth, C.L.; Obritzhauser, W.; Baumgartner, W.; Kasbohrer, A.; Pinior, B. A meta-analysis of bovine viral diarrhoea virus (BVDV) prevalences in the global cattle population. *Sci. Rep.* **2018**, *8*, 14420. [CrossRef]
17. Van Dyck, S.; Strahan, R. *The Mammals of Australia*; Reed New Holland: Sydney, NSW, Australia, 2008.
18. Aguirre, A.A.; Hansen, D.E.; Starkey, E.E.; Mclean, R.G. Serologic Survey of Wild Cervids for Potential Disease Agents in Selected National-Parks in the United-States. *Prev. Vet. Med.* **1995**, *21*, 313–322. [CrossRef]
19. Rodriguez-Prieto, V.; Kukielka, D.; Rivera-Arroyo, B.; Martinez-Lopez, B.; de las Heras, A.I.; Sanchez-Vizcaino, J.M.; Vicente, J. Evidence of shared bovine viral diarrhea infections between red deer and extensively raised cattle in south-central Spain. *BMC Vet. Res.* **2016**, *12*. [CrossRef]
20. Fernandez-Aguilar, X.; Lopez-Olvera, J.R.; Marco, I.; Rosell, R.; Colom-Cadena, A.; Soto-Heras, S.; Lavin, S.; Cabezon, O. Pestivirus in alpine wild ruminants and sympatric livestock from the Cantabrian Mountains, Spain. *Vet. Rec.* **2016**, *178*, U557–U586. [CrossRef] [PubMed]
21. Frolich, K. Bovine Virus Diarrhea and Mucosal Disease in Free-Ranging and Captive Deer (Cervidae) in Germany. *J. Wildl. Dis.* **1995**, *31*, 247–250. [CrossRef]
22. Duncan, C.; Van Campen, H.; Soto, S.; LeVan, I.K.; Baeten, L.A.; Miller, M.W. Persistent Bovine viral diarrhea virus infection in wild cervids of Colorado. *J. Vet. Diagn. Investig.* **2008**, *20*, 650–653. [CrossRef]
23. Passler, T.; Walz, P.H.; Ditchkoff, S.S.; Walz, H.L.; Givens, M.D.; Brock, K.V. Evaluation of hunter-harvested white-tailed deer for evidence of bovine viral diarrhea virus infection in Alabama. *J. Vet. Diagn. Investig.* **2008**, *20*, 79–82. [CrossRef] [PubMed]
24. Duncan, C.; Ridpath, J.; Palmer, M.V.; Driskell, E.; Spraker, T. Histopathologic and immunohistochemical findings in two white-tailed deer fawns persistently infected with Bovine viral diarrhea virus. *J. Vet. Diagn. Investig.* **2008**, *20*, 289–296. [CrossRef] [PubMed]
25. Passler, T.; Walz, P.H.; Ditchkoff, S.S.; Givens, M.D.; Maxwell, H.S.; Brock, K.V. Experimental persistent infection with bovine viral diarrhea virus in white-tailed deer. *Vet. Microbiol.* **2007**, *122*, 350–356. [CrossRef] [PubMed]
26. Collins, M.E.; Heaney, J.; Thomas, C.J.; Brownlie, J. Infectivity of pestivirus following persistence of acute infection. *Vet. Microbiol.* **2009**, *138*, 289–296. [CrossRef] [PubMed]
27. Howard, C.J. Immunological responses to bovine virus diarrhoea virus infections. *Rev. Sci. Tech.* **1990**, *9*, 95–103. [CrossRef]
28. Van Campen, H.; Williams, E.S.; Edwards, J.; Cook, W.; Stout, G. Experimental infection of deer with bovine viral diarrhea virus. *J. Wildl. Dis.* **1997**, *33*, 567–573. [CrossRef]
29. Tessaro, S.V.; Carman, P.S.; Deregt, D. Viremia and virus shedding in elk infected with type 1 and virulent type 2 bovine viral diarrhea virus. *J. Wildl. Dis.* **1999**, *35*, 671–677. [CrossRef]
30. Fredriksen, B.; Sandvik, T.; Loken, T.; Odegaard, S.A. Level and duration of serum antibodies in cattle infected experimentally and naturally with bovine virus diarrhoea virus. *Vet. Rec.* **1999**, *144*, 111–114. [CrossRef]
31. Hanon, J.B.; De Baere, M.; De la Ferte, C.; Roelandt, S.; Van der Stede, Y.; Cay, B. Evaluation of 16 commercial antibody ELISAs for the detection of bovine viral diarrhea virus-specific antibodies in serum and milk using well-characterized sample panels. *J. Vet. Diagn. Investig.* **2017**, *29*, 833–843. [CrossRef]
32. Morrondo, M.P.; Perez-Creo, A.; Prieto, A.; Cabanelas, E.; Diaz-Cao, J.M.; Arias, M.S.; Fernandez, P.D.; Pajares, G.; Remesar, S.; Lopez-Sandez, C.M.; et al. Prevalence and distribution of infectious and parasitic agents in roe deer from Spain and their possible role as reservoirs. *Ital. J. Anim. Sci.* **2017**, *16*, 266–274. [CrossRef]
33. Loeffen, W.L.; van Beuningen, A.; Quak, S.; Elbers, A.R. Seroprevalence and risk factors for the presence of ruminant pestiviruses in the Dutch swine population. *Vet. Microbiol.* **2009**, *136*, 240–245. [CrossRef] [PubMed]
34. Evans, C.A.; Lanyon, S.R.; Reichel, M.P. Investigation of AGID and two commercial ELISAs for the detection of Bovine viral diarrhea virus-specific antibodies in sheep serum. *J. Vet. Diagn. Investig.* **2017**, *29*, 181–185. [CrossRef] [PubMed]

35. Cherpes, T.L.; Meyn, L.A.; Hillier, S.L. Plasma versus serum for detection of herpes simplex virus type 2-specific immunoglobulin G antibodies with a glycoprotein G2-based enzyme immunoassay. *J. Clin. Microbiol.* **2003**, *41*, 2758–2759. [CrossRef]
36. Parra-Álvarez, S.; Coronel-Ruiz, C.; Castilla, M.G.; Velandia-Romero, M.L.; Castellanos, J.E. Alta correlación en la detección de anticuerpos y antígenos de virus del dengue en muestras de suero y plasma. *Rev. Fac. Med.* **2015**, *63*, 687–693. [CrossRef]
37. Lin, H.T.; Hsu, C.H.; Tsai, H.J.; Lin, C.H.; Lo, P.Y.; Wang, S.L.; Wang, L.C. Influenza A plasma and serum virus antibody detection comparison in dogs using blocking enzyme-linked immunosorbent assay. *Vet. World* **2015**, *8*, 580–583. [CrossRef]
38. Tavernier, P.; Sys, S.U.; De Clercq, K.; De Leeuw, I.; Caij, A.B.; De Baere, M.; De Regge, N.; Fretin, D.; Roupie, V.; Govaerts, M.; et al. Serologic screening for 13 infectious agents in roe deer (*Capreolus capreolus*) in Flanders. *Infect. Ecol. Epidemiol.* **2015**, *5*, 29862. [CrossRef]
39. Krametter, R.; Nielsen, S.S.; Loitsch, A.; Froetscher, W.; Benetka, V.; Moestl, K.; Baumgartner, W. Pestivirus exposure in free-living and captive deer in Austria. *J. Wildl. Dis.* **2004**, *40*, 791–795. [CrossRef]
40. Cuteri, V.; Diverio, S.; Carnieletto, P.; Turilli, C.; Valente, C. Serological survey for antibodies against selected infectious agents among fallow deer (*Dama dama*) in central Italy. *Zentralbl Veterinarmed B* **1999**, *46*, 545–549. [CrossRef]
41. Boadella, M.; Carta, T.; Oleaga, A.; Pajares, G.; Munoz, M.; Gortazar, C. Serosurvey for selected pathogens in Iberian roe deer. *BMC Vet. Res.* **2010**, *6*, 51. [CrossRef]
42. Decaro, N.; Lucente, M.S.; Lanave, G.; Gargano, P.; Larocca, V.; Losurdo, M.; Ciambrone, L.; Marino, P.A.; Parisi, A.; Casalinuovo, F.; et al. Evidence for Circulation of Bovine Viral Diarrhoea Virus Type 2c in Ruminants in Southern Italy. *Transbound. Emerg. Dis.* **2017**, *64*, 1935–1944. [CrossRef]
43. Fernandez-Sirera, L.; Cabezon, O.; Dematteis, A.; Rossi, L.; Meneguz, P.G.; Gennero, M.S.; Allepuz, A.; Rosell, R.; Lavin, S.; Marco, I. Survey of Pestivirus infection in wild and domestic ungulates from south-western Italian Alps. *Eur. J. Wildlife Res.* **2012**, *58*, 425–431. [CrossRef]
44. Bianchi, M.V.; Silveira, S.; Mosena, A.C.S.; de Souza, S.O.; Konradt, G.; Canal, C.W.; Driemeier, D.; Pavarini, S.P. Pathological and virological features of skin lesions caused by BVDV in cattle. *Braz. J. Microbiol.* **2019**, *50*, 271–277. [CrossRef] [PubMed]
45. Grooms, D.L.; Brock, K.V.; Ward, L.A. Detection of bovine viral diarrhea virus in the ovaries of cattle acutely infected with bovine viral diarrhea virus. *J. Vet. Diagn. Investig.* **1998**, *10*, 125–129. [CrossRef] [PubMed]
46. Givens, M.D.; Heath, A.M.; Brock, K.V.; Brodersen, B.W.; Carson, R.L.; Stringfellow, D.A. Detection of bovine viral diarrhea virus in semen obtained after inoculation of seronegative postpubertal bulls. *Am. J. Vet. Res.* **2003**, *64*, 428–434. [CrossRef]
47. Ohmann, H.B. Pathogenesis of bovine viral diarrhoea-mucosal disease: Distribution and significance of BVDV antigen in diseased calves. *Res. Vet. Sci.* **1983**, *34*, 5–10. [CrossRef]
48. Thiry, J.; Keuser, V.; Muylkens, B.; Meurens, F.; Gogev, S.; Vanderplasschen, A.; Thiry, E. Ruminant alphaherpesviruses related to bovine herpesvirus 1. *Vet. Res.* **2006**, *37*, 169–190. [CrossRef]
49. Gu, X.; Kirkland, P.D. Infectious Bovine Rhinotracheitis. In *Australian and New Zealand Standard Diagnostic Procedures*; The Department of Agriculture and Water: Canberra, ACT, Australia, 2008; pp. 1–18.
50. Lillehaug, A.; Vikoren, T.; Larsen, I.L.; Akerstedt, J.; Tharaldsen, J.; Handeland, K. Antibodies to ruminant alpha-herpesviruses and pestiviruses in Norwegian cervids. *J. Wildl. Dis.* **2003**, *39*, 779–786. [CrossRef]
51. Inglis, D.M.; Bowie, J.M.; Allan, M.J.; Nettleton, P.F. Ocular disease in red deer calves associated with a herpesvirus infection. *Vet. Rec.* **1983**, *113*, 182–183. [CrossRef]
52. Ek-Kommonen, C.; Pelkonen, S.; Nettleton, P.F. Isolation of a herpesvirus serologically related to bovine herpesvirus 1 from a reindeer (*Rangifer tarandus*). *Acta Vet. Scand.* **1986**, *27*, 299–301.
53. Lyaku, J.R.S.; Nettleton, P.F.; Marsden, H. A Comparison of Serological Relationships among 5 Ruminant Alphaherpesviruses by Elisa. *Arch. Virol.* **1992**, *124*, 333–341. [CrossRef]
54. Martin, W.B.; Castrucci, G.; Frigeri, F.; Ferrari, M. A Serological Comparison of Some Animal Herpesviruses. *Comp. Immunol. Microb.* **1990**, *13*, 75–84. [CrossRef]
55. Nixon, P.; Edwards, S.; White, H. Serological Comparisons of Antigenically Related Herpesviruses in Cattle, Red Deer and Goats. *Vet. Res. Commun.* **1988**, *12*, 355–362. [CrossRef] [PubMed]
56. Walker, P.J.; Klement, E. Epidemiology and control of bovine ephemeral fever. *Vet. Res.* **2015**, *46*, 124. [CrossRef] [PubMed]

57. Weir, R.P.; Agnihotri, K. Epizootic Haemorrhagic Disease. In *Australian and New Zealand Standard Diagnostic Procedures*; The Department of Agriculture and Water: Canberra, ACT, Australia, 2014; pp. 1–14.
58. Australia Animal Health. National Arbovirus Monitoring Program 2015–2016 Report. Available online: https://www.animalhealthaustralia.com.au/wp-content/uploads/2015/09/NAMP-Annual-Report_FA_print.pdf (accessed on 1 April 2020).

viruses

MDPI

Article

Prolonged Detection of Bovine Viral Diarrhoea Virus Infection in the Semen of Bulls

Andrew J. Read, Sarah Gestier, Kate Parrish, Deborah S. Finlaison, Xingnian Gu, Tiffany W. O'Connor and Peter D. Kirkland *

Virology Laboratory, Elizabeth Macarthur Agriculture Institute, Woodbridge Road, Menangle, New South, Wales 2568, Australia; andrew.j.read@dpi.nsw.gov.au (A.J.R.); sarah.gestier@dpi.nsw.gov.au (S.G.); kate.parrish@dpi.nsw.gov.au (K.P.); deborah.finlaison@dpi.nsw.gov.au (D.S.F.); xingnian.gu@dpi.nsw.gov.au (X.G.); tiffany.o'connor@dpi.nsw.gov.au (T.W.O.)

* Correspondence: peter.kirkland@dpi.nsw.gov.au; Tel.: +61-2-4640-6331

Received: 28 May 2020; Accepted: 16 June 2020; Published: 22 June 2020

Abstract: Infection of bulls with bovine viral diarrhoea virus (BVDV) can result in the development of virus persistence, confined to the reproductive tract. These bulls develop a normal immune response with high neutralizing antibody titres. However, BVDV can be excreted in the semen for a prolonged period. Although relatively rare, in this study we describe six separate cases in bulls being prepared for admission to artificial breeding centres. Semen samples were tested in a pan-Pestivirus-reactive real-time PCR assay and viral RNA was detected in semen from five of the bulls for three to eight months after infection. In one bull, virus was detected at low levels for more than five years. This bull was found to have one small testis. When slaughtered, virus was only detected in the abnormal testis. The low levels of BVDV in the semen of these bulls were only intermittently detected by virus isolation in cell culture. This virus-contaminated semen presents a biosecurity risk and confirms the need to screen all batches of semen from bulls that have been previously infected with BVDV. The use of real-time PCR is recommended as the preferred laboratory assay for this purpose.

Keywords: bovine viral diarrhoea virus; Pestivirus; persistent testicular infection; prolonged testicular infection; bovine; testes; semen

1. Introduction

Pestiviruses are a group of viruses that belong to the family Flaviviridae and are of animal health and economic importance worldwide [1]. The taxonomy of the Pestivirus genus has been reorganised to 11 species based on nucleotide and deduced amino acid sequence relatedness, antigenic similarity and host of origin [2,3] Bovine viral diarrhoea virus type 1 (BVDV-1) has been taxonomically classified as *Pestivirus A*, and BVDV type 2 (BVDV-2) has been classified as *Pestivirus B*. These viruses are predominantly pathogens of cattle, though sheep, other ruminants and pigs may also become infected.

Transmission of BVDV occurs horizontally by direct and indirect contact between cattle, and vertically by transplacental infection of the foetus. The virus replicates in mucosal epithelium and associated lymphoid tissues. A viraemia follows with subsequent viral replication throughout the body. Post-natal transient infections are often subclinical, but can result in a range of clinical conditions including reproductive losses, immune suppression, respiratory disease and acute diarrhoea [4]. Periconceptional infections can result in conception failure and embryonic deaths. Gestational infections can result in foetal death, congenital abnormalities, or a calf that becomes persistently infected (PI) for life [4].

The detection of BVDV in the semen of bulls is well documented [5–10]. Contamination of semen can arise from three types of infection. Firstly, male calves exposed to BVDV in utero, before they become immunocompetent, are born PI and will shed virus from most organs including in the semen for life [11,12]. These calves are born immunotolerant to BVDV and are often weak, ill-thrifty and are frequently infertile. However, occasionally PI bulls may grow normally, have semen of normal quality and can even gain entry to an artificial breeding centre [13]. There is usually a high concentration of BVDV in the semen of PI bulls (10^4 to $10^{7.6}$ $TCID_{50}$/mL) [5,9,14]. Semen with high titres of virus can infect susceptible cows following insemination [15,16].

Secondly, post-pubertal bulls exposed to BVDV undergo a transient infection and may shed virus in semen for up to 14 days [9], or perhaps 28 days [17] following infection. Virus levels tend to be low ($10^{0.9}$ to $10^{1.8}$ $TCID_{50}$/mL) but can occasionally infect a susceptible female [9]. These bulls will eventually mount an immune response to the virus which prevents further shedding.

Thirdly, some post-pubertal bulls exposed to BVDV become acutely infected but develop a prolonged or perhaps persistent testicular infection. These bulls are not PI, become transiently infected and mount a normal antibody response. The virus is cleared systemically, yet the virus is excreted in semen for prolonged periods. For this study, we have defined this prolonged testicular infection as the detectable presence of BVDV RNA in semen more than 60 days from when the systemic infection occurred. Until recently, these prolonged testicular infections (PTI) were considered to be rare. The first case was reported in 1998 [18]. This bull was not viraemic and therefore not PI. Nevertheless, infectious BVDV was detected in his semen at a moderate level ($10^{3.3}$ $TCID_{50}$/mL) for the rest of his life.

PTIs were induced experimentally in 2003 by inoculating susceptible post-pubertal bulls with BVDV [14]. Following acute infection, the bulls appeared to mount a typical immune response with the cessation of detectable viraemia. BVDV RNA was detected in the semen of these bulls by reverse transcription-nested PCR for a prolonged period of time. Attempts to isolate BVDV from the semen of these bulls were unsuccessful in cell culture, however semen collected five months after inoculation established infection in a naïve calf when given intravenously. These results were replicated by the same research team [6]. While the aetiology of PTIs with BVDV is still poorly understood, administration of a live attenuated vaccine virus has also been shown to induce PTI [7].

The incidence of PTI is considered to be very low and few have been described [17–19], yet due to the economic significance of BVDV infection, the World Organization for Animal Health (OIE) recommends that semen from seropositive bulls be screened for the presence of BVDV [20]. Before entry to a semen collection centre, bulls are tested to demonstrate that they are not viraemic or PI. Bulls that give negative results are then held in isolation at least 28 days [21] to undertake pre-entry testing and to ensure that animals recently exposed to BVDV can be reliably identified by allowing time for seroconversion. Once admitted to the semen collection centre, semen collections from all bulls with neutralising antibodies to BVDV are tested by a real-time, reverse transcription PCR (qRT-PCR). Following the introduction of this requirement, we have identified six bulls that have exhibited prolonged or persistent testicular infection. The detailed investigation and subsequent longitudinal sampling of these bulls is described.

2. Materials and Methods

2.1. Specimens

Blood samples and commercially prepared, extended semen from six bulls were submitted to the Virology Laboratory at the Elizabeth Macarthur Agriculture Institute (EMAI), New South Wales, Australia by veterinarians from artificial breeding centres. These samples were tested to meet regulatory protocols for the export of semen, where the serological status of an animal is determined and, if seropositive, their semen was subjected to virus detection. Subsequent to testing of the initial samples from these bulls, both archival samples and prospective collections were requested.

Bull 2 was sent to an abattoir where the testes were collected and sent to EMAI. The testes were measured, weighed and dissected. Cut surfaces from various positions within the testes were swabbed using cotton tipped swabs which were placed into 3 mL of phosphate buffered gelatin saline (PBGS) and tested with a pan-Pestivirus-reactive qRT-PCR.

2.2. Serology

A virus neutralisation test [20,22] based on a non-cytopathogenic Australian reference strain of BVDV (Trangie, subtype 1c) [23,24] was used as the primary assay for the detection of anti-Pestivirus antibodies. In some instances, an agar gel immunodiffusion assay [22] had been used to test samples prior to the introduction of the international protocol for the screening of bulls entering semen collection centres.

2.3. Virus and Nucleic Acid Detection

Serum from each of the bulls was tested to determine if the bull was persistently infected with BVDV using a commercial antigen capture ELISA (IDEXX BVDV Ag/Serum Plus, IDEXX Laboratories, Liebefeld-Bern, Switzerland) run according to the manufacturer's instructions.

The detection of BVDV RNA was performed by qRT-PCR as described previously [25]. Briefly, total nucleic acid was extracted from semen samples with a magnetic particle handling system, using a volume of 25 μL of undiluted semen and 50 μL of semen diluted 1/4 in PBGS. An exogenous internal control was included in the extraction buffer for each sample. The qRT-PCR utilised pan-Pestivirus reactive primers and probe [26] run on an ABI 7500 thermocycler in standard mode for a total of 45 cycles. Cycling conditions were as specified by the Ag-Path master mix manufacturer (Ambion, Austin, TX, USA). The fluorescence threshold was set manually at 0.05 and background was automatically adjusted. qRT-PCR results were expressed as cycle threshold (Ct) values and classified as negative if no amplification was observed after 45 cycles.

Semen samples were processed [20] and virus isolation (VI) was conducted with primary bovine testis cells grown in Basal Medium Eagle (BME) with Hank's salts containing antibiotics and supplemented with serum from BVDV antibody- and virus-free donor animals. After primary culture, samples were passaged at weekly intervals for 2 additional sub-cultures and then screened for evidence of BVDV antigen by immunoperoxidase staining [22] using a mixture of anti-Pestivirus monoclonal antibodies (P1H11, P4G11, P1D8 and 2NB2) [27]. Virus isolation was attempted on the earliest available semen samples and then periodically throughout the sampling period when BVDV RNA had been detected by qRT-PCR.

2.4. Immunohistochemistry (IHC)

Sections (4 μm thickness) of formalin fixed testes from Bull 2 were cut, mounted on Superfrost Plus slides (Menzel Gläser, Thermo Fisher Scientific, Waltham, MA, USA) and oven-dried for 30 min. Sections were de-waxed in xylene and rehydrated in an ethanol series. Endogenous peroxides were then blocked with a 1.8% solution of peroxide in methanol for 20 min. Slides were washed briefly in distilled water followed by 2 × 5 min washes in phosphate buffered saline (PBS) with 0.05% Tween 20 (MP Biomedicals, Santa Ana, CA, USA). Antigens were unmasked at 37 °C for 15 min in undiluted Proteinase K (Dako, Agilent, Santa Clara, CA, USA). Sections were washed twice with distilled water and once with PBS/0.05% Tween 20 before being blocked for 1 h with 0.5% bovine serum albumin (Roche, Basel, Switzerland) and 1% normal goat serum in PBS. Following a rinse in PBS/0.05% Tween 20, 150 μL of mouse MAb 15C5 (IDEXX Laboratories, Westbrooke, ME, USA) at a 1/10 dilution was added to each section and incubated at room temperature (RT; approximately 25 °C) for 1 h. The sections were washed twice in PBS/0.05% Tween 20 and incubated in EnVision™ anti-mouse serum (Dako, Agilent, Santa Clara, CA, USA) at RT for 1 h. Following three PBS/0.05% Tween 20 washes, 200 μL Dakocytomation DAB+ (Dako, Agilent, Santa Clara, CA, USA) chromogenic solution was added to each section and developed for 10 min in the dark. Slides were rinsed with distilled water to stop the colour reaction, counterstained with Mayers haematoxylin, dehydrated and mounted with dibutylphthalate polystyrene xylene (DPX) mounting medium (Sigma-Aldrich, St. Louis, MO, USA).

3. Results

Bulls that enter semen collection centres are shown to not be viraemic for BVDV and neutralizing antibody titres are determined. A review of records held at EMAI indicated that during the period November 2012 to October 2019, 586 bulls were tested at EMAI for entry to artificial breeding centres. The majority of bulls in this group were *Bos taurus* breeds (95.9%), with *Bos taurus* × *Bos indicus* breeds (2.5%) and *Bos indicus* breeds (1.6%) making up the remainder. Neutralizing antibodies to BVDV were detected in 295 (50.3%) of these bulls at the time of screening. Semen samples from all bulls that had virus neutralizing antibodies to BVDV were tested by qRT-PCR.

Six bulls were identified to have PTIs because BVDV RNA was detected in semen more than 60 days after the first detection of infection. The 60-day limit provided sufficient time to elapse to exclude the possibility of residual RNA or virus from a recent acute infection being detected. All of these bulls gave negative results in the antigen ELISA, confirming that none was persistently infected. All bulls had high neutralizing antibody titres during the observation period. Live attenuated Pestivirus vaccines are not used in Australia and none of the bulls in this study had been vaccinated. Full details for the six infected bulls are described in Appendix A Tables A1–A6

Both beef (five) and dairy (one) bulls were involved, with ages at the time of first detection of virus infection ranging from 10 to 21 months (Table 1). BVDV RNA was detected in the semen of these bulls for periods ranging from 3 to 73 months after the latest date at which the bulls could have been infected. Due to the very low levels of RNA detected on many occasions, virus isolation was only attempted on a limited proportion of samples. BVDV was successfully isolated from 12.5% (4/32) of samples subjected to virus isolation.

Table 1. Summary of details for bulls studied.

Bull Number	Breed	Estimated Age at First Detection of Infection (Months) [a]	Estimated Duration of Shedding (Months) [b]	qRT-PCR Results in Semen (Ct)	Virus Isolation (Pos/Total Tested)
1	Main Anjou	21	3	32.8–35.8	0/6
2	Holstein	10	73	29.7–38.4	2/10
3	Angus	18	3	29.1–32.6	0/5
4	Angus	19	4	30.1–32.0	1/3
5	Angus	20	6	32.6–34.8	0/6
6	Wagyu	21	8	27.6–38.0	1/2 [c]

[a] Detection of seroconversion or detection of bovine viral diarrhoea virus (BVDV) RNA in semen, whichever occurred earlier. [b] Time between detection of infection and last detection of BVDV RNA in semen. [c] Virus isolation successful in the early phase of infection.

The data for Bull 2 (Table A2) provides an interesting insight into PTIs. BVDV was retrospectively detected in the semen of this bull over a period of more than five years. The first detection of BVDV RNA was in 2017 during the routine screening protocol for seropositive bulls. A review of the testing history for this bull indicated that it had been infected with BVDV prior to late January 2012. Fifteen batches of semen were available over a six-month period between April and October 2012. Interestingly, when this semen was tested in 2018, BVDV RNA was only detected in the last of these 15 samples (26 October 2012). After the detection of BVDV RNA in May 2017, 10 consecutive positive semen samples were identified in 2018, followed by two negative semen samples prior to this bull being sent to an abattoir. No semen was available for testing between October 2012 and May 2017.

Each of these bulls with a presumed PTI was clinically normal and without any physical abnormalities, with the exception of Bull 2. Soon after arriving at the artificial breeding centre, this bull had scrotal measurements recorded on two occasions, with a circumference of 31 cm on 20 February 2012 and 34 cm on 16 April 2012. These measurements were considered normal for a bull of his breed and age [28]. Six years later, Bull 2 was sent an abattoir in June 2018. The testes from this bull were forwarded to EMAI where they were measured, weighed and dissected. One testis was 20 cm long and weighed 720 g. The other was 16 cm long and weighed 431 g. Swabs were taken from freshly cut surfaces at various locations within each testis and qRT-PCR performed. The 16 swabs taken from the larger of the testes gave negative qRT-PCR results, while positive results were obtained from seven of the nine swabs taken from the smaller one. The highest levels of BVDV RNA (as indicated by the lowest Ct values) were found in swabs taken from the parenchyma of the testis (Figure 1). However, this result was not uniform across the testicle with one of the four parenchymal swabs yielding a very high Ct value (37.3). Lower levels of BVDV RNA were detected along the epididymis and no viral RNA was detected in the deferent duct or the rete testis (Figure 1). BVDV was isolated from the testicular parenchyma.

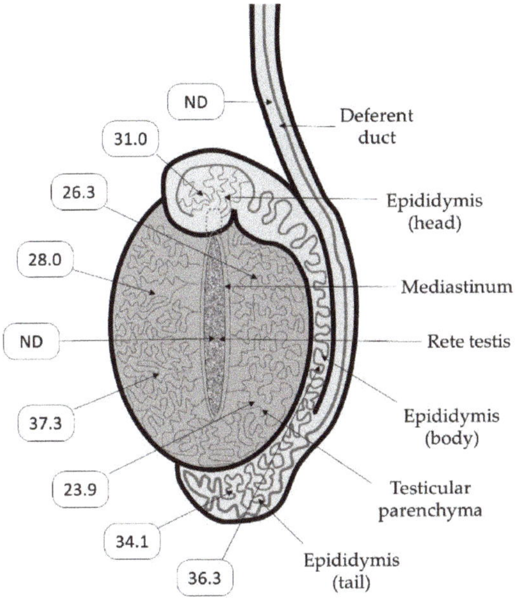

Figure 1. Schematic view of the affected testis of Bull 2. The numbers indicate the Ct values of swabs taken from the sites indicated by arrow heads. ND = BVDV RNA was not detected.

When examined microscopically, the abnormal testis from Bull 2 showed multifocal but locally extensive areas of seminiferous tubular atrophy (Figure 2A,C) characterised by reduced or absent germinal cell stages and aspermia. Affected tubules had either a reduced lumen diameter lined by a variably intact Sertoli cell layer or else were collapsed with complete or near total loss of Sertoli cells. Tubular atrophy was accompanied by moderate to marked tubular hyalinization and fibrosis and an increased prominence of Leydig cells. Immunohistochemical staining for BVDV antigen in these areas showed intense circumferential staining of the tubular luminal surface (Figures 2B and 3B). Within these tubules, there was also occasional staining of the subjacent thickened tubular wall mesenchyme and rarely within the cytoplasm of free intraluminal degenerative cells (Figure 3B). Adjacent to some areas of collapsed tubules were small clusters of less-affected tubules (Figure 2C) with low or no BVDV staining at the tubular luminal surface but containing numerous identifiable germinal cell stages which contained BVDV staining, including primary spermatocytes, round spermatids and elongate spermatids (Figure 3A). There was no evidence of inflammation associated with the affected tubules. Areas predominated by tubular lesions containing BVDV staining often intermingled with nearby histologically relatively normal tubules in which BVDV staining was not detected (Figure 2D). Representative sections of the epididymis were examined, found to be histologically unremarkable and BVDV IHC negative. The contralateral testicular tissue was histologically within normal limits (Figure 2E) and BVDV IHC negative (Figure 2F).

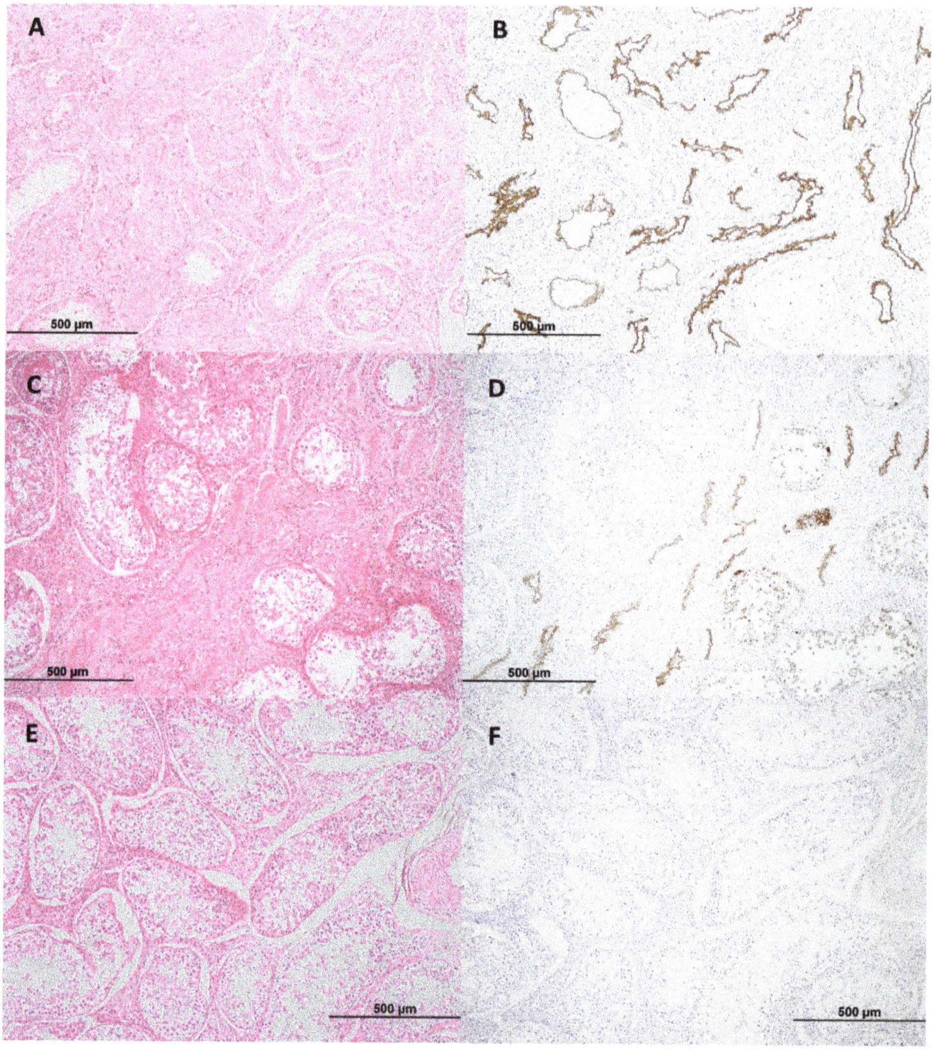

Figure 2. (**A**) Locally extensive areas of seminiferous tubular atrophy and fibrosis in a BVDV-infected bovine testicle. Many tubules are collapsed and shrunken with total loss of Sertoli cells, with a few tubules retaining some Sertoli cells but no germinal cells. The lower right quadrant contains a tubule with germinal cells. Haematoxylin and eosin (H&E) stain, 10×. (**B**) Non-serial section of A. Collapsed tubules have circumferential profiles of intense BVDV staining at the luminal surface. BVDV immunohistochemistry (IHC), 10×. (**C**) Collapsed seminiferous tubules intermingle with less-affected or relatively histologically normal tubules (lower right and upper left, respectively) in a BVDV-infected testicle. H&E stain, 10×. (**D**) Serial section of C; Collapsed tubules have intense BVDV staining while adjacent less-affected tubules (lower right) have BVDV staining within still-existing germinal cell stages and Sertoli cells. Histologically normal tubules (upper left) are BVDV antigen negative. BVDV IHC, 10×. (**E**) Histologically normal testicular tissue in the BVDV PCR negative contralateral testicle from Bull 2. H&E stain, 10×. (**F**) Serial section of E. The contralateral testicle is BVDV antigen negative. BVDV IHC, 10×.

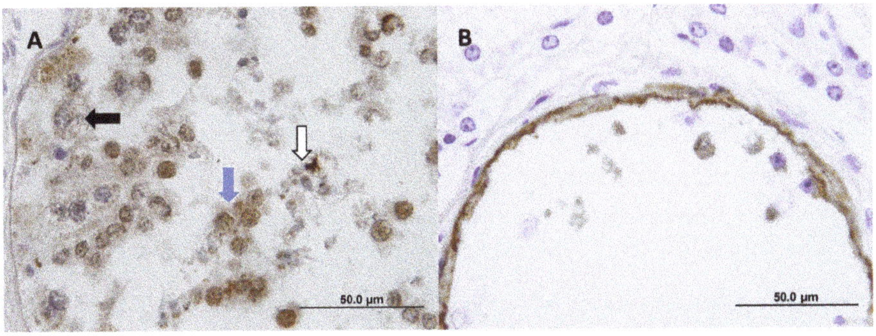

Figure 3. (**A**) A seminiferous tubule containing still-existing germinal cell stages in a BVDV-infected bovine testicle. BVDV staining is present in cells morphologically consistent with primary spermatocytes (black arrow), round spermatids (blue arrow) and elongated spermatids (white arrow). BVDV IHC, 100× (oil). (**B**) An atrophic seminiferous tubule in a BVDV-infected bovine testicle. The tubular luminal surface has intense BVDV staining which appears to occasionally extend into the superficial tubular mesenchyme and within degenerate free intraluminal cells. BVDV IHC, 100× (oil).

4. Discussion

International semen collection protocols and the associated health certification of donor bulls [21] are designed to ensure that bovine semen is free of contamination with BVDV. In addition to the exclusion of PI animals, measures are aimed at the exclusion of both transiently infected bulls and those with PTI. It was the adoption of these measures that resulted in the detection of the PTI bulls that are the subject of this report, and the detection of contaminated batches following retrospective testing. This study adds to the collective knowledge of prolonged or persistent testicular infection (PTI) in bulls through the identification and detailed investigation of six bulls. The longitudinal sampling, histopathology and IHC results have increased our understanding of the pathology and pathogenesis of PTI.

Studies of transiently infected bulls in the past have relied on virus isolation for the detection of BVDV in the semen. The use of qRT-PCR to detect BVDV in semen is recommended by the OIE [20], yet there is a lack of published literature on investigations of either transiently infected bulls or those with a PTI using qRT-PCR. It is not clear precisely how long BVDV RNA may be detected in semen following a transient infection. The possibility that some of these bulls were undergoing a transient infection when BVDV RNA was detected in the semen has to be considered due to the higher sensitivity of qRT-PCR, which may detect virus for a much longer time than virus isolation [26]. The date when Bull 5 was first infected can be estimated with the greatest precision due to the rising neutralizing antibody titre on the first two serum samples. Notably, the first semen sample to be tested gave a positive result and this sample was collected at least two months after the likely time of infection. Additionally, positive qRT-PCR results were obtained for semen samples for another two months. Therefore, it would seem unlikely that this RNA detection was the product of a transient infection. Similarly, with the exception of Bull 2, the other bulls were observed to shed viral RNA in semen for periods between three and eight months. Bull 2 shed virus for more than five years. Each of these bulls was likely to have experienced PTIs because the length of these shedding periods is in marked contrast to the 10 to 14 day periods reported for transiently infected bulls [9,10].

In 1998, Voges et al. described a post-pubertal bull that was shown to shed BVDV in semen over an 11-month period. This bull had a consistently high antibody titre against BVDV and yet was not viraemic and virus could not be detected in any organ apart from the testes. This bull was described as being in good health until his slaughter at 22 months of age. His growth rate and testicular development were unremarkable. Two other cases of PTI in bulls have been described in 2005 and 2014. [17,19]. In the latter report, a post-pubertal bull was also shown to be not PI. Infectious virus and viral RNA were detected in this bull up to 42 months of age. Virus was not detected for the last six months of this bull's life. IHC on this bull's testes at 48 months of age failed to detect viral antigen.

Voges et al. [18] proposed that for PTI to occur, BVDV infection should occur shortly before puberty, however, the pathogenesis of prolonged or persistent testicular infection is not clear. In cattle, puberty is evaluated on the basis of testicular growth and quantitative measures of sperm production [29]. Published estimates for the average age at puberty in bulls put this at approximately 10.5 months, ranging from 9.4 to 12.1 months [30–32]. It is of interest to note that five of the six bulls in our study were in the range of 17–21 months old when virus infection was first detected. The age at which three bulls seroconverted could be determined relatively accurately (Bulls 4, 5 and 6). These bulls were between 18 and 20 months of age when infection was first detected. Bull 2 was less than 10 months of age when he was exposed to BVDV. The age at which the other two bulls seroconverted to BVDV could not be accurately determined but their virus detection results indicate an upper age limit (18–21 months) at which infection may have occurred and is similar to the age at which the other three bulls became infected. The results of the three bulls whose age at infection could be estimated with some reliability challenges the notion that infection of PTI bulls only occurs at or around puberty.

The localization of BVDV in the reproductive tract of Bull 2 was different to what had been observed for transiently infected bulls. BVDV RNA was only detected in the smaller of the testes. Relatively high levels of RNA and antigen were detected in the parenchyma of this infected testis and BVDV was also isolated in cell culture. This is in contrast to what had been observed for transiently infected bulls, where virus was not isolated from the testicular parenchyma. In this bull, trace levels of BVDV RNA were also detected in the epididymis but, unlike observations of transiently infected bulls, no antigen was detected. Unfortunately, the accessory glands of Bull 2 were not examined, preventing further comparisons with the observations made in transiently infected bulls where virus was only isolated from the accessory glands.

The pathology and IHC staining of Bull 2 clearly demonstrated that the persistence of the BVDV infection was localized within the immune-privileged seminiferous tubule compartment. The bovine testis consists of two compartments, the peritubular compartment (Leydig cells and testicular macrophages) and the seminiferous tubule compartment (germ cells protected by Sertoli cells). The Sertoli cells span the basal to adluminal sections of the tubule and tight junctions between these cells form the blood–testes barrier [33]. This barrier protects developing germ cells from stimulating the adaptive immune system while allowing nutritional and structural support [34]. The peritubular compartment is exposed to the full immune response of the animal, while the seminiferous tubule compartment is in an immune-privileged site [35].

The pattern of pathology detected in this testis shows a continuum of tubular degenerative change correlating closely with a progression of BVDV staining, suggesting that the infection in this case was confined to seminiferous tubules and chronicity resulted in the eventual destruction of the normal tubule structure. It is also noteworthy that the concentration of BVDV RNA was not uniform across the testis. Within one area of the testicular parenchyma, the qRT-PCR gave a Ct value of 37.3. This result indicates that the level of BVDV RNA present was very close to the limit of detection for this assay. Yet, in the other areas of the testicular parenchyma tested, the Ct values ranged from 23.9 to 28.0. These values are typically seen in the semen of PI cattle. We speculate that the infection in PTI remains largely hidden from the immune system and manages to smoulder within the seminiferous tubules for a variable amount of time. The case of the bull from which the testes were collected at the abattoir supports this notion as there was a lack of interstitial or tubular inflammatory cellular infiltrates in response to the widespread presence of BVDV within the tubules, even in areas where BVDV tubule staining appears to extend superficially into the subjacent thickened peritubular mesenchyme of intensely stained tubules This latter observation may be a by-product of excessive antigen accumulation in these tubules, or perhaps the peritubular fibrosis assisted to shelter the infection from the peritubular compartment immune response. We speculate that the destruction of the tubules contributed to the small overall size of this testis. It was not possible to determine if the uninfected testis was ever infected, though no histological signs of past infection were detected. BVDV RNA was only detected in one of the early semen collections in 2012. It is not clear if BVDV RNA levels were too low to detect in these collections, or if there was deterioration in the RNA over the six years between when the RNA was shed and when it was tested by qRT-PCR. Additionally, the final two semen collections for this bull gave negative results in the qRT-PCR. Nevertheless, the ability to isolate virus from the testis of this bull 73 months after the initial infection demonstrates the capacity of the infection to persist in the testis for long periods.

The time course over which a PTI may be detected appears to range from a few months to many years. Eventually infection may be cleared, as appeared to occur in Bull 5 and as described by Newcomer et al., [17], or infection may remain persistent as demonstrated by Bull 2. Distinctions between persistent testicular infections and prolonged testicular infections have been proposed based on the ability to infect susceptible animals [17]. This distinction has not been applied to this series of bulls. BVDV was isolated from three of the six bulls (Bulls 2,4 and 6). Moreover, BVDV was isolated from the semen of Bull 2 on two occasions but was not isolated on eight. It is probable that differences between the ability to detect BVDV by qRT-PCR and by VI reflect shedding of the virus in semen, which in turn may vary between semen batches and progress of the infection.

Virus isolation has been considered the gold standard for BVDV detection when screening semen for the presence of BVDV-1. However, the results of the current investigations demonstrate the benefits of using qRT-PCR, which offers both higher analytical sensitivity and speed for the detection of BVDV infection. It is important that qRT-PCR assays employ broadly reactive pan-Pestivirus primers and probes to ensure that the potential genetic variability of all strains of Pestiviruses are reliably detected [20]. Many of the Ct values recorded for these positive bull semen samples were relatively high. The work of Hoffman et al. demonstrated that, for qRT-PCR assays, Ct values greater than 38 represent less than 10 genome copies per 5 μL of sample [26]. However, the volume of sample should not be overlooked because detection of 10 copies in a qRT-PCR assay represents approximately 1000 copies of RNA in a single 500 μL straw of semen. Parallel qRT-PCR and virus isolation experiments demonstrated that qRT-PCR had 100 to 1000 times greater sensitivity in contrast to the limit of detection for virus isolation methods. While qRT-PCR does not provide an indication of whether the sample contains infectious virus, a positive result in this assay does identify a batch as presenting a risk that infectious virus may be present. This risk is exemplified by experimental work [14] that demonstrated that, despite BVDV being detected by PCR, virus isolation attempts in cell culture with the semen of PTI bulls were unsuccessful. This same semen was capable of producing a systemic infection when injected into calves. Therefore, as recommended by the OIE [20], a reliable qRT-PCR should be the preferred diagnostic assay for BVDV detection for the screening of semen. Diagnostic laboratories are encouraged to undertake regular evaluation of any qRT-PCR assays that are being used to detect BVDV. This evaluation should involve comparison of published assays considering both the nucleic acid extraction and qRT-PCR components, using suitable samples from the country of origin, assessing the limit of detection and participating in proficiency testing.

With an increasing number of countries progressing towards freedom from BVDV, consideration may need to be given to screening of batches of stored semen. Most collections of semen from bulls that experience a PTI have very low levels of infectious virus and perhaps often at levels that would be lower than an infective dose following insemination. Nevertheless, this semen does present a biosecurity risk. Indeed, it had been shown that use of semen from the bull with the first recognized PTI was able to establish a systemic infection in an inseminated heifer [36], thereby confirming this risk. Contaminated semen can be used many years after collection and is often transported long distances internationally. These factors, combined with the potential for exposure of a large number of breeding females to a single batch of semen, increase the potential for the introduction of BVDV into free herds and countries, or potentially the introduction of other species of Pestivirus (such as BVDV-2 and Hobi-like Pestiviruses) into countries where these viruses are absent and where appropriately protective vaccines are not used.

In conclusion, although prolonged or persistent testicular infections are a relatively rare event, the findings of this study emphasize the recommendations [20] for all batches of semen from seropositive bulls to be screened to exclude the possibility of BVDV. Further, it is recommended that this screening is more effectively carried out by qRT-PCR rather than by virus isolation methods.

Author Contributions: Conceptualization, A.J.R. and P.D.K.; methodology and investigation, A.J.R., S.G., K.P., D.S.F., X.G., T.W.O. and P.D.K.; data curation, A.J.R.; writing—Original draft preparation, A.J.R., S.G. and P.D.K.; writing—Review and editing, A.J.R., S.G., K.P., D.S.F., X.G., T.W.O. and P.D.K. All authors have read and agreed to the published version of the manuscript.

Funding: This research received no external funding.

Acknowledgments: We acknowledge contributions of Megan Beca, Pennie Harkness, Steve Williams and Reon Holmes for supplying additional information about the bulls and providing additional samples. We are indebted to the staff of the Virology Laboratory at EMAI for their invaluable assistance during the testing of the samples described in this study and to the technical staff of the EMAI pathology laboratory for preparation of the tissue sections. Roger Cook provided helpful comments on the histological findings.

Conflicts of Interest: The authors declare no conflict of interest.

Appendix A

Table A1. Serological and virological test results for Bull 1.

Sample Date	Sample Type	Age (Months)	qRT-PCR (Ct)	Virus Isolation	Antigen ELISA	VNT (Titre)
28/08/2012	Serum	21			Negative	512
22/10/2012	Serum	23		Negative		≥512
12/11/2012	Extended semen	23	Positive (35.4)	Negative		
23/11/2012	Extended semen	24	Positive (35.4)	Negative		
26/11/2012	Extended semen	24	Positive (35.8)	Negative		
28/11/2012	Extended semen	24	Positive (32.8)	Negative		
30/11/2012	Extended semen	24	Positive (35.5)	Negative		
6/12/2012	Serum	24				≥512

Breed: Maine-Anjou. Date of birth: December 2010. Time of infection: prior to 28 August 2012. VNT—virus neutralisation titre.

Table A2. Serological and virological test results for Bull 2.

Sample Date	Sample Type	Age (Months)	qRT-PCR (Ct)	Virus Isolation	Antigen ELISA	AGID	VNT (Titre)
23/01/2012	Serum	10			Negative	3	
13/03/2012	Serum	12			Negative	2	
17/04/2012	Extended semen	13	Negative [a]	Negative			
17/04/2012	Extended semen	15	Negative [a]				
19/06/2012	Extended semen	15	Negative [a]				
29/06/2012	Extended semen	16	Negative [a]				
20/07/2012	Extended semen	17	Negative [a]				
8/08/2012	Extended semen	17	Negative [a]				
20/08/2012	Extended semen	17	Negative [a]				
24/08/2012	Extended semen	17	Negative [a]				
31/08/2012	Extended semen	18	Negative [a]				
4/09/2012	Extended semen	18	Negative [a]				
7/09/2012	Extended semen	18	Negative [a]				
11/09/2012	Extended semen	18	Negative [a]				
21/09/2012	Extended semen	18	Negative [a]				
28/09/2012	Extended semen	19	Negative [a]				
26/10/2012	Extended semen	20	Positive (36.5) [a]				
13/11/2012	Serum	26			Negative	>3	
7/05/2013	Serum	38			Negative	3	
5/05/2014	Serum	50			Negative	3	
4/05/2015	Serum	62			Negative	3	
2/05/2016	Serum	74			Negative		≥512
1/05/2017	Extended semen	82	Positive (29.7)	Positive			
18/01/2018	Extended semen	82	Positive (32.6)	Negative			
23/01/2018	Extended semen	82	Positive (32.4)	Negative			
25/01/2018	Extended semen	83	Positive (30.9)	Positive			

Table A2. *Cont.*

Sample Date	Sample Type	Age (Months)	qRT-PCR (Ct)	Virus Isolation	Antigen ELISA	AGID	VNT (Titre)
1/02/2018	Extended semen	83	Positive (35.9)	Negative			
6/02/2018	Extended semen	83	Positive (32.5)	Negative			
8/02/2018	Serum	83			Negative		
12/02/2018	Extended semen	83	Positive (33.3)	Negative			
13/02/2018	Extended semen	83	Positive (34.4)	Negative			
15/02/2018	Extended semen	83	Positive (31.58)	Negative			
16/02/2018	Extended semen	83	Positive (38.46)				
20/02/2018	Extended semen	83	Positive (36.82)				
22/02/2018	Extended semen	83	Negative				
27/02/2018	Extended semen	84	Negative				

Breed: Holstein. Date of birth: March 2011. Time of infection: prior to 23 January 2012. [a] Tested in 2018. AGID—agar gel immunodiffusion assay

Table A3. Serological and virological test results for Bull 3.

Sample Date	Sample Type	Age (Months)	qRT-PCR (Ct)	Virus Isolation	Antigen ELISA	VNT(Titre)
20/02/2013	Serum	15			Negative	
22/05/2013	Extended semen	18	Positive (29.2)			
24/05/2013	Extended semen	18	Positive (29.7)	Negative		
28/05/2013	Extended semen	18	Positive (29.3)	Negative		
30/05/2013	Extended semen	18	Positive (29.1)			
3/06/2013	Extended semen	18	Positive (32.6)			
7/06/2013	Extended semen	18	Positive (30.0)			
10/06/2013	Serum	18		Negative		≥512
12/06/2013	Extended semen	18	Positive (31.0)			
14/06/2013	Extended semen	19	Positive (30.5)			
18/06/2013	Extended semen	19	Positive (29.5)			
20/06/2013	Extended semen	19	Positive (30.0)			
24/06/2013	Extended semen	19	Positive (30.4)			
26/06/2013	Extended semen	19	Positive (30.5)			
3/07/2013	Extended semen	19	Positive (30.5)			
8/07/2013	Extended semen	19	Positive (32.0)			
12/07/2013	Extended semen	19	Positive (31.0)			
12/07/2013	Extended semen	19	Positive (30.5)	Negative		
16/07/2013	Extended semen	20	Positive (30.4)			
18/07/2013	Extended semen	20	Positive (31.4)			
23/07/2013	Extended semen	20	Positive (32.2)			
15/08/2013	Extended semen	21	Positive (30.7)	Negative		

Breed: Angus. Date of birth: December 2011. Time of infection: prior to 22 May 2013.

Table A4. Serological and virological test results for Bull 4.

Sample Date	Sample Type	Age (Months)	qRT-PCR (Ct)	Virus Isolation	Antigen ELISA	AGID	VNT (Titre)
3/09/2013	Serum	18			Negative	Negative	
8/10/2013	Serum	19			Negative	>3	
22/10/2013	Extended semen	20	Positive (31.7)	Positive			
24/10/2013	Extended semen	20	Positive (32.0)	Negative			
16/12/2013	Seminal fluid	22	Positive (30.2)	Negative			
16/02/2014 [a]	Extended semen	24	Positive (30.1)				
16/02/2014 [a]	Serum	24					≥512

Breed: Angus. Date of birth: March 2012. Time of infection: between late August and September 2013. [a] Slaughtered immediately after this collection.

Table A5. Serological and virological test results for Bull 5.

Sample Date	Sample Type	Age (Months)	qRT-PCR (Ct)	Virus Isolation	Antigen ELISA	VNT(Titre)
31/03/2014	Serum	20			Negative	4
12/05/2014	Serum	21		Negative		256
15/05/2014	Extended semen	21	Positive (32.6)			
24/05/2014	Neat semen	22	Positive (34.2)	Negative		
11/06/2014	Extended semen	22	Positive (34.8)	Negative		
2/07/2014	Neat semen	23	Positive (34.1)	Negative		
17/07/2014	Extended semen	23	Positive (33.9)	Negative		
20/02/2015	Serum	31			Negative	≥512
8/04/2015	Extended semen	32	Negative			
20/04/2015	Serum	32		Negative		≥512
22/04/2015	Extended semen	33	Negative			
27/04/2015	Extended semen	33	Negative			
14/05/2015	Extended semen	33	Negative			
19/05/2015	Extended semen	33	Negative			
21/05/2015	Extended semen	34	Negative			
25/05/2015	Extended semen	34	Negative			
28/05/2015	Extended semen	34	Negative			
2/06/2015	Extended semen	34	Negative			
4/06/2015	Extended semen	34	Negative			
8/06/2015	Extended semen	34	Negative			
12/06/2015	Extended semen	34	Negative			
16/06/2015	Extended semen	34	Negative			
28/06/2015	Extended semen	35	Negative			
18/07/2015	Extended semen	35	Negative			
17/08/2015	Extended semen	36	Negative			
21/08/2015	Extended semen	37	Negative			
24/08/2015	Extended semen	37	Negative			
31/08/2015	Extended semen	38	Negative			
6/02/2017	Serum	54			Negative	512
3/04/2017	Serum	56				128
1/05/2017	Extended semen	57	Negative			
4/05/2017	Extended semen	57	Negative			

Breed: Angus. Date of birth: August 2012. Time of infection: probably infected mid-March 2014.

Table A6. Serological and virological test results for Bull 6.

Sample Date	Sample Type	Age (Months)	qRT-PCR (Ct)	Virus Isolation	Antigen ELISA	VNT
14/12/2018	Extended semen	15	Negative			
6/04/2019 [a]	Serum	19			Negative	
21/05/2019	Extended semen	20	Positive (27.6)	Positive		
1/09/2019	Serum	24			Negative	>512
24/09/2019	Serum	24				>512
17/10/2019	Extended semen	25	Positive (38.0)	Negative		
22/11/2019	Extended semen	26	Positive (36.4)			
8/01/2020	Extended semen	28	Positive (36.2)			
23/01/2020	Extended semen	28	Negative			

Breed: Wagyu. Date of birth: September 2017. Time of infection: estimated infected between mid-April and mid-May 2019. [a] A BVDV antibody ELISA was performed on serum taken on 6 April 2019 with negative results (data not shown).

References

1. Richter, V.; Lebl, K.; Baumgartner, W.; Obritzhauser, W.; Kasbohrer, A.; Pinior, B. A systematic worldwide review of the direct monetary losses in cattle due to bovine viral diarrhoea virus infection. *Vet. J.* **2017**, *220*, 80–87. [CrossRef] [PubMed]
2. Smith, D.B.; Meyers, G.; Bukh, J.; Gould, E.A.; Monath, T.; Scott Muerhoff, A.; Pletnev, A.; Rico-Hesse, R.; Stapleton, J.T.; Simmonds, P.; et al. Proposed revision to the taxonomy of the genus Pestivirus, family Flaviviridae. *J. Gen. Virol.* **2017**, *98*, 2106–2112. [CrossRef] [PubMed]

3. Walker, P.J.; Siddell, S.G.; Lefkowitz, E.J.; Mushegian, A.R.; Dempsey, D.M.; Dutilh, B.E.; Harrach, B.; Harrison, R.L.; Hendrickson, R.C.; Junglen, S.; et al. Changes to virus taxonomy and the International Code of Virus Classification and Nomenclature ratified by the International Committee on Taxonomy of Viruses. *Arch. Virol.* **2019**, *164*, 2417–2429. [CrossRef] [PubMed]
4. Walz, P.H.; Grooms, D.L.; Passler, T.; Ridpath, J.F.; Tremblay, R.; Step, D.L.; Callan, R.J.; Givens, M.D.; American College of Veterinary Internal, M. Control of bovine viral diarrhea virus in ruminants. *J. Vet. Intern. Med.* **2010**, *24*, 476–486. [CrossRef] [PubMed]
5. Bielanski, A.; Dubuc, C.; Hare, W.C.D. Failure to Remove Bovine Diarrhea Virus (BVDV) from Bull Semen by Swim up and other Separatory Sperm Techniques Associated with in vitro Fertilization. *Reprod. Domest. Anim.* **1992**, *27*, 303–306. [CrossRef]
6. Givens, M.D.; Riddell, K.P.; Edmondson, M.A.; Walz, P.H.; Gard, J.A.; Zhang, Y.; Galik, P.K.; Brodersen, B.W.; Carson, R.L.; Stringfellow, D.A. Epidemiology of prolonged testicular infections with bovine viral diarrhea virus. *Vet. Microbiol.* **2009**, *139*, 42–51. [CrossRef] [PubMed]
7. Givens, M.D.; Riddell, K.P.; Walz, P.H.; Rhoades, J.; Harland, R.; Zhang, Y.; Galik, P.K.; Brodersen, B.W.; Cochran, A.M.; Brock, K.V.; et al. Noncytopathic bovine viral diarrhea virus can persist in testicular tissue after vaccination of peri-pubertal bulls but prevents subsequent infection. *Vaccine* **2007**, *25*, 867–876. [CrossRef]
8. Kirkland, P.D.; McGowan, M.R.; Mackintosh, S.G.; Moyle, A. Insemination of cattle with semen from a bull transiently infected with pestivirus. *Vet. Rec.* **1997**, *140*, 124–127. [CrossRef]
9. Kirkland, P.D.; Richards, S.G.; Rothwell, J.T.; Stanley, D.F. Replication of bovine viral diarrhoea virus in the bovine reproductive tract and excretion of virus in semen during acute and chronic infections. *Vet. Rec.* **1991**, *128*, 587–590. [CrossRef]
10. Whitmore, H.L.; Gustafsson, B.K.; Havareshti, P.; Duchateau, A.B.; Mather, E.C. Inoculation of bulls with bovine virus diarrhea virus: Excretion of virus in semen and effects on semen quality. *Theriogenology* **1978**, *9*, 153–163. [CrossRef]
11. McClurkin, A.W.; Coria, M.F.; Cutlip, R.C. Reproductive performance of apparently healthy cattle persistently infected with bovine viral diarrhea virus. *J. Am. Vet. Med. Assoc.* **1979**, *174*, 1116–1119. [PubMed]
12. Barlow, R.M.; Nettleton, P.F.; Gardiner, A.C.; Greig, A.; Campbell, J.R.; Bonn, J.M. Persistent bovine virus diarrhoea virus infection in a bull. *Vet. Rec.* **1986**, *118*, 321–324. [CrossRef] [PubMed]
13. Howard, T.H.; Bean, B.; Hillman, R.; Monke, D.R. Surveillance for persistent bovine viral diarrhea virus infection in four artificial insemination centers. *J. Am. Vet. Med. Assoc.* **1990**, *196*, 1951–1955. [PubMed]
14. Givens, M.D.; Heath, A.M.; Brock, K.V.; Brodersen, B.W.; Carson, R.L.; Stringfellow, D.A. Detection of bovine viral diarrhea virus in semen obtained after inoculation of seronegative postpubertal bulls. *Am. J. Vet. Res.* **2003**, *64*, 428–434. [CrossRef] [PubMed]
15. Kirkland, P.; Mackintosh, S.; Moyle, A. The outcome of widespread use of semen from a bull persistently infected with pestivirus. *Vet. Rec.* **1994**, *135*, 527–529. [CrossRef] [PubMed]
16. Meyling, A.; Jensen, A.M. Transmission of bovine virus diarrhoea virus (BVDV) by artificial insemination (AI) with semen from a persistently-infected bull. *Vet. Microbiol.* **1988**, *17*, 97–105. [CrossRef]
17. Newcomer, B.W.; Toohey-Kurth, K.; Zhang, Y.; Brodersen, B.W.; Marley, M.S.; Joiner, K.S.; Zhang, Y.; Galik, P.K.; Riddell, K.P.; Givens, M.D. Laboratory diagnosis and transmissibility of bovine viral diarrhea virus from a bull with a persistent testicular infection. *Vet. Microbiol.* **2014**, *170*, 246–257. [CrossRef] [PubMed]
18. Voges, H.; Horner, G.W.; Rowe, S.; Wellenberg, G.J. Persistent bovine pestivirus infection localized in the testes of an immuno-competent, non-viraemic bull. *Vet. Microbiol.* **1998**, *61*, 165–175. [CrossRef]
19. Dunser, M.; Altmann, M.; Dengg, J.; Eichinger, M.; Loitsch, A.; Revilla-Fernandez, S.; Schweighardt, H. Localised persistent infection of the genital tract with bovine viral diarrhoea virus in an immunocompetent bull. *Wien. Tierarztl. Mon.* **2005**, *92*, 227–232.
20. International Office of Epizootics (OIE). Chapter 3.4.7. Bovine viral diarrhoea. In *Manual of Diagnostic Tests and Vaccines for Terrestrial Animals*; Office international des épizooties: Paris, France, 2019; pp. 1075–1096.
21. International Office of Epizootics (OIE). Chapter 4.7. Collection and processing of bovine, small ruminant and porcine semen. In *Terrestrial Animal Health Code*; Office international des épizooties: Paris, France, 2019.
22. Kirkland, P.D.; MacKintosh, S.G. Ruminant Pestivirus Infections. In *Australia and New Zealand Standard Diagnostic Procedures*; Department of Agriculture, Australian Government: Canberra, Australia, 2006; pp. 1–30.

23. Mahony, T.J.; McCarthy, F.M.; Gravel, J.L.; Corney, B.; Young, P.L.; Vilcek, S. Genetic analysis of bovine viral diarrhoea viruses from Australia. *Vet. Microbiol.* **2005**, *106*, 1–6. [CrossRef]
24. Gong, Y.; Trowbridge, R.; Macnaughton, T.B.; Westaway, E.G.; Shannon, A.D.; Gowans, E.J. Characterization of RNA synthesis during a one-step growth curve and of the replication mechanism of bovine viral diarrhoea virus. *J. Gen. Virol.* **1996**, *77 Pt 11*, 2729–2736. [CrossRef]
25. Gu, X.; Davis, R.J.; Walsh, S.J.; Melville, L.F.; Kirkland, P.D. Longitudinal study of the detection of Bluetongue virus in bull semen and comparison of real-time polymerase chain reaction assays. *J. Vet. Diagn. Investig.* **2014**, *26*, 18–26. [CrossRef] [PubMed]
26. Hoffmann, B.; Depner, K.; Schirrmeier, H.; Beer, M. A universal heterologous internal control system for duplex real-time RT-PCR assays used in a detection system for pestiviruses. *J. Virol. Methods* **2006**, *136*, 200–209. [CrossRef] [PubMed]
27. Brown, L.M.; Papa, R.A.; Frost, M.J.; Mackintosh, S.G.; Gu, X.; Dixon, R.J.; Shannon, A.D. A single amino acid is critical for the expression of B-cell epitopes on the helicase domain of the pestivirus NS3 protein. *Virus Res.* **2002**, *84*, 111–124. [CrossRef]
28. Hueston, W.D.; Monke, D.R.; Milburn, R.J. Scrotal circumference measurements on young Holstein bulls. *J. Am. Vet. Med. Assoc.* **1988**, *192*, 766–768.
29. Wolf, F.R.; Almquist, J.O.; Hale, E.B. Prepuberal Behavior and Puberal Characteristics of Beef Bulls on High Nutrient Allowance. *J. Anim. Sci.* **1965**, *24*, 761–765. [CrossRef] [PubMed]
30. Brito, L.F.; Barth, A.D.; Rawlings, N.C.; Wilde, R.E.; Crews, D.H., Jr.; Mir, P.S.; Kastelic, J.P. Effect of nutrition during calfhood and peripubertal period on serum metabolic hormones, gonadotropins and testosterone concentrations, and on sexual development in bulls. *Domest. Anim. Endocrinol.* **2007**, *33*, 1–18. [CrossRef] [PubMed]
31. Dance, A.; Thundathil, J.; Wilde, R.; Blondin, P.; Kastelic, J. Enhanced early-life nutrition promotes hormone production and reproductive development in Holstein bulls. *J. Dairy Sci.* **2015**, *98*, 987–998. [CrossRef]
32. Harstine, B.R.; Maquivar, M.; Helser, L.A.; Utt, M.D.; Premanandan, C.; DeJarnette, J.M.; Day, M.L. Effects of dietary energy on sexual maturation and sperm production in Holstein bulls. *J. Anim. Sci.* **2015**, *93*, 2759–2766. [CrossRef]
33. Fawcett, D.W.; Leak, L.V.; Heidger, P.M., Jr. Electron microscopic observations on the structural components of the blood-testis barrier. *J. Reprod Fertil Suppl.* **1970**, *10*, 105–122.
34. Mital, P.; Hinton, B.T.; Dufour, J.M. The blood-testis and blood-epididymis barriers are more than just their tight junctions. *Biol. Reprod.* **2011**, *84*, 851–858. [CrossRef] [PubMed]
35. Setchell, B.P.; Pollanen, P.; Zupp, J.L. Development of the blood-testis barrier and changes in vascular permeability at puberty in rats. *Int. J.* **1988**, *11*, 225–233. [CrossRef] [PubMed]
36. Niskanen, R.; Alenius, S.; Belak, K.; Baule, C.; Belak, S.; Voges, H.; Gustafsson, H. Insemination of susceptible heifers with semen from a non-viraemic bull with persistent bovine virus diarrhoea virus infection localized in the testes. *Reprod. Domest. Anim.* **2002**, *37*, 171–175. [CrossRef] [PubMed]

viruses MDPI

Article

Real Time Analysis of Bovine Viral Diarrhea Virus (BVDV) Infection and Its Dependence on Bovine CD46

Christiane Riedel [1,*], Hann-Wei Chen [1], Ursula Reichart [2], Benjamin Lamp [3], Vibor Laketa [4,5] and Till Rümenapf [1]

1 Institute of Virology, Vetmeduni Vienna, 1210 Vienna, Austria; Hann-Wei.Chen@vetmeduni.ac.at (H.-W.C.); Till.Ruemenapf@vetmeduni.ac.at (T.R.)
2 VetCore Facility for Research, Vetmeduni Vienna, 1210 Vienna, Austria; ursula.reichart@vetmeduni.ac.at
3 Institute of Virology, Faculty of Veterinary Medicine, Justus-Liebig University, 35392 Gießen, Germany; Benjamin.J.Lamp@vetmed.uni-giessen.de
4 Department of Infectious Diseases, Virology, University of Heidelberg, 69120 Heidelberg, Germany; vibor.laketa@med.uni-heidelberg.de
5 German Center for Infection Research, 69120 Heidelberg, Germany
* Correspondence: christiane.riedel@vetmeduni.ac.at

Received: 16 December 2019; Accepted: 14 January 2020; Published: 17 January 2020

Abstract: Virus attachment and entry is a complex interplay of viral and cellular interaction partners. Employing bovine viral diarrhea virus (BVDV) encoding an mCherry-E2 fusion protein ($BVDV_{E2\text{-}mCherry}$), being the first genetically labelled member of the family *Flaviviridae* applicable for the analysis of virus particles, the early events of infection—attachment, particle surface transport, and endocytosis—were monitored to better understand the mechanisms underlying virus entry and their dependence on the virus receptor, bovine CD46. The analysis of 801 tracks on the surface of SK6 cells inducibly expressing fluorophore labelled bovine CD46 ($CD46_{fluo}$) demonstrated the presence of directed, diffusive, and confined motion. 26 entry events could be identified, with the majority being associated with a $CD46_{fluo}$ positive structure during endocytosis and occurring more than 20 min after virus addition. Deletion of the $CD46_{fluo}$ E2 binding domain ($CD46_{fluo}\Delta$E2bind) did not affect the types of motions observed on the cell surface but resulted in a decreased number of observable entry events (2 out of 1081 tracks). Mean squared displacement analysis revealed a significantly increased velocity of particle transport for directed motions on $CD46_{fluo}\Delta$E2bind expressing cells in comparison to $CD46_{fluo}$. These results indicate that the presence of bovine CD46 is only affecting the speed of directed transport, but otherwise not influencing BVDV cell surface motility. Instead, bovine CD46 seems to be an important factor during uptake, suggesting the presence of additional cellular proteins interacting with the virus which are able to support its transport on the virus surface.

Keywords: *Pestivirus*; BVDV; CD46; life cell imaging; attachment; surface transport

1. Introduction

Before entering a host cell, viruses have to establish contact with the cell and travel to sites that are suitable for entry. This process of attachment can involve different receptor molecules and transport mechanisms, which can be classified into diffusion, drifts, and confinement [1]. Diffusive motion can be caused by interaction of viruses with a number of receptor molecules that is too low to cause confinement or can support the screening of the cell surface for suitable sites of endocytosis. Directed motion is the result of interaction of the virus with cellular proteins or lipid structures that are linked to F-actin. Understanding the extracellular movement of virus particles, as well as the kinetics of entry and protein expression, provides important insights into the key players involved in attachment and entry and in infection dynamics in general.

For the family *Retroviridae*—facilitated by the ease of generation of genetically labelled virus particles—attachment and entry dynamics have been studied excessively. Virus surfing—the directed, actin-dependent transport of virus particles on the outside of filopodia, cytonemes, and retraction fibres—was also discovered employing a member of the *Retroviridae*, namely murine leukemia virus (MLV) [2]. This mode of extracellular, directed transport has since been described for members of the *Adenoviridae* [1], *Herpesviridae* [3], and *Papillomaviridae* [4].

For the medically relevant family *Flaviviridae*, the lack of genetically labelled virus particles has been overcome by the utilization of lipophilic dyes or covalently linked fluorophores [5–8]. These studies demonstrated the diffusive movement of dengue virus along the cell surface to clathrin coated pits and the subsequent association with specific markers of endocytosis [8] and helped in the identification of T cell immunoglobulin mucin-1 as a dengue virus receptor [9]. Also, labelled hepatitis C virus (HCV) particles demonstrated the involvement of actin in cell surface and intracellular virus transport [7].

The genus *Pestivirus* is also part of the *Flaviviridae* and characterized by the presence of three viral surface glycoproteins—E^{rns}, E1, and E2—and an additional N-terminal protease, N^{pro}. Members of this genus are pathogens of cloven-hoofed animals, including the highly economically relevant pathogens bovine viral diarrhea virus (BVDV) and classical swine fever virus (CSFV). In recent years, genetically labelled BVDV and CSFV clones have been constructed based on N-terminal fusion of luciferase or fluorophores to either E^{rns} or E2 [10–13]. For a BVDV E2-fluorophore fusion protein, visualization of purified particles in fluorescence microscopy based on the specific fluorescence signal has been reported [13]. BVDV enters host cells after initial interactions of E^{rns} with heparan sulphates [14] and subsequent binding of E2 to its cellular receptor, bovine CD46 [15,16], by clathrin mediated endocytosis [17,18] and fusion occurs after endosomal acidification [18,19]. Interestingly, recent evidence implies that BVDV is preferably transmitted by direct cell-to-cell spread in a CD46 independent manner [12], indicating the involvement of additional factors in BVDV spread. The BVDV-1 clone employed in [12] also encodes for an mCherry-E2 fusion protein, in the backbone of strain NADL, demonstrating the utilization of different transmission modes in the presence of an E2-fusion protein.

Bovine CD46 is an ubiquitously expressed, type I transmembrane glycoprotein and exists as different splice variants, affecting the length of its heavily O-glycosylated, membrane proximal regions (STP) and its cytoplasmic C-terminus. The membrane distal, extracellular part of the protein consists of four complement control protein modules (CCP) which have been implicated in the binding of a variety of pathogens, leading to its description as a "pathogens magnet" [20]. Physiologically, CD46 is a cofactor of the inactivation of complement components C3b and C4b and also involved in T cell regulation, modulation of autophagy and reproductive biology (reviewed in [21]). Its surface levels are regulated by clathrin-dependent endocytosis [21] or micropinocytosis after cross-linking [22]. BVDV E2 binding to CD46 is mediated by the 30 C-terminal amino acids of CCP-1 [16]. Interestingly, CD46's physiological ligands mainly interact with the membrane proximal CCPs 3–4, while pathogens mostly interact with the membrane distal CCPs 1–2 [23].

While the function of bovine CD46 as a receptor for BVDV is well documented in the literature, it is currently unclear what its exact functions are during virus attachment and entry. Also, it has been shown that bovine CD46 is not sufficient for entry. To determine which stages of the entry process—attachment, surface motion, or uptake—are affected by CD46, we conducted life cell imaging experiments employing purified, fluorophore-E2 labelled BVDV particles and SK6 cell lines inducibly expressing fluorophore labelled CD46 with and without the virus interacting CCP-1 module [16].

2. Materials and Methods

2.1. Viruses and Cells

For propagation, cells were cultured in DMEM (Capricorn, Ebsdorfergrund, Germany) supplemented with 10% FCS (Bio & Sell, Feucht/Nürnberg, Germany) and penicillin/streptomycin (Merck, Darmstadt, Germany) at 37 °C, 5% CO_2. SK6 tet-on cells inducibly expressing bovine CD46 labelled with mCherry or mClover ($CD46_{fluo}$) have been described in [13]. The E2 binding CCP of CD46—as reported by [16]—was deleted by PCR (Vazyme Biotech, Nanjing, China) in the previously described $CD46_{fluo}$-encoding construct employing the following primers (Eurofins, Ebersberg, Germany): forward: cgaCGGTGTCCTACCCTAGCTGATC; reverse: GGCATCGGAGGACGTGGGCAG, resulting in $CD46_{fluo}\Delta$E2bind. SK6 tet-on cells were transfected with $CD46_{fluo}\Delta$E2bind by electroporation and clonally selected as described in [13].

For the determination of susceptibility of SK6 $CD46_{fluo}$ or $CD46_{fluo}\Delta$E2bind cells to $BVDV_{E2\text{-}mCherry}$, 2×10^5 MDBK, SK6 $CD46_{fluo}$ or $CD46_{fluo}\Delta$E2bind cells or SK6 cells inducibly expressing GFP were seeded in each well of a 24-well plate 24 h before the start of the experiment. SK6 cells are of porcine origin and porcine CD46 is not a receptor for BVDV [16]. Cells were induced 16 h before infection with 2.5 µg/mL doxycycline (Merck, Darmstadt, Germany). Cells were infected with ten-fold dilutions of a BVDV-1 strain expressing mCherry fused to the N-terminus of E2 ($BVDV_{E2\text{-}mCherry}$) [13] for 4h. Thereafter, medium was exchanged to medium containing 1% carboxymethylcellulose to prevent virus transfer via the culture medium. 48 h after infection, infected foci were quantified by fluorescence microscopy (Olympus IX-70, Tokyo, Japan) to determine the titer and the susceptibility was calculated in % by dividing the titer determined on a given cell line through the titer determined on MDBK cells. $BVDV_{E2\text{-}mCherry}$ was propagated on MDBK cells in 5-layer tissue culture flasks (Corning, Corning, NY, USA) and purified and concentrated as described in [13]. The sequence of the plasmid encoding the full sequence of $BVDV_{E2\text{-}mCherry}$ is provided as supplementary data.

For life cell imaging, 2×10^4 SK6 $CD46_{fluo}$ or $CD46_{fluo}\Delta$E2bind cells were seeded in each well of an IBIDI µ slide 8 well chamber slide 24 h before the start of the experiment in medium devoid of phenol red. 16 h before the start of the experiment, expression of $CD46_{fluo}$ or $CD46_{fluo}\Delta$E2bind was induced by addition of 2.5 µg/mL doxycycline (Merck, Darmstadt, Germany).

2.2. Life Cell Imaging

All data was acquired on an Andor Revolution spinning disk confocal microscope (Oxford Instruments, Abingdon, UK) based on a Yokogawa CSU W1 with Nikon Ti2 stand equipped with two Andor DU-888 cameras. Imaging was performed in a humidified chamber at 37 °C and 5% CO_2. An APO TIRF NA 1.49 100× magnification oil immersion objective was used, resulting in a lateral pixel size of 0.13 µm. For the detection of $CD46_{fluo}$ and $CD46_{fluo}\Delta$E2bind, mClover was excited by a 515 nm laser for 0.1 s and the emitted signal was bandpass filtered with a 540/30 nm filter. E2-mCherry was excited by a 594 nm laser for 0.1 s and the emitted light was bandpass filtered with a 647/57 nm filter before detection. Excitation and detection of each fluorophore was performed consecutively.

For the examination of virus attachment, cell surface transport, and entry, one frame/10 s was recorded starting directly, 5, 10, 15, 20, or 25 min after virus addition (multiplicity of infection (MOI) = 10) and data was acquired for 5 or 10 min. Three focal planes (Z-step size = 0.8 µm) were acquired for each time point to compensate for potential focusing errors and to increase the area accessible for analysis. The z-level was chosen close to the cover slip to allow imaging of filopodia, retraction fibers, lamellipodia, and the lamella. In order to judge intracellular background levels during E2-mCherry excitation, one z-stack was acquired of each field of view to be employed in the experiment before addition of virus.

To follow the time course of infection, cells were imaged at one z-level every 10 min starting 60 min after infection until 16 h after infection. Two different MOIs (1 and 10) were employed to assess the effect of different MOIs on the development of E2-mCherry signal. Due to phototoxicity, it was not possible to acquire several z-levels.

2.3. Image Processing and Analysis

All image processing and particle tracking was performed in ImageJ [24]. Raw frames were filtered with a Gaussian filter (sigma = 1.4) to improve data visualization. For initial particle tracking, maximum intensity projections along the z-axis were generated and particles tracked employing ImageJ's [24] manual tracking plug-in. Subsequently, tracks were verified on the full z-stack and their localization (lamellipodium, lamella, cell body) as well as their direction of movement (towards, tangential, away, random, with reference to the cell body) and the association with $CD46_{fluo}$ or $CD46_{fluo}\Delta E2bind$ was documented. A python script (supplementary information) was used to analyze the output of the manual tracking plug-in with regard to direct and relative distance covered by a given particle, directionality (ratio direct to relative distance), average, minimum and maximum speed as well as the standard deviation of speed for a given track. The mean squared displacement (MSD) was calculated in R employing a customized script based on the code developed by [25]. Fits of the MSD curves were employed to calculate the velocity (V) of directed movement in case of an exponential slope using the following formula: $y = V^2x^2 + D_0x + a$ and the diffusion coefficient (D_0) was calculated using $y = 4D_0x + a$ in case of a linear slope. A two-tailed Student's t-test was performed to assess the likelihood of significant differences.

3. Results

3.1. BVDV Entry Is a Slow Process

To assess the infection cycle of BVDV in real time, we employed a previously described labelling strategy of the viral E2 surface glycoprotein [12,13]. SK6 cells inducibly expressing the fluorophore labelled, cellular surface receptor, bovine CD46 were chosen as system to study the infection cycle, as these cells demonstrated a high susceptibility to infection with BVDV after induction. SK6 cells not expressing bovine CD46 display a low susceptibility to BVDV infection, which is not mediated by porcine CD46, as porcine CD46 has previously been shown not to be involved in BVDV invasion [16]. After initial experiments employing either a system of mClover labelled BVDV and mCherry labelled CD46 or vice versa, we decided to employ mCherry labelled BVDV ($BVDV_{E2\text{-}mCherry}$) and mClover labelled CD46 ($CD46_{fluo}$), as strong photobleaching of mClover labelled BVDV rendered it inapplicable for life cell imaging.

Time series of 5 or 10 min duration with a frame rate of one frame/10 s were acquired at different time points (0–25 min in 5 min steps) after addition of $BVDV_{E2\text{-}mCherry}$ (MOI 10). Association of mCherry positive particles with the surface of SK6 $CD46_{fluo}$ cells could readily be observed and 801 particles were tracked by hand (n cells = 160). Due to the low signal to noise ratio, automatic tracking approaches did not improve the ease of analysis and were therefore not utilized. Entry events were rarely observed (n = 26, 2.8% of total particles tracked) and the majority occurred more than 20 min after virus addition (Figure 1A). In 65.5% of these entry events, the signal of the virus was associated with a $CD46_{fluo}$ positive vesicular structure. Association with such a vesicle was maintained after endocytosis in 76% of events. After endocytosis, particles could on average be tracked for 2.8 min and migrated on average 3 μm inside the cell (Figure 1A). Two examples of mCherry positive particles entering a cell are shown in Figure 1B,C and movies 1 and 2.

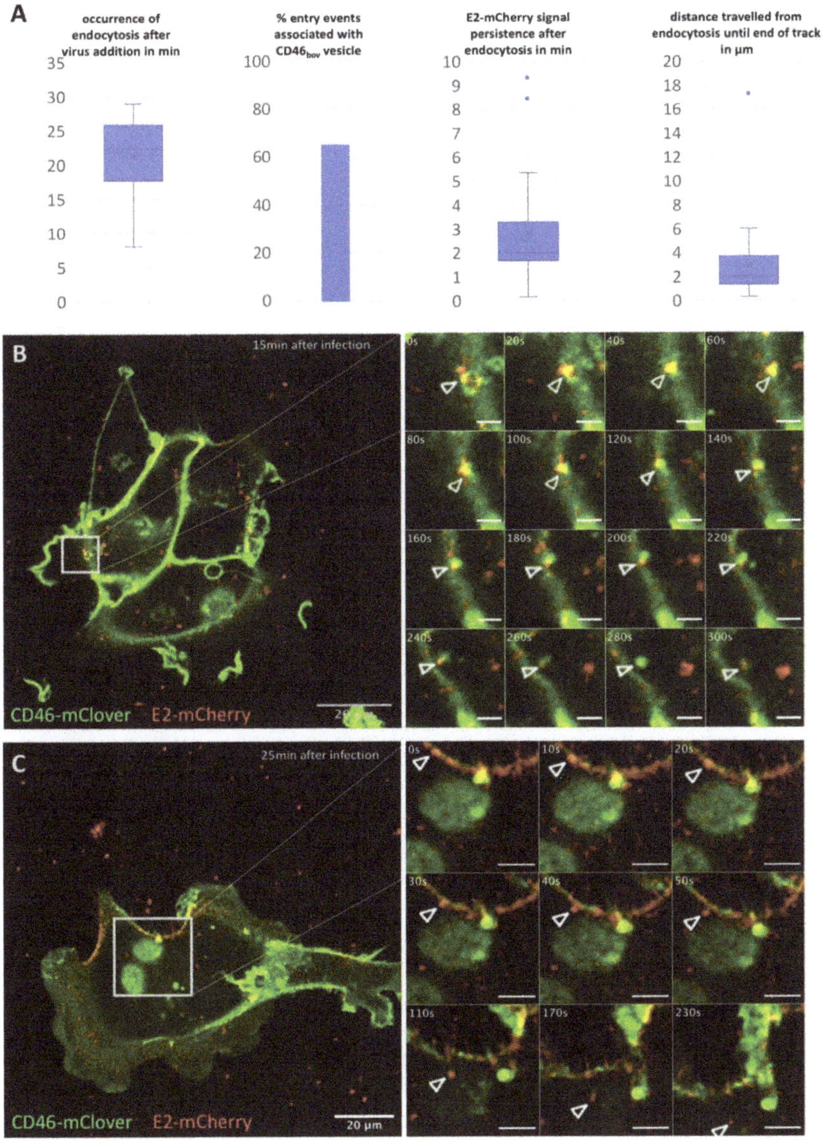

Figure 1. Characterization of $BVDV_{E2\text{-}mCherry}$ entry into SK6 $CD46_{fluo}$ cells. (**A**) Characterization of 26 entry events of $BVDV_{E2\text{-}mCherry}$ in SK6 $CD46_{fluo}$ cells with regard to occurrence after virus addition, association with $CD46_{fluo}$ signal, intracellular persistence of the E2-mCherry signal and intracellular distance travelled. Outliers (Q3 + 1.5-times interquartile range) are indicated by dots. (**B**,**C**) Examples of entry events of $BVDV_{E2\text{-}mCherry}$ (red) into SK6 $CD46_{bov}$ (green) cells. The full field of view at the time of the start of acquisition (time after virus addition is specified in the top right corner) is shown and the area of interest is indicated by grey squares. Frames as depicted in the detail images were acquired every 10 s for up to 10 min after the start of acquisition. Times indicated in s refer to the start of acquisition. The length of the scale bar in the detail images in (**B**) = 2.5 µm and in (**C**) = 5 µm.

Due to the intracellular background level in the mCherry emission range, the amount of intracellular, mCherry positive foci and their migration behavior was not analyzed as an unambiguous differentiation between background and specific signal was not possible.

Particles travelled with an average speed of 0.062 µm/s (s.d. 0.036 µm/s), an average maximum speed of 0.147 µm/s (s.d. 0.15) and an average directionality—defined as the ratio of direct distance versus real distance covered by a particle—of 0.34 (s.d. 0.22) on SK6 $CD46_{fluo}$ cells.

Directed transport of viruses along the outside of filopodia, cytonemes, and retraction fibers has been described for enveloped and non-enveloped viruses. The potential usurpation of this transport route was hence also examined in the context of $BVDV_{E2\text{-}mCherry}$. Virus surfing could be observed for 16 particles (2.0% of total tracks, movie 3). These particles migrated with an average velocity of 0.070 µm/s and an average directionality of 0.446, indicating an advantage of this type of transport regarding movement in a given direction. The average maximum speed of these particles was 0.176 µm/s, indicating a potential coupling to retrograde actin flow [2,26].

3.2. *First Detection of BVDV E2-mCherry Signal Is Depending on MOI*

In order to further analyze the progression of infection and to visualize the intracellular distribution of E2-mCherry over time, time series starting 60 min after virus addition and running for 16 h (1 frame every 10 min) were recorded. In all cells observed, an evenly distributed, slightly granular E2-mCherry signal was present from the initial detection of a specific signal (movies 4 and 5). This intracellular distribution pattern was reminiscent of an ER staining pattern, which is in good accordance with the localization of E2 in the ER lumen as already reported by [27,28]. Interestingly, the time after which an E2-mCherry signal could be resolved correlated with the MOI employed in a given experiment. For an MOI of 1, E2-mCherry could be detected on average 645 min after virus addition (n = 13) (Figure 2A, movie 4), whereas E2-mCherry could already be resolved on average 195 min (n = 16, movie 5) after infection if an MOI of 10 was used.

To gain further insights into the dynamics of E2-mCherry trafficking 20 h after infection, cells were imaged for 5 min with a frame rate of 1 frame/10 s. In addition to the diffuse, granular staining pattern of E2-mCherry, point-shaped, high signal intensities could be observed (Figure 2B, movie 6). They were partially associated with high $CD46_{fluo}$ signal intensities and this association was maintained during the whole course of the experiment. High intensity E2-mCherry foci could both be stationary or highly mobile (up to 0.5 µm/s), especially in the cell periphery (movie 6).

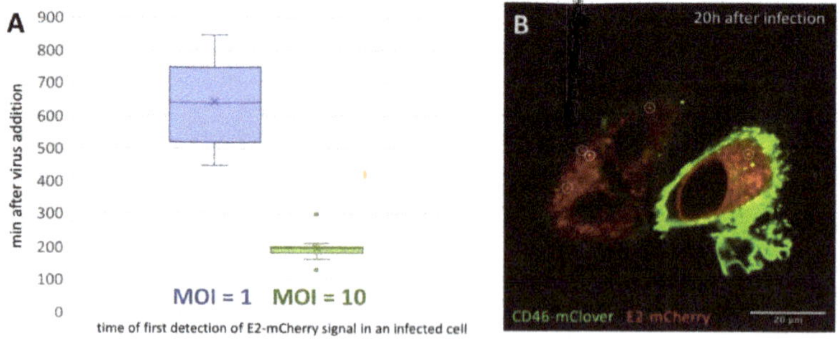

Figure 2. E2-mCherry signal development but not signal distribution is depending on the MOI. (**A**) Occurrence of E2-mCherry signal in min after addition of $BVDV_{E2\text{-}mCherry}$ at an MOI of 1 (blue) or 10 (green) to SK6 $CD46_{fluo}$ cells. SK6 $CD46_{fluo}$ cells were imaged recording one frame every 10 min at one z-level for 16 h. The mean is indicated by x and outliers (Q3 + 1.5-times interquartile range) are indicated by dots. (**B**) Distribution of E2-mCherry (red) and $CD46_{fluo}$ (green) signal 20 h after infection with $BVDV_{E2\text{-}mCherry}$ in SK6 $CD46_{fluo}$ cells. SK6 $CD46_{fluo}$ cells were imaged for 10 min recording 3 z-levels (0.8 µm) every 10 s (movie 6). Points of high E2-mCherry intensity that are colocalizing with $CD46_{fluo}$ are indicated by white circles.

3.3. *CD46$_{fluo}$ Decreases the Speed of Directed Surface Motion of BVDV$_{E2\text{-}mCherry}$*

In order to elucidate the effect of specific receptor binding on virus entry, SK6 cells inducibly expressing fluorophore-labelled CD46 with a deleted E2-binding domain (CD46$_{fluo}$ΔE2bind) were generated based on previous results by [16], reporting the CCP-1 as interaction partner. These cells showed a susceptibility to BVDV comparable to the susceptibility of SK6 cells inducibly expressing GFP (Figure 3). GFP expressing cells were chosen as control to account for a potential effect of induced expression on virus replication. The susceptibility of SK6 CD46$_{fluo}$ΔE2bind cells was more than 250-fold reduced in comparison to SK6 cells expressing unmodified CD46$_{fluo}$ after induction of expression.

This system was chosen over experiments employing inhibition of virus attachment by antibodies targeting neutralizing epitopes of E2 or functionally important domains of CD46 as it—in our opinion—provided the most direct and defined system to study the role of E2-CD46 interaction in the viral life cycle by life cell imaging. An inhibition of infection by BVDV tagged with a fluorophore at the E2 N-terminus employing before mentioned antibodies has already been demonstrated by [12] and this inhibition is comparable to BVDVs not bearing a tag at the E2 N-terminus. This clearly demonstrates the importance of the same domains in CD46-E2 interactions.

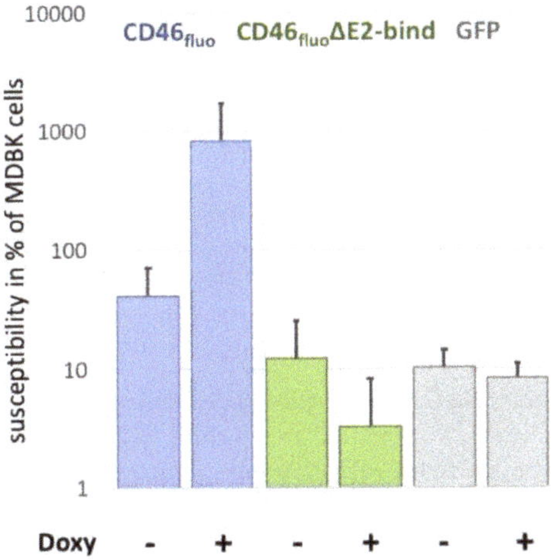

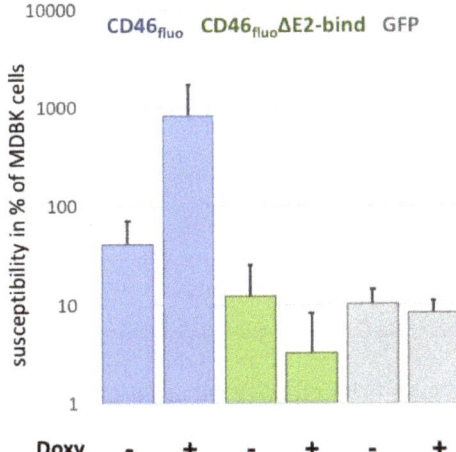

Figure 3. Deletion of CCP-1 reduces CD46$_{fluo}$ effect of the susceptibility of SK6 cells to the level of GFP-expressing controls. Susceptibility of SK6 CD46$_{fluo}$ (blue), CD46$_{fluo}$ΔE2bind (green) and GFP-expressing (grey) SK6 cells to infection with BVDV$_{E2\text{-}mCherry}$ in % of MDBK cells (=100%) with and without the induction of expression by the addition of doxycycline (Doxy) (n = 6). Cells were infected with serial dilutions of BVDV$_{E2\text{-}mCherry}$ for 4 h. Subsequently, medium was exchanged to medium containing 1% carboxymethycellulose and E2-mCherry positive foci were detected 48 h after infection by fluorescence microscopy to determine the focus forming units (ffu)/mL. Susceptibility of a given cell line was calculated as the percentage of ffu/mL determined for MDBK cells, which was set to 100%.

To assess the effect of receptor binding on the surface movement of BVDV$_{E2\text{-}mCherry}$, 1081 mCherry-positive particles were tracked on SK6 CD46$_{fluo}$ΔE2bind cells (n cells = 297). Of those 1081 particles, only 2 particles (0.2%) could be observed entering a cell, which is 7% of the entry events found for SK6 CD46$_{fluo}$ cells. One of these events was associated with a CD46$_{fluo}$ΔE2bind after uptake.

In order to better understand the movement of particles on the surface of SK6 CD46$_{fluo}$ and SK6 CD46$_{fluo}$ΔE2bind cells, particle trajectories were further analyzed. The average direct distance covered on the cell surface during a given observation period was 3.15 μm (s.d. 2.71 μm) for SK6 CD46$_{fluo}$ and 2.83 μm (s.d. 2.18 μm) for SK6 CD46$_{fluo}$ΔE2bind cells (Figure 4), while the average real distance covered was 10.9 μm (s.d. 8.64 μm) for $CD46_{bov}$ and 9.14 μm (s.d. 6.06 μm) for $CD46_{bov}\Delta$E2bind. The average directionality—defined as the quotient of direct versus real distance—was 0.34 (s.d. 0.22) for SK6 $CD46_{bov}$ and 0.37 (s.d. 0.23) for SK6 $CD46_{bov}\Delta$E2bind cells and therefore seems unaffected by CD46 binding. Particles on SK6 $CD46_{fluo}$ cells travelled with an average velocity of 0.062 μm/s (s.d. 0.036 μm/s) and an average maximum speed of 0.147 μm/s (s.d. 0.15). Particles on SK6 $CD46_{fluo}\Delta$E2bind cells were characterized by an average velocity of 0.063 μm/s (s.d. 0.028 μm/s) and an average maximum speed of 0.136 μm/s (s.d. 0.063 μm/s), indicating no effect of CD46 binding on observed velocities.

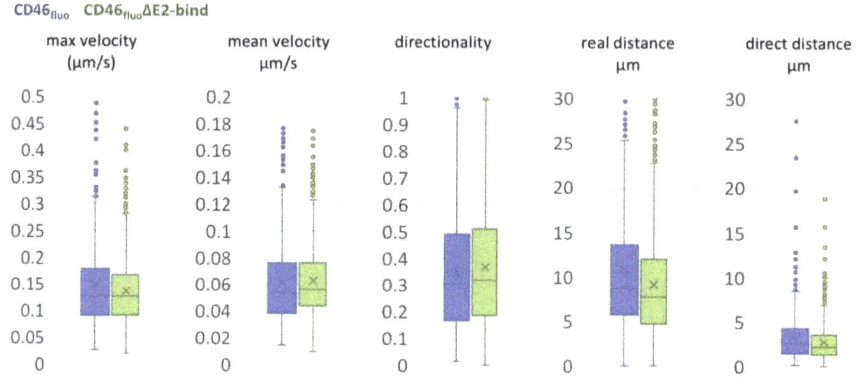

Figure 4. Movements of $BVDV_{E2\text{-}mCherry}$ on the surface of SK6 $CD46_{fluo}$ (blue) or $CD46_{fluo}\Delta$E2bind cells (green) are comparable. Box and Whisker blots of maximum and mean velocity, directionality, real distance and direct distance of particles tracked on the surface of SK6 $CD46_{fluo}$ (blue, n = 801) or $CD46_{fluo}\Delta$E2bind cells (green, n = 1081), respectively. The mean is indicated by x and outliers (Q3 + 1.5-times interquartile range) are indicated by dots.

On both, SK6 $CD46_{fluo}$ and SK6 $CD46_{fluo}\Delta$E2bind cells, the majority of particles was localized on the lamella or lamellipodium up to 10 min after infection; beginning with 15 min after infection, the majority of particles was bound to the cell body (Supplementary Figure S1). On average 51% of the particles on SK6 $CD46_{fluo}$ cells were associated with an increased CD46 signal intensity for at least 3 consecutive frames. On SK6 $CD46_{fluo}\Delta$E2bind cells, this percentage increased to an average of 74%. To assess potential differences in particle travelling directions, the overall direction of movement of a given particle was described as moving random or away from, tangential to or towards the cell body. Such a direction could unambiguously be assigned to 423 tracks on SK6 $CD46_{fluo}$ cells and 705 tracks on SK6 $CD46_{fluo}\Delta$E2bind cells (Supplementary Figure S2). 3% of particles tracked on SK6 $CD46_{fluo}$ cells and 12% of particles tracked on SK6 $CD46_{fluo}\Delta$E2bind cells showed a movement away from the cell body, whereas 57% and 42% respectively exhibited random movement. Tangential movement was observed for 17% of particles on SK6 $CD46_{fluo}$ cells and 18% of SK6 $CD46_{fluo}\Delta$E2bind cells. 23% and 29% of particles exhibited movement towards the cell body on SK6 $CD46_{fluo}$ cells and SK6 $CD46_{fluo}\Delta$E2bind cells, respectively. These results indicate that the major difference in the direction of particle transport is a four times reduced number of particles moving away from the cell body if CD46 is functional.

To further characterize the number of tracks following a certain type of motion on the cell surface and to determine transport velocities and the diffusion coefficients (D_0), MSD analysis was performed. Within both datasets, tracks exhibiting characteristics of directed motion, as well as limited diffusion or purely diffusive motility, could be identified. For SK6 $CD46_{fluo}$ cells, 86 tracks (10.7% of total tracks, 47.5% of tracks in MSD analysis) exhibited linear slopes, indicative of purely diffusive motion, while 46 tracks (4.3% of total tracks, 31.7% of tracks in MSD analysis) were identified for SK6 $CD46_{fluo}\Delta$E2bind cells (Figure 5A). The calculated diffusion coefficients were 0.020 μm²/s (s.d.

0.017 μm²/s) for SK6 $CD46_{bov}$ cells and 0.015 μm²/s (s.d. 0.019 μm²/s) for SK6 $CD46_{fluo}$ΔE2bind cells and were not significantly different ($p = 0.103$). 70 tracks exhibiting an exponential slope could be identified for SK6 $CD46_{bov}$ cells (8.7% of total tracks, 38.7% of tracks in MSD analysis) and 61 for SK6 $CD46_{bov}$ΔE2bind cells (5.6% of total tracks, 42.1% of tracks in MSD analysis) (Figure 5B). The calculated average particle velocity was 0.028 μm/s (s.d. 0.014 μm/s) for SK6 $CD46_{fluo}$ cells and 0.039 μm/s (s.d. 0.018 μm/s) for SK6 $CD46_{fluo}$ΔE2bind cells, suggesting a significantly increased directional transport speed on SK6 $CD46_{fluo}$ΔE2bind cells ($p < 0.001$). Twenty-five tracks on SK6 $CD46_{fluo}$ cells (3.1% of total tracks, 13.8% of tracks in MSD analysis) and 38 tracks on SK6 $CD46_{fluo}$ΔE2bind cells (3.5% of total tracks, 26.2% of tracks in MSD analysis) showed characteristics indicative of limited diffusion in MSD analysis.

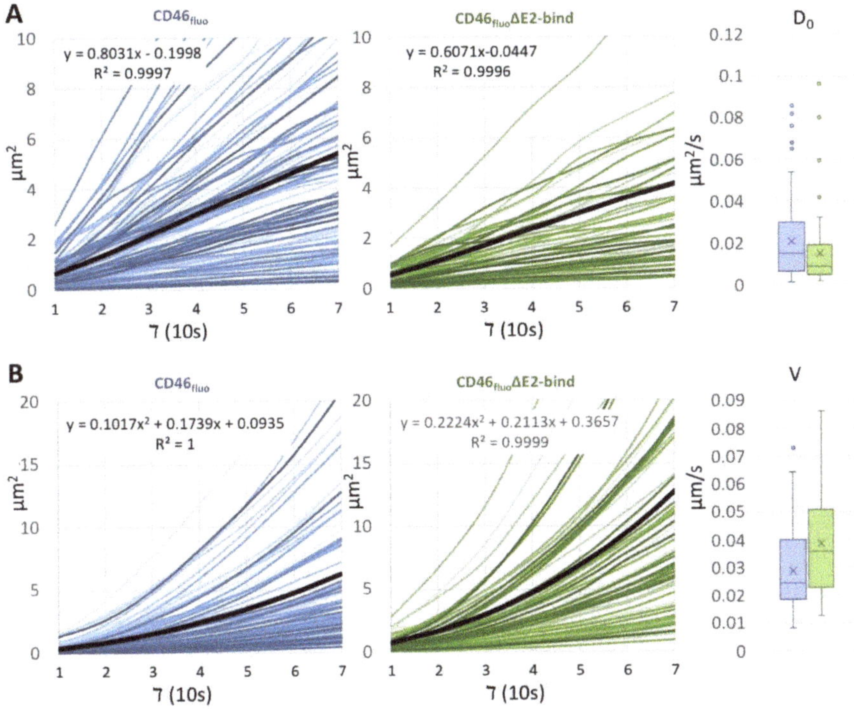

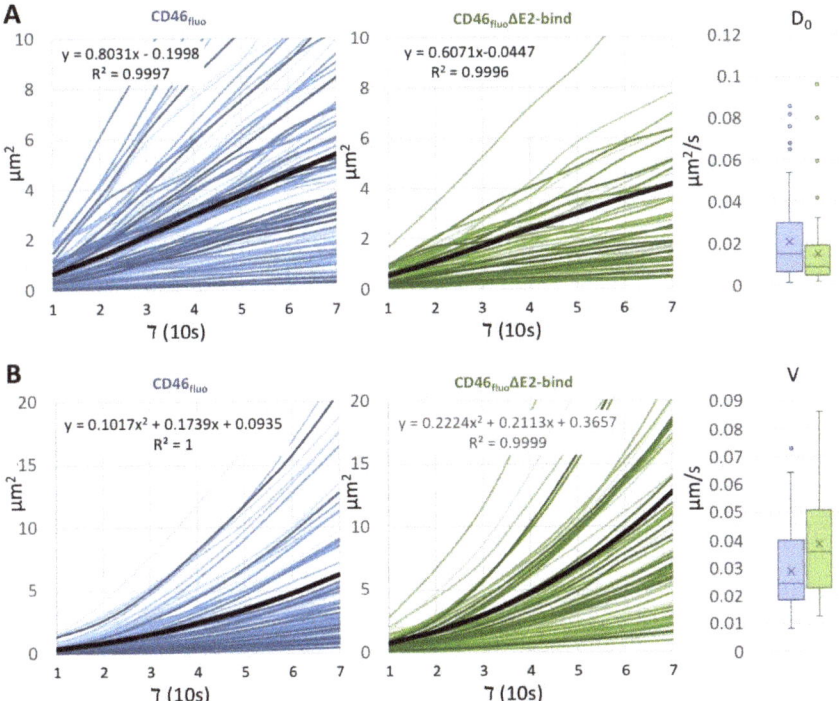

Figure 5. The average velocity of directed motion of $BVDV_{E2\text{-}mCherry}$ particles is increased on the surface of SK6 $CD46_{fluo}$ΔE2bind cells. MSD analysis of particle movement on SK6 $CD46_{fluo}$ and $CD46_{fluo}$ΔE2bind cells. (**A**) Slopes of tracks on SK6 $CD46_{fluo}$ (blue) and $CD46_{fluo}$ΔE2bind (green) cells displaying a linear increase and distribution of the calculated diffusion coefficient D_0 depicted as box and whisker blot. The mean is indicated by x and outliers (Q3 + 1.5 times interquartile range) are depicted as dots. Average slopes are depicted as thick black lines. (**B**) Slopes of tracks on SK6 $CD46_{fluo}$ (blue) and $CD46_{fluo}$ΔE2bind (green) cells displaying an exponential increase and distribution of the calculated particle velocities V, depicted as box and whisker blot. The mean is indicated by x and outliers (Q3 + 1.5 times interquartile range) are depicted as dots. Average slopes are depicted as thick black lines.

4. Discussion

Virus attachment and entry are governed by various affinities and highly dynamic processes. For BVDV, our findings suggest the presence of different types of motility on the cell surface, being diffusion, directed motion, and confined diffusion. This is different from observations of Dengue virus, as the main motion type until the association with a clathrin-coated pit is diffusion [8]. Virus surfing, as a movement frequently utilized by other viruses (e.g., MLV), was rarely observed, albeit the presence of filopodia, cytonemes, and retraction fibers, implying a minor role of this transport mechanism for BVDV, at least in this experimental setup. All motion types could be observed independent of the presence or absence of the BVDV-binding competent cellular receptor, CD46 and track properties were comparable. Similarly, the proportions of cellular compartments—lamellipodium, lamella, and cell body—with associated virus particles were not affected by CD46's ability to bind E2 when comparing different time points after infection. However, when analyzing directed motion in more detail by MSD analysis, track velocities were significantly increased if CD46 was not able to interact with E2. This might suggest that another interaction partner on the cellular surface—for example heparan sulphate or an as of yet unidentified attachment factor—might be responsible for coupling to the underlying actin network and that CD46 is rather involved in deceleration of particle transport on the cell surface or the anchoring of particles at potential sites of endocytosis. The importance of attachment factors such as heparan sulphate has also been demonstrated for other members of the *Flaviviridae* (reviewed in [29]). Also, CD46 is not sufficient to render cell lines derived from non-cloven-hoofed animals or humans susceptible to infection with BVDV. Instead, it increases the amount of cell-associated virus [15]. This indicates that additional cellular factors are required for successful virus infection after binding to CD46. Interestingly, overexpression of $CD46_{fluo}\Delta$E2bind resulted in a further drop of susceptibility to BVDV, which might indicate the sequestration of a potential additional entry factor by $CD46_{fluo}\Delta$E2bind.

In general, the number of observable entry events into SK6 $CD46_{fluo}$ was very low (2.8% of tracked particles). This might indicate that cellular uptake of BVDV is a rather inefficient process. However, during data acquisition, we were not able to sample the whole cellular surface due to the short frame interval and phototoxicity. The focus of acquisition was on the cellular surface in contact with the growth support, to clearly depict cellular protrusions, the lamellipodium and the lamella. Particles could frequently be observed exiting the acquisition planes to regions of the cell body that were inaccessible in the chosen acquisition scheme. It is therefore possible that the low amount of observed entry events is—at least partially—due to a preference of BVDV to enter cells at specific areas of the cell body. Also, the specific infectivity of $BVDV_{E2\text{-}mCherry}$ is below 1:10, meaning that more than 10 genome equivalents are needed for successful establishment of infection. It is currently unclear what particle defects are responsible for this phenomenon, but it has to be taken into consideration that a certain number of particles are already unable to evoke cellular uptake. In order to clarify these open questions and to also examine the entry process of BVDV into primary host cells, MDBK CD46 knock-out cell lines would be a valuable tool for future elucidation of the entry process.

Contrasting the increased speed of particle transport, cell entry events were very rare if CD46 was E2-binding incompetent. This, in combination with the rather rare CD46—virus co-migration on the cell surface, might implicate that CD46 is important for the signaling to evoke endocytosis, but not for surface movements towards potential sites of endocytosis. The frequently observed association of virus particles with CD46 positive vesicles after endocytosis and their comigration for several frames further supports this hypothesis, but additional experiments will be required for verification. Still, $BVDV_{E2\text{-}mCherry}$ is able to enter host cells independent of a functional CD46 molecule. This strongly indicates the potential to use other entry receptors, but with much lower efficiency. The importance of factors different from CD46 in BVDV propagation is also highlighted by recent results reporting that BVDV cell-to-cell spread is independent of CD46 [12].

The CD46 splice variant employed in this study is identical to the bovine CD46 variant initially reported by [15]. Unfortunately, we were not able to elucidate the effects of different allelic versions of CCP-1 and different CD46 splice variants on the dynamics of BVDV entry due to the huge efforts of data analysis. A different permissiveness to infection of these different allelic and splice variants has however already been demonstrated [30] and their effect on the dynamics of BVDV entry will be an interesting topic of future studies.

Taking advantage of the easy tracking of virus protein production in infected cells, the effect of different MOIs on the earliest time point of specific fluorescence detection was examined. The highly significant difference observed leads to the conclusion that signal development is depending on the virus dose applied and that superinfection exclusion is not acting fast enough to prevent the infection of one cell with several particles in the context of this experimental setup [31]. Surveillance of the signal development of $BVDV_{E2\text{-}mCherry}$ after infection revealed an initially finely granular, likely ER-associated [27,28] distribution throughout the whole cell, which intensified over time and finally also developed strong signal foci. Therefore, it seems likely that E2 is quickly distributed throughout the cell and not restricted to and radially expanding from a potential initial replication site. To better understand viral replication sites, additional studies are warranted to elucidate how the intracellular distribution of E2 is reflecting the overall distribution of non-structural proteins like NS3 and NS5B as part of the replication complex. This, in combination with ultrastructural studies, will provide further insights into the morphology of replication sites and allow comparison to the detailed morphological studies already performed on the membranous replication compartments of HCV [32] and flaviviruses [33,34]. The occasional association of CD46 positive intracellular foci with E2 mCherry positive foci suggests that CD46 is usually excluded from E2 mCherry containing compartments or only present in amounts below the detection limit of this experimental setup. Whether this association is unfavorable for the virus or serves an as of yet unknown purpose—like facilitation of exocytosis by taking advantage of CD46 transport to the cell surface—will need to be elucidated in the future.

In the presented study a large variety of BVDV surface motions has been documented. Interestingly, apart from a decrease in the velocity of directed motion, no effect of $CD46_{fluo}$ on the surface transport of BVDV could be identified, indicating a minor role during this process. First insights into the progression of virus infection indicate the quick distribution of BVDV protein throughout the ER in an initial virus dose dependent manner. It will be highly interesting to identify and characterize the additional factors orchestrating BVDV entry to develop a clear picture of this process.

Supplementary Materials: The following are available online at http://www.mdpi.com/1999-4915/12/1/116/s1, Figure S1: Particle localization on the cell surface depending on the time after virus addition, Figure S2: Directions of movements of virus particles on the cell surface with reference to the cell body. supplementary information: python script used for track analysis, legend of the movie files, nucleotide sequence of the plasmid encoding the genome of $BVDV_{E2\text{-}mCherry}$; Movie 1: Example 1 of an mCherry positive particle entering a cell, Movie 2: Example 2 of an mCherry positive particle entering a cell, Movie 3: Surfing of a $BVDV_{E2\text{-}mCherry}$ particle on the surface of a retraction fibre., Movie 4: Development of E2-mCherry (red) signal after infection of SK6 TO CD46fluo cells with an MOI of 1, Movie 5: Development of E2-mCherry (red) signal after infection of SK6 TO CD46fluo cells with an MOI of 10, Movie 6: Colocalization of E2-mCherry and $CD46_{fluo}$ inside the cell.

Author Contributions: C.R. designed the study. B.L., C.R., H.-W.C. and V.L. performed the experiments. C.R., T.R., U.R. and V.L. analysed the data. C.R. wrote the manuscript and all authors commented on it. All authors have read and agreed to the published version of the manuscript.

Funding: C.R. was supported by a Vetmeduni Start-up grant (PP23016276) and a CORBEL grant (PID 2375).

Acknowledgments: We would like to acknowledge the microscopy support from the Infectious Diseases Imaging Platform (IDIP) at the Center for Integrative Infectious Disease Research, Heidelberg, Germany. The authors thank Tobias Rasse and Christian Tischer for their support and the fruitful discussions. We would like to thank the University of Veterinary Medicine Vienna for Open Access Funding.

Conflicts of Interest: The authors declare no conflict of interest.

References

1. Burckhardt, C.J.; Greber, U.F. Virus movements on the plasma membrane support infection and transmission between cells. *PLoS Pathog.* **2009**, *5*, e1000621. [CrossRef] [PubMed]
2. Lehmann, M.J.; Sherer, N.M.; Marks, C.B.; Pypaert, M.; Mothes, W. Actin- and myosin-driven movement of viruses along filopodia precedes their entry into cells. *J. Cell Biol.* **2005**, *170*, 317–325. [CrossRef] [PubMed]
3. Oh, M.-J.; Akhtar, J.; Desai, P.; Shukla, D. A role for heparan sulfate in viral surfing. *Biochem. Biophys. Res. Commun.* **2010**, *391*, 176–181. [CrossRef] [PubMed]

4. Schelhaas, M.; Shah, B.; Holzer, M.; Blattmann, P.; Kühling, L.; Day, P.M.; Schiller, J.T.; Helenius, A. Entry of human papillomavirus type 16 by actin-dependent, clathrin- and lipid raft-independent endocytosis. *PLoS Pathog.* **2012**, *8*, e1002657. [CrossRef] [PubMed]
5. Ayala-Nuñez, N.V.; Wilschut, J.; Smit, J.M. Monitoring virus entry into living cells using DiD-labeled dengue virus particles. *Methods* **2011**, *55*, 137–143. [CrossRef]
6. Zhang, S.L.-X.; Tan, H.-C.; Hanson, B.J.; Ooi, E.E. A simple method for Alexa Fluor dye labelling of dengue virus. *J. Virol. Methods* **2010**, *167*, 172–177. [CrossRef]
7. Coller, K.E.; Berger, K.L.; Heaton, N.S.; Cooper, J.D.; Yoon, R.; Randall, G. RNA Interference and Single Particle Tracking Analysis of Hepatitis C Virus Endocytosis. *PLoS Pathog.* **2009**, *5*, e1000702. [CrossRef]
8. van der Schaar, H.M.; Rust, M.J.; Chen, C.; van der Ende-Metselaar, H.; Wilschut, J.; Zhuang, X.; Smit, J.M. Dissecting the cell entry pathway of dengue virus by single-particle tracking in living cells. *PLoS Pathog.* **2008**, *4*, e1000244. [CrossRef]
9. Dejarnac, O.; Hafirassou, M.L.; Chazal, M.; Versapuech, M.; Gaillard, J.; Perera-Lecoin, M.; Umana-Diaz, C.; Bonnet-Madin, L.; Carnec, X.; Tinevez, J.Y.; et al. TIM-1 Ubiquitination Mediates Dengue Virus Entry. *Cell Rep.* **2018**, *23*, 1779–1793. [CrossRef]
10. Tamura, T.; Fukuhara, T.; Uchida, T.; Ono, C.; Mori, H.; Sato, A.; Fauzyah, Y.; Okamoto, T.; Kurosu, T.; Setoh, Y.X.; et al. Characterization of Recombinant Flaviviridae Viruses Possessing a Small Reporter Tag. *J. Virol.* **2018**, *92*, e01582-17. [CrossRef]
11. Tamura, T.; Igarashi, M.; Enkhbold, B.; Suzuki, T.; Okamatsu, M.; Ono, C.; Mori, H.; Izumi, T.; Sato, A.; Fauzyah, Y.; et al. In vivo dynamics of reporter Flaviviridae viruses. *J. Virol.* **2019**. [CrossRef] [PubMed]
12. Merwaiss, F.; Czibener, C.; Alvarez, D.E. Cell-to-Cell Transmission Is the Main Mechanism Supporting Bovine Viral Diarrhea Virus Spread in Cell Culture. *J. Virol.* **2019**, *93*. [CrossRef] [PubMed]
13. Riedel, C.; Lamp, B.; Chen, H.-W.; Heimann, M.; Rümenapf, T. Fluorophore labelled BVDV: A novel tool for the analysis of infection dynamics. *Sci. Rep.* **2019**, *9*, 5972. [CrossRef]
14. Iqbal, M.; Flick-Smith, H.; McCauley, J.W. Interactions of bovine viral diarrhoea virus glycoprotein E(rns) with cell surface glycosaminoglycans. *J. Gen. Virol.* **2000**, *81*, 451–459. [CrossRef] [PubMed]
15. Maurer, K.; Krey, T.; Moennig, V.; Thiel, H.-J.; Rümenapf, T. CD46 is a cellular receptor for bovine viral diarrhea virus. *J. Virol.* **2004**, *78*, 1792–1799. [CrossRef] [PubMed]
16. Krey, T.; Himmelreich, A.; Heimann, M.; Menge, C.; Thiel, H.-J.; Maurer, K.; Rümenapf, T. Function of bovine CD46 as a cellular receptor for bovine viral diarrhea virus is determined by complement control protein 1. *J. Virol.* **2006**, *80*, 3912–3922. [CrossRef]
17. Grummer, B.; Grotha, S.; Greiser-Wilke, I. Bovine Viral Diarrhoea Virus is Internalized by Clathrin-dependent Receptor-mediated Endocytosis. *J. Vet. Med. Ser. B* **2004**, *51*, 427–432. [CrossRef]
18. Lecot, S.; Belouzard, S.; Dubuisson, J.; Rouillé, Y. Bovine viral diarrhea virus entry is dependent on clathrin-mediated endocytosis. *J. Virol.* **2005**, *79*, 10826–10829. [CrossRef]
19. Krey, T.; Thiel, H.-J.; Rümenapf, T. Acid-resistant bovine pestivirus requires activation for pH-triggered fusion during entry. *J. Virol.* **2005**, *79*, 4191–4200. [CrossRef]
20. Cattaneo, R. Four viruses, two bacteria, and one receptor: Membrane cofactor protein (CD46) as pathogens' magnet. *J. Virol.* **2004**, *78*, 4385–4388. [CrossRef]
21. Ni Choileain, S.; Astier, A.L. CD46 processing: A means of expression. *Immunobiology* **2012**, *217*, 169–175. [CrossRef]
22. Crimeen-Irwin, B.; Ellis, S.; Christiansen, D.; Ludford-Menting, M.J.; Milland, J.; Lanteri, M.; Loveland, B.E.; Gerlier, D.; Russell, S.M. Ligand binding determines whether CD46 is internalized by clathrin-coated pits or macropinocytosis. *J. Biol. Chem.* **2003**, *278*, 46927–46937. [CrossRef] [PubMed]
23. Persson, B.D.; Schmitz, N.B.; Santiago, C.; Zocher, G.; Larvie, M.; Scheu, U.; Casasnovas, J.M.; Stehle, T. Structure of the extracellular portion of CD46 provides insights into its interactions with complement proteins and pathogens. *PLoS Pathog.* **2010**, *6*, e1001122. [CrossRef] [PubMed]
24. Schneider, C.A.; Rasband, W.S.; Eliceiri, K.W. NIH Image to ImageJ: 25 years of image analysis. *Nat. Methods* **2012**, *9*, 671–675. [CrossRef] [PubMed]
25. Marini, F.; Mazur, J.; Binder, H. flowcatchR: A user-friendly workflow solution for the analysis of time-lapse cell flow imaging data. *F1000Research* **2015**, *4*. [CrossRef]
26. Kohler, F.; Rohrbach, A. Surfing along Filopodia: A Particle Transport Revealed by Molecular-Scale Fluctuation Analyses. *Biophys. J.* **2015**, *108*, 2114–2125. [CrossRef] [PubMed]

27. Schmeiser, S.; Mast, J.; Thiel, H.-J.; König, M. Morphogenesis of pestiviruses: New insights from ultrastructural studies of strain Giraffe-1. *J. Virol.* **2014**, *88*, 2717–2724. [CrossRef]
28. Gray, E.W.; Nettleton, P.F. The Ultrastructure of Cell Cultures Infected with Border Disease and Bovine Virus Diarrhoea Viruses. *J. Gen. Virol.* **1987**, *68*, 2339–2346. [CrossRef]
29. Gerold, G.; Bruening, J.; Weigel, B.; Pietschmann, T. Protein Interactions during the Flavivirus and Hepacivirus Life Cycle. *Mol. Cell. Proteomics* **2017**, *16*, S75–S91. [CrossRef]
30. Zezafoun, H.; Decreux, A.; Desmecht, D. Genetic and splice variations of Bos taurus CD46 shift cell permissivity to BVDV, the bovine pestivirus. *Vet. Microbiol.* **2011**, *152*, 315–327. [CrossRef]
31. Lee, Y.-M.; Tscherne, D.M.; Yun, S.-I.; Frolov, I.; Rice, C.M. Dual Mechanisms of Pestiviral Superinfection Exclusion at Entry and RNA Replication. *J. Virol.* **2005**, *79*, 3231–3242. [CrossRef] [PubMed]
32. Romero-Brey, I.; Merz, A.; Chiramel, A.; Lee, J.-Y.; Chlanda, P.; Haselman, U.; Santarella-Mellwig, R.; Habermann, A.; Hoppe, S.; Kallis, S.; et al. Three-Dimensional Architecture and Biogenesis of Membrane Structures Associated with Hepatitis C Virus Replication. *PLoS Pathog.* **2012**, *8*, e1003056. [CrossRef] [PubMed]
33. Welsch, S.; Miller, S.; Romero-Brey, I.; Merz, A.; Bleck, C.K.E.; Walther, P.; Fuller, S.D.; Antony, C.; Krijnse-Locker, J.; Bartenschlager, R. Composition and three-dimensional architecture of the dengue virus replication and assembly sites. *Cell Host Microbe* **2009**, *5*, 365–375. [CrossRef] [PubMed]
34. Cortese, M.; Goellner, S.; Acosta, E.G.; Neufeldt, C.J.; Oleksiuk, O.; Lampe, M.; Haselmann, U.; Funaya, C.; Schieber, N.; Ronchi, P.; et al. Ultrastructural Characterization of Zika Virus Replication Factories. *Cell Rep.* **2017**, *18*, 2113–2123. [CrossRef]

viruses

MDPI

Article

A CRISPR/Cas9 Generated Bovine CD46-knockout Cell Line—A Tool to Elucidate the Adaptability of Bovine Viral Diarrhea Viruses (BVDV)

Kevin P. Szillat, Susanne Koethe, Kerstin Wernike, Dirk Höper and Martin Beer *

Institute of Diagnostic Virology, Friedrich-Loeffler-Institut, Federal Research Institute for Animal Health, 17493 Greifswald-Insel Riems, Germany; Kevin.Szillat@fli.de (K.P.S.); Susanne.Koethe@fli.de (S.K.); Kerstin.Wernike@fli.de (K.W.); Dirk.Hoeper@fli.de (D.H.)

* Correspondence: Martin.Beer@fli.de; Tel.: +49-38351-71200

Received: 29 June 2020; Accepted: 2 August 2020; Published: 6 August 2020

Abstract: Bovine viral diarrhea virus (BVDV) entry into a host cell is mediated by the interaction of the viral glycoprotein E2 with the cellular transmembrane CD46 receptor. In this study, we generated a stable Madin–Darby Bovine Kidney (MDBK) CD46-knockout cell line to study the ability of different *pestivirus A* and *B* species (BVDV-1 and -2) to escape CD46-dependent cell entry. Four different BVDV-1/2 isolates showed a clearly reduced infection rate after inoculation of the knockout cells. However, after further passaging starting from the remaining virus foci on the knockout cell line, all tested virus isolates were able to escape CD46-dependency and grew despite the lack of the entry receptor. Whole-genome sequencing of the escape-isolates suggests that the genetic basis for the observed shift in infectivity is an amino acid substitution of an uncharged (glycine/asparagine) for a charged amino acid (arginine/lysine) at position 479 in the E^{RNS} in three of the four isolates tested. In the fourth isolate, the exchange of a cysteine at position 441 in the E^{RNS} resulted in a loss of E^{RNS} dimerization that is likely to influence viral cell-to-cell spread. In general, the CD46-knockout cell line is a useful tool to analyze the role of CD46 for pestivirus replication and the virus–receptor interaction.

Keywords: bovine viral diarrhea virus (BVDV); pestivirus; escape mutant; CD46; E^{RNS}; adaptation; CRISPR; knockout; MDBK; cell entry

1. Introduction

The genus *Pestivirus* belongs to the family *Flaviviridae* and contains several veterinary-relevant virus species with major animal welfare and economic importance like *pestivirus A* and *B* (bovine viral diarrhea virus types 1 and 2, BVDV-1 and -2), *pestivirus D* (border disease virus, BDV) and *pestivirus C* (classical swine fever virus, CSFV) [1–4]. The host range of the classical pestivirus species includes cloven-hoofed animals. However, in recent years, a continuously growing diversity of pestiviruses has been seen worldwide with atypical pestiviruses being isolated from host species like rat, bat and whale [5–8].

Pestiviruses are positive-sense, enveloped, single-stranded RNA viruses with a genome size of 12.3–13 kb [9]. The genome encodes eight nonstructural and four structural proteins in a single open reading frame (ORF) [10], of which the glycoproteins E^{RNS}, E1 and E2 play an important role in the initiation of BVDV uptake by the host cell [11,12]. It has been shown that E^{RNS} interacts with the cell surface heparan sulfate and E1–E2 heterodimers bind the cellular receptor CD46 and mediate clathrin-dependent endocytosis [11,13,14]. CSFV E^{RNS} interacts with the heparan sulfate of porcine cells and both CD46 and heparan sulfate, are major factors for the attachment of CSFV in vitro [15,16]. Interestingly, a shift to a dominant role of the E^{RNS}-mediated binding in CSFV was connected with a drastic in vivo attenuation [15]. Furthermore, dimerization of E^{RNS} is an important virulence factor and abrogation leads to attenuation [17]. E^{RNS} also plays a role in the control of the activation of beta interferon induced upon viral infection by inhibiting the double stranded RNA-induced response of cells [18]. Heparan sulfate has been further described to be important for cellular binding of different viruses, e.g., Schmallenberg virus, hepatitis E virus and rabies virus [19–22].

The binding partner of the pestivirus envelope protein E2, cellular CD46, is a type 1 transmembrane glycoprotein expressed on all nucleated cells that protects host cells from damage by the complement system by the inactivation of C3b and C4b complement products (reviewed in [23]). CD46 is known to serve as a binding partner for several human pathogens like certain adenoviruses, as well as for animal viruses like BVDV and CSFV [16,24]. In particular the two peptide domains $E_{66}QIV_{69}$ and $G_{82}QVLAL_{87}$ of the complement control protein module 1 (CCP1) are crucial for the binding of BVDV—preincubation of MDBK cells with an anti-CD46 serum leads to a strong reduction in infection efficiency [25]. In vitro studies have shown that BVDV spreads by direct cell-to-cell transmission from infected to uninfected cells, even in the presence of neutralizing antibodies against the virus and also if CD46 receptors are blocked by antibodies [26].

Interestingly, there is to our knowledge no bovine CD46-knockout cell line available, allowing e.g., the analysis of the interaction of bovine pestiviruses with receptor molecules. Due to rapid developments in the field of CRISPR/Cas9-mediated gene editing, genetically modified in vitro models like knockout cell lines can be generated in a straightforward manner in a relatively short amount of time [27]. By using ribonucleoprotein (RNP) complexes, target-specific guide RNAs (gRNAs) are complexed with the Cas9 protein and transfected into the target cell. Unlike delivery of mRNA or DNA, RNP-mediated editing does not depend on the host cell for the synthesis of Cas9 and gRNAs. Furthermore, the RNP-based approach reduces the risk of off-target effects and cell death due to the shortened half-life of the proteins inside the target cell [28,29]. We used the CRISPR/Cas9 RNP approach in this study to knockout the cellular receptor CD46 in the Madin–Darby Bovine Kidney (MDBK) cell line to establish a stable in vitro model. We used the generated cell line to passage different BVDV type 1 and 2 strains to study adaptation mechanisms of these viruses in a CD46-negative cellular environment.

2. Materials and Methods

2.1. Cells and Viruses

All cell lines used in this study were obtained from the Collection of Cell Lines in Veterinary Medicine (CCLV) at the Federal Research Institute for Animal Health, Insel Riems, Germany (FLI). Cells were cultured in minimal essential medium (MEM), supplemented with 10% fetal calf serum (FCS), at 37 °C and 5% CO_2. A Madin–Darby Bovine Kidney cell line (MDBK, RIE0261) was used to generate the CD46-knockout cell line, further referred to as CD46-MDBK. Bovine esophagus cells (KOP-R, RIE0244) were used to propagate and titrate virus stocks of the BVDV-1 strains Paplitz (1b, cytopathic), NADL (1a), D02/11-2 (1d) and BVDV-2 strain CS8644 (2a). All pestiviruses were obtained from the German National Reference Laboratory for BVD/MD (FLI).

2.2. Generation of a CD46-knockout Cell Line

2.2.1. Transfection of MDBK Cells and Clone Selection

The CRISPR/Cas9 RNP-mediated editing approach was used in this study to knockout the BVDV binding domains $E_{66}QIV_{69}$ and $G_{82}QVLAL_{87}$ of the cellular receptor CD46 in MDBK cells. CRISPR RNAs (crRNAs) were designed to introduce double-strand breaks (DBS) inside the CCP1 domain, spanning these binding domains (overview in Figure S1). Suitability of the crRNAs was confirmed by CRISPOR [30] and CHOPCHOP [31]. The crRNAs (crRNA-1: GGCTTCATAGAGACAAATCT and crRNA-3: TCATACACAATCTGCTCCCC) and all transfection reagents were purchased from Thermo Fisher Scientific (Waltham, MA, USA). Annealing of the crRNAs with the trans-activating crRNA (tracrRNA) was done following the manufacturer's protocol. MDBK cells were seeded one day prior to transfection and subsequently transfected according to the manufacturer's instructions, using Lipofectamine CRISPRMAX Cas9 transfection reagents (Thermo Fisher Scientific) and TrueCut Cas9 Protein v2 (Thermo Fisher Scientific). The nontarget control cell line (NTC) was generated under identical conditions, using nontarget control gRNAs (Thermo Fisher Scientific). Single cell dilution and subsequent polymerase chain reaction (PCR) of the monoclonal cell colonies were used to screen for a cell clone with the intended deletion.

2.2.2. Isolation of DNA, PCR and Sequencing

DNA was isolated using the QIAamp DNA Mini Kit (Qiagen, Hilden, Germany) according to the manufacturer's instructions. The PCR to confirm the knockout was performed using the QuantiTect Multiplex-PCR Kit (Qiagen) in combination with CD46-specific primers (forward primer CD46_CCP1_F: 5′-GAT GCT GTC TCT TCC ATT TAC T-3′; reverse primer CD46_CCP1_R: 5′-GCC TGA ATG CAT GGC TAT CT-3′) and the following conditions: 15 min, 95 °C; 45× (60 s, 95 °C; 30 s 58 °C and 30 s 72 °C); 5 min, 72 °C; 12 °C storage. The PCR amplicon of the wild-type MDBK cell line is 545 nucleotides long and covers the entire CCP1 domain.

PCR products were analyzed by gel electrophoresis and bands were extracted from the gel using the QIAquick Gel Extraction Kit (Qiagen) according to the manufacturer's instructions. Single cell colonies with the intended knockout mutation display a truncated amplicon with a size of around 400 nucleotides. The PCR gel extract was further sequenced using the IonTorrent platform, as described later.

2.3. *Immunofluorescence (IF) Staining, Plaque Assay and Fluorescence-Activated Cell Sorting (FACS)*

Expression of the CD46 receptor by NTC and CD46⁻MDBK was visualized by IF staining, using antibodies BVD/CA 26/2/5 and BVD/CA 17/2/1(1:16 dilution in tris-buffered saline with tween (TBST)). The antibodies are directed against CD46 and were kindly provided by Prof. Becher, Institute of Virology, University of Veterinary Medicine, Hannover [32]. Antimouse Alexa Fluor™488 F(ab′)2 (1:1000 in TBST, Life Technologies, Carlsbad, CA, USA) was used as conjugate. FACS and IF was used to study the impact of the CD46-knockout on pestivirus entry into the host cell. In brief, CD46⁻MDBK and NTC cells were seeded one day prior to infection. On the day of infection, the cells were incubated with the different pestivirus isolates for 1 h at 37 °C and 5% CO_2, multiplicity of infection (MOI) of 1. Confluent cells were used at this point to study the initial infectivity of the virus isolates and at later time points the cell-to-cell spread. Uninfected cells were used as a negative control and incubated with maintenance medium only (supplemented with penicillin and streptomycin). After the incubation period, all cells were washed and cultured for 24 h at 37 °C and 5% CO_2 in maintenance medium.

For the IF-staining of pestiviruses, cell supernatant was discarded, and cells were fixed at 80 °C for 2 h. Subsequently, fixed cells were incubated with a 1:500 dilution of antibody WB103/105 (APHA Scientific, Addlestone, UK) in TBST for 1 h at room temperature (RT). Cells were washed thrice with TBST and incubated with a 1:1000 dilution of antimouse Alexa Fluor™488 F(ab′)2 (Life Technologies) in TBST for 1 h at RT. Cells were washed thrice and finally covered with 1,4-Diazabicyclo [2.2.2]octane (DABCO) fluorescence preservation buffer (Sigma-Aldrich, St. Louis, MO, USA) containing propidium iodide (Sigma-Aldrich).

For the IF plaque assay, CD46⁻MDBK cells were seeded one day prior to infection into a 24-well plate. Subsequently, cells were infected with the stock virus (passage 0) and passage 15 of virus isolate Paplitz (BVDV-1b) at MOI = 0.1 and incubated for 1 h at 37 °C, 5% CO_2. Thereafter, the cells were washed once with maintenance medium and overlayed with a 1:1 mix of 4% agarose (Lonza, Basel, Switzerland) in aqua dest. and MEM (2× concentrated) supplemented with penicillin/streptomycin and gentamicin. The medium layer was removed after 48 h, cells were washed once with TBST and fixed with 4% paraformaldehyde (Sigma-Aldrich) for 20 min. Cells were washed once and permeabilized with 0.5% Triton X-100 in phosphate-buffered saline (PBS) for 5 min. Cells were washed again and treated as previously described for the IF staining of pestiviruses. The size of 50 virus plaques was measured manually by using the Nikon Eclipse Ti-U inverted microscope (Nikon GmbH, Düsseldorf, Germany) and NIS-Elements software (v. 4.50). The assay was performed in triplicate. The difference between the groups was statistically evaluated with a Mann-Whitney rank sum test as implemented in SigmaPlot (Version 11.0; Systat Software Inc., San Jose, CA, USA).

For FACS analysis, cell supernatant and trypsinized cells were collected in a FACS tube (Sarstedt, Nümbrecht, Germany) and centrifuged for 5 min at 800× *g*. Supernatant was discarded and cells fixed with 4% paraformaldehyde (Sigma-Aldrich) for 15 min at 4 °C. Cells were further permeabilized with 0.01% Digitonin (Sigma-Aldrich) for 10 min at 4 °C and washed with PBS. Cells were incubated with antibody WB103/105 (1:500 dilution in PBS) for 15 min at 4 °C. The cells were washed with PBS prior application of the anti-mouse Alexa Fluor™488 F(ab')2 (1:1000 dilution in PBS) for 15 min at 4 °C. Cells were washed and resuspended in PBS. A total of 10,000 events were acquired from each sample, using the FACSCalibur (BD Biosciences, Franklin Lakes, NJ, USA). Uninfected cells were stained and used as a negative control. They were included in each run separately to allow subtraction of the background signal (<2.0%). Experiments were performed in three independent replicates.

2.4. Pestivirus Passaging

The different pestivirus isolates were passaged on CD46⁻MDBK and NTC cells to study the ability of the viruses to escape CD46-dependent cell entry and as a control, respectively. Cells were seeded one day prior to infection into a 24-well plate and infected with passage 0 (MOI = 1) for 1 h at 37 °C and 5% CO_2. Passage 0 represents the virus stock grown from the initial virus isolates on KOP-R cells. Cells were washed once and incubated in 1 mL maintenance medium for 72 h before they were frozen at −20 °C. For the next virus passage, crude cell extract (noncleared mix of cells and supernatant) were thawed and 100 µL per well were used for infection of the next passage of CD46⁻MDBK and NTC cells. Virus titers of passage 15 were determined by endpoint titration on KOP-R cells and further tested in FACS and IF regarding their infectivity on CD46⁻MDBK.

2.5. Sequencing, Sequence Assembly and Sequence Comparison

Virus passages 0 and 15 of the virus isolates passaged on CD46⁻MDBK were processed using a modification of the protocol published by Wylezich and colleagues [33] and sequenced using the Ion Torrent S5XL platform. Briefly, RNA was extracted from the freeze-thawed crude cell extract using the RNAdvance kit (Beckman Coulter, Fullerton, CA, USA) following the manufacturer's instructions and further concentrated using the Agencourt RNAClean XP magnetic beads (Beckman Coulter). Afterwards, cDNA was synthesized using a combination of the SuperScript™ IV First-Strand cDNA Synthesis System (Thermo Fisher Scientific) and the NEBNext®Ultra™ II Non-Directional RNA Second Strand Synthesis Module (New England Biolabs, Ipswich, MA, USA). The generated cDNA was fragmented to 500 bp using the Covaris M220 Focused-Ultrasonicator (Covaris, Brighton, UK) before Ion Torrent-compatible libraries were generated using the GeneRead L Core Kit (Qiagen) and IonXpress Barcode Adapter (Thermo Fisher Scientific). After size selection, library quality was checked using the Agilent 2100 Bioanalyzer system (Agilent Technologies, Santa Clara, CA, USA). The library concentration was measured with the KAPA Library Quantification kit (Roche, Mannheim, Germany). Using an Ion Torrent S5 XL, libraries were sequenced on an Ion 530 chip in 400-bp mode according to the manufacturer's instructions (Thermo Fisher Scientific).

For sequence assembly, a random subset of 50,000 to 500,000 reads was assembled (Newbler v.3.0; 454/Roche). The complete data set was mapped against the assembled genome (Newbler v.3.0; 454/Roche) to confirm the sequence. Consensus sequences were annotated based on the NADL (BVDV-1a) reference strain (NC_001461) and amino acid (aa) sequences were annotated accordingly (E^{RNS}: aa 271-497, E1: aa 498-692 and E2: aa 693-1066). Sequences were aligned using the MAFFT alignment of the bioinformatic software Geneious Prime (v.2020.1.2; Biomatters Ltd., Auckland, New Zealand). The variant analysis for the detection of viral quasispecies was performed in Geneious Prime (v.2020.1.2; Biomatters Ltd., Auckland, New Zealand), using the 454Contigs.bam from the mapping of passage 0 and passage 15 reads against the consensus sequence of passage 0 (Newbler v.3.0./Roche). Single-nucleotide-polymorphisms were detected with a minimum variant frequency of 0.1.

2.6. Immunoblotting

Cell lysates of passage 15 (on CD46^{-}MDBK cells) and passage 0 (on NTC cells) were prepared 72 h after infection of the cells. For this purpose, cells were harvested and incubated for 30 min on ice in lysis buffer (Na_2HPO_4, 1% Triton), supplemented with protease inhibitor (Roche). Proteins were separated under nonreducing conditions on a 10% SDS-polyacrylamide gel for 75 min at 130 mA and subsequently blotted onto a GE Healthcare Amersham™ Protan™ nitrocellulose membrane (Thermo Fisher Scientific). The membrane was blocked with Roti®Block (Carl Roth, Karlsruhe, Germany) for 1 h. For the detection of pestivirus E^{RNS}, the membrane was incubated with monoclonal antibodies HC/TC169/2/3 and BVD/C46/2/1 for 1 h (1:100 in Roti®Block, kindly provided by Prof. Becher, Institute of Virology, University of Veterinary Medicine Hannover). The POD-antimouse antibody (Dianova, Hamburg, Germany) was used as a secondary antibody, diluted 1:20,000 in phosphate buffered saline with tween (PBST). Proteins were detected by chemiluminescence, using SuperSignal™ West Pico PLUS Chemiluminescent Substrate (Thermo Fisher Scientific).

3. Results

3.1. Characterization of CD46^{-}MDBK

After single cell dilution of the transfected cells, colonies were screened via PCR and the knockout cell was identified to carry a homozygous deletion in the target region. Deep sequencing of the PCR product confirmed the successful biallelic knockout of 120 and 134 nucleotides in the target region, spanning the BVDV binding domains $E_{66}QIV_{69}$ and $G_{82}QVLAL_{87}$ (Figure 1b). CD46^{-}MDBK were propagated and the deletion was confirmed by PCR up to the last passage used for passaging of the virus isolates (CD46^{-}MDBK passage 54) (Figure 1a). IF staining with the anti-CD46 antibodies BVD/CA 26/2/5 and BVD/CA 17/2/1 (data not shown) showed no fluorescence signal on CD46^{-}MDBK, compared to NTC cells (Figure 1c).

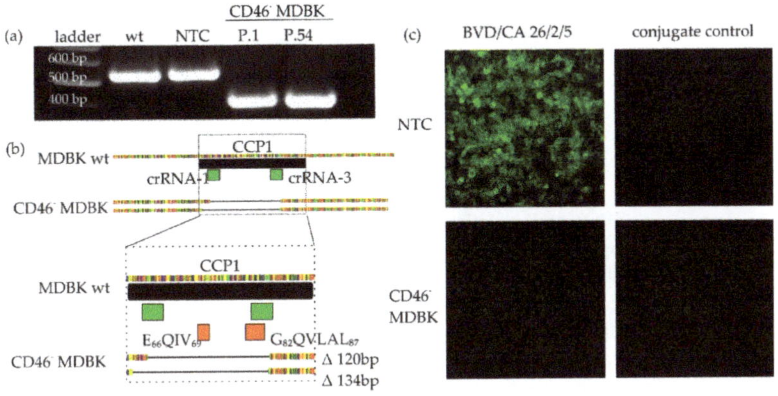

Figure 1. Characterization of CD46^{-}MDBK cell line. (**a**) PCR amplification of knockout target region of MDBK wild-type (wt), nontarget control (NTC), CD46^{-}MDBK passage 1 (P. 1) and passage 54 (P. 54) (**b**) Sequence comparison of wt and CD46^{-}MDBK reveals a biallelic deletion of 120 and 134 nucleotides in the CCP1 region (black box), including the bovine viral diarrhea virus (BVDV) binding sides $E_{66}QIV_{69}$ and $G_{82}QVLAL_{87}$ (orange box). crRNAs are depicted in green (**c**) Staining of CD46 receptor using the anti-CD46 BVD/CA 26/2/5 antibody and conjugate only control on NTC (top) and CD46^{-}MDBK (bottom). Magnification: 20×.

The functional knockout of the receptor was further confirmed via FACS and IF. The different BVDV isolates grew successfully on the NTC cells, whereas growth was strongly inhibited on the CD46⁻MDBK cells (Figure 2b,c). Based on the FACS data, the isolates used in this study (passage 0) infected 70–95% of the NTC cells, compared to less than 5% of the CD46⁻MDBK (Figure 2a).

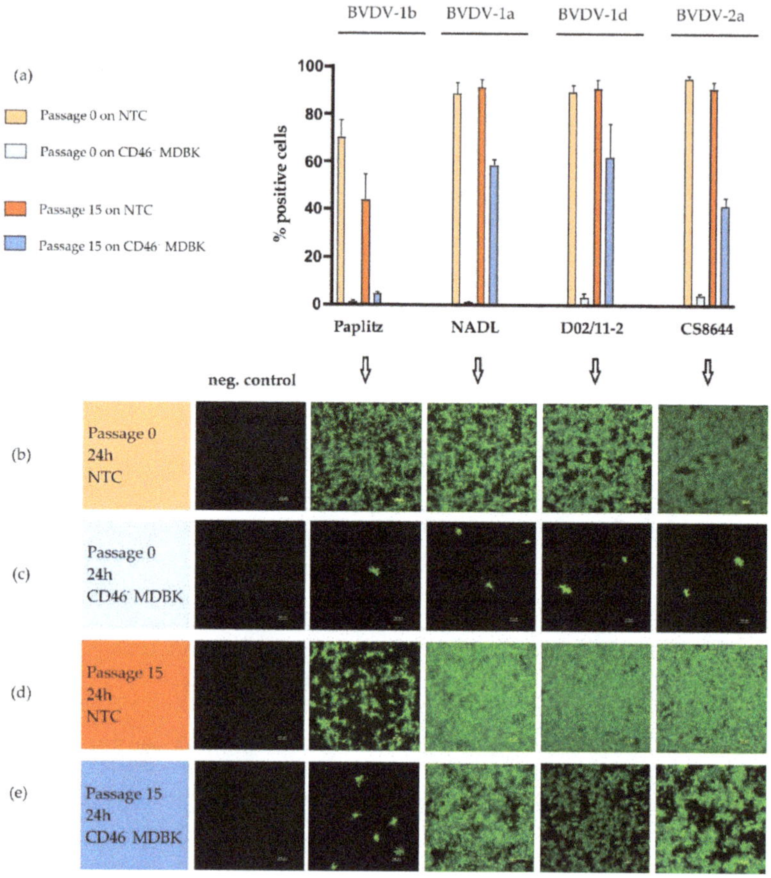

Figure 2. Detection of BVDV in the initial and the 15th passage on the indicated cell lines after incubation for 24 h. (**a**) Diagram depicting the percentages of positive cells in the fluorescence-activated cell sorting (FACS) analysis, conducted in triplicates. Immunofluorescence (IF) staining of (**b**) isolate passage 0 on non-target control (NTC), (**c**) passage 0 on CD46⁻MDBK (**d**) passage 15 on NTC and (**e**) passage 15 on CD46⁻MDBK. Scale bar 100 µm. Cells were incubated with the different pestivirus isolates (MOI = 1) for 1 h at 37 °C and subsequently washed and incubated for 24 h before analysis. Negative control: cells were incubated with maintenance medium only.

3.2. Adaption of Pestivirus Isolates by Passaging on CD46⁻MDBK Cells

The generated CD46⁻MDBK cell line was subsequently used to adapt different BVDV isolates to a host cell lacking the CD46 receptor. After 15 passages on the knockout cell line, the isolates NADL, D02/11-2 (BVDV-1d)and CS8644 (BVDV-2a) showed a strong increase in infectivity, from less than 5% positive cells (passage 0) to 40–60% (passage 15) on CD46⁻MDBKas measured by FACS analysis (Figure 2a). For all viruses, infectivity analyzed by IF staining was in line with the FACS data (Figure 2).

The cytopathic virus isolate Paplitz (cytopathic effect becomes apparent 3 to 5 days post infection) did not show a significant increase in CD46⁻MDBK cell infectivity after passaging based on the FACS results (from 1% positive cells with passage 0 virus to 4% with passage 15 virus, Figure 2a). Similarly, the IF staining did not show a strong increase in infectivity after incubation for 24 h on CD46⁻MDBK, when compared to the other isolates (Figure 2e). However, when the CD46⁻MDBK cells were incubated for 72 h, a complete infection of the cell layer was visible for all isolates (passage 15) (Figure 3d), that differed markedly from incubation for 72 h with the passage 0 virus (Figure 3b).

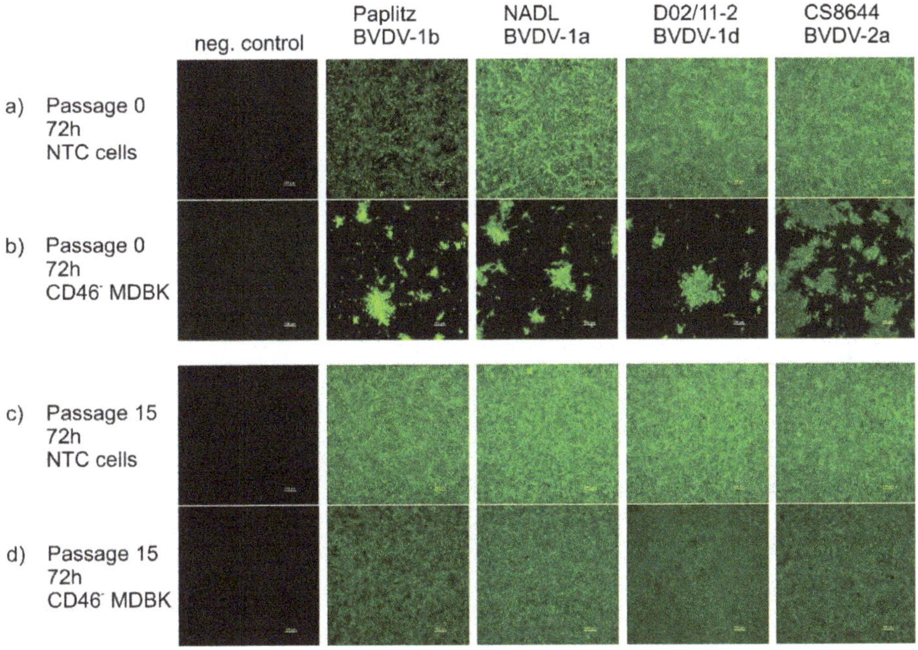

Figure 3. Comparison of the initial passage and passage 15 on non-target control (NTC) and CD46⁻MDBK after incubation for 72 h. Passage 0 on (**a**) NTC and (**b**) CD46⁻MDBK, passage 15 on (**c**) NTC and (**d**) CD46⁻MDBK. Cells were incubated with the designated pestivirus isolates (MOI = 1) for 1 h at 37 °C and subsequently washed and incubated for 72 h before analysis. Negative control: cells were incubated with maintenance medium only.

Therefore, passage 0 and 15 of virus isolate Paplitz (BVDV-1b) were further tested in a plaque assay, to investigate the underlying mechanism that led to a densely infected cell layer after 72 h. Virus plaques of passage 0 had an average area of 28,462 μm^2 (standard deviation (SD): 11,230 μm^2, median: 25,489 μm^2). Virus plaques of passage 15 were on average twice as big, with an average area of 55,985 μm^2 (SD: 21,349 μm^2, median: 53,725 μm^2) (Figure 4).

The difference between the plaque sizes of both groups was statistically significant ($p < 0.001$). In comparison, the plaque size of the other passage 0 virus isolates measures for NADL (BVDV-1a): 25,243 μm^2 (SD: 11,850 μm^2, median: 23,132 μm^2), D02/11-2 (BVDV-1d): 24,655 μm^2 (SD: 9181 μm^2, median: 24,011 μm^2) and for CS8644 (BVDV-2a): 52,072 μm^2 (SD: 21,839 μm^2, median: 50,417 μm^2). Based on these measurments, the Paplitz (BVDV-1b) passage 0 isolate is comparable to the plaque size of the other tested BVDV-1 strains.

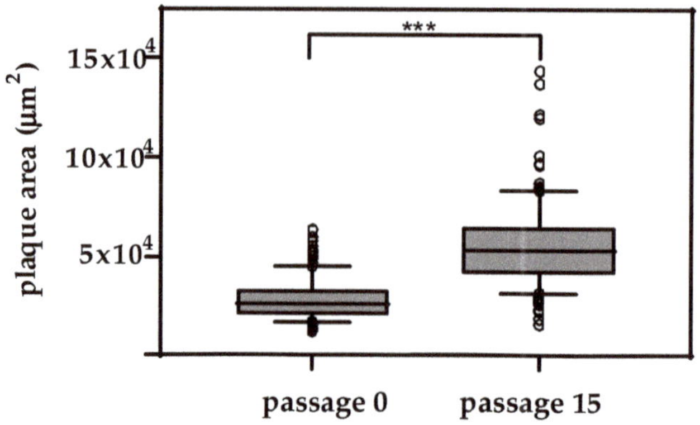

Figure 4. Comparison of plaque sizes of the initial passage and passage 15 of virus isolate Paplitz (BVDV-1b) on CD46⁻MDBK. Cells were incubated with a MOI = 0.1 for 1 h at 37 °C and subsequently washed and incubated for 48 h. The experiment was conducted in triplicate and 50 plaques were measured for each virus passage in each experiment (total $n = 150$ plaques/isolate). Whiskers indicate the 10% and 90% percentiles, respectively, and each outlier is depicted by a dot. The difference between the plaque sizes is statistically significant (*** $p < 0.001$).

3.3. *Sequence Analysis after Virus Passaging on CD46⁻MDBK*

In order to identify genetic adaptations that led to the observed shift in infectivity and virus-growth characteristics between the different virus passages, they were sequenced using next-generation sequencing. The genomic regions coding for the E^{RNS}, E1 and E2 proteins of the initial virus passage were compared to the 15th passage, since these regions are known to be involved in pestivirus entry into the host cell [11,14]. The graphical overview of the aligned sequences is depicted in Figure S2. After 15 passages, all virus isolates gained at least one aa substitution in the E^{RNS}-protein (Table 1).

Table 1. Overview of the amino acid substitution in E^{RNS}, E1 and E2 proteins based on the consensus sequences of the original BVDV isolates and their 15th passage on the CD46-deficient cell line. Amino acid positions with known effects are highlighted in bold.

Virus Strain	Amino Acid Position	Amino Acid Passage 0	Amino Acid Passage 15	Genome Region	Variant Frequency Passage 0	Variant Frequency Passage 15
Paplitz	**441**	**Cystein**	**Arginine**	**E^{RNS}**	**-**	**98.7%**
(BVDV-1b)	673	Alanine	Valine	E1	-	54.3%
	764	Glycine	Arginine	E2	-	99.7%
NADL	450	Histidine	Leucine	E^{RNS}	-	100%
(BVDV-1a)	**479**	**Glycine**	**Arginine**	**E^{RNS}**	**-**	**100%**
	677	Isoleucine	Methionine	E1	36.0%	100%
D02/11-2	**479**	**Asparagine**	**Lysine**	**E^{RNS}**	**42.1%**	**100%**
(BVDV-1d)	631	Alanine	Valine	E1	-	91.6%
CS8644	423	Glycine	Arginine	E^{RNS}	-	75.3%
(BVDV-2a)	477	Glutamine	Lysine	E^{RNS}	-	77.6%
	479	**Glycine**	**Arginine**	**E^{RNS}**	**-**	**99.4%**
	480	Isoleucine	Threonine	E^{RNS}	-	99.8%
	735	Histidine	Glutamine	E2	-	99.7%
	847	Isoleucine	Arginine	E2	-	99.4%
	880	Isoleucine	Valine	E2	-	95.8%

Virus isolates NADL (BVDV-1a), D02/11-2 (BVDV-1d) and CS8644 (BVDV-2a) displayed an aa exchange at the same position, namely 479 in the E^{RNS}. NADL (BVDV-1a) and CS8644 (BVDV-2a) exchanged the uncharged glycine with a charged arginine and D02/11-2 (BVDV-1d) exchanged the uncharged asparagine with a charged lysine. The cytopathic strain Paplitz (BVDV-1b) exchanged an uncharged cysteine for a charged arginine at position 441 in the E^{RNS} (Table 1). From the 15 mutations listed, two were already detected in the viral sequence reads from passage 0 as a minor population. Amino acid mutation in NADL (BVDV-1a) at position 677 and in D02/11-2 (BVDV-1d) at position 479 were identified with a sequence prevalence of 36% and 42%, respectively (Table 1). An important finding is that the minor population at aa position 479 of D02/11-2 (BVDV-1d) passage 0 is linked with another minor population in passage 0 that substitutes the original lysine at position 480 with an asparagine with a frequency of 39.9% (Table S1). During analysis of the actual reads of passage 0, it becomes apparent that reads are coding either for a lysine-asparagine or an asparagine-lysine (aa position 479 and 480) but not for a lysine-lysine or asparagine-asparagine (visualized in Figure S3).

3.4. E^{RNS} Dimerization of Virus Isolate Paplitz (BVDV-1b)

The formation of E^{RNS} dimers of virus isolate Paplitz (BVDV-1b) was analyzed, since the exchanged aa cysteine at position 441 was previously described to be important for dimerization of E^{RNS} [17]. Immunoblot analysis confirmed formation of E^{RNS} dimers (88-96 kDa) in virus passage 0, compared to monomer formation (48 kDa) of virus passage 15 (Figure 5).

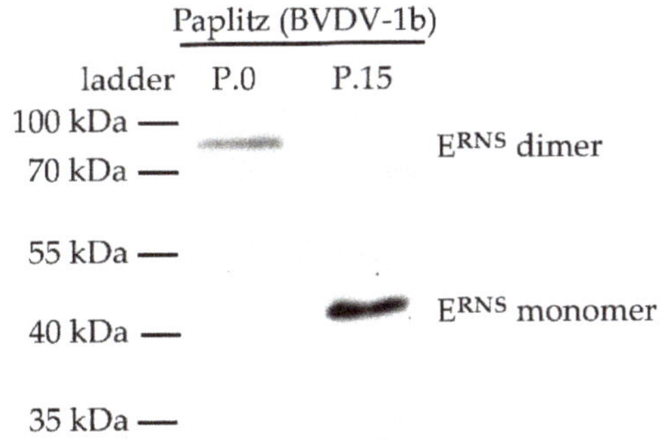

Figure 5. E^{RNS} formation of virus isolate Paplitz (BVDV-1b). Detection of E^{RNS} in the lysates of infected cells 72 h post infection using immunoblotting. Proteins were separated under nonreducing conditions. E^{RNS} of passage 0 (P.0) forms dimers (88–96 kDa), whereas passage 15 (P.15) forms monomers (44–48 kDa).

4. Discussion

The importance of bovine CD46 for pestivirus entry into a host cell has been already described in detail in the past and the two peptide domains $E_{66}QIV_{69}$ and $G_{82}QVLAL_{87}$ have been identified to be crucial [25,34,35]. However, for further analysis, bovine CD46-knock-out cells were needed. We therefore successfully knocked out the aforementioned domains and showed that this leads to a significant decrease in infectivity of the tested BVDV-1 and -2 isolates.

The strong reduction in infectivity on $CD46^{-}$MDBK is consistent with previous findings of Krey et al. [25], who showed reduced binding of pestivirus isolates in vitro by at least 70% (in 29 out of 30 isolates tested) after preincubation of MDBK cells with an anti-CD46 serum. Interestingly, we could show that the knockout of CD46 did not result in a complete loss of infectivity. A small percentage of less than 5% of the cells were still able to get infected by the tested BVDV isolates (passage 0). Therefore, a small fraction of the viruses most likely uses a CD46-independent route to enter the respective host cells.

After passaging the four different virus isolates 15 times on $CD46^{-}$MDBK, infectivity of the virus isolates NADL (BVDV-1a), D02/11-2 (BVDV-1d) and CS8644 (BVDV-2a) increased significantly when compared to passage 0 after incubation for 24 h. Sequencing of these passages showed that the isolates share the same aa exchange at position 479 after passaging. This aa exchange is of importance, since the same position (aa 476 in *pestivirus C* (CSFV) correlates to aa 479 in *pestivirus A/B*) has been shown to be crucial for CSFV interaction with membrane-associated heparan sulfate [36]. The exchange of an uncharged aa (glycine) to a positively charged aa (arginine) at this position has been described to increase virus replication in vitro of CSFV variants carrying this mutation [15,36]. It has been also suggested that this aa substitution (position 476 in CSFV; position 479 in BVDV-1 and -2) increases the positive charge of the E^{RNS} region and that this particular aa is exposed to the surface and involved into direct binding to the negatively charged heparan sulfate [36]. Reimann and colleagues [37] also associated the increased virus infectivity in vitro with the same aa substitution of glycine to arginine at position 479 in CP7_E2alf, a chimeric pestivirus constructed from a BVDV-1 backbone (strain CP7) and E2 from CSFV (strain Alfort). Overall, these findings indicate very clearly that the loss of the host cells CD46 favours the introduction of a positively charged aa at position 479 of the E^{RNS} in the pestivirus isolates studied. Even though the aa position 479 is in the amphipathic C-terminal end of the E^{RNS} that is important for membrane binding, it is likely that this aa substitution has no or only minor effects on lipid binding. Research has shown that the amphipathic helix has a robust lipid affinity that cannot be disturbed by mutating single amino acids [38].

The aa substitution at position 479 occurred most likely de novo in virus isolate NADL (BVDV-1a) and CS8644 (BVDV-2a), however, very low frequencies might be missed due to the sequencing depths. In case of virus isolate D02/11-2 (BVDV-1d), the minor variant of the aa substitution (asparagine for lysine) at position 479 can be detected already in passage 0 with a proportion of 42%. Therefore, we would expect the phenotype of this passage 0 virus isolate to show a higher infectivity than actually detected, since we postulate that this mutation is important for the change to a phenotype with higher infectivity. However, when we looked into the other minor variants of passage 0 virus isolate D02/11-2 (BVDV-1d), we found that a lysine at aa position 480 is substituted with an asparagine with a similar proportion of 39.9%. Furthermore, when looking into the actual reads of passage 0, we could see that the reads are coding either for a lysine-asparagine (in consensus) or an asparagine-lysine (in minor variant) at position 479–480 but not for a lysine-lysine or an asparagine-asparagine. We therefore assume that in case of D02/11-2 (BVDV-1d), both lysine at position 479 and 480 are necessary to increase the infectivity of the virus. This has probably been selected for since neither the lysine-asparagine nor the asparagine-lysine variant could be detected in passage 15.

We suggest that the aa exchange at position 479 may allow the virus isolates NADL (BVDV-1a), D02/11-2 (BVDV-1d) and CS8644 (BVDV-2a) to compensate for the loss of the potential binding side CD46 by an increased binding of heparan sulfate. This finding provides therefore additional evidence for previous assumptions concerning the role of the heparan sulfate binding and CD46 independent entry of pestiviruses [25]. Furthermore, it would be interesting to passage these mutated virus isolates (passage 15) in vivo, to study if the mutation at position 479 is selected against in vivo like it was shown for CSFV [15]. We would speculate that this is the case, considering the similarities of BVDV and CSFV cell entry in vitro regarding their dependency on CD46 and heparan sulfate [11,16,34].

Nevertheless, the cytopathic strain Paplitz (BVDV-1b) is of special interest, since the virus shows an altered and unexpected growth behaviour after passaging. It seems that the selection pressure from CD46-MDBK cell passaging does not result in an improved initial cell entry of the virus but rather changed the growth behaviour of the virus towards a markedly improved cell-to-cell spread after initial cell entry. The lower infectivity of Paplitz (passage 15) compared to Paplitz (passage 0) on the NTC cells might be explained by the special way of adaption of this strain to the CD46 deficient environment and the subsequent loss of E^{RNS} dimers. It can be hypothesized that this trade-off favours cell spread in CD46-deficient cells for the cost of reduced binding to CD46 and growth in NTC cells. Interestingly, one mutation identified in the E^{RNS} of Paplitz (BVDV-1b) results in the exchange of a cysteine at position 441, which is important for the formation of a intermolecular disulphide bond in the E^{RNS} of CSFV [17]. Mutation of this cysteine, that occurred most likely de novo, leads to a loss of dimerization of E^{RNS}

that results in case of CSFV in an attenuated virus that is still able to grow in cell culture [17] but also with larger plaque sizes due the reduced binding to heparan sulfate [39]. We could also demonstrate here the loss of E^{RNS} dimerization in the passage 15 variant. Paplitz (BVDV-1b) passages 0 and 15 are not able to enter the host cell faster after passaging, but once inside the cell, the virus grows differently than the initial passage 0 and shows an enhanced cell-to-cell spread. This change is likely influenced by substitution of the cystein at position 441, that leads to monomeric E^{RNS}. Further research should study if the adaption described for virus isolate Paplitz (BVDV-1b) is typical for cytopathic pestiviruses and if the other mutations identified in E1 and E2 might also play a role. It could be hypothesized that dimerizaton of E^{RNS} could have a function concering cell-to-cell spread of pestiviruses like BVDV-1.

In summary, we describe here the succesful generation of a stable CD46-knockout cell line and we demonstrated that the infectivity of different BVDV isolates strongly depends on CD46. At the same time, we showed that a fraction of the virus particles is still able to enter the host cell even in the absence of the CD46 receptor. The study also shows that forced adaption of the pestiviruses studied leads to compensatory mutations in the E^{RNS}, affecting the virus-host interplay. Forced adaption of virus isolates NADL (BVDV-1a), and CS8644 (BVDV-2a) and D02/11-2 (BVDV-1d) led to a mutation at aa position 479 probably increasing heparan sulfate binding. In contrast, isolate Paplitz (BVDV-1b) substituted a cystein at position 441 that led to a loss of E^{RNS}-dimer formation, that is further suggested to increase cell-to-cell spread of the virus. The newly generated cell line is now available for future research to elucidate the role of the CD46 receptor in the context of the pestivirus biology and growth cycle.

Supplementary Materials: The following are available online at http://www.mdpi.com/1999-4915/12/8/859/s1. Supplementary Figure S1. Overview of the PCR amplicon, including the complement control protein 1 domain (CCP1, depicted in grey) with BVDV binding sides $E_{66}QIV_{69}$ and $G_{82}QVLA_{87}$ (both in orange), CRISPR RNAs (crRNA-1 and -3, depicted in yellow) and PCR primer (CD46_CCP1_F and _R, depicted in green). Supplementary Figure S2. Amino acid sequence alignment of passage 0 and 15 of the BVDV isolates. Differences between passages are depicted as a black gap in the alignment. The red box indicates the shared amino acid exchange at position 479 in virus passage 15 of NADL (BVDV-1a), D02/11-2 (BVDV-1b) and CS8644 (BVDV-2a). Supplementary Figure S3. Comparison of passage 0 and passage 15 sequence reads of D02/11-2 (BVDV-1b) at aa position 479 and 480 compared to consensus sequence passage 0. Two minor variants resulting in either a AAA-AAC or AAC-AAA codon (aa lysine-asparagine or asparagine-lysine). In comparison, passage 15 reads assemble only AAA-AAA (aa lysine-lysine) at position 479 and 480. The initial consensus sequence of passage 0 assembles AAC-AAA (aa lysine-asparagine). Supplementary Table S1. Overview of the variant analysis for passage 0 and 15 of virus isolate D02/11-2 (BVDV-1b). The variant of aa 479 and 480 is represented by nucleotide position 1437 and 1440 respectively and highlighted in bold.

Author Contributions: Conceptualization, M.B.; methodology, S.K., K.W., D.H. and M.B.; formal analysis, K.P.S., S.K., K.W., D.H. and M.B.; investigation, K.P.S., S.K., D.H. and K.W.; resources, D.H. and K.W.; writing—original draft preparation, K.P.S.; writing—review and editing, S.K., K.W., D.H. and M.B.; visualization, K.P.S. All authors have read and agreed to the published version of the manuscript.

Funding: This work was supported by the European Union's Horizon 2020 research and innovation program HONOURs, under grant agreement No 721367.

Acknowledgments: The authors are grateful to Andrea Aebischer, Patrick Zitzow, Doreen Schulz and Bianka Hillmann for excellent technical assistance and support.

Conflicts of Interest: The authors declare no conflict of interest.

References

1. Edwards, S.; Fukusho, A.; Lefevre, P.C.; Lipowski, A.; Pejsak, Z.; Roehe, P.; Westergaard, J. Classical swine fever: The global situation. *Vet. Microbiol.* **2000**, *73*, 103–119. [CrossRef]
2. Richter, V.; Lebl, K.; Baumgartner, W.; Obritzhauser, W.; Käsbohrer, A.; Pinior, B. A systematic worldwide review of the direct monetary losses in cattle due to bovine viral diarrhoea virus infection. *Vet. J.* **2017**, *220*, 80–87. [CrossRef] [PubMed]
3. Braun, U.; Hilbe, M.; Peterhans, E.; Schweizer, M. Border disease in cattle. *Vet. J.* **2019**, *246*, 12–20. [CrossRef] [PubMed]

4. Smith, D.B.; Meyers, G.; Bukh, J.; Gould, E.A.; Monath, T.; Scott Muerhoff, A.; Pletnev, A.; Rico-Hesse, R.; Stapleton, J.T.; Simmonds, P.; et al. Proposed revision to the taxonomy of the genus Pestivirus, family Flaviviridae. *J. Gen. Virol.* **2017**, *98*, 2106–2112. [CrossRef] [PubMed]
5. Wu, Z.; Ren, X.; Yang, L.; Hu, Y.; Yang, J.; He, G.; Zhang, J.; Dong, J.; Sun, L.; Du, J.; et al. Virome analysis for identification of novel mammalian viruses in bat species from Chinese provinces. *J. Virol.* **2012**, *86*, 10999–11012. [CrossRef] [PubMed]
6. Firth, C.; Bhat, M.; Firth, M.A.; Williams, S.H.; Frye, M.J.; Simmonds, P.; Conte, J.M.; Ng, J.; Garcia, J.; Bhuva, N.P.; et al. Detection of zoonotic pathogens and characterization of novel viruses carried by commensal Rattus norvegicus in New York City. *mBio* **2014**, *5*, e01933–e02014. [CrossRef]
7. Blome, S.; Beer, M.; Wernike, K. New leaves in the growing tree of Pestiviruses. *Adv. Virus Res.* **2017**, *99*, 139–160.
8. Jo, W.K.; van Elk, C.; van de Bildt, M.; van Run, P.; Petry, M.; Jesse, S.T.; Jung, K.; Ludlow, M.; Kuiken, T.; Osterhaus, A. An evolutionary divergent pestivirus lacking the N^{pro} gene systemically infects a whale species. *Emerg. Microbes Infect.* **2019**, *8*, 1383–1392. [CrossRef]
9. Simmonds, P.; Becher, P.; Bukh, J.; Gould, E.A.; Meyers, G.; Monath, T.; Muerhoff, S.; Pletnev, A.; Rico-Hesse, R.; Smith, D.B.; et al. ICTV Virus Taxonomy Profile: Flaviviridae. *J. Gen. Virol.* **2017**, *98*, 2–3. [CrossRef]
10. Tautz, N.; Tews, B.A.; Meyers, G. The Molecular Biology of Pestiviruses. *Adv. Virus Res.* **2015**, *93*, 47–160.
11. Iqbal, M.; Flick-Smith, H.; McCauley, J.W. Interactions of bovine viral diarrhoea virus glycoprotein E^{rns} with cell surface glycosaminoglycans. *J. Gen. Virol.* **2000**, *81 Pt 2*, 451–459. [CrossRef]
12. Liang, D.; Sainz, I.F.; Ansari, I.H.; Gil, L.; Vassilev, V.; Donis, R.O. The envelope glycoprotein E2 is a determinant of cell culture tropism in ruminant pestiviruses. *J. Gen. Virol.* **2003**, *84 Pt 5*, 1269–1274. [CrossRef]
13. Grummer, B.; Grotha, S.; Greiser-Wilke, I. Bovine viral diarrhoea virus is internalized by clathrin-dependent receptor-mediated endocytosis. *J. Vet. Med. B Infect. Dis. Vet. Public Health* **2004**, *51*, 427–432. [CrossRef] [PubMed]
14. Ronecker, S.; Zimmer, G.; Herrler, G.; Greiser-Wilke, I.; Grummer, B. Formation of bovine viral diarrhea virus E1-E2 heterodimers is essential for virus entry and depends on charged residues in the transmembrane domains. *J. Gen. Virol.* **2008**, *89 Pt 9*, 2114–2121. [CrossRef]
15. Hulst, M.M.; van Gennip, H.G.; Vlot, A.C.; Schooten, E.; de Smit, A.J.; Moormann, R.J. Interaction of classical swine fever virus with membrane-associated heparan sulfate: Role for virus replication in vivo and virulence. *J. Virol.* **2001**, *75*, 9585–9595. [CrossRef] [PubMed]
16. Dräger, C.; Beer, M.; Blome, S. Porcine complement regulatory protein CD46 and heparan sulfates are the major factors for classical swine fever virus attachment in vitro. *Arch. Virol.* **2015**, *160*, 739–746. [CrossRef] [PubMed]
17. Tews, B.A.; Schürmann, E.M.; Meyers, G. Mutation of cysteine 171 of pestivirus E^{rns} RNase prevents homodimer formation and leads to attenuation of classical swine fever virus. *J. Virol.* **2009**, *83*, 4823–4834. [CrossRef] [PubMed]
18. Iqbal, M.; Poole, E.; Goodbourn, S.; McCauley, J.W. Role for bovine viral diarrhea virus E^{rns} glycoprotein in the control of activation of beta interferon by double-stranded RNA. *J. Virol.* **2004**, *78*, 136–145. [CrossRef] [PubMed]
19. Murakami, S.; Takenaka-Uema, A.; Kobayashi, T.; Kato, K.; Shimojima, M.; Palmarini, M.; Horimoto, T. Heparan sulfate proteoglycan is an important attachment factor for cell entry of Akabane and Schmallenberg viruses. *J. Virol.* **2017**, *91*, e00503–e00517. [CrossRef]
20. Kalia, M.; Chandra, V.; Rahman, S.A.; Sehgal, D.; Jameel, S. Heparan sulfate proteoglycans are required for cellular binding of the hepatitis E virus ORF2 capsid protein and for viral infection. *J. Virol.* **2009**, *83*, 12714–12724. [CrossRef]
21. Sasaki, M.; Anindita, P.D.; Ito, N.; Sugiyama, M.; Carr, M.; Fukuhara, H.; Ose, T.; Maenaka, K.; Takada, A.; Hall, W.W.; et al. The role of heparan sulfate proteoglycans as an attachment factor for rabies virus entry and infection. *J. Infect. Dis.* **2018**, *217*, 1740–1749. [CrossRef] [PubMed]
22. Thamamongood, T.; Aebischer, A.; Wagner, V.; Chang, M.W.; Elling, R.; Benner, C.; Garcia-Sastre, A.; Kochs, G.; Beer, M.; Schwemmle, M. A genome-wide CRISPR-Cas9 screen reveals the requirement of host cell sulfation for Schmallenberg virus infection. *J. Virol.* **2020**. [CrossRef] [PubMed]

23. Liszewski, M.K.; Atkinson, J.P. Complement regulator CD46: Genetic variants and disease associations. *Hum. Genom.* **2015**, *9*, 7. [CrossRef]
24. Cattaneo, R. Four viruses, two bacteria, and one receptor: Membrane cofactor protein (CD46) as pathogens' magnet. *J. Virol.* **2004**, *78*, 4385–4388. [CrossRef] [PubMed]
25. Krey, T.; Himmelreich, A.; Heimann, M.; Menge, C.; Thiel, H.J.; Maurer, K.; Rümenapf, T. Function of bovine CD46 as a cellular receptor for bovine viral diarrhea virus is determined by complement control protein 1. *J. Virol.* **2006**, *80*, 3912–3922. [CrossRef] [PubMed]
26. Merwaiss, F.; Czibener, C.; Alvarez, D.E. Cell-to-cell transmission is the main mechanism supporting bovine viral diarrhea virus spread in cell culture. *J. Virol.* **2019**, *93*, e01776–e01818. [CrossRef]
27. Doudna, J.A.; Charpentier, E. Genome editing. The new frontier of genome engineering with CRISPR-Cas9. *Science* **2014**, *346*, 1258096. [CrossRef]
28. Lin, S.; Staahl, B.T.; Alla, R.K.; Doudna, J.A. Enhanced homology-directed human genome engineering by controlled timing of CRISPR/Cas9 delivery. *Elife* **2014**, *3*, e04766. [CrossRef]
29. DeWitt, M.A.; Corn, J.E.; Carroll, D. Genome editing via delivery of Cas9 ribonucleoprotein. *Methods* **2017**, *121*, 9–15. [CrossRef] [PubMed]
30. Haeussler, M.; Schönig, K.; Eckert, H.; Eschstruth, A.; Mianne, J.; Renaud, J.B.; Schneider-Maunoury, S.; Shkumatava, A.; Teboul, L.; Kent, J.; et al. Evaluation of off-target and on-target scoring algorithms and integration into the guide RNA selection tool CRISPOR. *Genome Biol.* **2016**, *17*, 148. [CrossRef]
31. Labun, K.; Montague, T.G.; Krause, M.; Torres Cleuren, Y.N.; Tjeldnes, H.; Valen, E. CHOPCHOP v3: Expanding the CRISPR web toolbox beyond genome editing. *Nucleic Acids Res.* **2019**, *47*, W171–W174. [CrossRef] [PubMed]
32. Schelp, C.; Greiser-Wilke, I.; Wolf, G.; Beer, M.; Moennig, V.; Liess, B. Identification of cell membrane proteins linked to susceptibility to bovine viral diarrhoea virus infection. *Arch. Virol.* **1995**, *140*, 1997–2009. [CrossRef] [PubMed]
33. Wylezich, C.; Papa, A.; Beer, M.; Höper, D. A versatile sample processing workflow for metagenomic pathogen detection. *Sci. Rep.* **2018**, *8*, 13108. [CrossRef] [PubMed]
34. Maurer, K.; Krey, T.; Moennig, V.; Thiel, H.J.; Rümenapf, T. CD46 is a cellular receptor for bovine viral diarrhea virus. *J. Virol.* **2004**, *78*, 1792–1799. [CrossRef] [PubMed]
35. Alzamel, N.; Bayrou, C.; Decreux, A.; Desmecht, D. Soluble forms of CD46 are detected in Bos taurus plasma and neutralize BVDV, the bovine pestivirus. *Comp. Immunol. Microbiol. Infect. Dis.* **2016**, *49*, 39–46. [CrossRef] [PubMed]
36. Hulst, M.M.; van Gennip, H.G.; Moormann, R.J. Passage of classical swine fever virus in cultured swine kidney cells selects virus variants that bind to heparan sulfate due to a single amino acid change in envelope protein E^{rns}. *J. Virol.* **2000**, *74*, 9553–9561. [CrossRef] [PubMed]
37. Reimann, I.; Depner, K.; Trapp, S.; Beer, M. An avirulent chimeric Pestivirus with altered cell tropism protects pigs against lethal infection with classical swine fever virus. *Virology* **2004**, *322*, 143–157. [CrossRef]
38. Aberle, D.; Oetter, K.M.; Meyers, G. Lipid binding of the amphipathic helix serving as membrane anchor of Pestivirus glycoprotein E^{rns}. *PLoS ONE* **2015**, *10*, e0135680. [CrossRef]
39. van Gennip, H.G.; Hesselink, A.T.; Moormann, R.J.; Hulst, M.M. Dimerization of glycoprotein E^{rns} of classical swine fever virus is not essential for viral replication and infection. *Arch. Virol.* **2005**, *150*, 2271–2286. [CrossRef]

viruses

MDPI

Article

Fetal Lymphoid Organ Immune Responses to Transient and Persistent Infection with Bovine Viral Diarrhea Virus

Katie J. Knapek [1,2,†,‡], Hanah M. Georges [1,†], Hana Van Campen [1,2], Jeanette V. Bishop [1], Helle Bielefeldt-Ohmann [3], Natalia P. Smirnova [1,§] and Thomas R. Hansen [1,*]

1 Department of Biomedical Sciences, Colorado State University, Fort Collins, CO 80523, USA; kjknapek@central.uh.edu (K.J.K.); h.georges@colostate.edu (H.M.G.); hana.van_campen@colostate.edu (H.V.C.); jeanette.bishop@colostate.edu (J.V.B.); natalia.smirnova@sidelinesoft.com (N.P.S.)

2 Department of Microbiology, Immunology and Pathology, Colorado State University, Fort Collins, CO 80523, USA

3 Australian Infectious Diseases Research Centre and School of Veterinary Science, The University of Queensland, St. Lucia, QLD 4072, Australia; h.bielefeldtohmann1@uq.edu.au

* Correspondence: thomas.hansen@colostate.edu; Tel.: +1-970-988-4582

† These authors contributed equally to this work.

‡ Current Address: Department of Animal Care Operations, University of Houston, Houston, TX 77004, USA.

§ Current Address: Beckman Coulter Life Sciences, Loveland, CO 80538, USA.

Received: 5 June 2020; Accepted: 24 July 2020; Published: 28 July 2020

Abstract: Bovine Viral Diarrhea Virus (BVDV) fetal infections occur in two forms; persistent infection (PI) or transient infection (TI), depending on what stage of gestation the fetus is infected. Examination of lymphoid organs from both PI and TI fetuses reveals drastically different fetal responses, dependent upon the developmental stage of the fetal immune system. Total RNA was extracted from the thymuses and spleens of uninfected control, PI, and TI fetuses collected on day 190 of gestation to test the hypothesis that BVDV infection impairs the innate and adaptive immune response in the fetal thymus and spleen of both infection types. Transcripts of genes representing the innate immune response and adaptive immune response genes were assayed by Reverse Transcription quitatitative PCR (RT-qPCR) ($2^{-\Delta\Delta Cq}$; fold change). Genes of the innate immune response, interferon (IFN) inducible genes, antigen presentation to lymphocytes, and activation of B cells were downregulated in day 190 fetal PI thymuses compared to controls. In contrast, innate immune response genes were upregulated in TI fetal thymuses compared to controls and tended to be upregulated in TI fetal spleens. Genes associated with the innate immune system were not different in PI fetal spleens; however, adaptive immune system genes were downregulated, indicating that PI fetal BVDV infection has profound inhibitory effects on the expression of genes involved in the innate and adaptive immune response. The downregulation of these genes in lymphocytes and antigen-presenting cells in the developing thymus and spleen may explain the incomplete clearance of BVDV and the persistence of the virus in PI animals while the upregulation of the TI innate immune response indicates a more mature immune system, able to clear the virus.

Keywords: bovine viral diarrhea virus; fetus; thymus; immune response

1. Introduction

Bovine viral diarrhea viruses (BVDV) cause significant economic losses in all sectors of cattle production worldwide [1–4]. BVDVs are small single-stranded RNA viruses belonging to the Pestivirus genus in the family Flaviviridae [5]. Isolates of BVDV are classified into two genotypes, type 1 and type 2, with several subtypes and two biotypes: cytopathic (cp) and noncytopathic (ncp) [6]. Acute BVDV infection of immunocompetent cattle results in diverse clinical presentations, including subclinical infection, fever, nasal and/or ocular discharge, pneumonia, severe systemic disease, hemorrhage, and peracute death [7]. Importantly, BVDV infection of pregnant cattle results in fetal infection and reproductive losses, including early embryonic death, abortion, and stillbirth [8–11]. If maternal infection with ncp BVDV occurs prior to 125 days of gestation, the fetus becomes persistently infected (PI) with the virus and is born without BVDV-specific antibodies [12,13]. PI animals will shed BVDV throughout life, serving as the main source of infection for other cattle. BVDV infection of pregnant cows after 150 days of gestation results in a transient infection (TI) of the fetus. These TI calves are born with BVDV-specific antibodies indicative of a functional adaptive immune response and clearance of the virus [9,14,15]. Virus-specific immune responses in the bovine fetus develops between days 125 and 150; therefore, BVDV infection during this time may result in either a PI or TI, depending on the individual fetuses [16]. The differences in the outcomes of PI and TI fetal infections have been attributed to the maturation and function of the fetal immune system at the time of infection [17].

At least two explanations have been tendered to explain the persistence of BVDV in PI cattle. First, ncp BVDV has been shown to inhibit type I interferon (IFN) induction in vitro through the actions of virally encoded RNase (E^{rns}) and the N-terminal protease (N^{pro}) (reviewed in [18–21]). The latter N^{pro} degrades *interferon regulatory factor 3* (IRF3), thus inhibiting the transcription of IFNB and its antiviral activity in adjacent cells [22–24]. E^{rns} binds to and degrades double-stranded RNA (dsRNA), preventing its binding to cells and induction of IFNs [25]. Inhibition of the IFN response by N^{pro} and E^{rns} allows BVDV to persist and replicate in cells. Conversely, in an in vivo model, mutations of these two viral genes affecting their functions negate persistent infection [26]. In another in vivo model using BVDV strain Pe515nc, fetuses were inoculated in utero (amniotic fluid) and collected on days 3, 5, and 7 post-inoculation [27]. BVDV was confirmed in fetal tissues; however, an innate immune response was not seen, suggesting an inhibition of the fetal innate immune response to BVDV [27]. A second explanation of BVDV persistence in fetal infections might be that BVDV does not inhibit the IFN response, but instead, another unknown mechanism is responsible for persistence. Previously, we observed that bovine fetuses whose dams were inoculated with ncp BVDV 96B2222 on day 75 of gestation had a peak in BVDV RNA in fetal blood on day 97 of gestation, 21 days post-maternal inoculation (dpmi), followed by a 10-fold decrease in viral RNA on days 192 and 245 of gestation, suggesting partial viral clearance by an active immune response in the PI fetus, or possibly a decrease in infected cell types/numbers [28–30]. Interferon stimulated genes (ISG), such as *ISG15, protein kinase RNA-activated* (*PKR*), as well as the RNA helicases: *DEAD Box Protein 58* (*DDX58*), also known as *retinoic acid-inducible gene* (*RIGI*), *melanoma differentiation-associated protein 5* (*MDA5*); and *DExH-box helicase 58* (*DHX58*) were also shown to be chronically upregulated in the PI animal postnatally [31]. These results indicate that PI fetuses and placenta respond to BVDV with an innate immune response, albeit somewhat reduced compared to TI fetuses [28,29,31]. These findings indicate that the fetus responds to ncp BVDV infection with an innate immune response, much like the innate response to ncp BVDV infection in postnatal calves shown by Palomares et al. (2013); hence, inhibition of the innate immune response by viral proteins does not entirely explain viral persistence in vivo [32]. Another in vivo postnatal model used BVDV strain 11,249, which was previously shown to inhibit the IFN response in vitro [33]. The study revealed that in vivo, this BVDV strain does induce IFN responses [33]. The stimulation of an IFN response to BVDV has been exhibited in in vitro studies, in vivo fetal infection studies, and in vivo postnatal infections. Differences in these theories could be explained by differences in BVDV strains used and/or differences in experimental models. It is important to examine these differences and theories when considering the results of the present study.

An additional explanation for viral persistence is that the presence of BVDV during the development of the T cell repertoire permits its antigens to be accepted as "self" antigens, resulting in a state referred to as immunotolerance in which elements of the adaptive immune system do not respond to viral antigens and do not clear the virus from fetal tissues [34]. We hypothesize that ncp BVDV fetal infection early in gestation (<125 days) interferes with T cell development in the bovine fetal thymus during a critical period in which T cells are selected based on their recognition of self-antigens, affecting T cell response to BVDV in the spleen, resulting in immunotolerance to the virus and persistent infection in the calf. BVDV fetal infection later in gestation (>150 days) results in an active innate immune response, which clears the virus infection and causes a delay in antigen-specific T and B cell responses. To examine the effect of ncp BVDV infection on the bovine fetal thymus and spleen, transcripts of genes representative of the innate and adaptive immune response pathways were compared with control fetuses collected on day 190 of gestation.

2. Materials and Methods

2.1. Animals

Animal experiments were performed as previously described [28,34]. All animal experiments were approved by the Institutional Animal Care and Use Committees at the University of Wyoming, approval 05-265A-02 (19/10/2005) and 08-16A-01 (11/01/09). BVDV antigen negative and seronegative yearling Hereford heifers were synchronized for estrus (ovulation) and artificially inseminated with BVDV-free semen. Pregnancy was confirmed by ultrasound examination on days 35 to 40 and on day 70 of gestation.

2.2. Experimental Design: BVDV Inoculation and Fetal Collections

A power analysis was performed, and a group of 6 animals per treatment group (control, TI, and PI) was determined to be appropriate for a power of 1. Therefore, 18 unvaccinated pregnant heifers were randomly placed into treatment groups and inoculated intranasally with 2 mL culture media to generate sham-treated controls ($n = 6$) or with media containing 4.4 $\log_{10}$ $TCID_{50}/_{mL}$ of ncp BVDV2 strain 96B2222 on day 75 of gestation to generate PI ($n = 6$) fetuses and on day 175 to generate TI ($n = 6$) fetuses [34]. Treatment groups were kept in widely separated pens and fed at the end of feeding rounds to minimize viral transmission. To capture the fetal immune response during maternal seroconversion (in the PI group), day 190 of gestation was chosen for fetal collections. Eighteen fetuses were collected by Cesarean section and necropsied on day 190 of gestation. Samples of fetal thymuses and spleens were frozen in liquid nitrogen and stored at −80 °C.

2.3. RNA Extraction and RT-qPCR

Total RNA from 70 mg of frozen thymus and 50 mg of frozen spleen was isolated using TRIzol reagent according to the manufacturer's instructions (Invitrogen ThermoFisher, Rockford, IL, USA). The isolated RNA was treated with DNase I (Qiagen, Germantown, MD, USA) and purified using the RNeasy MiniElute Cleanup Kit (Qiagen, Germantown, MD, USA). RNA concentration and 260/280 and 260/230 ratios were measured using the NanoDrop 1000 Spectrophotometer (ThermoScientific, Rockford, IL, USA). The primer sequences and gene accession numbers are listed in Table 1. One µg of RNA was reverse transcribed to synthesize cDNA using iScript™ Reverse Transcription Supermix (Bio-Rad, Hercules, CA, USA). Reverse transcription quantitative polymerase chain reaction (RT-qPCR) was performed with iQ™ SYBR® Green Supermix (Bio-Rad, Hercules, CA, USA). Each cDNA reaction was diluted (1:5 for thymus and 1:10 for spleen) with RNase-free water. Primers were used at 3 µM concentration. Each sample was assayed in duplicate wells on a 384-well plate. Four biological replicates of RT-qPCR plates were performed at one cycle of 95 °C for 3 min, 40 cycles of 95 °C for 30 s, 58 °C for 30 s, and 72 °C for 15 s with a final 5 min elongation in a LightCycler-480 Instrument (Roche, Basel, Switzerland). Upon completion of RT-qPCR, melting curve analysis was performed to assess the quality of amplification.

Table 1. Primers utilized for RT-qPCR.

Gene	Sequence	Accession	Efficiency
18S	FW: GAACGAGACTCTGGCATGC REV: CTGAACGCCACTTGTCCCTC	NR_036642	102%
DDX58	FW: GAGCACTGGTGGATGCCTTA REV: GCTGTCTCTGTTGGTTCGGA	XM_024996055.1	157%
IRF7	FW: GCCTCCTGGAAAACCAACTT REV: CCTTATGAGGGTCGGTAGGGG	NM_001105040.1	127%
NFKB	FW: CGAGGTTCGGTTCTACGAGG REV: TGCAGGAACACGGGTTACAGG	NM_001102101.1	134%
IFNB	FW: TCCAGCACATCTTCGGCATT REV: TTCCCTAGGTGGGGAACGAT	NM_174350.1	104%
ISG15	FW: GGTATCCGAGCTGAAGCAGTT REV: ACCTCCCTGCTGTCAAGGT	NM_174366	115%
STAT4	FW: TTCTTCCCATGTCGCCAAGT REV: AACCAGATGTGATTGTTGGCA	NM_001083692.2	104%
IFI30	FW: GCATGCAGCTCTTGCACATC REV: GGCCCCAAGAGTTCTTACCC	NM_001101251.2	131%
PSMB9	FW: ATCTACCTGGCCACCATCAC REV: AGGAGAGTCCGAGGAAGGAG	NM_001034388	115%
PSMB8	FW: ACTGGAAGGCAGCACAGAGT REV: ATTGTGCTTAGTGGGGCATC	NM_001040480	124%
B2M	FW: AFTAAGCCGCAGTGGAGGT Rev: CGCAAAACACCCTGAAGACT	AC_000167.1	123%
CXCL10	FW: ACACCGAGGCACTACGTTCT Rev: TAAGCCCAGAGCTGGAAAGA	CB533091	121%
CXCL16	FW: CTTGTGAGGGCAGATTGTGA Rev:GGTCAATAGCTGGTTAGTTGTGAA	CK770974	105%
CD4	FW: GGGCAGAACGGATGTCTCAA REV: ATAGGTCTTCTGGAGCCGGT	NM_001103225.1	110%
CD8a	FW: TACATCTGGGCTCCCTTGGT REV: CCACAGGCCTGGGACATTTG	NM_174015.1	130%
CD8b	FW: AGCTGAGTGTGTTGATGTTCT REV: TTCTGAGTCACCTGGGTTGG	NM_001105344.2	93%
CD79b	FW: TGATTCCCGGGCTCAACAAC REV: CTGCCAGATCCGGGAACAAG	XM_002696068.6	160%

DDX58: DExD/H-Box Helicase 58; *IRF7*: Interferon Regulatory Factor 7; *NFKB*: Nuclear Factor Kappa-Light-Chain -Enhancer of Activated B Cells; *IFNB*: Interferon Beta; *ISG15*: Interferon Stimulated Gene 15; *STAT4*: Signal Transducer and Activator of Transcription 4; *IFI30*: IFI30 Lysosomal Thiol Reductase; *PSMB9*: Proteasome 20S Subunit Beta 9; *PSMB8*: Proteasome 20S Subunit Beta 8; *B2M*: Beta-2-Microglobulin; *CXCL10*: C-X-C Motif Chemokine Ligand 10; *CXCL*: C-X-C Motif Chemokine Ligand 16. FW: Forward primer; REV: Reverse Primer.

2.4. RT-qPCR Targets and Validation

RT-qPCR targets for each tissue were chosen based on current knowledge and available tissue. Targets shared between the thymus and spleen were chosen to understand basic changes in both the innate and adaptive branches of the immune system. For the innate immune branch, targets chosen were *DDX58*, *NFKB*, *IRF7*, *IFNB*, and *ISG15*. For the adaptive branch, targets chosen were *PSMB9*, *IFI30*, *CD4*, *CD8A*, *CD8B*, and *CD79B*. Thymic tissue samples were more abundant than the spleen, allowing for additional targets to be studied in the thymus; *STAT1*, *IFI6*, *CXCL10*, *CXCL16*, *CXCR6*, *TAP1*, *B2M*, *CIITA*, and *CD46*. These results can be found in Table A1. A previous study which used a microarray to study BVDV-infected spleens indicated *STAT4* and *PSMB8* as targets of interest specific for splenic samples [35]. Therefore, both *STAT4* and *PSMB8* were additional targets for splenic samples in this study. BVDV viral RNA was also targeted to confirm tissue infection in TI and PI fetuses. These results can also be found in Table A1. Standard RT-qPCR primers were validated and evaluated based on MIQE guidelines [36].

2.5. Morphogenesis of Thymus during Bovine Fetal Development and Effect of In Utero BVDV Infection

Fetal thymus and spleen samples were fixed for 48 h in 10% neutral-buffered formalin, transferred into 70% ethanol, and paraffin-embedded. Four to 5 µm sections were cut and stained with eosin and hematoxylin (H&E). Microscopic assessment was performed with special attention to the components of the reticulo-endothelial network, myeloid, and lymphoid cells, including where the latter two cell types appear in the organ primordium. The relative cortical-to-medulla ratio was assessed, including full encirclement of the medulla by cortical thymocytes. Additionally, attention was paid to myoid cells and Hassall's corpuscles in terms of appearance, frequency, and size.

2.6. Statistical Analysis

RT-qPCR data are presented according to MIQE guidelines and statistically analyzed using the $2^{-\Delta\Delta Cq}$ method [37]. Briefly, the Δ quantification cycle (C_q; also known as C_t) was calculated by subtracting the mean C_q of the reference gene (18S rRNA), within the treatment group, from the C_q of the target gene (same treatment group as the reference). The average of the ΔC_q for controls for the target gene were then subtracted from each infection treatment ΔC_q, for the same target gene as controls, to calculate the fold change ($2^{-\Delta\Delta Cq}$). Statistical analysis of data obtained by RT-qPCR ($2^{-\Delta\Delta Cq}$) was performed in GraphPad Prism 8 (GraphPad Software, San Diego, CA, USA). The data were checked for normality using Shapiro-Wilks test; normally distributed data were analyzed with a one-way ANOVA and Dunnett's multiple comparisons test, while non-normally distributed data were analyzed with a Kruskal-Wallis test and Dunn's multiple comparisons test. Significant differences were at $p < 0.05$ and tendencies/trends were at $p < 0.10$. Data are presented as the mean ± SEM. Graphical data represents $2^{-\Delta\Delta Cq}$ of which statistical analysis was run. In the text, fold change is presented for ease of understanding up- and down-regulation. For $2^{-\Delta\Delta Cq}$ values > 1, fold changes were reported as averaged $2^{-\Delta\Delta Cq}$ values. For $2^{-\Delta\Delta Cq}$ values between 0 and 1 (down regulation), the negative inverse ($-1/(2^{-\Delta\Delta Cq})$) was calculated and reported as fold change.

3. Results

3.1. Detection of BVDV RNA Expression in Thymus and Spleen and BVDV Receptor CD46 in Fetal Thymuses

BVDV RNA was not detected in any of the control fetal thymus RNA samples and was present in 6 of 6 PI and 5 of 6 TI fetal thymic RNA samples with PI thymuses having 9.7- to 68.8-fold higher BVDV RNA concentrations compared to TI thymuses (Figure 1A). PI fetal spleens had a significantly increased amount of BVDV RNA ($p < 0.001$, $2^{-\Delta\Delta Cq} = 3037$-fold greater) compared to controls ($2^{-\Delta\Delta Cq} = 1.0$) and very low amounts of BVDV RNA in TI fetal spleens (Figure 1B). The concentration of BVDV receptor *CD46* mRNA in PI fetal thymuses was decreased 6.2-fold ($p < 0.001$) and increased in TI fetal thymuses ($p < 0.05$) compared to controls on day 190 (Figure 1C).

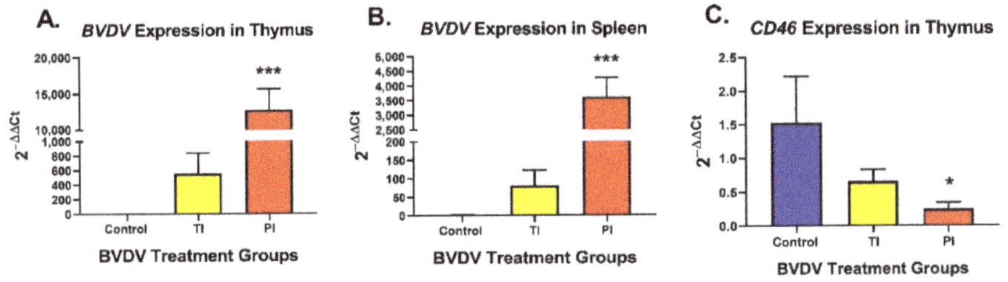

Figure 1. Bovine Viral Diarrhea Virus (BVDV) expression in (**A**) thymus and (**B**) spleen samples (Control, TI, and PI). Expression of BVDV was significantly upregulated in PIs compared to controls in both thymic and splenic samples. Expression of BVDV in TIs was higher than controls, but not significant due to individual variation. (**C**) BVDV receptor CD46 expression in thymus (Control, TI, and PI). *CD46* expression was significantly down-regulated in PI thymuses. Control: Uninfected, blue; TI: Transiently Infected, yellow; PI: Persistently Infected, red. * $p < 0.05$, *** $p < 0.001$.

3.2. Thymic Responses

3.2.1. Innate Immune Responses in TI and PI Fetal Thymuses

In TI fetal thymuses, mRNA concentrations were significantly increased ($p < 0.01$) for the innate immune response genes *NFKB* (1.5-fold) and *IRF7* (4.5-fold) compared to controls on day 190 (15 dpmi). Neither *DDX58, IFNB*, nor *ISG15* were significantly different than the controls (Figure 2).

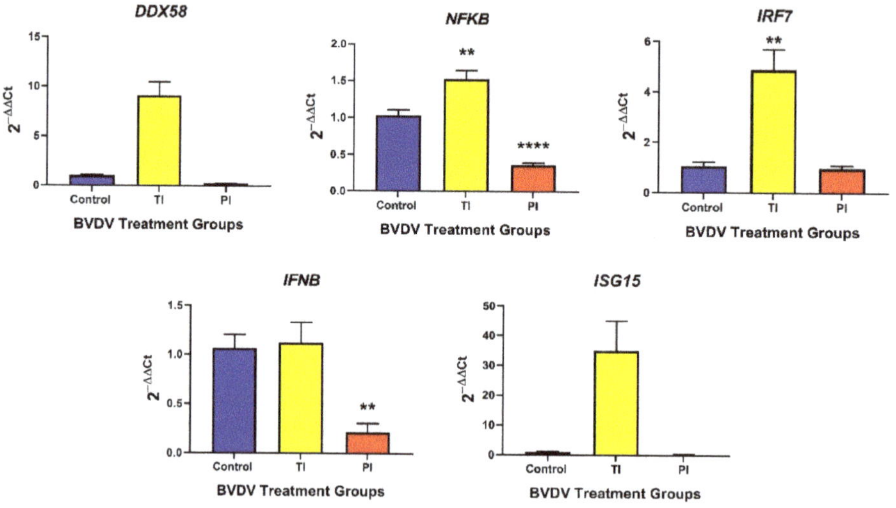

Figure 2. Control, TI, and PI thymus transcripts associated with the innate immune response. Interferon transcriptional regulators *NFKB* and *IRF7* were significantly upregulated in TI fetuses compared to controls. In PI fetal thymuses, only IFN transcriptional regulator *NFKB* and *IFNB* were down-regulated compared to control fetuses. TI: Transiently Infected, PI: Persistently Infected, ** $p < 0.01$, **** $p < 0.0001$.

In the PI fetal thymuses, the mRNAs of the innate response genes *NFKB* (−2.8-fold) and *IFNB* (6.0-fold) were significantly downregulated in PI thymuses compared to controls on day 190 ($p < 0.01$) (Figure 2); whereas *DDX58, IRF7*, and *ISG15* concentrations remained at control levels and were not significantly different between control and PI thymuses. A trending decrease in the type I IFN stimulated transcripts *STAT1* (−22.7-fold; Table A1) was found in PI fetal thymuses compared to controls ($p < 0.10$), which was consistent with the observed downregulation of *IFNB*. Type II IFN stimulated transcripts, transitional responses from the innate to adaptive immune response, *IFI16* (−11.3-fold, trend $p < 0.10$) and *CXCL16* (−22.4-fold, $p < 0.05$) were similarly down-regulated (Table A1). *CXCR6* mRNA concentration was not different between control and PI fetal thymuses (Table A1).

3.2.2. Adaptive Immune Responses in TI and PI Fetal Thymuses

In TI fetal thymuses, the expression of *MHC I* mRNA *PSMB9* (Figure 3), *TAP1* (Table A1), and *B2M* (Table A1) were not significantly different from controls. T cell markers *CD8A, CD8B*, and *CD4* and activated B cell marker *CD79B* mRNAs were not significantly different in TI fetal thymuses compared to controls.

The relative expression of MHC I antigen presentation pathway genes *PSMB9* (−5.2-fold), *TAP1* (−5.5-fold), and *B2M* (−27.9-fold) transcripts were significantly decreased in PI fetal thymuses relative to controls on day 190 of gestation ($p < 0.05$) (Figure 3, Table A1 respectively). CTL co-receptors *CD8A* (−16.2-fold) and *CD8B* (−9.6-fold) mRNA concentrations were significantly decreased in PI fetal thymuses compared to the controls on day 190 of gestation ($p < 0.05$) (Figure 3). *CIITA* (−3.4-fold, trend $p < 0.10$) and *CD4* (−8.1-fold, $p < 0.0001$) mRNA associated with MHC II antigen presentation were significantly decreased in PI fetal thymuses on day 190 (Table A1 and Figure 3, respectively). Activated B cell marker, *CD79B*, mRNA concentration was significantly decreased (−3.8-fold) in PI fetal thymuses compared to control ($p < 0.001$) thymuses at day 190 of gestation (Figure 3).

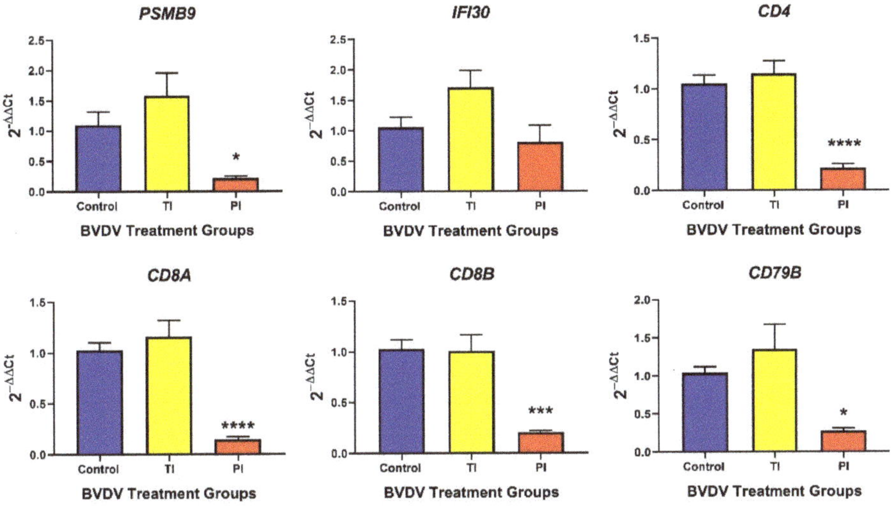

Figure 3. Thymic (Control, TI, and PI) transcript expression associated with the adaptive immune response. No significant differences were found in TI fetal thymuses compared to controls. However, all adaptive genes, with the exception of *IFI30*, were significantly down-regulated in PI fetal thymuses compared to controls. TI: Transiently Infected, PI: Persistently Infected. * $p < 0.05$, *** $p < 0.001$, **** $p < 0.0001$.

3.2.3. Morphogenesis and Histology of the Thymus during Bovine Fetal Development and the Effect of in Utero BVDV Infection

There were no discernible differences between the control or PI fetuses with regard to the thymus morphology at day 190. The morphology of the fetal thymus had attained almost a neonatal appearance, with the cortex having expanded to completely surround the medulla, and the cortico-medullary ratio being ≥2. The medulla contained large, often complex Hassall's corpuscles, large numbers of myoid cells, and eosinophilic granulocytes. The latter could be seen throughout the medulla, but appeared in highest frequency in the cortico-medullary border zone. The myoid cells were large, round cells with a deeply eosinophilic cytoplasm, either homogenous or with finely fibrillar patterning. The almost euchromatic nucleus was round and centrally located, or more oval and acentric. A small amount of chromatin clumping along the nuclear membrane and occasionally a small nucleolus was present. These cells may, in some cases, be considered "single-cell" Hassall's corpuscles.

3.3. Splenic Responses

3.3.1. Innate Immune Responses in TI and PI Fetal Spleens

DDX58 (25-fold; $p < 0.05$), IRF7 (22-fold; $p < 0.05$), and ISG15 (160-fold; $p < 0.05$) mRNA were upregulated in TI fetal spleens compared to controls, whereas *NFKB* and *IFNB* mRNA concentrations remained at basal levels and were not significantly different (Figure 4). The innate immune response in PI fetal spleens was similar to controls with only *NFKB* being significantly downregulated in PI fetal spleens compared to controls (0.1532-fold, $p < 0.05$, Figure 4).

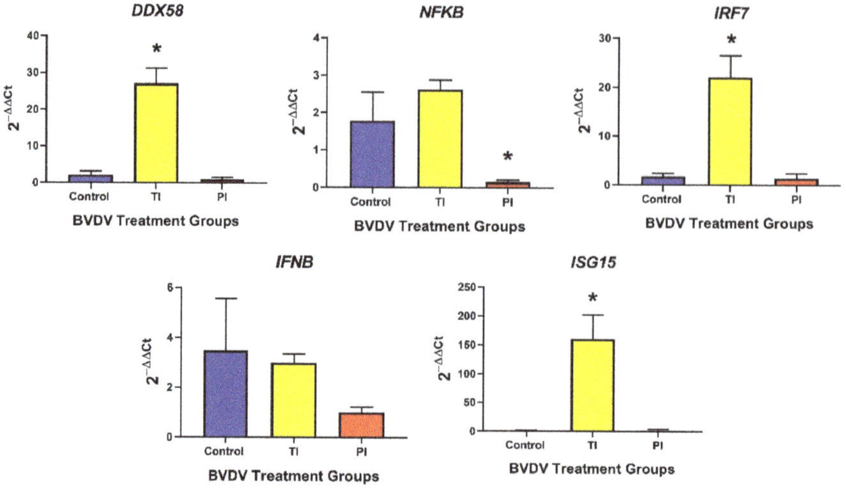

Figure 4. Spleen (Control, TI, and PI) expression of transcripts associated with the innate immune response. TI fetal spleens exhibited an upregulation of *DDX58*, *IRF7*, and *ISG15* compared to controls, indicating an active type I IFN response. In PI fetal spleens, only *NFKB* was downregulated compared to controls. TI: Transiently Infected, PI: Persistently Infected. * $p < 0.05$.

3.3.2. Adaptive Immune Responses in TI and PI Fetal Spleens

Adaptive immune response genes, such as *STAT4* (Figure 5), *PSMB9* (MHC class I, Table A1), and *IFI30* (MHC class II; Figure 5) were significantly upregulated in TI spleens compared to controls (5.54-fold, $p < 0.01$; 12.9-fold, $p < 0.05$ respectively); however, *PSMB8* (MHC class I) was not significantly different between the two groups (Figure 5). T and B lymphocyte markers, *CD4*, *CD8A*, *CD8B*, and *CD79B*, were not different in TI compared to the control spleen (Figure 5).

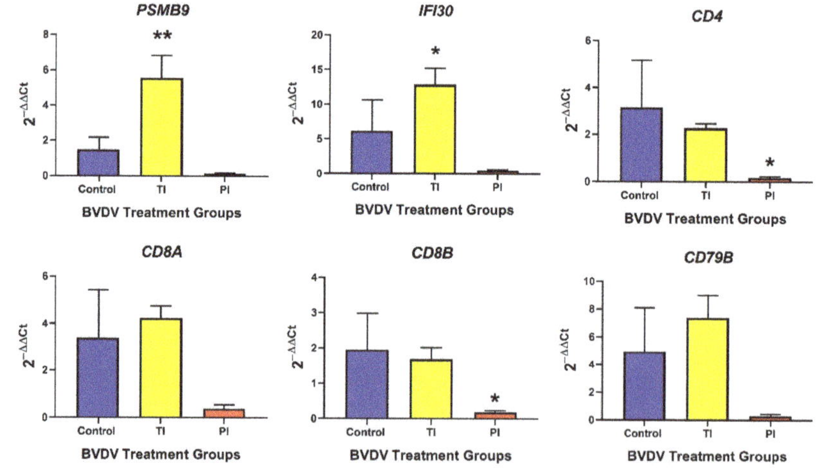

Figure 5. Splenic (Control, TI, and PI) expression of transcripts associated with the adaptive immune response. In TI fetal spleens, only *PSMB9* and *IFI30*, associated with antigen presentation, were significantly different to the controls. In PI fetal spleens, T cell markers *CD4* and *CD8A* were significantly downregulated compared to controls. TI: Transiently Infected, PI: Persistently Infected. * $p < 0.05$, ** $p < 0.01$.

In PI fetal spleens, *STAT4* (Table A1), *PSMB8/9* (Table A1 and Figure 5, respectively), and *IFI30* (Figure 5) were not differentially expressed compared to control fetal spleens. PI fetal spleens exhibited a downregulation of T lymphocyte markers mRNA (*CD4*, and *CD8B*) compared to controls (0.1465-fold, $p < 0.5$; 0.4005-fold, $p < 0.05$, respectively) (Figure 5) but T lymphocyte marker *CD8A* and B lymphocyte marker *CD79B* were not significantly different between the two groups.

4. Discussion

4.1. Transient Infection on Day 175 of Gestation Elicits an Immunocompetent Response in the Bovine Fetal Thymus and Spleen on Day 190

BVDV takes approximately 7–14 days to cross the placenta and infect the fetus following intranasal inoculation of the dam [38]. The TI fetal spleens and thymuses collected at day 190 of gestation, 15 days post-maternal infection were positive for BVDV RNA, indicating that the fetuses were infected for 1–7 days, which is sufficient time to have elicited an innate immune response in all TI fetuses and initial adaptive responses in some TI fetuses.

The rudimentary thymus is first detected around day 27 of gestation in ruminants with colonization and lymphoid development occurring at day 42 of gestation. Day 42 progenitor cells come from the fetal liver until the bone marrow develops and matures enough (around day 70) to take over as the primary source of thymocytes [39]. The bovine fetal thymus contains a discernable cortico-medullary delineation by day 42, Hassall's corpuscles by day 65, and IgM positive cells by day 70 to 100 [40]. In addition to T lymphocytes and prothymocytes of T cell lineage, the postnatal thymus contains thymic epithelial cells, myoid cells, fibroblasts, dendritic cells, macrophages and miscellaneous neutrophils, eosinophils and B cells [41,42]. These cell types would be present in the thymus at the time of maternal inoculation with BVDV (75 dpmi and 175 dpmi) and subsequent fetal infection in this experimental model. In utero BVDV infections have been shown to result in thymic hypoplasia in bovine fetuses characterized by marked lymphocyte depletion in the cortex, accompanied by hypertrophy of the epithelial stroma and infiltration by monocytes and mature macrophages with BVDV antigens primarily localized to macrophages [43–45].

The innate immune gene transcripts, *NFKB*, and *IRF7* were significantly upregulated in TI fetal thymuses indicative of a robust activation of the innate immune response to BVDV, which agrees with previous findings [34,35,38]. Significant changes were not observed in adaptive response genes, including the T cell marker transcripts, *CD8A*, *CD8B*, *CD4*, or B cell activation marker *CD79B* compared to controls, as expected due to the short course between fetal infection and tissue collection. In addition, the thymus is a primary lymphoid organ responsible for the selection and maturation of T cells but with limited antigen-specific immune responses. The decreased BVDV RNA concentrations in TI thymuses compared to PI thymuses, however, is supportive evidence that TI fetuses were able to mount an effective immune reaction against the virus. Despite the ability of TI fetuses to clear the virus, BVDV fetal infection during day 175–190 of gestation occurs during a critical stage of thymus and spleen development, as well as T cell selection and maturation in the thymus. BVDV fetal infection during this stage of fetal development may alter the animal's ability to fight other infections postnatally.

The spleen is a secondary lymphoid organ responsible for rapid immune responses to bloodborne pathogens like BVDV. During embryonic development, mesenchymal progenitor cells differentiate into marginal reticular cells, fibroblastic reticular cells, and follicular dendritic cells (reviewed in [46]). These cells will contribute to the spleen stroma and begin to form the white pulp, a lymphocyte-rich area which is the site of antigen-specific immune responses to blood-borne pathogens [46]. White pulp establishment occurs at approximately 15 weeks of gestation in humans, assumed to be a similar time point for cattle [40,46]. At this early time point, few B cells cluster around splenic arterioles, and 3 weeks later, are joined by T cells [46]. The T cells form the periarteriolar sheath (PALS), while the B cells form a follicle around the PALS [46]. Lymphocyte infiltration is believed to occur well after persistent infection with BVDV. However, TI fetuses were infected following the establishment of the white pulp and lymphocyte infiltration of the spleen. This may play a large role in the transient and persistent infection of BVDV in fetuses, as discussed later.

In TI fetal spleens, the cytoplasmic sensor mRNA *DDX58*, transcriptional factor *IRF7*, and interferon-stimulated gene *ISG15* were upregulated indicating the detection of BVDV RNA by the immune system of TI fetuses and activation of the innate immune response. Transcripts for genes associated with antigen presentation, *PSMB9* and *IFI30*, were upregulated in the TI fetal spleen, suggesting that the antigen processing mechanisms of cells had been stimulated. However, transcripts for T and B cell markers were not increased as viral peptide presentation in the context of MHC I and MHC II would not have stimulated the proliferation of T and B cells in the short time period between fetal infection and tissue collection. In addition, T and B cell activation may also have been delayed in the TI fetuses. BVDV is known to inhibit lymphocyte activation, resulting in delayed T and B cell responses during acute infections of postnatal animals [47,48]. The low amounts of BVDV RNA detected in the TI fetal spleen compared to PI fetal spleens suggest the innate immune mechanisms effectively inhibited viral replication prior to the elaboration of an adaptive immune response. Additionally, the fetal spleen is assumed to have both T and B cells on day 105 of gestation. It can be assumed these lymphocytes in the spleen, once activated, contribute to the clearance of the virus, seen in TI postnatal animals [9,15].

4.2. PI Thymus and Spleen Exhibit Diminished or Inhibited Immune Responses at Day 190 of Gestation

The mechanism of BVDV PI is controversial. Some of the first in vitro studies on BVDV infection concluded that BVDV infection blocked the host's interferon response contributing to the persistence of BVDV in fetuses infected before day 120 of gestation [21,23,24]. In a paper published in 2001, the authors infected fetuses in utero (through amniotic fluid) on day 60 of gestation and collected fetuses 3, 5, and 7 days post-maternal inoculation [27]. It was concluded that ncp BVDV inhibits the fetal IFN response [27]. Subsequently, using a model of maternal infection, the fetal IFN response was stimulated by BVDV 22 days post-maternal infection [28,31,34,35,38]. However, it is important to note the differences in BVDV strains used and the differences in experimental models. This current study follows several others with similar experimental models and the same BVDV strain that have demonstrated an active innate immune response in PI fetuses soon after the establishment of fetal infection. The discussion below highlights these previous findings [28,31,34,35,38].

Previously, it was established that the early gestation bovine fetus (day 75) could respond to ncp BVDV infection with an innate immune response indicated by the upregulation of cytosolic dsRNA sensors in fetal blood and spleen, type I IFNs in cotyledons, and ISGs in fetal tissues, including the liver and spleen [28,31,38]. Moreover, the IFNG protein, an important bridge between the innate and the adaptive immune response, was detected in the amniotic fluid and fetal blood, and *IFNG* mRNA was upregulated in liver, spleen, and thymus of PI fetuses 14–22 dpmi (89–97 days of gestation) [34]. The activation of these genes corresponded with a reduction in BVDV titer in PI fetal blood at day 190 and 245, suggesting partial clearance of the virus by an active innate immune response; however, BVDV continued to be present in fetal tissues, albeit at a reduced level (reviewed in [14,28,34]). We hypothesized that the persistence of ncp BVDV was due to exhaustion of the initial innate immune response, and impaired steps in the adaptive immune response leading to incomplete viral clearance in the PI fetus. In support of this hypothesis, two innate immune response mRNAs, *NFKB* and *IFNB*, the bridging mRNA *CXCL16*, and the adaptive immune response mRNAs, *PSMB9, TAP1, B2M, CD8A, CD8B, CD4*, and *CD79B*, were found to be significantly decreased on day 190 in PI fetal thymuses compared to controls, and T cell marker mRNAs (*CD4* and *CD8B*) were significantly downregulated in the PI spleen. However, as discussed below, it is believed attenuation of the immune system may be caused by Treg development, instead of our original hypothesis of immune system exhaustion [35,49–51].

DDX58 binds to and recognizes viral RNA in the innate immune system pathway. A reduction in *DDX58* concentrations could negatively impact the ability of fetal cells to respond to BVDV RNA and subsequently impair the induction of IFN antiviral activities. Persistent inhibition of NFKB has been associated with inappropriate immune cell development or delayed cell growth. For example, mice null for NFKB signaling molecules have impaired thymic medullary epithelial cell formation and autoimmune disease [52]. The transcription factor NFKB is both induced by type 1 IFNs and is a transcriptional regulator of cytokine expression, including IFNB and IFNG [53]. The decreased *IFNB* mRNA observed in the PI fetal thymus on day 190 coincided with the decreased expression of *NFKB* and the IFNI-inducible gene mRNAs, *ISG15* and *CXCL16*. Reduced *IFNB* mRNA may be explained by a paucity of dendritic cells in the PI fetal thymus at this stage of development; however, differences in the dendritic cell population between PI and control thymuses collected on day 190 were not observed histologically. IFI16 is a cytosolic sensor for viral RNA/DNA, which transcriptionally regulates the expression of type I IFNs in a positive manner [54]. The trended downregulation of *IFI16* observed in the PI thymus could explain the decreased expression of *IFNB*.

The PI fetal thymus exhibited differences in the innate immune response genes at day 190 compared to those previously observed in the fetal spleen and blood. *DDX58, MDA5*, and *ISG15* were previously observed to be increased in PI fetal blood from day 97 to day 192, and *ISG15* was increased in the PI fetal spleen, bone marrow, caruncle, and cotyledon, reflecting an active innate response in these tissues [28,31]. In contrast, *NFKB* and *IFNB* were significantly downregulated in PI fetal thymuses on day 190, but only *NFKB* was downregulated in spleens. A previous study by our group indicated a significant down-regulation of *IFNB* in day 190 PI fetal spleens, but no significant change in *NFKB* [35]. These studies were independent, and although results differ slightly, they both indicate an inhibition of the innate immune response 115 days post-maternal infection. Regulatory T cells (Tregs) originate in the thymus during fetal development and create tolerance of self in the developing fetus to prevent autoimmune disorders [49,50]. Due to the fetal BVDV infection occurring early in fetal development, prior to lymphocyte infiltration in the spleen, it is possible the Treg cells are creating tolerance to the BVDV viral antigen, allowing the virus to continue replication without an immune response from the host [49,50]. The development of Tregs in the thymus could be contributing to the difference in immune gene expression between the thymus and spleen [49,50]. On day 190 of gestation in PI animals, Tregs may be inhibiting an immune response in the thymus first before the Treg cells migrate to the periphery and reach the spleen.

Thymic concentrations of adaptive immune response transcripts, such as MHC I antigen presentation (*PSMB9, TAP1*, and *B2M*), CTL co-receptors *CD8A* and *CD8B*, and T cell marker CD4 were also decreased in PI fetal thymuses compared to controls consistent with an impaired adaptive immune response. The reduction in antigen presentation gene transcripts in PI fetal thymuses were associated with reduced *CD79B*, a component of antigen recognition and activation in B cells. In this study, *PSMB9* expression in PI thymuses was decreased at day 190 of gestation, suggesting a reduction in the normal function of the immunoproteasome [55]. Following proteolysis of viral proteins by the immunoproteasome, TAP1 transports peptides to the endoplasmic reticulum (ER) where peptides bind to MHC class I molecules for presentation on the cell membrane. Decreased *TAP1* concentrations could negatively impact antigen presentation to immune cells. In addition, *B2M* mRNA in PI thymuses was significantly decreased at day 190 compared to controls. The B2M protein assembles with the MHC I molecule in the ER to form a stable MHC I complex necessary for antigen binding. When B2M is not present, MHC I remains in the ER, and the MHC I complex is not expressed on the cell surface [56]. Decreased expression of *B2M* would result in additional dysfunction of the endogenous viral antigen presentation mechanism. In PI fetal spleens, *PSMB9* and *IFI30* were not significantly different compared to controls. This contradicts a previous study in which PSMB9 and IFI30 were significantly downregulated in day 190 PI spleens compared to controls [35]. This difference between the studies reflect increased individual variability between controls in the present study or possibly due to slight variation in infection/fetal collection times. The difference in the expression of antigen presentation mRNAs between the spleen and thymus may be indicative of APC and lymphocyte migration from the primary lymphoid organ, that is, the thymus, to secondary lymphoid organs, such as the spleen. In secondary lymphoid organs, such as the spleen, APCs reside and actively survey the blood and lymph for antigens; thus, the secondary lymphoid organs will have higher gene expression of mature APCs and lymphocyte markers [57].

Cytotoxic T cells recognize peptides presented in the context of the MHC I and B2M complex on target cells by the T cell receptor (TCR) and co-receptor, CD8. Antigen recognition by the CD8-TCR complex triggers the release of perforin, granzymes and cytokines from the lytic granules of CTL. *CD8A* and *CD8B* mRNA concentrations in PI fetal thymuses and *CD8B* in spleens were significantly decreased at day 190, suggesting either a reduced number of CTLs or decreased CD8A/B expressed per cell. Similarly, *CD4*, the co-receptor of the T cell receptor on T helper cells, mRNA was decreased in PI fetal thymuses and spleens at day 190 compared to controls and may reflect a decrease in T helper cells or the number of co-receptors per T helper cell. Supporting the results in a previous study, a reduction in BVDV-specific CTLs, T helper cells, or in antigen recognition would negatively impact the clearance of BVDV-infected cells in the thymus, spleen, and other fetal tissues, thus contributing to viral persistence [35].

The B-cell antigen receptor complex-associated protein Beta chain, *CD79B*, is part of the B lymphocyte antigen receptor complex and is required for T cell-dependent activation of B cells. In cooperation with CD79A, CD79B activates B cells through the interaction with CD4 T cells. After this interaction, B cells then have the capability to proliferate, secrete antibodies, cytokines, and chemokines, and initiate memory cell formation [58]. CD79B gene expression was decreased in PI thymuses on day 190 compared to controls, but were not downregulated in PI fetal spleens. However, a previous study by our group did find a decrease in CD79B expression in PI fetal spleens [35]. This difference may be due to individual variation of animals in the current study or slight differences in infection/collection timing. A sustained decrease in CD79B expression could result in the depletion of B cells later in gestation or postnatally, contributing to the overall deficit in adaptive immunity, failed viral clearance, and BVDV persistence. The outcome would be apparent immunotolerance to BVDV and the inability of PI animals to combat secondary viral and bacterial infections, such as bovine respiratory disease in postnatal life.

CD46 has been shown to be a receptor for BVDV entry [59,60]. CD46 is a regulator of complement activation through its interaction with plasma serine protease factor I, which cleaves complement factors C3b and C4b to prevent cell injury from complement attack. The concentration of *CD46* was significantly decreased in PI fetal thymuses compared to controls at day 190. The reduced concentration of *CD46* in the PI thymuses may reflect a general diminishment of constitutively expressed cellular mRNAs due to viral mechanisms that usurp cellular transcriptional machinery in favor of the production of viral proteins [61]. Alternatively, expression of the BVDV receptor might be inhibited in response to the binding of BVDV to its receptor and the presence of high viral titers during persistent infection. The downregulation of *CD46* in PI thymuses might also reduce the regulatory activity of CD46 on the complement system and increase the vulnerability of host cells to complement-mediated injury. A soluble form of CD46 which neutralizes BVDV in vitro is present in adult bovine plasma and may play a role in viral clearance in postnatal acute infections. However, soluble CD46 is not found in fetal or neonatal serum, and is therefore unlikely to impact PI fetal BVDV infections [62].

The comparison of TI and PI fetal immune responses in developing lymphoid organs has not been previously examined in experimental settings. Unfortunately, bovine-specific antibodies for the genes of interest are unavailable, and limited this study to RNA expression only. The inability to obtain protein data does limit interpretation. However, the results of this transcriptome study exhibit a drastic difference in immune response competence and maturity between PI and TI fetuses. The upregulation of the innate immune response of TI lymphoid organs, along with the inability of BVDV levels to be significantly different than controls indicate that the TI fetus is effectively fighting the virus, unlike the PI fetus. The timepoint of TI fetal collections may have been too early to capture the activation of the adaptive immune response, or the virus is possibly able to temporarily inhibit the adaptive immune response. It is further hypothesized that although the TI fetuses are able to respond appropriately to the virus, this viral insult during fetal development epigenetically alters the TI animals' DNA and causes further immune-related issues postnatally. This will require further studies postnatally.

5. Conclusions

Although animals infected with BVDV in utero may be born morphologically normal, ncp BVDV fetal infection has significant effects on the expression of innate and adaptive immune responses. In TI animals infected late in gestation, the immune system is developed enough to fight and clear the virus prior to or following parturition. We interpret these RT-qPCR data to reflect an active and upregulated innate immune response in TI thymuses and spleens, seen as an increase in type I IFN transcriptional regulators (Figure 6). In the case of PI fetuses, the virus infects the fetus prior to complete immune system development, ultimately causing drastic attenuation of the immune system, possibly due to the fetus identifying the virus as "self" (Figure 6). Specifically, the downregulation of genes, including type I IFN transcriptional regulators, antigen presentation, and T cell markers in PI fetal tissues was found. These trends in TI and PI fetuses were seen in both the thymic and splenic tissues, suggesting that these changes occur simultaneously in tissues critical for immune system education and development (Figure 6). These types of changes are believed to affect the infected animals postnatally, possibly through epigenetic changes.

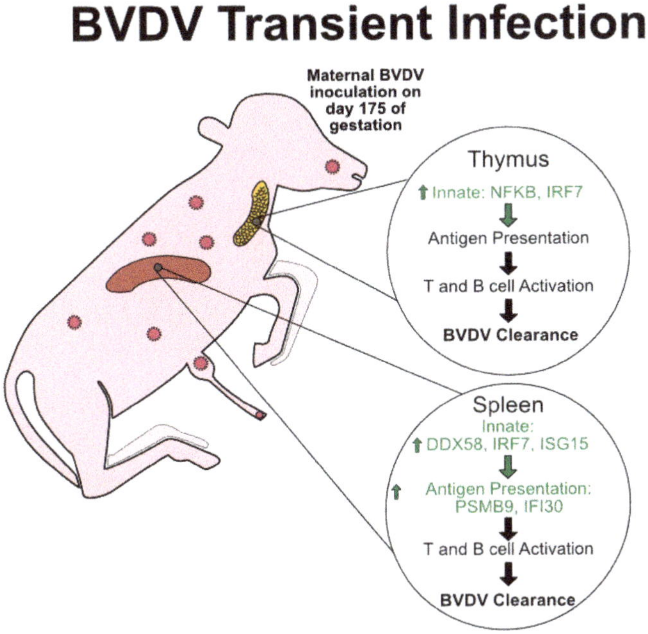

Figure 6. *Cont.*

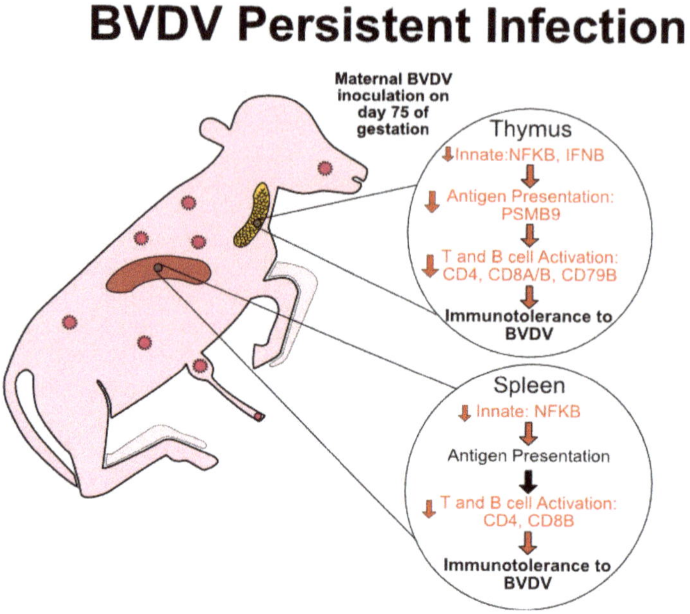

Figure 6. Graphic summarizing significant changes in fetal immune responses based on type of infection and tissue, as interpreted by RT-qPCR analysis. BVDV transient infection represents a fetal infection after gestational day 160 in which the fetus is able to clear the virus. Trends include an upregulation in genes associated with the innate immune response. Day 190 of collections may have been too early to capture an adaptive immune response in TI fetuses. However, we hypothesize that these pathways contribute to BVDV clearance in TI animals. BVDV-persistent infection represents a fetal infection before gestational day 120 in which the fetus is unable to clear the virus. Trends include a drastic down-regulation of the innate and adaptive branches of the immune system, and may lead to immunotolerance of the fetus to BVDV. In future studies, TI BVDV clearance and PI immunotolerance will be studied further with protein targets and flow cytometry. Green text, up arrow left of text, and listed genes represent pathways and genes which were significantly upregulated, as determined by RT-qPCR. Red text, down arrow left of text, and listed genes represent pathways and genes which were significantly downregulated as determined by RT-qPCR.

Author Contributions: Conceptualization, K.J.K. and T.R.H.; methodology, N.P.S., K.J.K.; validation, K.J.K., H.M.G., N.P.S.; formal analysis K.J.K., H.M.G., H.V.C.; investigation, K.J.K., H.M.G., H.V.C., H.B.-O.; resources, J.V.B.; writing—original draft preparation, K.J.K., H.M.G.; writing—review and editing, H.V.C., H.B.-O., T.R.H.; visualization, K.J.K., H.M.G.; supervision, J.V.B., H.B.-O.; project administration, T.R.H.; funding acquisition, T.R.H., H.M.G. All authors have read and agreed to the published version of the manuscript.

Funding: This work was supported by USDA AFRI NIFA Predoctoral Fellowship 2019-67011-29539/1019321 to H.M.G.; AFRI NIFA National Needs Fellowships Grant Program 2016-38420-25289, USDA-AFRI Grants 2004-35204-17005 and 2008-35204-04652, and W3112 Reproduction in Domestic Ruminants HATCH project #1011648 from the USDA National Institute of Food and Agriculture to T.R.H.

Acknowledgments: The authors would like to thank Ann Hess for her assistance with statistical analysis.

Conflicts of Interest: The authors declare no conflict of interest. The funders had no role in the design of the study; in the collection, analyses, or interpretation of data; in the writing of the manuscript, or in the decision to publish the results.

Appendix A

Table A1. Table of genes and their expression in respective tissues. These genes were not measured in both tissues, only one tissue.

Tissue	Target	Control Mean	Control SEM	TI Mean	TI SEM	PI Mean	PI SEM	TI v. C *p* value	PI v. C *p* value
Thymus	*STAT1*	1.32	0.48	2.29	0.21	0.04	0.01	0.58	0.07
Thymus	*IFI6*	1.10	0.20	2.09	0.49	0.10	0.02	0.09	0.09
Thymus	*CXCL10*	1.24	0.35	3.10	1.20	0.27	0.12	0.99	0.11
Thymus	*CXCL16*	1.22	0.34	1.21	0.27	0.05	0.01	0.99	0.01
Thymus	*CXCR6*	1.40	0.57	1.64	0.41	0.48	0.15	0.99	0.53
Thymus	*TAP1*	1.09	0.20	1.07	0.21	0.20	0.06	0.99	0.01
Thymus	*B2M*	1.17	0.31	1.33	0.24	0.04	0.01	0.89	0.01
Thymus	*CIITA*	1.19	0.27	1.35	0.35	0.35	0.08	0.99	0.08
Spleen	*STAT4*	1.62	0.83	2.25	0.51	0.32	0.04	0.43	0.08
Spleen	*PSMB8*	1.66	0.79	6.09	1.40	0.06	0.02	0.21	0.07

SEM: Standard Error Mean; TI: Transient Infection; PI: Persistent Infection; C: Control; STAT1: Signal Transducer and Activator of Transcription 1; IFI6: Interferon Alpha Inducible Protein 6; CXCL10: C-X-C Motif Chemokine Ligand 10, CXCL16: C-X-C Motif Chemokine Ligand 16; TAP1: Transporter 1 ATP Binding Cassette Subfamily B Member; B2M: Beta-2-Microglobulin; CIITA: Class II Major Histocompatibility Complex Transactivator; STAT4: Signal Transducer and Activator of Transcription 4; PSMB8: Proteasome 20S Subunit Beta 8.

References

1. Moennig, V.; Houe, H.; Lindberg, A. BVD control in Europe: Current status and perspectives. *Anim. Health Res. Rev.* **2005**, *6*, 63–74. [CrossRef] [PubMed]
2. Larson, R.; Pierce, V.; Grotelueschen, D.; Wittum, T. Economic evaluation of beef cowherd screening for cattle persistently-infected with bovine viral diarrhea virus. *Bov. Pract.* **2002**, *36*, 106–112.
3. Pinior, B.; Firth, C.L.; Richter, V.; Bakran-Lebl, K.; Trauffler, M.; Dzieciol, M.; Hutter, S.E.; Burgstaller, J.; Obritzhauser, W.; Winter, P.; et al. A systematic review of financial and economic assessments of bovine viral diarrhea virus (BVDV) prevention and mitigation activities worldwide. *Prev. Vet. Med.* **2017**, *137*, 77–92. [CrossRef]
4. Richter, V.; Bakran-Lebl, K.; Baumgartner, W.; Obritzhauser, W.; Käsbohrer, A.; Pinior, B. A systematic worldwide review of the direct monetary losses in cattle due to bovine viral diarrhoea virus infection. *Vet. J.* **2017**, *220*, 80–87. [CrossRef]
5. Simmonds, P.; Becher, P.; Bukh, J.; Gould, E.A.; Meyers, G.; Monath, T.; Muerhoff, S.; Pletnev, A.; Rico-Hesse, R.; Smith, D.B.; et al. ICTV Virus Taxonomy Profile: Flaviviridae. *J. Gen. Virol.* **2017**, *98*, 2–3. [CrossRef]
6. Tautz, N.; Tews, B.A.; Meyers, G. The Molecular Biology of Pestiviruses. *Nat. Eng. Resist. Plant Viruses Part II* **2015**, *93*, 47–160. [CrossRef]
7. Bielefeldt-Ohmann, H. The Pathologies of Bovine Viral Diarrhea Virus Infection. *Vet. Clin. N. Am. Food Anim. Pract.* **1995**, *11*, 447–476. [CrossRef]
8. Brownlie, J.; Clarke, M.C.; Howard, C.J.; Pocock, D.H. Pathogenesis and epidemiology of bovine virus diarrhoea virus infection of cattle. *Ann. Rech. Vet. Ann. Vet. Res.* **1987**, *18*, 157–166.
9. Kendrick, J.W. Bovine viral diarrhea-mucosal disease virus infection in pregnant cows. *Am. J. Vet. Res.* **1971**, *32*, 533–544.
10. Lanyon, S.; Hill, F.I.; Reichel, M.P.; Brownlie, J. Bovine viral diarrhoea: Pathogenesis and diagnosis. *Vet. J.* **2014**, *199*, 201–209. [CrossRef] [PubMed]
11. Olafson, P.; Maccallum, A.D.; Fox, F.H. An apparently new transmissible disease of cattle. *Cornell Vet.* **1946**, *36*, 205–213.
12. Coria, M.F.; McClurkin, A.W. Duration of active and colostrum-derived passive antibodies to bovine viral diarrhea virus in calves. *Can. J. Comp. Med. Rev. Can. Med. Comp.* **1978**, *42*, 239–243.
13. McClurkin, A.W.; Littledike, E.T.; Cutlip, R.C.; Frank, G.H.; Coria, M.F.; Bolin, S.R. Production of cattle immunotolerant to bovine viral diarrhea virus. *Can. J. Comp. Med. Rev. Can. Med. Comp.* **1984**, *48*, 156–161.
14. Hansen, T.R.; Smirnova, N.P.; Webb, B.T.; Bielefeldt-Ohmann, H.; Sacco, R.E.; Van Campen, H. Innate and adaptive immune responses to in utero infection with bovine viral diarrhea virus. *Anim. Health Res. Rev.* **2015**, *16*, 15–26. [CrossRef] [PubMed]
15. Brownlie, J.; Clarke, M.; Howard, C. Experimental production of fatal mucosal disease in cattle. *Vet. Rec.* **1984**, *114*, 535–536. [CrossRef]

16. Van Campen, H. Bovine Viral Diarrhea. In Proceedings of the Range Beef Cow Symposium, Rapid City, SD, USA, 9–11 December 1997; University of Nebraska-Lincoln Commons: Rapid City, SD, USA, 1997.
17. Potgieter, L.N. Immunology of Bovine Viral Diarrhea Virus. *Vet. Clin. N. Am. Food Anim. Pract.* **1995**, *11*, 501–520. [CrossRef]
18. Schweizer, M.; Peterhans, E. Pestiviruses. *Annu. Rev. Anim. Biosci.* **2014**, *2*, 141–163. [CrossRef]
19. Peterhans, E.; Schweizer, M. BVDV: A pestivirus inducing tolerance of the innate immune response. *Biology* **2013**, *41*, 39–51. [CrossRef]
20. Baigent, S.J.; Goodbourn, S.; McCauley, J.W. Differential activation of interferon regulatory factors-3 and -7 by non-cytopathogenic and cytopathogenic bovine viral diarrhoea virus. *Vet. Immunol. Immunopathol.* **2004**, *100*, 135–144. [CrossRef]
21. Baigent, S.J.; Zhang, G.; Fray, M.D.; Flick-Smith, H.; Goodbourn, S.; McCauley, J.W. Inhibition of Beta Interferon Transcription by Noncytopathogenic Bovine Viral Diarrhea Virus Is through an Interferon Regulatory Factor 3-Dependent Mechanism. *J. Virol.* **2002**, *76*, 8979–8988. [CrossRef]
22. Chen, Z.; Rijnbrand, R.; Jangra, R.K.; Devaraj, S.G.; Qu, L.; Ma, Y.; Lemon, S.M.; Li, K. Ubiquitination and proteasomal degradation of interferon regulatory factor-3 induced by Npro from a cytopathic bovine viral diarrhea virus. *Virology* **2007**, *366*, 277–292. [CrossRef] [PubMed]
23. Gil, L.H.; Ansari, I.H.; Vassilev, V.; Liang, D.; Lai, V.C.; Zhong, W.; Hong, Z.; Dubovi, E.J.; Donis, R.O. The amino-terminal domain of bovine viral diarrhea virus Npro protein is necessary for alpha/beta interferon antagonism. *J. Virol.* **2006**, *80*, 900–911. [CrossRef] [PubMed]
24. Gil, L.H.V.G.; Van Olphen, A.L.; Mittal, S.K.; Donis, R.O. Modulation of PKR activity in cells infected by bovine viral diarrhea virus. *Virus Res.* **2006**, *116*, 69–77. [CrossRef] [PubMed]
25. Magouras, I.; Mätzener, P.; Rümenapf, T.; Peterhans, E.; Schweizer, M. RNase-dependent inhibition of extracellular, but not intracellular, dsRNA-induced interferon synthesis by Erns of pestiviruses. *J. Gen. Virol.* **2008**, *89*, 2501–2506. [CrossRef]
26. Meyers, G.; Ege, A.; Fetzer, C.; Von Freyburg, M.; Elbers, K.; Carr, V.; Prentice, H.; Charleston, B.; Schürmann, E.-M. Bovine Viral Diarrhea Virus: Prevention of Persistent Fetal Infection by a Combination of Two Mutations Affecting Erns RNase and Npro Protease. *J. Virol.* **2007**, *81*, 3327–3338. [CrossRef]
27. Charleston, B.; Fray, M.D.; Baigent, S.; Carr, B.V.; Morrison, W.I. Establishment of persistent infection with non-cytopathic bovine viral diarrhoea virus in cattle is associated with a failure to induce type I interferon. *J. Gen. Virol.* **2001**, *82*, 1893–1897. [CrossRef]
28. Smirnova, N.P.; Webb, B.T.; Bielefeldt-Ohmann, H.; Van Campen, H.; Antoniazzi, A.Q.; Morarie, S.E.; Hansen, T.R. Development of fetal and placental innate immune responses during establishment of persistent infection with bovine viral diarrhea virus. *Virus Res.* **2012**, *167*, 329–336. [CrossRef]
29. Smirnova, N.P.; Ptitsyn, A.; Austin, K.J.; Bielefeldt-Ohmann, H.; Van Campen, H.; Han, H.; Van Olphen, A.L.; Hansen, T.R. Persistent fetal infection with bovine viral diarrhea virus differentially affects maternal blood cell signal transduction pathways. *Physiol. Genom.* **2008**, *36*, 129–139. [CrossRef]
30. Ohmann, H.B.; Rønsholt, L.; Bloch, B. Demonstration of Bovine Viral Diarrhoea Virus in Peripheral Blood Mononuclear Cells of Persistently Infected, Clinically Normal Cattle. *J. Gen. Virol.* **1987**, *68*, 1971–1982. [CrossRef]
31. Shoemaker, M.L.; Smirnova, N.P.; Bielefeldt-Ohmann, H.; Austin, K.J.; Van Olphen, A.; Clapper, J.A.; Hansen, T.R. Differential Expression of the Type I Interferon Pathway during Persistent and Transient Bovine Viral Diarrhea Virus Infection. *J. Interf. Cytokine Res.* **2009**, *29*, 23–36. [CrossRef]
32. Palomares, R.A.; Walz, H.G.; Brock, K.V. Expression of type I interferon-induced antiviral state and pro-apoptosis markers during experimental infection with low or high virulence bovine viral diarrhea virus in beef calves. *Virus Res.* **2013**, *173*, 260–269. [CrossRef] [PubMed]
33. Charleston, B.; Brackenbury, L.S.; Carr, B.V.; Fray, M.D.; Hope, J.C.; Howard, C.J.; Morrison, W.I. Alpha/Beta and Gamma Interferons are Induced by Infection with Noncytopathic Bovine Viral Diarrhea Virus In Vivo. *J. Virol.* **2002**, *76*, 923–927. [CrossRef] [PubMed]
34. Smirnova, N.P.; Webb, B.; McGill, J.L.; Schaut, R.G.; Bielefeldt-Ohmann, H.; Van Campen, H.; Sacco, R.E.; Hansen, T.R. Induction of interferon-gamma and downstream pathways during establishment of fetal persistent infection with bovine viral diarrhea virus. *Virus Res.* **2014**, *183*, 95–106. [CrossRef] [PubMed]
35. Georges, H.M.; Knapek, K.J.; Bielefeldt-Ohmann, H.; Van Campen, H.; Hansen, T.R. Attenuated lymphocyte activation leads to the development of immunotolerance in bovine fetuses persistently infected with BVDV[†]. *Biol. Reprod.* **2020**. [CrossRef]

36. Bustin, S.A.; Benes, V.; Garson, J.A.; Hellemans, J.; Huggett, J.F.; Kubista, M.; Mueller, R.; Nolan, T.; Pfaffl, M.W.; Shipley, G.L.; et al. The MIQE Guidelines: Minimum Information for Publication of Quantitative Real-Time PCR Experiments. *Clin. Chem.* **2009**, *55*, 611–622. [CrossRef] [PubMed]
37. Livak, K.J.; Schmittgen, T.D. Analysis of Relative Gene Expression Data Using Real-Time Quantitative PCR and the $2^{-\Delta\Delta CT}$ Method. *Methods* **2001**, *25*, 402–408. [CrossRef]
38. Smirnova, N.P.; Bielefeldt-Ohmann, H.; Van Campen, H.; Austin, K.J.; Han, H.; Montgomery, D.L.; Shoemaker, M.L.; Van Olphen, A.L.; Hansen, T.R. Acute non-cytopathic bovine viral diarrhea virus infection induces pronounced type I interferon response in pregnant cows and fetuses. *Virus Res.* **2008**, *132*, 49–58. [CrossRef]
39. Goddeeris, B.M.; Morrison, W.I. *Cell-Mediated Immunity in Ruminants*; CRC Press: Boca Raton, FL, USA, 1994.
40. Schultz, R.D.; Dunne, H.W.; Heist, C.E. Ontogeny of the Bovine Immune Response 1. *Infect. Immun.* **1973**, *7*, 981–991. [CrossRef]
41. Haynes, B.F. The Human Thymic Microenvironment. *Adv. Immunol.* **1984**, *36*, 87–142. [CrossRef]
42. Lobach, D.F.; Haynes, B.F. Ontogeny of the human thymus during fetal development. *J. Clin. Immunol.* **1987**, *7*, 81–97. [CrossRef]
43. Ohmann, H.B. Experimental fetal infection with bovine viral diarrhea virus. II. Morphological reactions and distribution of viral antigen. *Can. J. Comp. Med. Rev. Can. Med. Comp.* **1982**, *46*, 363–369.
44. Done, J.; Terlecki, S.; Richardson, C.; Harkness, J.; Sands, J.; Patterson, D.; Sweasey, D.; Shaw, I.; Winkler, C.; Duffell, S. Bovine virus diarrhoea-mucosal disease virus: Pathogenicity for the fetal calf following maternal infection. *Vet. Rec.* **1980**, *106*, 473–479. [CrossRef]
45. Falkenberg, S.M.; Bauermann, F.V.; Ridpath, J.F. Characterization of thymus-associated lymphoid depletion in bovine calves acutely or persistently infected with bovine viral diarrhea virus 1, bovine viral diarrhea virus 2 or HoBi-like pestivirus. *Arch. Virol.* **2017**, *162*, 3473–3480. [CrossRef] [PubMed]
46. Golub, R.; Tan, J.; Watanabe, T.; Brendolan, A. Origin and Immunological Functions of Spleen Stromal Cells. *Trends Immunol.* **2018**, *39*, 503–514. [CrossRef] [PubMed]
47. Ellis, J.A.; Davis, W.C.; Belden, E.L.; Pratt, D.L. Flow Cytofluorimetric Analysis of Lymphocyte Subset Alterations in Cattle Infected with Bovine Viral Diarrhea Virus. *Vet. Pathol.* **1988**, *25*, 231–236. [CrossRef] [PubMed]
48. Bolin, S.R.; McClurkin, A.W.; Coria, M.F. Effects of bovine viral diarrhea virus on the percentages and absolute num- bers of circulating B and T lymphocytes in cattle. *Am. J. Vet. Res.* **1988**, *46*, 884–886.
49. Burt, T.D. Fetal regulatory T cells and peripheral immune tolerance in utero: Implications for development and disease. *Am. J. Reprod. Immunol.* **2013**, *69*, 346–358. [CrossRef]
50. Mold, J.E.; Venkatasubrahmanyam, S.; Burt, T.D.; Michaëlsson, J.; Rivera, J.M.; Galkina, S.A.; Weinberg, K.; Stoddart, C.A.; McCune, J.M. Fetal and Adult Hematopoietic Stem Cells Give Rise to Distinct T Cell Lineages in Humans. *Science* **2010**, *330*, 1695–1699. [CrossRef]
51. Collen, T.; Douglas, A.J.; Paton, D.J.; Zhang, G.; Morrison, W. Single Amino Acid Differences are Sufficient for CD4+ T-Cell Recognition of a Heterologous Virus by Cattle Persistently Infected with Bovine Viral Diarrhea Virus. *Virology* **2000**, *276*, 70–82. [CrossRef]
52. Zhu, M.; Fu, Y. The complicated role of NF-kappaB in T-cell selection. *Cell Mol. Immunol.* **2010**, *7*, 89–93. [CrossRef]
53. Leonidas, C.P. Mechanisms of type-I- and type-II-interferon-mediated signalling. *Nat. Rev. Immunol.* **2005**, *5*, 375–386.
54. Thompson, M.R.; Sharma, S.; Atianand, M.; Jensen, S.B.; Carpenter, S.; Knipe, D.M.; Fitzgerald, K.A.; Kurt-Jones, E.A. Interferon γ-inducible Protein (IFI) 16 Transcriptionally Regulates Type I Interferons and Other Interferon-stimulated Genes and Controls the Interferon Response to both DNA and RNA Viruses*. *J. Biol. Chem.* **2014**, *289*, 23568–23581. [CrossRef] [PubMed]
55. Harwig, A.; Landick, R.; Berkhout, B. The Battle of RNA Synthesis: Virus versus Host. *Viruses* **2017**, *9*, 309. [CrossRef] [PubMed]
56. Benedictus, L.; Luteijn, R.D.; Otten, H.; Lebbink, R.J.; Van Kooten, P.J.S.; Wiertz, E.J.H.J.; Rutten, V.P.M.G.; Koets, A. Pathogenicity of Bovine Neonatal Pancytopenia-associated vaccine-induced alloantibodies correlates with Major Histocompatibility Complex class I expression. *Sci. Rep.* **2015**, *5*, 12748. [CrossRef]
57. Cyster, J.G. Chemokines and Cell Migration in Secondary Lymphoid Organs. *Science* **1999**, *286*, 2098–2102. [CrossRef] [PubMed]

58. Bishop, G.A.; Haxhinasto, S.A.; Stunz, L.L.; Hostager, B.S. Antigen-specific B-lymphocyte activation. *Crit. Rev. Immunol.* **2003**, *23*, 149–197. [CrossRef]
59. Maurer, K.; Krey, T.; Moennig, V.; Thiel, H.-J.; Rümenapf, T. CD46 Is a Cellular Receptor for Bovine Viral Diarrhea Virus. *J. Virol.* **2004**, *78*, 1792–1799. [CrossRef]
60. Krey, T.; Himmelreich, A.; Heimann, M.; Menge, C.; Thiel, H.-J.; Maurer, K.; Rümenapf, T. Function of Bovine CD46 as a Cellular Receptor for Bovine Viral Diarrhea Virus Is Determined by Complement Control Protein 1. *J. Virol.* **2006**, *80*, 3912–3922. [CrossRef]
61. Neill, J.D.; Ridpath, J.F. Gene expression changes in BVDV2-infected MDBK cells. *Biology* **2003**, *31*, 97–102. [CrossRef]
62. Alzamel, N.; Bayrou, C.; Decreux, A.; Desmecht, D. Soluble forms of CD46 are detected in Bos taurus plasma and neutralize BVDV, the bovine pestivirus. *Comp. Immunol. Microbiol. Infect. Dis.* **2016**, *49*, 39–46. [CrossRef]

viruses

MDPI

Article

The Effect of Bovine Viral Diarrhea Virus (BVDV) Strains and the Corresponding Infected-Macrophages' Supernatant on Macrophage Inflammatory Function and Lymphocyte Apoptosis

Karim Abdelsalam [1,2,*], Mrigendra Rajput [1], Gamal Elmowalid [2], Jacob Sobraske [1], Neelu Thakur [1], Hossam Abdallah [2], Ahmed A. H. Ali [2] and Christopher C. L. Chase [1,*]

1 Department of Veterinary and Biomedical Sciences, South Dakota State University, Brookings, SD 57007, USA; mrajput@atu.edu (M.R.); jacob.sobraske@sdstate.edu (J.S.); neelu.thakur@sdstate.edu (N.T.)

2 Faculty of Veterinary Medicine, Zagazig University, Zagazig 44519, Egypt; gelmowalid@yahoo.com (G.E.); bactine_ho@yahoo.com (H.A.); ahhmht35@yahoo.com (A.A.H.A.)

* Correspondence: karim.abdelsalam@sdstate.edu (K.A.); christopher.chase@sdstate.edu (C.C.L.C.)

Received: 4 May 2020; Accepted: 23 June 2020; Published: 29 June 2020

Abstract: Bovine viral diarrhea virus (BVDV) is an important viral disease of cattle that causes immune dysfunction. Macrophages are the key cells for the initiation of the innate immunity and play an important role in viral pathogenesis. In this in vitro study, we studied the effect of the supernatant of BVDV-infected macrophage on immune dysfunction. We infected bovine monocyte-derived macrophages (MDM) with high or low virulence strains of BVDV. The supernatant recovered from BVDV-infected MDM was used to examine the functional activity and surface marker expression of normal macrophages as well as lymphocyte apoptosis. Supernatants from the highly virulent 1373-infected MDM reduced phagocytosis, bactericidal activity and downregulated MHC II and CD14 expression of macrophages. Supernatants from 1373-infected MDM induced apoptosis in MDBK cells, lymphocytes or BL-3 cells. By protein electrophoresis, several protein bands were unique for high-virulence, 1373-infected MDM supernatant. There was no significant difference in the apoptosis-related cytokine mRNA (IL-1beta, IL-6 and TNF-a) of infected MDM. These data suggest that BVDV has an indirect negative effect on macrophage functions that is strain-specific. Further studies are required to determine the identity and mechanism of action of these virulence factors present in the supernatant of the infected macrophages.

Keywords: bovine viral diarrhea virus; cytopathic BVDV; immunosuppression; lymphocyte apoptosis; monocyte-derived macrophages; non-cytopathic BVDV

1. Introduction

Bovine viral diarrhea virus (BVDV) causes immune dysfunction in cattle and other ruminants, resulting in severe economic losses. The immune dysfunction associated with BVDV is believed to be, in part, a consequence of lymphoid depletion that could be mild or severe based on the virulence of the BVDV strains [1–4]. Furthermore, BVDV induces multiple abnormalities to immune cell function and immune mediators that could potentiate viral infection and pathogenesis [5,6]. BVDV infects neutrophils and decreases their microbicidal activity [7]. It also infects alveolar macrophage and decreases the expression of complement Fc receptors and chemokine production, reducing their ability to engulf opsonized pathogens [8].

Although only cytopathogenic (Cp) strains can induce direct apoptosis to peripheral blood mononuclear cells (PBMCs) [9,10], other studies have suggested that Ncp BVDV induces apoptosis in vitro in PBMCs isolated from BVDV-infected animals [11]. Ncp BVDV causes the most severe

BVDV immune dysfunction and is associated with severe lymphoid depletion [2,3,12–15]. Some Ncp BVDV strains have been categorized under a new biotype, lymphocytopathogenic, as they can induce apoptosis in lymphocytes while not effecting epithelial cells [3,16]. In addition, Ncp BVDV has been reported to induce apoptosis in vitro in PBMCs isolated from BVDV-infected animals [11].

Factors associated with Cp BVDV-induced apoptosis include: (i) oxidative stress and (ii) soluble factor(s) produced by monocytes that induced apoptosis in uninfected cells after lipopolysaccharide (LPS) treatment, [9,10,17,18]. However, the exact mechanism by which Ncp/lymphocytopathogenic BVDV induced the apoptotic effect is not well understood. Several studies suggest the role of macrophages in lymphoid depletion and immune dysfunction associated with BVDV. These studies demonstrated the importance of monocytes/macrophage in lymphocyte apoptosis in a mixed population of PBMCs [9,13]. However, the role of apoptotic-associated cytokines, which are important for classic swine fever virus (CSFV)-induced lymphocyte apoptosis, do not appear to be important for BVDV [13]. BVDV and CSFV are classified under the same genus: *Pestivirus*, Family: *Flavivirdae*. It has also been hypothesized that the *Pestivirus'* specific secreted Erns glycoprotein could play a role in the apoptosis mechanism associated with CSFV [19,20].

Immune mediators potentiate viral infection and pathogenesis. CSFV [21], and Cp BVDV [17], prime uninfected cells for activation-induced apoptosis. The mechanism for activation-induced apoptosis is virus-specific, where direct contact between infected and uninfected cells may also be involved in CSFV [21]. CSFV infection of bone marrow hematopoietic cells does not induce apoptosis in the majority of infected cells but does cause apoptosis in the uninfected cells, suggesting an apoptotic role for soluble factors released from infected cells. Cp BVDV induces apoptosis, while Ncp BVDV inhibits apoptosis [9,10,17,18,22,23].

Cp BVDV induces apoptosis in epithlial cellls in cell culture and in PBMCs, making it different from Lymphocytopathogenic BVDV, which can only induce apoptosis in lymphocytes. The soluble factor(s) involved in apoptosis has not been well characterized in any of these studies. Several studies have identified cytokines that prime for apoptosis, including IL-4 [24], TNF-α, IFN-γ [25–27] and IL1beta [28]. However, these cytokines are produced by lymphocytes rather than macrophages [23].

To investigate the indirect effect of different strains of BVDV on macrophage activity and lymphocytes apoptosis, both peripheral blood lymphocytes and a BVDV-free BL-3 cell line were used [3]. The direct apoptotic effect of BVDV was examined by the infection of lymphocytes with different BVDV strains. The indirect effect was investigated by infection of the MDM with one of two different strains of BVDV, followed by treatment of lymphocytes with the irradiated virus-free supernatant from BVDV-infected macrophages [29].

2. Materials and Methods

2.1. Viruses and Cells

Two Ncp strains of BVDV 2 were used in the current study: a high virulent 1373 and, a low virulent 28508-5, in addition to Cp 296C strain as control for apoptosis [30]. All strains were titrated using TCID50 titration assay [31].

BVDV-free Madin Darby bovine kidney MDBK cells (passage 113–135) were grown in minimal essential medium (MEM, Gibco BRL, Grand Island, NY, USA) (pH 7–7.4) supplemented with 10% BVDV-free fetal bovine serum (FBS) (Hyclone Laboratories, Logan, UT, USA), penicillin (100 U/mL) and streptomycin (100 µg/mL). MDBK cells were maintained at 37 °C with 5% CO_2 in humidified incubator and were used for viral titration and propagation as well as apoptosis experiments.

Bovine lymphosarcoma cell line (Bl-3) is a non-adherent, BVDV-free B cell line clone derived from BL-3.1 cell line that was obtained from ATCC, Manassas, VA, USA. BL-3 cell line was cultured as previously described [3] and they were used to examine the direct and indirect apoptotic effect of different BVDV strains.

2.2. Isolation of Monocytes and Differentiation to Monocyte Derived Macrophage (MDM)

PBMCs were isolated as previously described [32,33] with modifications. Briefly, the heparinized blood was diluted 1:1 with PBS and overlaid over histopaque 1083 gradient using SepMate™ 50 mL tubes (Stemcell Technologies, Cambridge, MA, USA) and centrifuged at 1200× *g* for 20 min at room temperature (RT). The buffy-coat was then transferred to a clean 50 mL conical tube and washed five times with PBS, followed by centrifugation at 120× *g* for 10 min at RT. The viability of PBMCs was determined by trypan blue exclusion assay according to Strobber [34]. The PBMCs were suspended in RPMI 1640 medium (GE Healthcare, Hyclone Laboratories, Logan, UT, USA) supplemented with 10% FBS, penicillin (100 U/mL) and streptomycin (100 μg/mL) to achieve a final concentration of 1×10^6 cells/mL. The cells were incubated in T175 flasks for 3 h at 37 °C. Then, the adherent monocyte was washed with PBS five times and detached by incubation for 10 min with Accutase™ (eBioscience, San Diego, CA, USA). Detached monocytes were PBS-washed twice to get rid of Accutase. The isolated monocyte cultured in complete RPMI 1640, as described by Elmowalid [29], at a concentration of 10^5 cells/well in 48-well plate, followed by incubation for 5 days at 37 °C. The incubated cells were fed every other day by replacing half of the conditioned media with fresh complete RPMI. At day 5, the MDM were characterized phenotypically as MHCI-, MHCII-, CD11b- and CD14-positive cells.

2.3. Production and Inactivation of Infected MDM Supernatant

The MDM were infected with 100 μL of 10^5 TCID50 BVDV strains at MOI of 1 in triplicate as described by Elmowalid [29], with modifications. The infected cells were incubated for 1 h, then washed to remove the excess unbound virus, and 500 μL of complete RPMI 1640 medium was added to each well in a 24-well plate. At least one column of the plate was mock-infected with complete RPMI 1640 medium as a negative control. The infected MDM were incubated for 12, 24 or 48 h at 37 °C in CO_2 incubator. The BVDV-infected macrophage supernatants were collected at 12, 24 and 48 hpi (hours post-infection) and centrifuged at 1000× *g* for 10 min at RT to remove cellular debris. The supernatant was UV-inactivated for 20 min on ice to exclude the direct virus effect [35]. The absence of any infectious viral particles in the treated supernatants was confirmed by inoculation on MDBK cells followed by a 5-day incubation and BVDV specific immune-staining of the inoculated MDBK cells using both immune-peroxidase and immunofluorescence. The positive control was 1373 infected MDBK cells.

2.4. Phagocytosis

One mL of virus-free (UV-inactivated) supernatants collected at 24 or 48 hpi from Ncp1373 or 28508-5 BVDV strains or mock-infected MDM were used to treat MDM, cultured on 24-well plates (approximately 5×10^5 cells/well whose viability was >92%), for 24 h. This was followed by washing the treated cells twice with PBS and then exposed to 250 μL containing approximately 2×10^7 of TRITC-labeled *C. albicans* (50 yeast/macrophage) and incubated for 30 min at 37 °C. Finally, the cells were washed twice in cold PBS and re-suspended in 200 μL/well freshly prepared paraformaldehyde (PFA) to be examined under UV-microscopy. The number of yeast/cell counted and the cells were classified into two groups: cells that contained >20 TRITC labeled yeast/cell that indicated normal phagocytic activity and cells that contained <20 TRITC labeled yeast/cell that indicated insufficient phagocytic activity. A total number of 100 MDM containing yeast were counted and the percentage of phagocytic activity was calculated according to the following formula: Phagocytic% = (number of MDM containing >20 yeasts/total number of MDM) × 100.

2.5. Bactericidal Activity

MDM were treated with 1 mL/well of UV-inactivated supernatant from MDM infected with 1373, 28508-5 or mock-infected 24 and 48 h supernatant, and incubated for 24 h at 37 °C. The cells were then washed twice in PBS and incubated for 30 min or more (up to 120 min) with 1×10^7 *E. coli* (20 bacteria/macrophage) suspended in RPMI 1640 supplemented with 5% FBS and no antibiotics to allow bacterial phagocytosis. One group of the cells were washed three times with cold PBS 30 post-incubation to stop phagocytosis and to remove excess bacteria (time 0) and were lysed and the lysates were centrifuged at 800× *g* for 10 min and the supernatants were aspirated, while the pellets were suspended in LB broth and stored at 4 °C. The other group of cells were incubated in 5% FBS, and no antibiotics for extra 90 min to allow killing (total 120 min) followed by washing and lysis, as described above. Cell lysates of both 0 and 90 min timepoints were placed in 96-well plates and incubated for 18 h at 37 °C, followed by adding 30 μL/well of of 50 mg/mL 3-(4,5-dimethylthiazol-2-yl)-2,5-diphenyl tetrazolium bromide (MTT, Sigma, St. Louis, MO, USA) and the plates were incubated at 37 °C for another 6–8 h. Optical density was measured at 540 nm wavelengths in ELISA reader (Biotek, ELx808, Winooski, VT, USA). The percentage of killing was calculated by the following formula: Killing% = (absorbance after incubation for 90 min/Absorbance at time 0) × 100.

2.6. Nitric Oxide Production

Nitric oxide (NO) assay was done in 24-well plates. MDM were treated by adding 1 mL/well of UV-treated supernatant from 1373 or by direct infection with 1373 strain of BVDV (MOI of 1) kept with the cells. Mock-infected supernatant or RPMI 1640 media containing 5% FBS were used as a control. The cells were incubated for 24 h at 37 °C. The cells were then stimulated with lipopolysaccharide (25 μg/mL) (LPS, Sigma, St. Louis, MO, USA) for another 24 h. Then, supernatants were collected and nitric oxide (NO) concentration was determined using Griess reagent (Sigma, St. Louis, MO, USA) using Nitrite (NO_2) measurement as an indicator of NO production according to the protocol of Blond et al., 2000: briefly, 100 μL of culture supernatant was added to 100 μL of Griess reagent, made of a 1/1 mixture of 1% (wt/vol) sulfanilamide and 0.5% (wt/vol) *N*-(1-naphthyl)ethylenediamine dihydrochloride (Sigma, St. Louis, MO, USA) in 30% acetic acid, in each well of a 96-well plate. Reactions were performed in triplicate at RT for 10 min. Chromophore absorbance was then measured at 550 nm in a microplate reader (Bio-TEK Instruments model EL 311; OSI, Paris, France). Nitrite concentration was evaluated by comparison with the nitrate standard curve (Sigma, St. Louis, MO, USA). The lower limit of detection for nitrite is 250 nM.

2.7. Immunostaining and Flow Cytometry of CD14 and MHC II

To examine the effect of BVDV direct infection or its supernatant on macrophage surface marker expression, the MDM were either infected or treated with different BVDV-48 h-supernatants: including 1373 and 28508-5 strains. This was in addition to the supernatant from mock-infected MDM. Surface marker expression was investigated 48 hpi, starting with using Accutase™ (eBioscience, San Diego, CA, USA) to detach the MDMs that were washed twice in PBS and plated in a U-bottom-96-well plate by adding100 μL of cells (2 × 10^5 cell) per well. Cells were then suspended in blocking buffer (50 μL/well) (PBS containing 2% FCS and 0.1% sodium azide) for 20 min at RT. 50 μL/well of mouse anti-bovine CD14, or MHC II (VMRD, Pullman, WA, USA) diluted at 1:200 in PBS was added and the cells incubated for 20 min at 4 °C, followed by washing three times in PBS. Subsequently, the cells were incubated with FITC-conjugated goat anti-mouse Ig G (Ig G whole molecule, ICN/Cappel laboratories, Santa Mesa, CA, USA) diluted at 1:1000 in staining buffer for 20 min in dark at 4 °C, followed by washing three times in PBS. Fifty (50) μL/well of propidium iodide (1 μg/mL) were added directly before analysis to stain the dead cells. Both cell control (without staining) and the FITC control (just FITC staining without the primary anti-bovine antibodies) were included to eliminate nonspecific background. For each sample, mean fluorescent intensity of immune-stained cells was estimated and analyzed on FACScan flow cytometry (Becton Dickinson, Mountain View, CA, USA) using CELL Quest software (BD Biosciences).

2.8. Examining the Indirect or Direct Effect of BVDV Strains on Peripheral Blood Lymphocyte Population and MDBK

To examine the indirect effect of BVDV strains, peripheral blood lymphocytes, BL-3 or MDBK were treated with various time-point irradiated supernatants from BVDV-infected macrophages at a dilution of 1:1 and incubated for 12, 24 or 48 h post treatment, as previously described [36]. To examine the direct effect of BVDV strains, peripheral blood lymphocytes or BL3 were infected with BVDV strain at MOI of 1, as described by Ridpath et al. [37]. The effect of the direct infection of lymphocyte with Cp 296C strain was included as a positive apoptosis control.

2.9. Chromatin Condensation

The apoptotic effect of the infected MDM supernatant on epithelial cells was investigated by incubating MDBK cells for 12, 24, or 48 h at 37 °C with 1 mL/well of UV-treated supernatant from 1373, or 28508-5 as well as the non-infected supernatant. The cells were then washed twice in PBS, and then stained using 4,6-diamidino-2-phenylindole (DAPI, Sigma, St Louis, MO, USA) according to the protocol of Mi Hyeon et al., 2002. Simply, one volume of pre-staining solution [citric acid (2.1 g), Tween 20 (0.5 mL), and distilled H_2O (100 mL)] was added and the cells were incubated for 5–7 min at RT in the dark. After the incubation, another six volumes of staining solution [(citric acid (2.1 g), DAPI (0.2 mg/mL), 1 mg DNAse, and distilled H_2O (100 mL)] were added for another 5–7 min. Cells were washed three times in PBS and examined under fluorescence microscopy. Two hundred cells (200) were counted and the mean value was calculated. The apoptotic index was the mean of three independent experiments.

2.10. Annexin V Staining

Both direct BVDV infection or the indirect effect of its supernatant on bovine lymphocytes and MDBK cells was investigated using the AnnexinV staining kit (eBioscience, San Diego, CA, USA). Briefly the infected or supernatant-treated lymphocytes were washed once with PBS and the cells were re-suspended in 100 µL of 1x binding buffer supplied by the kit. Five µL of FITC conjugated annexin v antibody added to the lymphocytes suspended in 100 µL of binding buffer and incubated for 10 min in the dark. This was followed by washing the cells twice by suspending the cells in 1x binding buffer followed by adding another five µL of propidium iodide to exclude the dead cells. Finally, the cells were re suspended in 200 µL of 1x BD FACSTM lysing solution followed by analysis with BD Accuri™ C6 Plus Flow Cytometer (BD Biosciences, San Jose CA, USA). The result was expressed as % of apoptosis.

2.11. Protein Electrophoresis

The BVDV-infected MDM supernatant was fractionated on 10% sodium dodecyl sulphate polyacrylamide gel (SDS-PAGE) according to the protocol of Schagger and von Jagow, 1987. Briefly, 40 µL of the supernatant (around 50 µg) were added to 10 µL of 5x sample buffer (65 mM Tris-HCL, pH 7.0, 2% SDS, 10% glycerol, 5% β mercaptoethanol, and 0.001% bromophenol blue) with heating at 100 °C for 3–5 min. Protein electrophoresis was then done by loading the treated sample (50 µL) on the SDS-PAGE gel at 25 mA current for 45–60 min followed by staining with Coomassie blue overnight (in some experiments for 4 h) with shaking at RT. Finally, the gel was immersed in de-staining buffer (40% methanol and 10% glacial acetic acid) with shaking for 15 min at RT until the protein bands become clear [38].

2.12. Quantification of Apoptosis-Related Cytokines by qRT-PCR of BVDV-Infected MDMs

Both the infected and mock-infected MDM were pelleted at 500× *g* for 8 min, then washed twice in PBS. The infected MDM were lysed and nucleic acid (NA) extraction done using RNeasy extraction kit (Qiagen, Valencia, CA, USA). The extracted NA of different samples was normalized to 5 ng/µL using Nanodrop ND-1000 Spectrophotometer (Fisher Scientific, Portsmouth, NH, USA). Nucleic acid extracted from mock-infected MDM with complete RPMI 1640 medium was used as a negative control while NA extracted from MDM treated with Concanavalin A (ConA) or lipopolysaccharide (LPS) (Sigma-Aldrich, St. Louis, MO, USA) was used as a positive control. Around 5 μL of the normalized NA was used for relative quantification of apoptosis-related cytokines; TNF-α, IL-1α, IL-1β and IL-6 for each sample in duplicates using quantitative reverse transcriptase PCR (qRT-PCR) and Power SYBR® Green RNA-to-CtTM 1-Step Kit (Thermo Fisher Scientific, Millersburg PA, USA). Additional cytokines were also measured that include: IFN–α, β and γ, IL-4, 8, 10, and 12. The cytokine primers used were described in Table 1 [39]. The relative expression of mRNA was standardized using beta actin and GAPDH as housekeeping genes. Each qPCR experiment was followed by heat dissociation curve step to exclude nonspecific amplification. qRT-PCR results were analyzed using relative expression software tool (REST©2009 software) [40].

Table 1. The set of primers used for quantifications of apoptosis-related cytokine mRNA.

Cytokine	Forward Primer	Reverse Primer	Annealing Temp °C
TNF-α	AGACCCCAGCACCCAGGACTCG	GGAGATGCCATCTGTGTGAGTG	55
IL-1α	GATGCCTGAGACACCCAA	GAAAGTCAGTGATCGAGGG	53
IL-1β	CAAGGAGAGGAAAGAGACA	TGAGAAGTGCTGATGTACCA	53
IL-6	TCCAGAACGAGTATGAGG	CATCCGAATAGCTCTCAG	52
β-actin	CGCACCACTGGCATTGTC	TCCAAGGCGACGTAGCAG	55
IFN-α	ACACACACCTGGT	GATGACAGCAGAAATGA	52
IFN-β	RTCTGSAGCCAAT	CAGGCACACCTGT	52
IFN-γ	ATAACCAGGTCATTCAAAGG	ATTCTGACTTCTCTTCCGCT	55
IL-8	TGGGCCACACTGTGAAAAT	TCATGGATCTTGCTTCTCAGC	53
IL-10	TGCTGGATGACTTTAAGGG	AGGGCAGAAAGCGATGACA	53
IL-12	GAGGCCTGTTTACCACTGGA	CTCATAGATACTTCTAAGGCACAG	58
IL-4	TGCATTGTTAGCGTCTCCTG	AGGTCTTTCAGCGTACTTGT	56

2.13. Supernatant Neutralization

To exclude the apoptotic effect induced due to possible viral secreted proteins, 24 or 48 h supernatants from the BVDV-infected macrophages with 1373 strain were incubated at 37 °C for 1 h with Erns specific 15c5 monoclonal antibody (IDEXX Laboratories, Westbrook, ME, USA) or with BVDV-polyclonal antibody (kindly supplied by Dr. Robert Fulton, Food Animal Research, Oklahoma State University). The mixture of the infected supernatant and antibody was prepared as two parts of infected supernatant incubated with one part of concentrated antibody. Three (3) parts of the supernatant/antibody mixture were added to one part of BL-3 cell line seeded at 1×10^5 cells/well, followed by incubation at 37 °C for 36 h post-treatment.

2.14. Statistical Analysis

Data were analyzed using a Student's *t*-test (Microsoft EXCEL, MAC 2011) to assess the significance of the differences between mean values of treated and control samples at each timepoint. Differences were considered significant at $p < 0.05$, however some treatments showed very significant difference with p value of <0.01. Every experiment was achieved using at least three different animals and each experiment was done in triplicates. The variations in results were calculated by standard deviations at each timepoint. For cytokine analysis, REST© 2009 [40], that is based on analysis of variance (AVOVA).

3. Results

3.1. Inactivation of Infected Supernatant

The irradiated supernant-treated MDBK cells had no immune-peroxidase stain for the UV-irradiated supernant (Figure 1A) compared to the positive intracytoplasmic staining of 1373-infected-MDBK non-irradiated control (Figure 1B). There was no fluorescence signal for the irradiated strain (Figure 1C) compared to the non-irradiated infection control (Figure 1D).

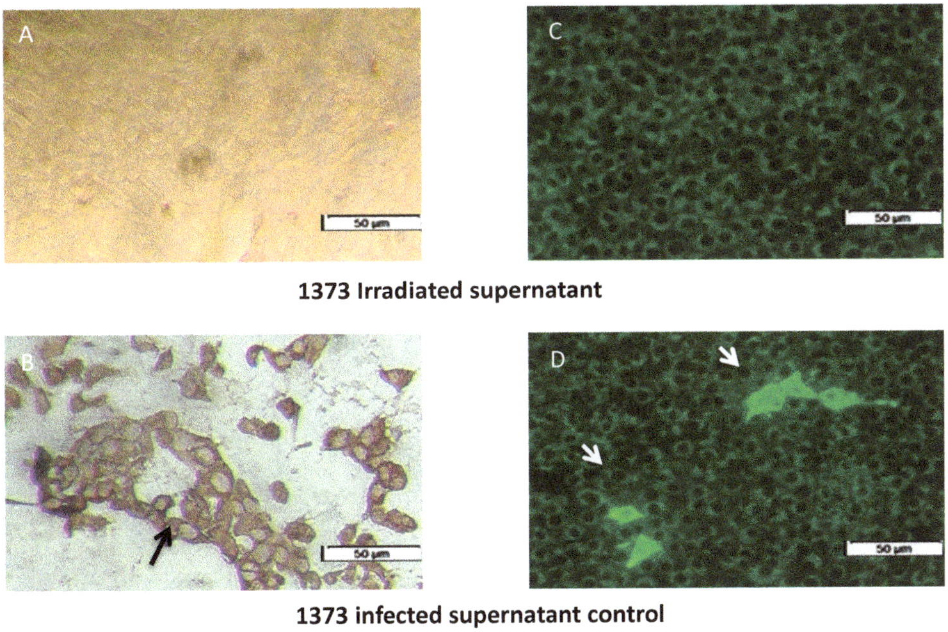

Figure 1. Successful inactivation of 1373-infected monocyte-derived macrophages (MDM) supernatant by UV-irradiation. (**A**,**C**) irradiated 1373-supernatant that show no evidence of intra-cytoplasmic replication by immune-peroxidase (IP) and immune-fluorescence (IF). (**B**,**D**) intra-cytoplasmic replication signal of the infected MDM supernatant, indicated by black and white arrows in the IP and IF assay, respectively (**B**,**D**).

3.2. Effect of Supernatant from BVDV Infected MDM on Macrophage Phagocytic Activity Compared to the Direct BVDV Effect

At 24 h post-treatment, the supernatant from Ncp 1373 significantly reduced MDM phagocytic activity by 38.5%, which continued to decrease to 51% at 48 h post-treatment compared to the supernatant from Ncp 28508-5, which was similar to the mock-infected control at any timepoint (Figure 2). Direct infection of MDM with the highly virulent 1373 strain, but not 28508-5, diminished phagocytic activity of MDM as early as 12 h post-treatment by 27.5% (Figure 2). We also found that heat-treated supernatant from 1373, or 28508-5 BVDV-infected supernatant had no effect on MDM phagocytic ability (data not shown).

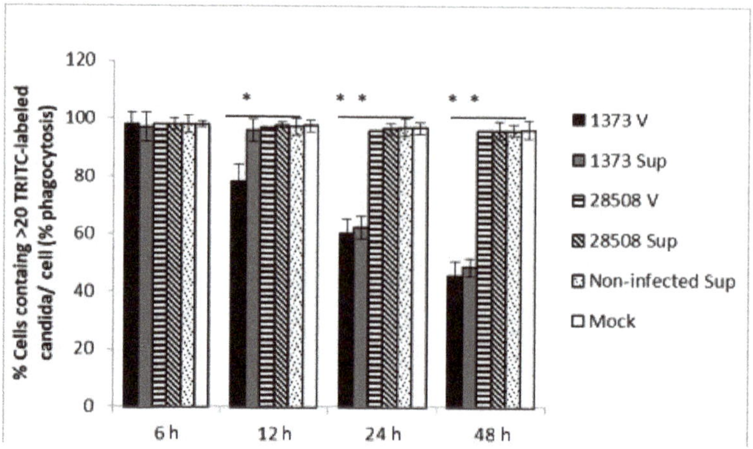

Figure 2. The effect of BVDV strains and the corresponding infected-MDM supernatants on macrophage phagocytic activity. Phagocytic index is defined as the % of MDMs that contains >20 TRITC-labeled *C. albicans*. h: hours post-treatment, V: direct virus infection, no supernatant, Sup: BVDV-infected macrophage supernatant, non-infected Sup: non-infected macrophage supernantant, Mock: culture medium *: $p < 0.05$ (>95% confidence), **: $p < 0.01$ (>99% confidence).

3.3. *Effect of Supernatant from BVDV-Infected MDM on Macrophage Bactericidal Activity Compared to the Direct BVDV Effect*

No significant effect was observed at 6 or 12 h post-treatment, however, the supernatant from Ncp 1373 strain significantly ($p < 0.05$) reduced MDM bactericidal activity by 37.7% and 51.4% at 24 and 48 h post-treatment, respectively (Figure 3). Direct infection of MDM with 1373 decreased the bactericidal activity by 22%, which began as early as 12 hpi, unlike the late indirect effect. The inhibition of bactericidal activity was time-dependent, and continued to increase to reach 54.7% by 48 hpi. There was no significant effect with 28508-5 strain, its corresponding infected-MDM supernatant, mock-infected, or the non-infected supernatant on bactericidal activity (Figure 3).

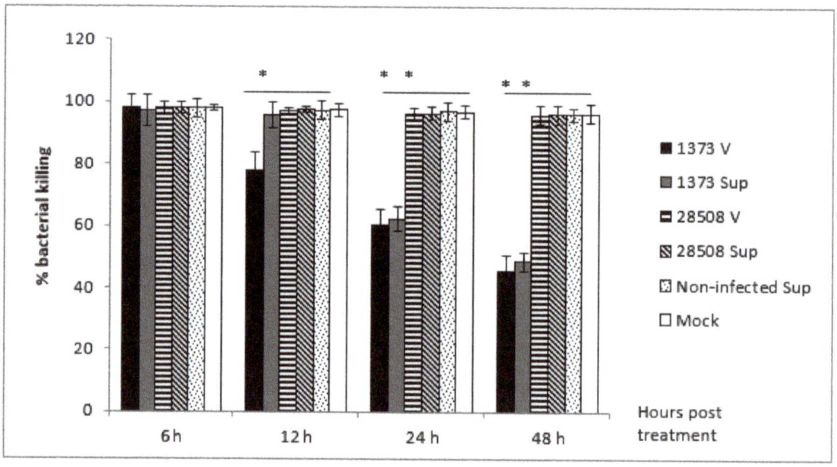

Figure 3. The effect of BVDV strains and the corresponding infected-MDM supernatants on macrophage bactericidal activity for 48 h post-treatment. Sup: supernatant (indirect effect), V: virus (direct effect). *: $p < 0.05$ (>95% confidence), **: $p < 0.01$ (>99% confidence).

3.4. Effect of Supernatant from BVDV Infected MDM on Macrophage Nitric Oxide Production Compared to the Direct BVDV Effect

The direct infection of MDM with both BVDV strains significantly increased NO production (158–164 μmol/mL), while neither 1373- nor 28508-5-infected MDM supernatant affected NO production in uninfected MDM (Figure 4). Both supernatant treatments, as well as non-infected supernatant and mock-infected controls, showed low nitric oxide production (93.3 μmol/mL) (Figure 4).

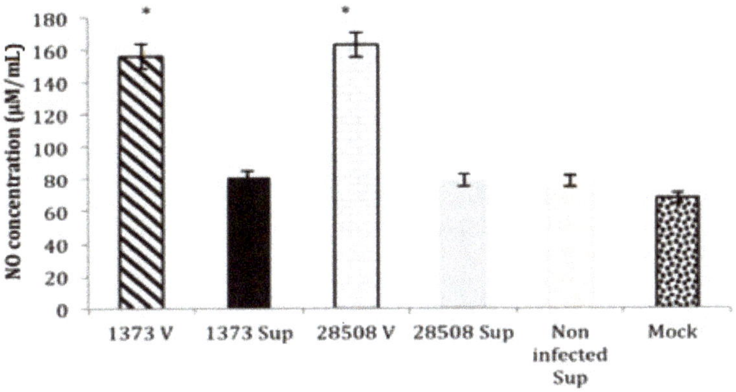

Figure 4. Comparison between the direct effect of BVDV strains and the corresponding infected-MDM supernatants (the indirect effect) on macrophage nitric acid production 24 h post-treatment. Sup: supernatant (indirect effect), V: virus (direct effect). *: $p < 0.05$ (>95% confidence).

3.5. Effect of Supernatant on MHC II Expression Compared to the Direct BVDV Effect

At 24 h post-treatment, cells treated with 1373 supernatant had significant reduction in MHC II expression to 61.0% (Figure 5A). At 48 h, MHC II expression was further reduced to 50.0% (Figure 5A) In contrast, 28508-5 or mock-infected supernatant had no significant reduction in MHC II expression and the expression level was greater than 94.0% at 24 or 48 h (Figure 5A).

3.6. Effect of Supernatant on CD14 Expression Compared to the Direct BVDV Effect

At 24 h post-treatment, the CD14 expression decreased in MDM infected with Ncp 1373 strain or it's supernatant (Figure 5B). At 36 h post-treatment, CD14 expression further decreased to 44.6% and 49.7% for Ncp 1373 and its supernatant, respectively (Figure 5B). The heat-treated supernatant or supernatant of 28508-5 did not show a significant decrease in the percentage of cells expressing CD14 and the average percentage was over 93% at 24 or 36 h (Figure 5B).

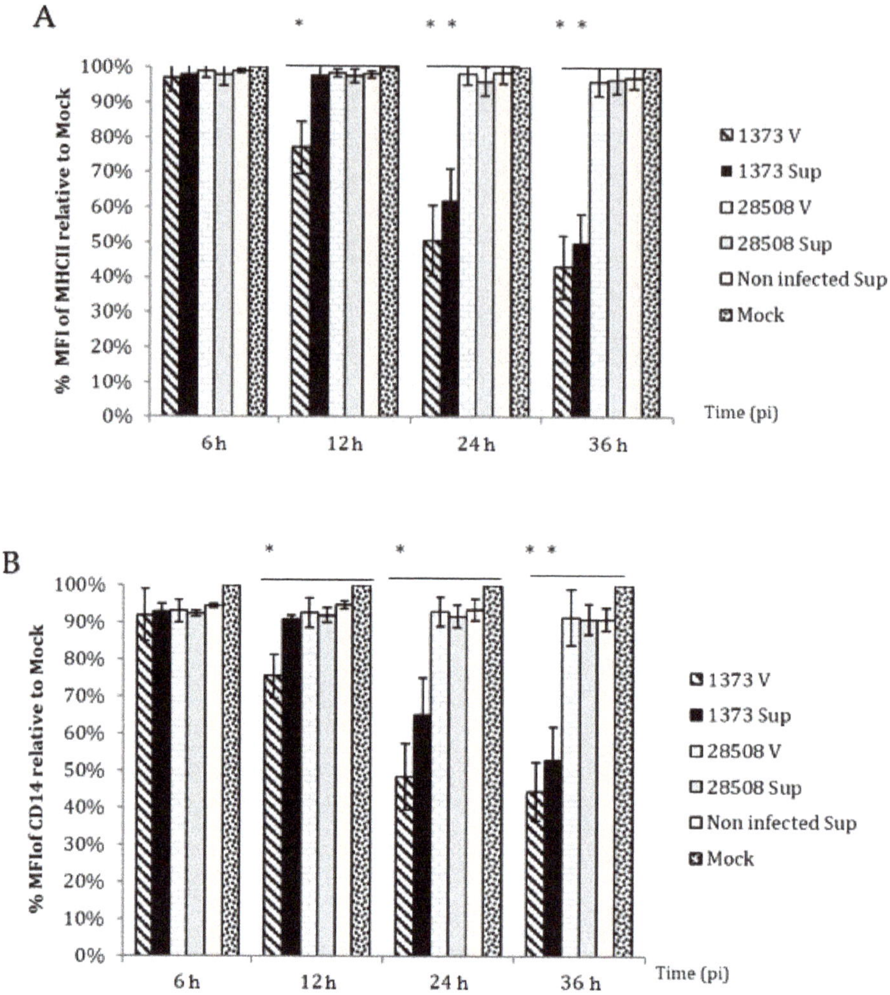

Figure 5. The effect of BVDV strains and the corresponding infected-MDM supernatants (indirect effect) on surface marker expression. (**A**) MHC II and (**B**) CD14 surface marker expression. Sup: supernatant (indirect effect), V: virus (direct effect). *: $p < 0.05$ (>95% confidence), **: $p < 0.01$ (>99% confidence).

3.7. Effect of Supernatant from BVDV Infected Macrophages on MDBK Cells

Chromatin condensation was detected after 24 h in treated MDBK with 1373 supernatant, while no chromatin condensation was detected in MDBK treated with 28508-5 nor non-infected supernatant (Figure 6A). The percentage of apoptotic cells was not significant until 24 h post-infection using Annexin V assay. It increased from 36.3% (24 h post-treatment) to reach 51.3% (48 h post-treatment) in contrast to less than 3% in case of mock-infected or treated MDBK with either 28508-5 or non-infected supernatant (Figure 6B).

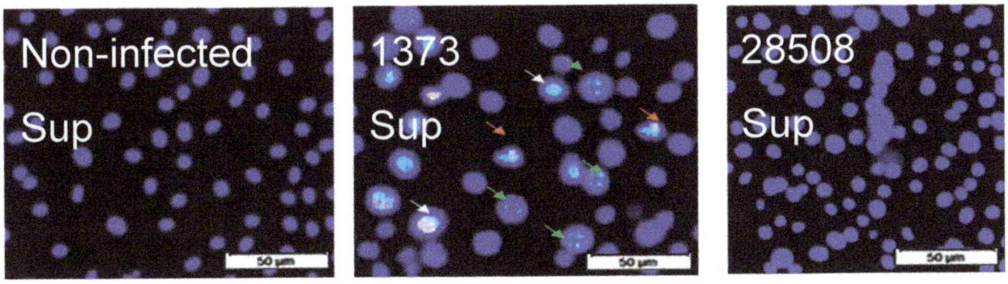

(A)

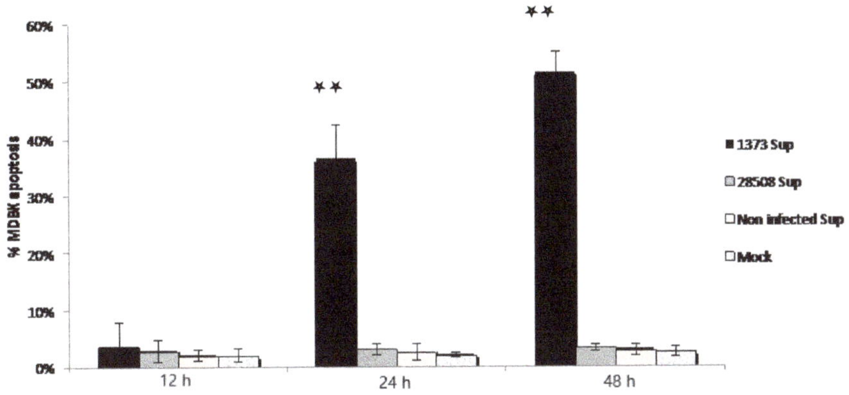

(B)

Figure 6. The indirect (infected MDM 24 h supernatant) effect of different BVDV strains on MDBK apoptosis. (**A**) Chromatin condensation by 1373 strain in contrast to 28508-5 and non-infected supernatant. (**B**) The % of MDBK apoptosis in response to different supernatants. 1373 Sup: BVDV 1373-infected macrophage supernatant, 28508-5 Sup: BVDV 28508-5-infected macrophage supernatant Non-infected Sup: non-infected macrophage supernantant, Mock: culture medium **: $p < 0.01$ (>99% confidence).

3.8. Effect of Supernatant from BVDV-Infected Macrophages on Different Lymphocyte Populations

The 12 h supernatants did not induce apoptosis compared to the mock-infected control (Figure 7). Both 24 and 48 h supernatants of the highly virulent 1373 resulted in an apoptotic effect on either freshly isolated (data not shown) or BL-3 lymphocytes. All supernatants of the low virulent Ncp 28508-5 at different timepoints did not induce apoptotic changes compared to the mock-infected control for either BL-3 lymphocytes or peripheral blood total lymphocyte population (Figure 7A,B).

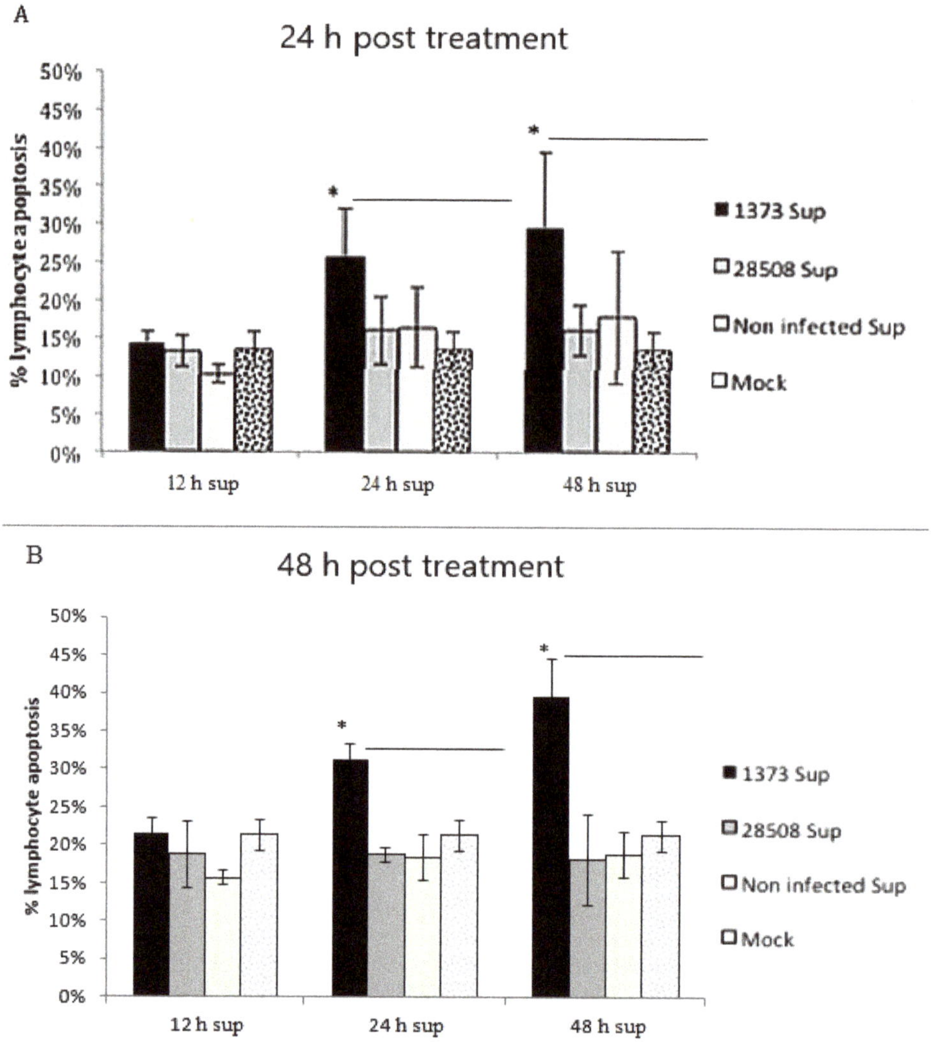

Figure 7. The indirect (infected MDM 12, 24 and 48 h supernatant) and the direct effect of different strains of BVDV on lymphocyte apoptosis. (**A**) The indirect effect 24 h post-treatment (**B**) The indirect effect 48 h post-treatment 1373 Sup: BVDV 1373- infected macrophage supernatant, 28508-5 Sup: BVDV 28508-5-infected macrophage supernatant non-infected Sup: non-infected macrophage supernatant, Mock: culture medium, *: $p < 0.05$ (>95% confidence).

3.9. Effect of Direct Infection of Lymphocyte with BVDV Strains

Only the Cp 296C strain was able to induce significant lymphocyte apoptosis as early as 12 h post-infection compared to the non-infected/mock control and the Ncp BVDV strains (Figure 8A) The direct infection of lymphocytes with 1373 was compared to the indirect effect of this virulent BVDV strain. Interestingly, the direct infection of lymphocytes did not induce significant changes compared to the mock-infected control until 48 h post-infection with 21.8% as the maximum apoptotic effect. The indirect effect of 1373 was tow times greater than the direct infection of lymphocyte at 24 and 48 hpi, with a maximum apoptotic effect of 35% at 48 h post-treatment (Figure 8B).

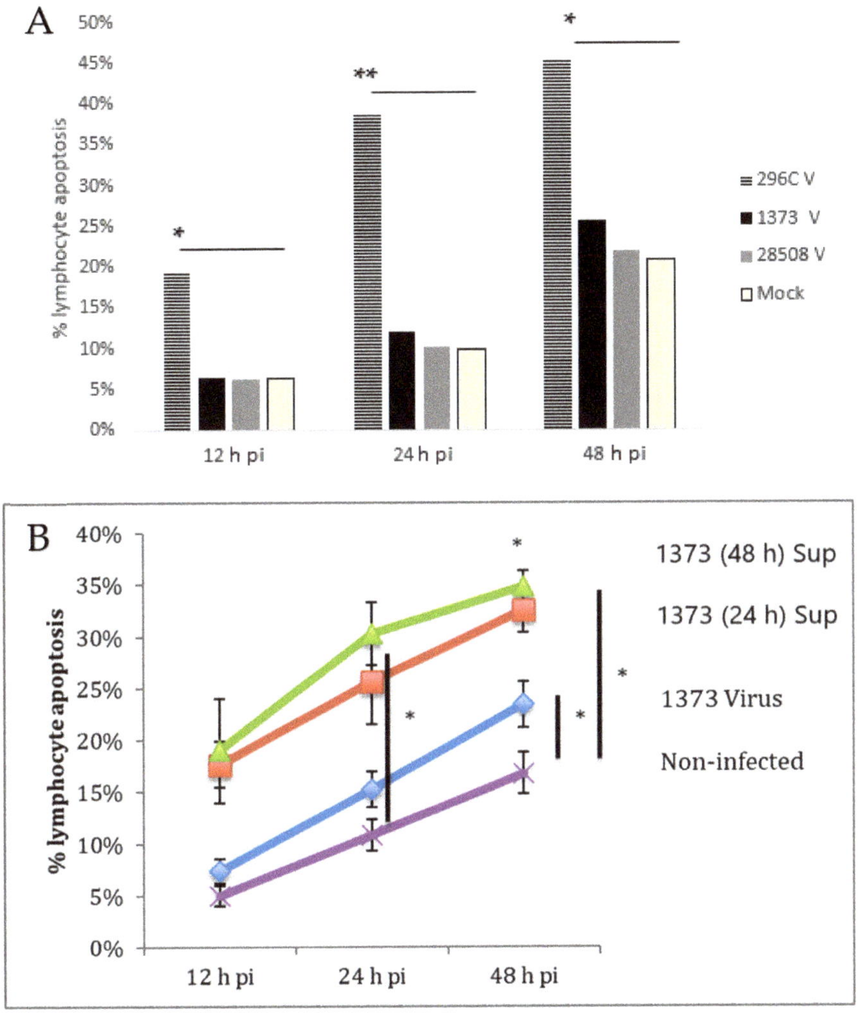

Figure 8. (**A**) The direct effect of BVDV strains on lymphocyte apoptosis (**B**) Comparison between the direct infection and the indirect (infected MDM 24 and 48 h supernatant) effect of the highly virulent 1373 BVDV strain on lymphocyte apoptosis for up to 48 pi. 296C V: direct lymphocyte infection with Cp 296C strain, 1373 V: direct lymphocyte infection with Ncp 1373 strain, 28,508 V: direct lymphocyte apoptosis with Ncp 28508-5 strain, Mock: non-infected lymphocyte, 1373 (48 h) Sup: supernatant from 1373-infected macrophage at 48 h post-treatment, 1373 (24 h) Sup: supernatant from 1373-infected macrophage at 24 h post-treatment, 1373 Virus: direct infection of macrophage with 1373 strain (not supernatant), Non-infected: supernatant from macrophage that was not infected, *: $p < 0.05$ (>95% confidence), **: $p < 0.01$ (>99% confidence).

3.10. Supernatant Protein Analysis

No viral proteins were detected in the supernatants at 6 hpi (Figure 9A), however, faint bands of 40–45 KD started to appear in the 1373-infected MDM supernatant at 12 hpi (Figure 9B). By 24 hpi, bands of 30 and 120 KD were observed in supernatant from Ncp1373. MDM culture infected with 28508-5 showed only one dense band of about 40–45 KD (Figure 9C). At 48 hpi, the band numbers and densities increased in MDM infected with the Ncp 1373. In contrast, only two bands of 40 to 45 KD were observed in supernatant from 28508-5 Ncp BVDV strains by that time (Figure 9D). No bands were observed in supernatant from mock-infected MDM at 6, or 12 h. At 24 or 48 hpi, only one band of 40–45 KD was observed in the mock-infected cells supernatant (Figure 9).

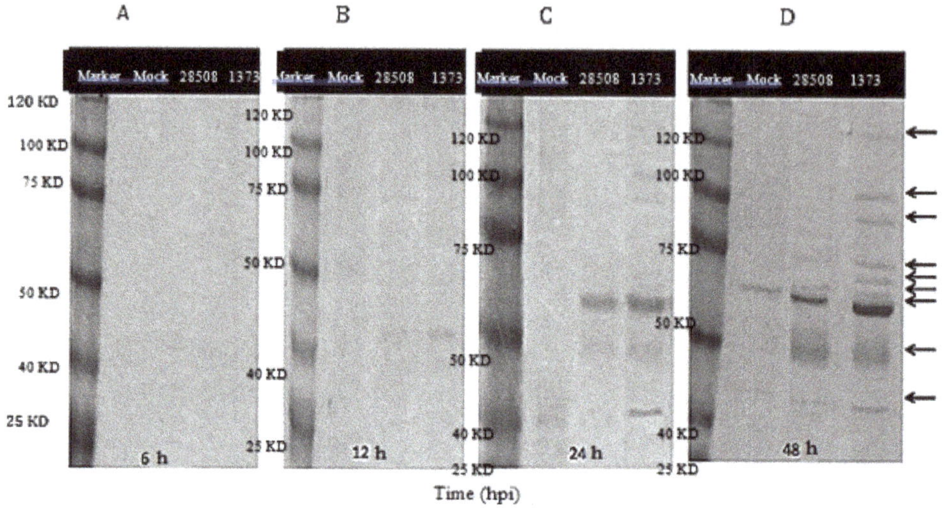

Figure 9. Protein electrophoresis of supernatants from BVDV infected and mock-infected MDM at (**A**) 6 hpi, (**B**) 12 hpi, (**C**) 24 hpi and (**D**) 48 hpi. Marker: protein marker that range from 25 to 120 KD, Mock: non-infected macrophage control, 28508: supernatant from 28508-5-infected macrophage at different timepoints, 1373: supernatant from 1373-infected macrophage at different timepoints, 6, 12, 24, 48 hpi: the corresponding timepoints post treatment of macrophage with supernatant, arrows refer to the different band intensity between strains.

3.11. Role of Cytokines in the Indirect Lymphocyte Apoptosis

Apoptosis-related cytokines including TNF-α, IL-1β and IL-6, did not show significant change at any of the three timepoints compared to mock-infected control. Other cytokines like IFN-γ, IL-10, IFN-α, IFN-β, IL-12, IL-4, IL-8 and TGF-β have also shown no significant difference between groups. The positive control mixture of Concavalin A and LPS upregulated all of the cytokines (data not shown).

3.12. Role of Viral Factors in the Indirect Lymphocyte Apoptosis

There was no significant difference in the percentage of lymphocyte apoptosis between the neutralized supernatant with either mAb or polyclonal Ab and the un-neutralized supernatant (data not shown).

4. Discussion

This study sheds light on the role of macrophages in the pathogenesis of BVDV infection. Phagocytosis is an important immune response of the macrophage and constitutes one of the first defenses against microbial invasion. Successful phagocytic activity leads to the initiation of successful immune response. In this study, supernatant from the highly virulent 1373 strain reduced macrophage phagocytic activity. In a previous study, Ncp BVDV infection of alveolar macrophage decreased expression of complement and Fc receptors and chemokine production, reducing their ability to engulf opsonized pathogens [8]. IL-2 inhibitory substances are produced in BVDV-infected PBMCs cultures that bind to receptors on lymphocytes, macrophages and dendritic cells that activate protein kinase C, resulting in the phosphorylation of proteins and inhibition of basic metabolic activities [41], that may reduce cell movement and the capacity to engulf organisms. A reduction in macrophage phagocytic capacity may be due to the presence of chemokine inhibitory mediators or the decrease in cytokine-induced chemotaxis [42] that lead to the decreased capacity of phagocytic cells for chemotaxis and phagocytosis.

Microbicidal activities are one of the main functions of macrophage to control and clear microbial infection. In vivo experiments suggest that BVDV immunosuppression resulted in subsequent secondary microbial infection. The supernatant from low virulent 28508-5-infected macrophage had no effect on bactericidal activity, while the highly virulent 1373 supernatant decreased the bactericidal activity of macrophages significantly with more than 50% by 48 hpi ($p < 0.05$). Considerable evidence has accumulated that BVDV may be a key component in multiple-etiology diseases [43].

Nitric oxide production was not affected by the supernatant exposure. Direct infection of MDM with highly virulent 1373 strain significantly induced NO production; however, 1373-infected macrophage supernatant (indirect) had no significant effect on NO production in MDM following stimulation with LPS. Another flavivirus, the hepatitis C virus (HCV) ERNS protein and HIV infection was associated with increased NO synthase (NOi) production in mice livers [44,45]. Type I IFN is a potent inducer of NO; however, it is downregulated by Ncp BVDV strains [25,46]. Taking these results together, Ncp BVDV might induce NO production directly through ERNS protein activation to NOi, but not indirectly due to the lack of secreted IFN in the MDM supernatant.

Microbial recognition, processing and presentation to lymphocytes is one of the unique functions of antigen-presenting cells for initiation of immune response. The recognition of pathogens is achieved through surface markers, which initiate the immune response. Our results indicated a significant reduction in surface marker expression of MHC II and CD14, associated with the highly virulent 1373-infected MDM supernatant. MHC II is an important marker for the initiation of immune response and antigen presentation. The current study indicated that the highly virulent 1373 reduced the surface expression of MHC II molecules. A previous study demonstrated that MHC II expression decreased in PBMCs infected with Ncp BVDV [47]. Another study reported that BVDV infection downregulated MHC I and II in monocytes [48]. This is consistent with what we found but in macrophage.

CD14 is important for toll-like receptors and gram-negative recognition and phagocytosis. Our findings indicated that low virulent 28508-5 supernatant had no effect on CD14 expression, while the highly virulent 1373 supernatant significantly decreased CD14 expression. CD14 downregulation was associated with the reduction in the phagocytic activity of neutrophils in HCV patients [49]. In human and mouse models, downregulation of CD14 was associated with suppression of antigen-specific lymphocyte proliferation [50]. Impairment of macrophage inflammatory function is a likely consequence of this surface reduction with subsequent secondary microbial invasion and the immunosuppressive effect of BVDV on the infected animals.

The findings of this study indicate the importance of soluble factors induced in BVDV-infected macrophages in viral pathogenesis. These results demonstrate that only the highly virulent Ncp BVDV strain induced the production of soluble factors capable of impairing macrophage inflammatory functions and surface marker expression as early as 12 h post-infection. These mediators and factors also induced epithelial cell and lymphocyte apoptosis. The release of several BVDV induced-soluble factors release has been reported [8,51–53]. On gel electrophoresis, several proteins ranging from 30 to 120 kDa were produced in 1373-infected MDM supernatants that were either less intense or not present at all in 28508-5-infected MDM supernatant. In a previous study, inflammatory mediators of 40–75 KD play a pivotal role in viral infection and pathogenesis [54].

Another study indicated that PMNC infected with Ncp or Cp released unknown factor(s) in the culture media that could be of cellular or viral origin [52,55]. Previously, secreted Erns viral glycoprotein played an important role in lymphocyte apoptosis induced by BVDV and CSFV [19]. In our study, we found no association of Erns glycoprotein or any other viral factors in the induction of apoptosis. Neither anti-ERNS mAb or a polyclonal BVDV Ab treatment of BVDV-infected MDM supernantant reduced lymphocyte apoptosis. The difference between BVDV and CSFV may be due to the use of a cloned Erns glycoprotein of CSFV that would be expressed at higher levels versus constitutive expression of BVDV Erns in BVDV-infected cells.

In the current study, we found that a supernatant from the highly virulent 1373 infected MDM induced apoptosis in MDBK as early as 24 h post-treatment, while low virulent 28508-5 infected MDM supernatant did not induce apoptosis. Previous work has shown that Ncp 28508-5 BVDV strain did not cause an inflammatory response in vivo, while Ncp 1373 caused inflammation and depletion of lymphocytes [56]. Interestingly, the Ncp 28508-5 strain is much more successful at establishing persistent infection. This lack of an inflammatory response may be one evasion mechanism to avoid the immune response and establish persistent infection in the developing fetus.

Furthermore, we found that the virulence of the BVDV strain is very important in determining the severity of the corresponding lymphoid depletion and subsequent immune suppression. The low virulence of the 28508-5 supernatant failed to induce apoptosis on the freshly isolated lymphocytes, while the highly virulent 1373 supernatant induced significant lymphocyte and somatic cell apoptosis. These results support the association between virulence and the degree of immune suppression that is associated with lymphoid depletion in vivo [1–3,57].

The role of macrophages in lymphocyte apoptosis was investigated in this study. Our results suggested that the supernatant of infected macrophages with only virulent BVDV strains induced lymphocyte apoptosis. These results provide an explanation for the observation that there is an increased number of macrophages in lymph nodes 3 days post-infection prior to lymphoid depletion [13]. Moreover, this indirect effect mediated by macrophages was not significant until 24 h post-infection with virulent BVDV strains, suggesting that rapid treatment strategies may help to limit the infection and would decrease or prevent the severe immune suppression induced by BVDV.

Our data support the results of Pedrera et al., 2009 [13] that did not relate the apoptotic effect to the biotype of the strain but to its virulence. Interestingly, we noted that the highly virulent 1373 strain of BVDV induced a more marked and faster apoptotic effect on lymphocytes indirectly as compared to the direct effect. These results suggest that there is a more severe and rapid immune suppression associated with the highly virulent 1373 that is likely due to the indirect effect mediated by macrophages.

Apoptosis-related cytokines produced by BVDV-infected macrophages did not play a role in lymphocyte apoptosis. There was no significant change in TNF-α, IL-1α, IL-1β or IL-6 transcriptional levels. These results were different from those described for CSFV, where TNF-a was involved in the indirect lymphocyte apoptosis activated by the virus [36]. Although TNF-α induction appears to be included in the mechanism of CSFV-induced lymphocyte apoptosis, this does not appear to be true for BVDV. The lack of TNF-α induced apoptosis is consistent with previous research with BVDV lymphocyte apoptosis [13]. Another study suggested that a role of IFN-γ in inducing apoptosis, unlike our research, was done using monocytes and a different BVDV strain [26].

The BL-3 cell line gave similar results, as seen with the peripheral blood lymphocytes. BL-3 cell findings were consistent with previously reported apoptotic effect of the highly virulent BVDV strains on the BL-3 cell line [30]. The degree of apoptosis induced in the lymphocytes of both peripheral blood or BL-3 cell line was similar. This finding leads us to speculate that B cells are more susceptible than T lymphocytes, but this hypothesis needs to be tested.

Another study suggested the identity of some possible factors that mediate the lymphocyte apoptosis post-BVDV infection, like micro RNA and programmed death-1 (PD-1) [58,59]. Finally, the identification and characterization of these factors and the molecular mechanisms involved in their induction and modes of action will be important goals for future studies.

5. Conclusions

BVDV causes immune dysfunction that leads to vaccination failure and secondary bacterial infection with lymphoid depletion associated with BVDV clinical cases. It was important for our study to focus on immune dysfunction caused by BVDV through investigating the indirect effect of an infected macrophage supernatant of various BVDV strains on macrophage inflammatory function as well as lymphocyte apoptosis, these two cells that are important for clearance of the virus from the body of infected animals. Our data suggested that the immune dysfunction associated with the highly virulent Ncp 1373 strain was mainly due to the indirect effect mediated by macrophage-secreted mediators. These mediators may impair macrophage surface marker expression, which in turn disrupt the general macrophage inflammatory function, including phagocytic and bactericidal activity but not NO production. These soluble factors also enhance lymphocyte apoptosis with subsequent lymphoid depletion associated with BVDV cases. Unlike CSFV, neither TNF-a nor secreted Erns glycoprotein induces BVDV lymphocyte apoptosis. This led us to the hypothesis that host but not virus factors likely mediated the indirect lymphocyte apoptotic effect.

Further investigations of the factors present in the supernatant of the infected macrophages with the highly virulent strains need to be conducted so that we have a better understanding of the disease mechanism. This can lead to finding a way to block these factors that can be used as a therapeutic practice in acutely infected animals to help improve recovery. Understanding these factors better would also help to improve the current control strategies and decrease the risk and the economic losses of BVDV. Future studies that include other BVDV strains need to be conducted to further confirm our findings.

Author Contributions: Conceptualization, K.A., G.E., and C.C.L.C.; writing-original draft preparation, K.A., M.R.; writing-review and editing, K.A., G.E., and A.A.H.A.; visualization N.T., J.S., and H.A. All authors have read and agreed to the published version of the manuscript. K.A. and M.R. have the same contribution as first author.

Funding: This research received no external funding.

Acknowledgments: We would like to thank the support from Shollie Falkenberg for providing the BL-3 cell line. Robert Fulton for providing the polyclonal antibodies for BVDV testing. Adam Hoppe help and guidance with using the flow cytometry machine. The Department of Veterinary and Biomedical Sciences/Animal. Disease Research and Diagnostic Laboratory, SDSU, Brookings, SD, USA; SDSU. Functional Genomic Core Facility (FGCF), SDSU.

Conflicts of Interest: The authors declare no conflict of interest.

References

1. Ammari, M.; McCarthy, F.; Nanduri, B.; Pinchuk, G.; Pinchuk, L. Understanding the Pathogenesis of Cytopathic and Noncytopathic Bovine Viral Diarrhea Virus Infection Using Proteomics. In *Proteomic Applications in Biology*; IntechOpen: London, UK, 2012.
2. Bittar, J.H.; Palomares, R.A.; Hurley, D.J.; Hoyos-Jaramillo, A.; Rodriguez, A.; Stoskute, A.; Hamrick, B.; Norton, N.; Adkins, M.; Saliki, J.T.; et al. Immune response and onset of protection from Bovine viral diarrhea virus 2 infection induced by modified-live virus vaccination concurrent with injectable trace minerals administration in newly received beef calves. *Vet. Immunol. Immunopathol.* **2020**, *225*, 110055. [CrossRef] [PubMed]
3. Falkenberg, S.M.; Johnson, C.; Bauermann, F.V.; McGill, J.; Palmer, M.V.; Sacco, R.E.; Ridpath, J.F. Changes observed in the thymus and lymph nodes 14 days after exposure to BVDV field strains of enhanced or typical virulence in neonatal calves. *Vet. Immunol. Immunopathol.* **2014**, *160*, 70–80. [CrossRef] [PubMed]
4. Liebler-Tenorio, E.M.; Ridpath, J.F.; Neill, J.D. Distribution of viral antigen and development of lesions after experimental infection of calves with a BVDV 2 strain of low virulence. *J. Vet. Diagn. Investig.* **2003**, *15*, 221–232. [CrossRef] [PubMed]
5. Potgieter, L.N. Immunology of bovine viral diarrhea virus. *Vet. Clin. N. Am. Food Anim. Pract.* **1995**, *11*, 501–520. [CrossRef]
6. Sopp, P.; Hooper, L.B.; Clarke, M.C.; Howard, C.J.; Brownlie, J. Detection of bovine viral diarrhoea virus p80 protein in subpopulations of bovine leukocytes. *J. Gen. Virol.* **1994**, *75*, 1189–1194. [CrossRef]
7. Roth, J.A.; Kaeberle, M.L. Isolation of neutrophils and eosinophils from the peripheral blood of cattle and comparison of their functional activities. *J. Immunol. Methods* **1981**, *45*, 153–164. [CrossRef]
8. Welsh, M.D.; Adair, B.M.; Foster, J.C. Effect of BVD virus infection on alveolar macrophage functions. *Vet. Immunol. Immunopathol.* **1995**, *46*, 195–210. [CrossRef]

9. Lambot, M.; Hanon, E.; Lecomte, C.; Hamers, C.; Letesson, J.-J.; Pastoret, P.-P. Bovine viral diarrhoea virus induces apoptosis in blood mononuclear cells by a mechanism largely dependent on monocytes. *J. Gen. Virol.* **1998**, *79*, 1745–1749. [CrossRef] [PubMed]
10. Zhang, G.; Aldridge, S.; Clarke, M.C.; McCauley, J.W. Cell death induced by cytopathic bovine viral diarrhoea virus is mediated by apoptosis. *J. Gen. Virol.* **1996**, *77*, 1677–1681. [CrossRef] [PubMed]
11. Risalde, M.A.; Romero-Palomo, F.; Lecchi, C.; Ceciliani, F.; Bazzocchi, C.; Comazzi, S.; Besozzi, M.; Gómez-Villamandos, J.C.; Luzzago, C. BVDV permissiveness and lack of expression of co-stimulatory molecules on PBMCs from calves pre-infected with BVDV. *Comp. Immunol. Microbiol. Infect. Dis.* **2020**, *68*, 101388. [CrossRef] [PubMed]
12. Chase, C.C.; Thakur, N.; Darweesh, M.F.; Morarie-Kane, S.E.; Rajput, M.K. Immune response to bovine viral diarrhea virus—looking at newly defined targets. *Anim. Health Res. Rev.* **2015**, *16*, 4–14. [CrossRef] [PubMed]
13. Pedrera, M.; Gomez-Villamandos, J.C.; Romero-Trevejo, J.L.; Risalde, M.A.; Molina, V.; Sanchez-Cordon, P.J. Apoptosis in lymphoid tissues of calves inoculated with non-cytopathic bovine viral diarrhea virus genotype 1: Activation of effector caspase-3 and role of macrophages. *J. Gen. Virol.* **2009**, *90*, 2650–2659. [CrossRef] [PubMed]
14. Romero–Palomo, F.; Risalde, M.A.; Gómez–Villamandos, J.C. Immunopathologic Changes in the Thymus of Calves Pre–infected with BVDV and Challenged with BHV–1. *Transbound. Emerg. Dis.* **2017**, *64*, 574–584. [CrossRef]
15. Roth, J.A.; Bolin, S.R.; Frank, D.E. Lymphocyte blastogenesis and neutrophil function in cattle persistently infected with bovine viral diarrhea virus. *Am. J. Vet. Res.* **1986**, *47*, 1139–1141.
16. Ridpath, J.F.; Bendfeldt, S.; Neill, J.D.; Liebler-Tenorio, E. Lymphocytopathogenic activity in vitro correlates with high virulence in vivo for BVDV type 2 strains: Criteria for a third biotype of BVDV. *Virus Res.* **2006**, *118*, 62–69. [CrossRef]
17. Adler, B.; Adler, H.; Pfister, H.; Jungi, T.W.; Peterhans, E. Macrophages infected with cytopathic bovine viral diarrhoea virus release a factor(s) capable of priming uninfected macrophages for activation-induced apoptosis. *J. Virol.* **1997**, *71*, 3255–3258. [CrossRef]
18. Schweizer, M.; Peterhans, E. Oxidative stress in cells infected with bovine viral diarrhoea virus: A crucial step in the induction of apoptosis. *J. Gen. Virol.* **1999**, *80*, 1147–1155. [CrossRef] [PubMed]
19. Bruschke, C.J.M.; Hulst, M.M.; Moormann, R.J.M.; Rijn, P.A.V.; Oirschot, J.T.V. Glycoprotein Erns of Pestiviruses Induces Apoptosis in Lymphocytes of Several Species. *J. Virol.* **1997**, *71*, 6692–6696. [CrossRef] [PubMed]
20. Fetzer, C.; Tews, B.A.; Meyers, G. The carboxy-terminal sequence of the pestivirus glycoprotein Erns represents an unusual type of membrane anchor. *J. Virol.* **2005**, *79*, 11901–11913. [CrossRef]
21. Summerfield, A.; Zingle, K.; Inumaru, S.; McCullough, C. Induction of apoptosis in bone marrow neutrophil-lineage by classical swine fever. *J. Gen. Virol.* **2001**, *82*, 1309–1318. [CrossRef]
22. Hoff, H.S.; Donis, R.O. Induction of apoptosis and cleavage of poly (ADP-ribose) polymerase by cytopathic bovine viral diarrhea virus infection. *Virus Res.* **1997**, *49*, 101–113. [CrossRef]
23. Schweizer, M.; Peterhans, E. Noncytopathic bovine viral diarrhea virus inhibits double-stranded RNA-induced apoptosis and interferon synthesis. *J. Virol.* **2001**, *75*, 4692–4698. [CrossRef] [PubMed]
24. Mangan, D.F.; Robertson, B.; Wahl, S.M. IL-4 enhances programmed cell death (apoptosis) in stimulated human monocytes. *J. Immunol.* **1992**, *148*, 1812–1816. [PubMed]
25. Adler, H.; Jungi, T.W.; Pfister, H.; Strasser, M.; Sileghem, M.; Peterhans, E. Cytokine regulation by virus infection: Bovine viral diarrhea virus, a flavivirus, downregulates production of tumor necrosis factor alpha in macrophages in vitro. *J. Virol.* **1996**, *70*, 2650–2653. [CrossRef]
26. Choi, K.S. The effect of bovine viral diarrhea virus on bovine monocyte phenotype. *Iran. J. Vet. Res.* **2017**, *18*, 13.
27. Lopez, B.I.; Santiago, K.G.; Lee, D.; Ha, S.; Seo, K. RNA Sequencing (RNA-Seq) Based Transcriptome Analysis in Immune Response of Holstein Cattle to Killed Vaccine against Bovine Viral Diarrhea Virus Type, I. *Animals* **2020**, *10*, 344. [CrossRef]
28. Aldo, M.A.; Yahel, L.R.; Montserrat, R.H.; Elena, S.S.; Lourdes, A.P.; Alejandro, B.G. The NADL strain of bovine viral diarrhea virus induces the secretion of IL-1β through caspase 1 in bovine macrophages. *Res. Vet. Sci.* **2020**. [CrossRef]

29. Elmowalid, G. Unmasking the Effect of Bovine Viral Diarrhea Virus on Macrophage Inflammatory Functions. Ph.D. Thesis, South Dakota State University, Brookings, SD, USA, 2003.
30. Schaut, R.G.; Ridpath, J.F.; Sacco, R.E. Bovine Viral Diarrhea Virus Type 2 Impairs Macrophage Responsiveness to Toll-Like Receptor Ligation with the Exception of Toll-Like Receptor 7. *PLoS ONE* **2016**, *11*, e0159491. [CrossRef]
31. Reed, L.J.; Muench, H. A simple method of estimating fifty per cent endpoints. *Am. J. Epidemiol.* **1938**, *27*, 493–497. [CrossRef]
32. Mwangi, W.; Brown, W.C.; Splitter, G.A.; Zhuang, Y.; Kegerreis, K.; Palmer, G.H. Enhancement of antigen acquisition by dendritic cells and MHC class II-restricted epitope presentation to CD4+ T cells using VP22 DNA vaccine vectors that promote intercellular spreading. *J. Leukoc. Biol.* **2005**, *78*, 401–411. [CrossRef]
33. Rajput, M.K.; Darweesh, M.F.; Park, K.; Braun, L.J.; Mwangi, W.; Young, A.J.; Chase, C.C. The effect of bovine viral diarrhea virus (BVDV) strains on bovine monocyte-derived dendritic cells (Mo-DC) phenotype and capacity to produce BVDV. *Virol. J.* **2014**, *11*, 1. [CrossRef] [PubMed]
34. Strober, W. Trypan blue exclusion test of cell viability. *Curr. Protoc. Immunol.* **2001**, *111*, A3-B1–A3-B3.
35. Perler, L.; Schweizer, M.; Jungi, T.W.; Peterhans, E. Bovine viral diarrhoea virus and bovine herpesvirus-1 prime uninfected macrophages forlipopolysaccharide-triggered apoptosis by interferon-dependent and-independent pathways. *J. Gen. Virol.* **2000**, *81*, 881–887. [CrossRef] [PubMed]
36. Choi, C.; Hwang, K.K.; Chae, C. Classical swine fever virus induces tumor necrosis factor-alpha and lymphocyte apoptosis. *Arch. Virol.* **2004**, *149*, 875–889. [CrossRef] [PubMed]
37. Ridpath, J.F.; Driskell, E.; Chase, C.C.; Neill, J.D. Characterization and detection of BVDV related reproductive disease in white tail deer. In Proceedings of the 13th International World Association of Veterinary Laboratory Diagnosticians Symposium, Melbourne, Australia, 11–14 November 2007; p. 34.
38. Gels, T.; Schagger, H.; von Jagow, G. Tricine-sodium dodecyl sulfate-polyacrylamide gel electrophoresis for the separation of proteins in the range from 1 to 100 kDa. *Anal. Biochem.* **1987**, *166*, 368–379.
39. Konnai, S.; Usui, T.; Ohashi, K.; Onuma, M. The rapid quantitative analysis of bovine cytokine genes by real-time RT-PCR. *Vet. Microbiol.* **2003**, *94*, 283–294. [CrossRef]
40. Pfaffl, M.W.; Horgan, G.W.; Dempfle, L. Relative expression software tool (REST©) for group-wise comparison and statistical analysis of relative expression results in real-time PCR. *Nucleic Acids Res.* **2002**, *30*, e36. [CrossRef]
41. Hou, L.; Wilkerson, M.; Kapil, S.; Mosie, R.D.; Shuman, W.; Reddy, J.R.; Loughin, T.; Minocha, H.C. The effect of different bovine viral diarrhea virus genotypes and biotypes on the metabolic activity and activation status of bovine peripheral blood mononuclear cells. *Viral Immunol.* **1998**, *11*, 233–244. [CrossRef]
42. Ketelsen, A.T.; Johnson, D.W.; Muscoplat, C.C. Depression of bovine monocyte chemotactic responses by bovine viral diarrhea virus. *Infect. Immun.* **1979**, *25*, 565–568. [CrossRef]
43. Potgieter, L.N.; McCracken, M.D.; Hopkins, F.M.; Walker, R.D.; Guy, J.S. Experimental production of bovine respiratory tract disease with bovine viral diarrhea virus. *Am. J. Vet. Res.* **1984**, *45*, 1582–1585.
44. Blond, D.; Raoul, H.; Le Grand, R.; Dormont, D. Nitric oxide synthesis enhances human immunodeficiency virus replication in primary human macrophages. *J. Virol.* **2000**, *74*, 8904–8912. [CrossRef]
45. Lasarte, J.J.; Sarobe, P.; Boya, P.; Casares, N.; Arribillaga, L.; de Cerio, A.L.D.; Gorraiz, M.; Borrás-Cuesta, F.; Prieto, J. A recombinant adenovirus encoding hepatitis C virus core and E1 proteins protects mice against cytokine–induced liver damage. *Hepatology* **2003**, *37*, 461–470. [CrossRef] [PubMed]
46. Charleston, B.; Hope, J.C.; Carr, B.V.; Howard, C.J. Masking of two in vitro immunological assays for Mycobacterium bovis (BCG) in calves acutely infected with non-cytopathic bovine viral diarrhoea virus. *Vet. Rec.* **2001**, *149*, 481–484. [CrossRef] [PubMed]
47. Archambault, D.; Béliveau, C.; Couture, Y.; Carman, S. Clinical response and immunomodulation following experimental challenge of calves with type 2 noncytopathogenic bovine viral diarrhea virus. *Vet. Res.* **2000**, *31*, 215–227. [CrossRef] [PubMed]
48. Lee, S.R.; Nanduri, B.; Pharr, G.T.; Stokes, J.V.; Pinchuk, L.M. Bovine viral diarrhea virus infection affects the expression of proteins related to professional antigen presentation in bovine monocytes. *Biochim. Et Biophys. Acta (Bba)-Proteins Proteom.* **2009**, *1794*, 14–22. [CrossRef] [PubMed]
49. Jirillo, E.; Greco, B.; Caradonna, L.; Satalino, R.; Pugliese, V.; Cozzolongo, R.; Cuppone, R.; Manghisi, O.G. Evaluation of cellular immune responses and soluble mediators in patients with chronic hepatitis C virus (cHCV) infection. *Immunopharmacol. Immunotoxicol.* **1995**, *17*, 347–364. [CrossRef] [PubMed]

50. Chen, Y.C.; Wang, S.Y.; King, C.C. Bacterial lipopolysaccharide inhibits dengue virus infection of primary human monocytes/macrophages by blockade of virus entry via a CD14-dependent mechanism. *J. Virol.* **1999**, *73*, 2650–2657. [CrossRef]
51. Jensen, J.; Schultz, R.D. Effect of infection by bovine viral diarrhea virus (BVDV) in vitro on interleukin-1 activity of bovine monocytes. *Vet. Immunol. Immunopathol.* **1991**, *29*, 251–265. [CrossRef]
52. Markham, R.J.; Ramnaraine, M.L. Release of immunosuppressive substances from tissue culture cells infected with bovine viral diarrhea virus. *Am. J. Vet. Res.* **1985**, *46*, 879–883.
53. Van Reeth, K.; Adair, B. Macrophages and respiratory viruses. *Pathol.-Biol.* **1997**, *45*, 184–192.
54. Janeway, C.A., Jr.; Travers, P.; Walport, M.; Shlomchik, M.J. *The Complement System and Innate Immunity*; Garland Science: New York, NY, USA, 2001.
55. Darweesh, M.F.; Rajput, M.K.; Braun, L.J.; Rohila, J.S.; Chase, C.C. BVDV Npro protein mediates the BVDV induced immunosuppression through interaction with cellular S100A9 protein. *Microb. Pathog.* **2018**, *121*, 341–349. [CrossRef] [PubMed]
56. Liebler-Tenorio, E.M.; Ridpath, J.F.; Neill, J.D. Distribution of viral antigen and development of lesions after experimental infection with highly virulent bovine viral diarrhea virus type 2 in calves. *Am. J. Vet. Res.* **2002**, *63*, 1575–1584. [CrossRef] [PubMed]
57. Jungmann, A.; Nieper, H.; Müller, H. Apoptosis is induced by infectious bursal disease virus replication in productively infected cells as well as in antigen-negative cells in their vicinity. *J. Gen. Virol.* **2001**, *82*, 1107–1115. [CrossRef] [PubMed]
58. Liu, Y.; Liu, S.; He, B.; Wang, T.; Zhao, S.; Wu, C.; Yue, S.; Zhang, S.; He, M.; Wang, L.; et al. PD-1 blockade inhibits lymphocyte apoptosis and restores proliferation and anti-viral immune functions of lymphocyte after CP and NCP BVDV infection in vitro. *Vet. Microbiol.* **2018**, *226*, 74–80. [CrossRef]
59. Shi, H.; Ni, W.; Sheng, J.; Chen, C. Lentivirus-mediated bta-miR-193a Overexpression Promotes Apoptosis of MDBK Cells by Targeting BAX and Inhibits BVDV Replication. *Kafkas Üniversitesi Vet. Fakültesi Derg.* **2017**, *23*, 587–593.

viruses

MDPI

Review

Bovine Viral Diarrhea Virus: Recent Findings about Its Occurrence in Pigs

Luís Guilherme de Oliveira [1,*], Marina L. Mechler-Dreibi [1], Henrique M. S. Almeida [1] and Igor R. H. Gatto [2]

[1] School of Agricultural and Veterinarian Sciences, São Paulo State University (Unesp), Jaboticabal. Via de Acesso Prof. Paulo Donato Castelanne s/n, Jaboticabal - SP 14884-900, Brazil; mlopesvet@gmail.com (M.L.M.-D.); henri_almeida2003@yahoo.com.br (H.M.S.A.)

[2] Ourofino Animal Health Ltda. Rodovia Anhanguera SP 330, Km 298. Distrito Industrial, Cravinhos – SP 14140-000, Brazil; igatto_10@hotmail.com

* Correspondence: luis.guilherme@unesp.br

Received: 23 April 2020; Accepted: 26 May 2020; Published: 31 May 2020

Abstract: Bovine viral diarrhea virus (BVDV) is an important pathogen belonging to the *Pestivirus* genus, *Flaviviridae* family, which comprises viral species that causes an economic impact in animal production. Cattle are the natural host of BVDV and the main source of infection for pigs and other animal species. Due to its antigenic and genetic similarity with other important pestiviruses such as Classical Swine Fever Virus (CSFV), several studies have been conducted to elucidate the real role of this virus in piglets, sows, and boars, not only in the field but also in experimental infections, which will be discussed in this paper. Although BVDV does not pose a threat to pigs as it does to ruminants, the occurrence of clinical signs is variable and may depend on several factors. Therefore, this study presents a survey of data on BVDV infection in pigs, comparing information on prevalence in different countries and the results of experimental infections to understand this type of infection in pigs better.

Keywords: BVDV; experimental infection; natural infection; pigs

---

1. Updates on BVDV Infection in Swine

Bovine viral diarrhea (BVD) is an infection caused by the bovine viral diarrhea virus (BVDV), belonging to the genus *Pestivirus*, family *Flaviviridae*, with single-stranded positive polarity RNA [1]. Viruses belonging to the *Pestivirus* genus infect hosts of several animal species and include viral agents of great impact for animal production [2,3]. The *Pestivirus* species have been recently named from A to K and, among them, the *Pestivirus* A (BVDV-1), *Pestivirus* B (BVDV-2), *Pestivirus* C (Classical Swine Fever Virus), and *Pestivirus* K (atypical porcine pestivirus) are the main viral species related to swine [4]. BVDV has two genotypes, type 1 and type 2, which are classified into sub-genotypes: BVDV-1 (1a to 1u), adding up to 21 sub-genotypes, and BVDV-2 (2a to 2d), with four sub-genotypes [5]. BVDV-1 is related to most reference strains, is commonly used for vaccine production, and was most frequently isolated from mild to moderate clinical cases in cattle. Conversely, BVDV-2 was isolated from acute disease outbreaks, also presenting strains of mild and moderate virulence [6].

Based on the effect of replication on cell culture, BVDV isolates can be divided into cytopathic (cp) and non-cytopathic (ncp), with the ncp isolates being responsible for most natural infections and persistent fetal infections, and cp isolates constituting a minority, which are isolated almost exclusively from cattle with mucosal disease [6].

Cattle are natural hosts of BVDV, considered the major source of infection for pigs and other animal species [7,8]. Usually, positive pig herds for BVDV occur when cattle and pigs are raised on the same farm, and the direct contact between these animal species is considered the main source of BVDV transmission for pigs [7]. Infection caused by BVDV in pigs has been reported in China [9], the Netherlands [10], Brazil [11–13], Austria [14], Germany [15], Norway [16], Ireland [17], Denmark [18] and others. These data were found not only in domestic pigs but also in wild boars [19], which raise concerns about risk factors involved in BVDV infection, the clinical form of the disease, and the existence of accurate diagnostic tests. In Brazil, BVDV-1d was frequently reported in cattle [20]. Mósena et al. [11] states that by the phylogenetic analysis of sequenced samples collected from backyard pigs, classified as BVDV-1d and BVDV-2a, it is possible that one of the obtained sequences originated from contact between cattle and pigs.

It is known that all pestiviruses are genetically and antigenically related, and BVDV infection in pigs may be presented with a great variability of clinical signs [21]. Even though BVDV infections in pigs are not as problematic as Classical Swine Fever Virus (CSFV) infections, it is believed that distinguishing these two diseases could be difficult due to the similar clinical signs when considering low pathogenicity strains [22]. Reports of clinical signs associated with the infection consisting of anemia, delayed development, rough hair, polyarthritis, congenital tremors (CT), petechiae on the skin, diarrhea, conjunctivitis, and cyanosis [23]. Clinical signs similar to CSF, and sudden death [24] were also observed when the BVDV strain was isolated from both pigs and cattle from the same farm [24]. On the other hand, several recent studies with experimental infection did not report the presence of clinical signs of infection [25–34]. This may occur due to an inadequate level of viremia or a low virulence strain, biotype of the virus, host adaptation and/or route of inoculation [31–35]. A possible explanation is that cases in which BVDV infection-induced large numbers of lesions in adult pigs have been caused by viral strains that passed along previous adaptations in this species [23].

BVDV has a predilection for replication in defense cells, mainly lymphocytes, but it also infects monocytes and dendritic cells. As antigen presenters, dendritic cells play an important role in cellular immunity by initiating the nonspecific immune response against various pathogens [35]. Its infection promotes lysis of monocytes as a mechanism for evading the immune system, affecting the recognition and subsequent development of a specific immune-humoral response [36].

BVDV can contaminate cell cultures and fetal calf serum [37]. In countries that promote CSFV vaccination, the BVDV prevalence found in swine herds has been associated with the widespread use of live vaccines for Classical Swine Fever (CSF), which were produced with bovine sera from positive Chinese bovine herds [9]. Batches of live CSFV vaccines used in China confirmed five BVDV-contaminated samples out of 23 collected for testing [38].

Serological diagnosis by enzyme-linked immunosorbent assay (ELISA) is more efficient, cheaper, and faster than molecular techniques [37]. The neutralizing antibody titers in the serum of animals previously exposed to a Pestivirus member are usually medium to high regarding the homologous viral species, and low (or non-reactive) regarding other species [6]. Anti-BVDV antibodies were shown to be able to protect pigs against CSFV infection and the manifestation of clinical signs, even though the anti-CSFV antibody titers were low, which could hinder CSFV outbreaks in herds with a high prevalence of anti-BVDV antibodies [22]. The same condition could occur in the presence of anti-Border Disease Virus (BDV) antibodies, as cross-reactions could affect the transmission of CSFV and should be evaluated for an accurate diagnosis of a CSFV infection and for implementing specific surveillance protocols in cases of outbreaks [10].

Reverse transcription-polymerase chain reaction (RT-PCR) is widely used for detecting the viral agent for differential diagnosis [39], since samples of blood, milk, saliva, and tissue can be successfully tested [40], and can be stored for a long time with minimal losses [41]. Researchers have adapted numerous variations of PCR methods for the detection of infectious agents, using DNA templates as well as RNA templates after an RT step [42], which has enabled more accurate, sensible and specific diagnostics. Direct sequencing of the RT-PCR product for fragments of 5′UTR and N-terminal autoprotease (N^{pro}) may also provide accurate differential diagnosis [19].

Given the antigenic and genetic similarity and the improvement in laboratory diagnostic methods, the comparison between results from recent and former studies should be cautious. Several studies that examine data collection in the field, as well as experimental infection with BVDV, have been conducted, and the results that contrast with the former data in the literature will be further discussed.

2. Data Collection from Backyard and Intensive Pig Herds

Aiming at collecting data on the occurrence of anti-BVDV antibodies in Brazilian swine herds, cross-sectional studies were carried out [12,13] in the backyard and intensive pig herds, respectively, located in the CSF-free zone of Brazil. For the first study, 56 pig herds from the northwest region of the state of São Paulo were evaluated, which are part of 11 municipalities. Blood samples were collected for serological testing by virus neutralization (VN), and titers higher than ten were considered positive. Out of the 360 serum samples, 4.72% (17/360) were reactive to BVDV in VN, which is 1.94% (7/360) reactive to BVDV-1 (Singer strain), with antibody titers ranging from 10 to 640, and 3.06% (11/360) reactive to BVDV-2 (VS-253 strain), with antibody titers ranging from 10 to 80, and only one reactive sample against both genotypes. Regarding herds, 27% (15/56) presented at least one animal positive to any of the genotypes. The prevalence of BVDV in bovine herds in the same region where this study was conducted was 56.49% [43], which may have resulted in the highest prevalence values of swine in the region when compared to previous reports [10]. Most of the farms evaluated in this study had cattle and pigs in close contact. As ruminants are the main source of BVDV infection for pigs [7,8], the prevalence of the disease in cattle herds is closely related to the presence of infections and influences the prevalence of the disease in pigs [9,10,44].

On the other hand, in a cross-sectional study carried out in 33 commercial pig herds, collected 1705 blood samples for analysis [13]. Samples were also tested by VN, and 5.34% (91/1705) of the samples were sero-reactive to BVDV with antibody titers ranging from 10 to 80. Of these, 3% (51/1075) were positive for reference strains of BVDV-1 (Singer strain) and 2.35% (40/1705) for reference strains of BVDV-2 (VS-253 strain), with 0.1% (2/1705) of samples with cross-reactions between both genotypes. Herds were sampled from 27 municipalities, located in seven Brazilian states, which are part of three different regions (South, Southeast, and Midwest). In 64% of herds (21/33), there was at least one positive sample for any of the BVDV genotypes in VN. As the presence of anti-BVDV antibodies in swine serum can lead to false-positive results in serological tests for the diagnosis of CSFV, the positive samples from both studies were sent for anti-CSFV antibodies detection, and were both negative. A survey carried out in The Netherlands [10] on commercial farm animals found a prevalence of 2.5% for gilts and 0.42% for finishing animals via ELISA. The difference in prevalence found in finishing animals from these two studies can be explained not only by the sensitivity of the techniques used but also by the different levels of biosecurity in the farms studied.

In a more recent study [11], swine sera were collected from 320 backyard pig herds in southern Brazil. Serum samples were tested by VN against BVDV-1a, -1b, and -2 strains, resulting in 4.2% (27/639) positive samples. Of those, 16 samples presented the highest titers against BVDV-1a (2 samples), BVDV-1b (5 samples), and BVDV-2 (9 samples). These studies confirm that ruminant Pestiviruses have been circulating in swine herds and must be considered in future Pestivirus control programs conducted in Brazil.

The low prevalence of BVDV in pigs was also found by other authors in Norway, Ireland, Denmark, and The Netherlands [16–18] (Table 1). Not only domestic pigs are a concern when it comes to Pestivirus infections since wild boar have been described as important reservoirs or transmitters of pathogens in nature due to their ability to reach long distances and transmit diseases to domestic swine. Other studies [45–47] have reported a low prevalence of BVDV in wild boar in Germany, the Czech Republic, and Eastern Serbia, respectively. Weber [19] was the first to detect BVDV RNA in wild boars' blood samples, and Gatto [48] first reported the presence of anti-BVDV antibodies in Thayasuids. In general, the low prevalence of anti-BVDV antibodies in pigs may be associated not only with the level of interaction between pigs and ruminants but also with the host-pathogen specificity, which seems to be lower in pigs compared to cattle. In intensive pig farming, biosecurity measures reduce or eliminate the presence of some infectious agents. Some researchers [10,44] attributed the low prevalence of BVDV in swine herds in their studies to the high specialization of agriculture, in which interspecific contact was reduced due to single zootechnical breeding on the property.

Table 1. Bovine viral diarrhea virus prevalence in intensive pig farming, backyard pig herds, and wild boars observed by several authors in different regions worldwide.

References	Type of Pig Production	Animal Category	Number of Samples/Number of Herds	Type of Sample	Diagnostic Tool	Region	BVDV Prevalence
Loken et al., 1991	-	Adult pigs	1317/887	Serum	Virus neutralization	Norway	2.2%
Graham et al., 2001	-	-	660/46	Serum	ELISA*	Northern Ireland	0.15%
Loeffen et al., 2009	Intensive pig farming	Finishing pigs	1890/189	Serum	ELISA and Virus neutralization	The Netherlands	0.42%
Loeffen et al., 2009	Intensive pig farming	Sows	6020/616	Serum	ELISA and Virus neutralization	The Netherlands	2.5%
Deng et al., 2012	-	Pigs with clinical signs	511/11	Serum and tissue homogenate	Nested-RT-PCR	China (herds from 11 provinces)	26.8%
Gatto et al., 2017	Intensive pig farming		1705/33	Serum	Virus neutralization	Brazil (herds from six states)	5.34%
Almeida et al., 2017		Piglets and adults	360/56	Serum	Virus neutralization	Northwestern region of São Paulo State, Brazil	4.72%
Mósena et al., 2020	Backyard pig herds	Male and female animals 6–72 mo old	639/320	Serum	RT-PCR and virus neutralization	Rio Grande do Sul, Brazil	4.2%
Dahle et al., 1993	Wild boars	-	841 samples	Serum	Direct Neutralizing Peroxidase linked antibody assay (NPLA)	Northern Germany	0.83%
Sedlak et al., 2008	Wild boars	-	352 samples	Serum	ELISA	Czech Republic	1%
Milicevik et al., 2018	Wild boars	-	50 samples	Spleen	qRT-PCR	Eastern Serbia	8%
Weber et al., 2016	Farmed wild boars	-	40 samples	Lung	RT-PCR	Rio Grande do Sul, Brazil	2.5%
Gatto et al., 2020	Farmed white-lipped peccaries	-	72 samples	Serum	Virus neutralization	Midwest Brazil	1.38%

*ELISA: enzyme-linked immunosorbent assay.

In China, swine samples with clinical signs such as diarrhea, miscarriage, and death, between 2007 and 2010, were tested for BVDV by nested RT-PCR. Unlike that heretofore described, the observed prevalence was 23.1% in 2007, 27.7% in 2008, 33.6% in 2009, and 23.6% in 2010, showing a high prevalence of BVDV-1 infection [9] when compared to the abovementioned study. These numbers should be analyzed carefully since only samples from clinical cases compatible with BVDV infection were analyzed. As stated by these authors, the use of live vaccines against CSFV may also be directly related to this higher prevalence of BVDV in pig herds from China, since, in general, the prevalence of BVDV in pig herds is low.

3. Experimental Infection with BVDV in Pigs: Routes of Transmission and Disease Development

The lack of studies concerning the routes of transmission of BVDV between piglets highlighted the need to develop researches clarifying this information (Table 2). Three studies were separately conducted with three groups of two weaned piglets separated by isolation cabinets: the challenged, sentinel, and control groups. The isolation cabinets were arranged to allow only a specific route of transmission, namely airborne, nose-to-nose [25], and by back pond water [26]. Although all challenged piglets shed the virus and seroconverted, only transmission by the back-pond water was confirmed since sentinel animals also shed the virus and seroconverted in this study. An interesting fact regarding the viral shedding observed in these studies was the intermittent pattern of nasal shedding. Challenged piglets shed the virus between 5- and 24-days post-inoculation, intermittently, detected by RT-PCR of nasal swabs. In the challenged groups, clinical signs such as diarrhea, rough hair, and oculo-nasal discharge were observed about 15 dpi, when the piglets started seroconversion. Despite other pathogens that have not been searched for, these clinical signs may be suggestive of BVDV infection since it was not observed in piglets from the control group.

Table 2. Results obtained in experimental inoculation studies of bovine viral diarrhea virus in swine.

References	BVDV Strain	Route of Inoculation	Animal Category	BVDV-Shedding	Viremia	Seroconversion (Dpi / % of Positive Pigs)	Consequences of Infection
Stewart et al., 1980	-	Intranasal-oral	Pregnant sows	-	7 dpi	21 dpi / 100%	Intrauterine infection in one litter, fewer fetuses than corpora lutea. Not observed in fetuses
Terpstra and Wensvoort, 1988	76/4 and 77/5 strains	Natural infection	Sows and piglets	-	-	35 dpi / (NA)	Piglets with clinical signs similar to CSF
Paton and Done, 1994	91/1 and 87/6 strains	Intrauterine	Pregnant sows	only congenital persistently infected fetus, for 2.5 years	4–6 dpi in fetuses	variable in piglets, but all sows seroconverted	Intrauterine infection, some fetuses with no clinical signs, some with persistent congenital infection
Terpstra and Wensvoort, 1997	Van Ee, Appel and Toering	Natural infection	Piglet	urine, oropharyngeal fluids, and semen of persistently infected boar	Observed	30 dpi - 8 mo / (NA)	Congenital persistent infection, leukopenia, viral replication in several organs, clinical signs resembling chronic CSF
Kulcsar et al., 2001	Oregon C24V strain	Intranasal and subcutaneous	Pregnant sows	Not observed	-	28 dpi / 100%	No clinical signs on sows, birth of weak piglets with ruffled hair coat, splay leg, trembling, myoclonia, diarrhea, fever, death
Makoschey et al., 2002	BVDV-2	-	Piglets	-	-	Observed / 100%	Inapparent infection with a slight increase in body temperature in some piglets, viral replication in cells/organs. Slight leukopenia and/or thrombocytopenia
Walz et al., 2004	BVDV-1 ncp	Intranasal inoculation	Pregnant gilts	-	5–7 dpi	21 dpi / 100%	No clinical signs observed in gilts or piglets; transplacental infection occurred in only one fetus from one gilt. No antibodies were found in piglets
Wieringa- Jelsma et al., 2006	St. Oedenrode strain	Intranasal inoculation	Weaned pigs	2–11 dpi	7 dpi	21 dpi / 100%	Not observed
Dos Santos et al., 2017	BVDV-1 (Singer) cp	Oronasal inoculation	Weaned piglets	10–25 dpi	Not observed	25 dpi / 62.5%	Diarrhea, rough hair, nasal discharge
Nascimento et al., 2017	BVDV-1 (Singer) cp	Oronasal inoculation	Weaned piglets	5–24 dpi	Not observed	25 dpi / 100%	Diarrhea, shivering, nasal discharge
Mechler et al., 2018	BVDV-2 (VS-260) ncp	Oronasal and intrauterine inoculation	Pregnant gilts	No	Not observed	20 dpi / 100%	Not observed
Pereira et al., 2018	BVDV-2 (VS-253) cp	Oronasal inoculation	Pregnant gilts	6–24 dpi	3–12 dpi	12–33 dpi / 70%	Thrombocytopenia
Gomes et al., 2019	BVDV-2 (VS 260) ncp	Oronasal inoculation	Pregnant gilts	No	Not observed	20 dpi / 100%	Not observed
Storino et al., 2020	BVDV-2 (LVB 16557/15) ncp	Oronasal, intramuscular and intravenous inoculation	Boars	No	Not observed	20–40 dpi / 12.5%	Lymphocytosis and monocytopenia

Dpi: days post-inoculation; cp: cytopathic; ncp: non-cytopathic.; NA = not applicable.

These studies proved that BVDV can infect weaned piglets, which shed the virus by the nasal route, presented clinical signs, and seroconverted. The presence of BVDV in nasal secretions indicates that pigs can be a source of infection for other animals, especially if piglets become infected by high virulent strains of BVDV. In the literature, BVDV shedding by a persistently infected boar has also been reported [49], with the detection of the virus in oropharyngeal fluid, urine, and semen.

Although BVDV infection in young pigs can occur without clinical manifestation of the disease [32], other authors have reported the appearance of reproductive problems in pregnant gilts, such as abortions, birth of small piglets, stillborn and congenital persistently infected animals (PI) [22]. In cattle, during acute infection, viremia and viral shedding are usually transient and at low titers, but even so, they can result in vertical transmission [50]. BVDV-2 infection can lead to the occurrence of fetal malformations in cattle; however, if the fetus survives the infection long enough, non-specific changes in maturation may occur in the lymphoid tissues [51]. In bovine herds infected by BVDV, several fetal malformations were described comprising cerebellar hypoplasia, poor myelinization of the spinal cord, hydrocephalus, microcephaly, retinal atrophy or dysplasia, and many others [6]. Diseases affecting myelin sheath formation or nerve synapses alter electrical impulses in neurons, which may lead to tremors [52].

Congenital infection of piglets born from gilts infected with BVDV has been reported in some studies [22,23], in which piglets died between the 2nd and the 16th week of life with signs similar to that of CSF, in addition to showing growth retardation. In a study in which pregnant sows were inoculated with BVDV on the 35th and 45th day of gestation by intrauterine and intranasal route, respectively, nine piglets born from females infected by the intrauterine route and five born from animals infected by intranasal route were born persistently infected [22]. Other studies have also shown transplacental infection, with the virus isolated in at least one of the fetuses [32,53,54].

Conversely, Pereira [27] inoculated BVDV-2 in groups of pregnant gilts at different stages of gestation and before artificial insemination (AI). Seroconversion and a transient viremia were detected, but reproductive losses and clinical signs of the disease in gilts and piglets were not observed. Other recent studies analyzed groups of gilts inoculated by BVDV-2 by oronasal [28,29] and intrauterine [28] routes on the 45th day of pregnancy, before the fetal immunocompetence period. No transplacental transmission was observed since piglets from oro-nasally inoculated gilts were born BVDV-free; no anti-BVDV antibodies were detected in piglets at birth but were acquired by colostral passive transfer. Congenital persistent infection was not observed since piglets did not shed BVDV at any moment. Rates of BVDV transmission between pigs under field conditions is very low, and under experimental conditions it would be even more limited [33].

Regarding intrauterine inoculation [28], piglets were born with no clinical signs of infection and no signs of hypomyelination or CT. Surprisingly, high anti-BVDV-2 antibody titers were found. Serological investigations in bovine fetuses experimentally inoculated with the virus also indicated the development of specific immune competence before the period already established in the literature [51]. The average period of seroconversion of the gilts challenged with the virus was 20 days [28], varying between 12 and 33 days [27]. Other studies have described BVDV inducing viremia seven days post-infection and seroconversion three weeks after experimental inoculation in pigs [55–57]. BVDV was discarded as an etiological agent of CT [28], differently from atypical porcine pestivirus (APPV), which was linked to CT-disease in experimental and natural infection conditions [58–61].

Understanding the role of BVDV in the reproductive system of boars is valuable information, considering that biotechnological procedures have the expressive potential of spreading diseases to free herds [62]. When it comes to boars, the presence of agents of the genus Pestivirus has been confirmed in porcine semen. Shedding of CSFV in porcine semen was the first to be reported under natural and experimental infection [63], as well as virus transmission to sows and fetuses by AI and transplacentally, respectively [64]. Recently, APPV was also detected in the semen and preputial fluid of naturally infected boars, with a high viral load in semen [65]. Experimental infection with BVDV-2 did not result in changes in the post-period of pre-inoculation in most of the seminal characteristics evaluated, and no viral shedding was detected in semen or preputial fluid, but lymphocytosis and monocytopenia were observed [30]. Considering a mild and transient viremia, the likelihood that the circulating virus in the blood reaches different organs was low. Also, the blood-testis barrier would decrease the chance of reaching semen, which may explain the absence of viral RNA detection in the reproductive tract of the inoculated boars [30]. A BVDV persistently infected boar presented viral shedding in the ejaculate, which contained no sperm cells [49]. Possibly, BVDV transmission by semen occurs in atypical cases of congenital persistent infection in pigs [49].

4. Final Consideration

The course of BVDV infection in pigs will depend on the virulence of the viral strain and the pig immune response [66] and may be limited [44]. Even so, the presence of the virus in the nasal secretions of infected animals demonstrated that pigs could act as a source of infection, thus facilitating the spread in the herd [26,27]. Although BVDV does not pose the same threat to pig herds as it poses to ruminants, it may lead to the development of a range of clinical signs and culminate in a serological cross-reaction with the CSFV, interfering negatively in classical swine fever monitoring and surveillance programs, and misleading diagnosis of the disease [10].

Author Contributions: All authors have made substantial contributions to this paper, including manuscript organization, writing, editing, and approving the final version. All authors have read and agreed to the published version of the manuscript.

Funding: We are grateful for the grants #2014/13590-3 and #2016/21421-2, São Paulo Research Foundation (FAPESP); grant 409435/2016-3 of National Council for Scientific and Technological Development (CNPq) and an M.L.M-D Master's scholarship #2016/02982-3, São Paulo Research Foundation (FAPESP).

Acknowledgments: The authors appreciate the support provided by Prof. Eduardo Furtado Flores Federal University of Santa Maria (UFSM) and Dra. Edviges Maristela Pituco (Biological Institute of São Paulo).

Conflicts of Interest: The authors declare no conflict of interest. At the time the studies were conducted, IRHG was part of the Graduate Program in Veterinary Medicine at the School of Agricultural and Veterinarian Sciences, and started working at Ourofino Animal Health after completing his Ph.D.

References

1. King, A.M.; Lefkowitz, E.; Adams, M.J.; Carstens, E.B. (Eds.) *Virus Taxonomy: Ninth Report of the International Committee on Taxonomy of Viruses*; Elsevier: Amsterdam, The Netherlands, 2011; Volume 9.
2. Houe, H. Economic impact of BVDV infection in dairies. *Biologicals* **2003**, *31*, 137–143. [CrossRef]
3. Moennig, V.; Becher, P. Pestivirus control programs: How far have we come and where are we going? *Anim. Health Res. Rev.* **2015**, *16*, 83–87. [CrossRef] [PubMed]
4. King, A.M.; Lefkowitz, E.J.; Mushegian, A.R.; Adams, M.J.; Dutilh, B.E.; Gorbalenya, A.E.; Kropinski, A.M. Changes to taxonomy and the International Code of Virus Classification and Nomenclature ratified by the International Committee on Taxonomy of Viruses (2018). *Arch. Virol.* **2018**, *163*, 2601–2631. [CrossRef]
5. Yeşilbağ, K.; Alpay, G.; Becher, P. Variability and global distribution of subgenotypes of bovine viral diarrhea virus. *Viruses* **2017**, *9*, 128. [CrossRef] [PubMed]
6. Ridpath, J.; Bauermann, F.V.; Flores, E.F. *Flaviviridae*. In *Virologia Veterinária*, 2nd ed.; Flores, E.F., Ed.; Editora UFSM: Santa Maria, Rio Grande do Sul, Brazil, 2007.
7. Kirklant, P.; Le Potier, M.F.; Vannier, P.; Sinlaison, D. *Pestiviruses*. In *Diseases of Swine*, 10th ed.; Zimmerman, J.J., Karriker, L., Ramirez, A., Schwartz, K.J., Stevenson, G.W., Eds.; Blackwell: Oxford, UK, 2012.
8. Ridpath, J.F. Bovine viral diarrhea virus: Global status. *Vet. Clin. N. Am. Food Anim. Pract.* **2010**, *26*, 105–121. [CrossRef] [PubMed]
9. Deng, Y.; Sun, C.Q.; Cao, S.J.; Lin, T.; Yuan, S.S.; Zhang, H.B.; Wen, X.T.; Tong, G.Z. High prevalence of bovine viral diarrhea virus 1 in Chinese swine herds. *Vet. Microbiol.* **2012**, *159*, 490–493. [CrossRef]
10. Loeffen, W.L.A.; Van Beuningen, A.; Quak, S.; Elbers, A.R.W. Seroprevalence and risk factors for the presence of ruminant pestiviruses in the Dutch swine population. *Vet. Microbiol.* **2009**, *136*, 240–245. [CrossRef]
11. Mosena, A.C.; Weber, M.N.; Cibulski, S.P.; Silva, M.S.; Paim, W.P.; Silva, G.S.; Silveira, S. Survey for pestiviruses in backyard pigs in southern Brazil. *J. Vet. Diag. Investig.* **2020**, *32*, 136–141. [CrossRef]

12. Almeida, H.M.S.; Gatto, I.R.H.; dos Santos, A.C.R.; Ferraudo, A.S.; Samara, S.I.; de Oliveira, L.G. A Cross-Sectional and Exploratory Geospatial Study of Bovine Viral Diarrhea Virus (BVDV) Infections in Swines in the São Paulo State, Brazil. *Pak. Vet. J.* **2017**, *37*, 470–474.
13. Gatto, I.R.H.; Linhares, D.C.L.; de Souza Almeida, H.M.; Mathias, L.A.; de Medeiros, A.S.R.; Poljak, Z.; de Oliveira, L.G. Description of risk factors associated with the detection of BVDV antibodies in Brazilian pig herds. *Trop. Anim. Health Prod.* **2018**, *50*, 773–778. [CrossRef]
14. Liess, B.; Moennig, V. Ruminant pestivirus infection in pigs. *Rev. Sci. Tech.* **1990**, *9*, 151–161. [CrossRef] [PubMed]
15. O'Connor, M.; Lenihan, P.; Dillon, P. Pestivirus antibodies in pigs in Ireland. *Vet. Rec.* **1991**, *129*, 269. [CrossRef] [PubMed]
16. Løken, T.; Krogsrud, J.; Larsen, I.L. Pestivirus infections in Norway. Serological investigations in cattle, sheep and pigs. *Acta. Vet. Scand.* **1991**, *32*, 27–34. [PubMed]
17. Graham, D.A.; Calvert, V.; German, A.; McCullough, S.J. Pestiviral infections in sheep and pigs in Northern Ireland. *Vet. Rec.* **2001**, *148*, 69–72. [CrossRef] [PubMed]
18. Jensen, M.H. Screening for neutralizing antibodies against hog cholera-and/or bovine viral diarrhea virus in Danish pigs. *Acta. Vet. Scand.* **1985**, *26*, 72–80.
19. Weber, M.N.; Pino, E.H.M.; Souza, C.K.; Mósena, A.C.S.; Sato, J.P.H.; de Barcellos, D.E.S.N.; Canal, C.W. First evidence of bovine viral diarrhea virus infection in wild boars. *Acta Sci. Vet.* **2016**, *44*, 1–5. [CrossRef]
20. Weber, M.N.; Silveira, S.; Machado, G.; Groff, F.H.; Mósena, A.C.; Budaszewski, R.F.; Dupont, P.M.; Corbellini, L.G.; Canal, C.W. High frequency of bovine viral diarrhea virus type 2 in Southern Brazil. *Virus Res.* **2014**, *19*, 117–124. [CrossRef]
21. Becher, P.; Ramirez, R.A.; Orlich, M.; Rosales, S.C.; König, M.; Schweizer, M.; Thiel, H.J. Genetic and antigenic characterization of novel pestivirus genotypes: Implications for classification. *Virology* **2003**, *311*, 96–104. [CrossRef]
22. Paton, D.J.; Done, S.H. Congenital infection of pigs with ruminant-type pestiviruses. *J. Comp. Pathol.* **1994**, *111*, 151–163. [CrossRef]
23. Terpstra, C.; Wensvoort, G. Natural infections of pigs with bovine viral diarrhea virus associated with signs resembling swine fever. *Res. Vet. Sci.* **1988**, *45*, 137–142. [CrossRef]
24. Paton, D.J.; Lowings, J.P.; Barrett, A.D.T. Epitope mapping of the gp53 envelope protein of bovine viral diarrhea virus. *Virology* **1992**, *190*, 763–772. [CrossRef]
25. Santos, A.C.R.; Nascimento, K.A.; Mechler, M.L.; Almeida, H.M.S.; Gatto, I.R.H.; Carnielli, L.G.F.; Pollo, A.S.; De Oliveira, L.G. Experimental infection and evaluation of airborne transmission and nose-to-nose contact of bovine viral diarrhea virus among weaned piglets. *AJBAS* **2017**, *11*, 12–19.
26. Nascimento, K.A.; Mechler, M.L.; Gatto, I.R.; Almeida, H.M.S.; Pollo, A.S.; Sant'Ana, F.J.; Oliveira, L.G.D. Evidence of bovine viral diarrhea virus transmission by back pond water in experimentally infected piglets. *Pesq. Vet. Bras.* **2018**, *38*, 1896–1901. [CrossRef]
27. Pereira, D.A.; Peron, J.B.; de Souza Almeida, H.M.; Baraldi, T.G.; Gatto, I.R.H.; Kasmanas, T.C.; de Oliveira, L.G. Experimental inoculation of gilts with bovine viral diarrhea virus 2 (BVDV-2) does not induce transplacental infection. *Vet. Microbiol.* **2018**, *225*, 25–30. [CrossRef]
28. Mechler, M.L.; dos Santos Gomes, F.; Nascimento, K.A.; de Souza-Pollo, A.; Pires, F.F.B.; Samara, S.I.; Pituco, M.E.; de Oliveira, L.G. Congenital tremor in piglets: Is bovine viral diarrhea virus an etiological cause? *Vet. Microbiol.* **2018**, *220*, 107–112. [CrossRef]
29. Gomes, F.S.; Mechler-Dreibi, M.L.; Gatto, I.R.H.; Storino, G.Y.; Pires, F.F.B.; Xavier, E.B.; Samara, S.I. Congenital persistent infection with bovine viral diarrhea virus not observed in piglets. *Can. Vet. J.* **2019**, *10*, 1220–1222.
30. Storino, G.Y.; Xavier, E.B.; Mechler-Dreibi, M.L.; Simonatto, A.; Gatto, I.R.H.; Oliveira, M.E.F.; de Oliveira, L.G. No effects of noncytopathic bovine viral diarrhea virus type 2 on the reproductive tract of experimentally inoculated boars. *Vet. Microbiol.* **2020**, *240*, 108512. [CrossRef]
31. Walz, P.H.; Baker, J.C.; Mullaney, T.P.; Kaneene, J.B.; Maes, R.K. Comparison of type I and type II bovine viral diarrhea virus infection in swine. *Can. J. Vet. Res.* **1999**, *63*, 119.
32. Walz, P.H.; Baker, J.C.; Mullaney, T.P.; Maes, R.K. Experimental inoculation of pregnant swine with type 1 bovine viral diarrhoea virus. *J. Vet. Med. Ser. B* **2004**, *51*, 191–193. [CrossRef]
33. Wieringa-Jelsma, T.; Quak, S.; Loeffen, W.L.A. Limited BVDV transmission and full protection against CSFV transmission in pigs experimentally infected with BVDV type 1b. *Vet. Microbiol.* **2006**, *118*, 26–36. [CrossRef]

34. Langohr, I.M.; Stevenson, G.W.; Nelson, E.A.; Lenz, S.D.; Wei, H.; Pogranichniy, R.M. Experimental co-infection of pigs with Bovine viral diarrhea virus 1 and Porcine circovirus-2. *J. Vet. Diag. Investig.* **2012**, *24*, 51–64. [CrossRef] [PubMed]
35. Chase, C.C. The impact of BVDV infection on adaptive immunity. *Biologicals* **2013**, *41*, 52–60. [CrossRef] [PubMed]
36. Iwasaki, A.; Medzhitov, R. Regulation of adaptive immunity by the innate immune system. *Science* **2010**, *327*, 291–295. [CrossRef] [PubMed]
37. Vilček, Š.; Nettleton, P.F. Pestiviruses in wild animals. *Vet. Microbiol.* **2006**, *116*, 1–12. [CrossRef]
38. Fan, X.Z.; Ning, Y.B.; Wang, Q.; Xu, L.; Shen, Q.C. Detection of bovine viral diarrhea virus as contaminant in classical swine fever virus live vaccine with RT-PCR. *Chin. J. Vet. Med.* **2010**, *46*, 8–10.
39. Houe, H.; Lindberg, A.; Moennig, V. Test strategies in bovine viral diarrhea virus control and eradication campaigns. *Eur. J. Vet. Diag. Investig.* **2006**, *18*, 427–436. [CrossRef]
40. Kliučinskas, R.; Lukauskas, K.; Milius, J.; Vyšniauskis, G.; Kliučinskas, D.; Šalomskas, A. Detection of bovine viral diarrhoea virus in saliva samples. *Bull. Vet. Inst. Pulawy* **2008**, *52*, 31–37.
41. Vilček, Š.; Strojny, L.; Ďurkovič, B.; Rossmanith, W.; Paton, D. Storage of bovine viral diarrhoea virus samples on filter paper and detection of viral RNA by a RT-PCR method. *J. Virol. Methods* **2001**, *92*, 19–22. [CrossRef]
42. Blank, W.A.; Henderson, K.S.; White, L.A. Virus PCR assay panels: An alternative to the mouse antibody production test. *Lab Anim.* **2004**, *33*, 26–32. [CrossRef]
43. Samara, S.I.; Dias, F.C.; Moreira, S.P.G. Ocorrência da diarréia viral bovina nas regiões sul do Estado de Minas Gerais e nordeste do Estado de São Paulo. *Braz. J. Vet. Res. Anim. Sci.* **2004**, *41*, 396–403. [CrossRef]
44. O'Sullivan, T.; Friendship, R.; Carman, S.; Pearl, D.L.; McEwen, B.; Dewey, C. Seroprevalence of bovine viral diarrhea virus neutralizing antibodies in finisher hogs in Ontario swine herds and targeted diagnostic testing of 2 suspect herds. *Can. Vet. J.* **2011**, *52*, 1342–1344. [PubMed]
45. Dahle, J.; Patzelt, T.; Schagemann, G.; Liess, B. Antibody prevalence of hog cholera, bovine viral diarrhoea and Aujeszky's disease virus in wild boars in northern Germany. *DTW* **1993**, *100*, 330–333.
46. Sedlak, K.; Bartova, E.; Machova, J. Antibodies to selected viral disease agents in wild boars from the Czech Republic. *J. Wildl. Dis.* **2008**, *44*, 777–780. [CrossRef] [PubMed]
47. Milićević, V.; Maksimović-Zorić, J.; Veljović, L.; Kureljušić, B.; Savić, B.; Cvetojević, Đ.; Radosavljević, V. Bovine viral diarrhea virus infection in wild boar. *Res. Vet. Sci.* **2018**, *119*, 76–78. [CrossRef]
48. Gatto, I.R.; Di Santo, L.G.; Storino, G.Y.; Sanfilippo, L.F.; Ribeiro, M.G.; Mathias, L.A.; De Oliveira, L.G. Serological survey of bovine viral diarrhea (BVDV-1), brucellosis, and leptospirosis in captive white-lipped peccaries (Tayassu pecari) from the Midwest region in Brazil. *Austral J. Vet. Sci.* **2020**, *52*, 37–42. [CrossRef]
49. Terpstra, C.; Wensvoort, G. A congenital persistent infection of bovine virus diarrhoea virus in pigs: Clinical, virological and immunological observations. *Vet. Q.* **1997**, *19*, 97–101. [CrossRef]
50. Thurmond, M.C. Virus Transmission. In *Bovine Viral Diarrhea Virus: Diagnosis, Management, and Control*; Goyal, S.M., Ridpath, J.F., Eds.; John Wiley & Sons: Hoboken, NJ, USA, 2008.
51. Ohmann, H.B. Experimental fetal infection with bovine viral diarrhea virus. II. Morphological reactions and distribution of viral antigen. *Can. J. Comp. Med.* **1982**, *46*, 363.
52. Scarratt, W.K. Cerebellar disease and disease characterized by dysmetria or tremors. *Vet. Clin. N. Am. Food Anim. Pract.* **2004**, *20*, 275–286. [CrossRef]
53. Grooms, D.L. Reproductive losses caused by bovine viral diarrhea virus and leptospirosis. *Theriogenology* **2006**, *66*, 624–628. [CrossRef]
54. Bachofen, C.; Vogt, H.R.; Stalder, H.; Mathys, T.; Zanoni, R.; Hilbe, M.; Peterhans, E. Persistent infections after natural transmission of bovine viral diarrhoea virus from cattle to goats and among goats. *Vet. Res.* **2013**, *44*, 32. [CrossRef]
55. Stewart, W.C.; Miller, L.D.; Kresse, J.I.; Snyder, M.L. Bovine viral diarrhea infection in pregnant swine. *Am. J. Vet. Res.* **1980**, *41*, 459–462. [PubMed]
56. Kulcsar, G.; Soos, P.; Kucsera, L.; Glavits, R.; Pálfi, V. Pathogenicity of a bovine viral diarrhoea virus strain in pregnant sows. *Acta. Vet. Hung.* **2001**, *49*, 117–120. [CrossRef] [PubMed]
57. Makoschey, B.; Liebler-Tenorio, E.M.; Biermann, Y.M.; Goovaerts, D.; Pohlenz, J.F. Leukopenia and thrombocytopenia in pigs after infection with bovine viral diarrhoea virus-2 (BVDV-2). *DTW* **2002**, *109*, 225–230.

58. Arruda, B.L.; Arruda, P.H.; Magstadt, D.R.; Schwartz, K.J.; Dohlman, T.; Schleining, J.A.; Victoria, J.G. Identification of a divergent lineage porcine pestivirus in nursing piglets with congenital tremors and reproduction of disease following experimental inoculation. *PLoS ONE* **2016**, *11*, e0150104. [CrossRef]
59. De Groof, A.; Deijs, M.; Guelen, L.; Van Grinsven, L.; Os-Galdos, V.; Vogels, W.; Suijskens, J. Atypical porcine pestivirus: A possible cause of congenital tremor type A-II in Newborn piglets. *Viruses* **2016**, *8*, 271. [CrossRef]
60. Schwarz, L.; Riedel, C.; Högler, S.; Sinn, L.J.; Voglmayr, T.; Wöchtl, B.; Rümenapf, T. Congenital infection with atypical porcine pestivirus (APPV) is associated with disease and viral persistence. *Vet. Res.* **2017**, *48*, 1. [CrossRef]
61. Gatto, I.R.H.; Harmon, K.; Bradner, L.; Silva, P.; Linhares, D.C.L.; Arruda, P.H.; Arruda, B.L. Detection of atypical porcine pestivirus in Brazil in the central nervous system of suckling piglets with congenital tremor. *Transbound. Emerg. Dis.* **2018**, *65*, 375–380. [CrossRef]
62. Maes, D.; Van Soom, A.; Appeltant, R.; Arsenakis, I.; Nauwynck, H. Porcine semen as a vector for transmission of viral pathogens. *Theriogenology* **2016**, *85*, 27–38. [CrossRef]
63. Choi, C.; Chae, C. Detection of classical swine fever virus in boar semen by reverse transcription–polymerase chain reaction. *J. Vet. Diag. Investig.* **2003**, *15*, 35–41. [CrossRef]
64. De Smit, A.J.; Bouma, A.; Terpstra, C.; Van Oirschot, J.T. Transmission of classical swine fever virus by artificial insemination. *Vet. Microbial.* **1999**, *67*, 239–249. [CrossRef]
65. Gatto, I.R.H.; Arruda, P.H.; Visek, C.A.; Victoria, J.G.; Patterson, A.R.; Krull, A.C.; Arruda, B.L. Detection of atypical porcine pestivirus in semen from commercial boar studs in the United States. *Transbound. Emerg. Dis.* **2018**, *65*, e339–e343. [CrossRef] [PubMed]
66. Penrith, M.L.; Vosloo, W.; Mather, C. Classical swine fever (hog cholera): Review of aspects relevant to control. *Transbound. Emerg. Dis.* **2011**, *58*, 187–196. [CrossRef] [PubMed]

viruses

MDPI

Article

Clinical and Serological Evaluation of LINDA Virus Infections in Post-Weaning Piglets

Alexandra Kiesler [1,†], Kerstin Seitz [1,†], Lukas Schwarz [2,†], Katharina Buczolich [1], Helga Petznek [1], Elena Sassu [2], Sophie Dürlinger [2], Sandra Högler [3], Andrea Klang [3], Christiane Riedel [1], Hann-Wei Chen [1], Marlene Mötz [1], Peter Kirkland [4], Herbert Weissenböck [3], Andrea Ladinig [2], Till Rümenapf [1] and Benjamin Lamp [1,*,‡]

1 Institute of Virology, Department for Pathobiology, University of Veterinary Medicine, Veterinaerplatz 1, 1210 Vienna, Austria; Alexandra.Kiesler@vetmeduni.ac.at (A.K.); Kerstin.Seitz@vetmeduni.ac.at (K.S.); katharina.buczolich@gmx.net (K.B.); Helga.Petznek@vetmeduni.ac.at (H.P.); Christiane.Riedel@vetmeduni.ac.at (C.R.); Hann-Wei.Chen@vetmeduni.ac.at (H.-W.C.); Marlene.Moetz@vetmeduni.ac.at (M.M.); Till.Ruemenapf@vetmeduni.ac.at (T.R.)

2 Department for Farm Animals and Veterinary Public Health, University Clinic for Swine, University of Veterinary Medicine, Veterinaerplatz 1, 1210 Vienna, Austria; Lukas.Schwarz@vetmeduni.ac.at (L.S.); Elena.Sassu@vetmeduni.ac.at (E.S.); Sophie.Duerlinger@vetmeduni.ac.at (S.D.); Andrea.Ladinig@vetmeduni.ac.at (A.L.)

3 Institute of Pathology and Forensic Veterinary Medicine, Department of Pathobiology, University of Veterinary Medicine, Veterinaerplatz 1, 1210 Vienna, Austria; Sandra.Hoegler@vetmeduni.ac.at (S.H.); Andrea.Klang@vetmeduni.ac.at (A.K.); Herbert.Weissenboeck@vetmeduni.ac.at (H.W.)

4 Virology Laboratory, Elizabeth Macarthur Agriculture Institute, Woodbridge Rd, Menangle, New South Wales 2568, Australia; Peter.Kirkland@dpi.nsw.gov.au

* Correspondence: Benjamin.J.Lamp@vetmed.uni-giessen.de; Tel.: +0049-641-99-383-56

† These authors contributed equally to the work.

‡ Current affiliation: Institute of Virology, Faculty of Veterinary Medicine, Justus-Liebig-Universität, Schubertstrasse 81, 35392 Giessen, Germany; Benjamin.J.Lamp@vetmed.uni-giessen.de.

Received: 17 September 2019; Accepted: 21 October 2019; Published: 23 October 2019

Abstract: The novel pestivirus species known as lateral-shaking inducing neuro-degenerative agent (LINDA) virus emerged in 2015 in a piglet-producing farm in Austria. Affected piglets showed strong congenital tremor as a result of severe lesions in the central nervous system. Here, we report the results of a controlled animal infection experiment. Post-weaning piglets were infected with LINDA to determine the susceptibility of pigs, the clinical consequences of infection and the humoral immune response against LINDA. No clinically overt disease signs were observed in the piglets. Viremia was hardly detectable, but LINDA was present in the spleen and several lymphatic organs until the end of the experiment on day 28 post-infection. Oronasal virus shedding together with the infection of one sentinel animal provided additional evidence for the successful replication and spread of LINDA in the piglets. Starting on day 14 post-infection, all infected animals showed a strong humoral immune response with high titers of neutralizing antibodies against LINDA. No cross-neutralizing activity of these sera with other pestiviral species was observed. According to these data, following postnatal infection, LINDA is a rather benign virus that can be controlled by the pig's immune system. However, further studies are needed to investigate the effects of LINDA on the fetus after intrauterine infection.

Keywords: Linda virus; serological profile; virus neutralization assay; virus pathogenicity; humoral immune response

1. Introduction

The genus *Pestivirus* within the family Flaviviridae currently comprises 11 different species—recently termed *Pestivirus A–K* [1]. In addition to the long known classical swine fever virus (CSFV, *Pestivirus C*), a number of other pestivirus species have been identified in the porcine host in recent years, such as border disease virus (BDV, *Pestivirus D*), bovine viral diarrhea virus (BVDV-1, *Pestivirus A*), Bungowannah virus (BUNGO, *Pestivirus F*), and atypical porcine pestivirus (APPV, *Pestivirus K*) [1–5]. In 2015, we detected a yet unknown pestivirus species in a piglet-producing farm in Austria, which was termed lateral-shaking inducing neuro-degenerative agent (LINDA) virus [6]. Since the nucleotide sequence of LINDA shows a significant divergence of over 30% compared to the accepted pestivirus species, we proposed it as the new species *Pestivirus L* [1,6].

Pestiviruses are small enveloped viruses with a positive-sense, single-stranded, non-segmented RNA genome with a length of about 12 to 13 kilobases (kb) [7]. The genome consists of one large open reading frame (ORF), flanked by 5′- and 3′-non-coding regions [7]. This single ORF encodes a hypothetical polyprotein, that is co- and post-translationally processed into non-structural and structural proteins by viral and cellular proteases [8]. The three structural glycoproteins, termed E^{rns}, E1 and E2, and the nucleocapsid protein named Core are generated by cellular proteases [9,10]. The generation of the non-structural proteins N^{pro}, p7, NS2, NS3, NS4A, NS4B, NS5A and NS5B is very complex. Multiple processing steps mediated by autoproteases (N^{pro} and NS2) and the major NS3/4A protease yield partially processed precursors, mature proteins and enzymatically active protein fragments [8,11–13]. The presence of the autoprotease N^{pro} and the envelope glycoprotein E^{rns} are recognized as characteristic of the genus *Pestivirus* [1,7]. Since the corresponding proteins have been found in the genome of LINDA, it can undoubtedly be classified in the genus *Pestivirus* [6].

CSFV is listed by the World Organization for Animal Health (OIE) as an economically important pig pathogen [14]. The clinical signs of classical swine fever (CSF) vary significantly depending on the virulence of the virus strain as well as the age and susceptibility of the infected pigs. CSF is usually characterized by fever, skin lesions, convulsions and, especially in young animals, death within a few days [15]. BUNGO emerged on a pig farm in Australia in 2003, causing an increased rate of stillbirths, mummification and sudden deaths of piglets [2,16]. Experimental studies were conducted to investigate the pathogenicity of BUNGO in weaner pigs and porcine fetuses under laboratory conditions. Despite the low pathogenicity of the virus in weaned piglets, a long-lasting viremia, efficient virus shedding and rapid seroconversion were detected [17]. In contrast, a multifocal non-suppurative myocarditis with myonecrosis was observed following direct fetal exposure to BUNGO mimicking intrauterine infection [18]. APPVs were discovered in the United States in 2015 by next-generation sequencing [4], and subsequently detected in many countries around the world [19–23]. A close correlation between intrauterine APPV infections and the occurrence of congenital tremor (CT) type A-II in newborn piglets was reported [24]. The simultaneous detection of nucleic acids of APPV and hypomyelination in the central nervous system of these piglets implied a causative role of APPV for the appearance of the so-called shaking piglet syndrome [20]. This causal relationship is further supported by the birth of shaking piglets after inoculation of pregnant sows with APPV-containing material [24].

LINDA was discovered during the investigation of an outbreak of CT in a piglet-producing farm. We identified the agent, isolated the virus, sequenced its genome and established a RT-PCR assay as well as serological reagents for its detection [6]. Since then, LINDA has not been found in any other farm in Austria or elsewhere in the world [25]. To gain a deeper insight into the biology of this virus, we infected weaned piglets with LINDA under controlled experimental conditions. The aim of this small-scale animal experiment was the determination of susceptibility, pathogenicity and virulence of LINDA in the immunocompetent porcine host. Sera from the experimentally infected piglets were further used to characterize the humoral immune response against LINDA and to study the induction of cross-neutralizing antibodies against other pestiviruses.

2. Materials and Methods

2.1. Cells and Viruses

SK-6 cells [26] and MDBK cells (ATCC® CCL-22™) [27] were grown in Dulbecco's modified Eagle's medium (DMEM) supplemented with 10% fetal bovine serum (FCS, Bio and Sell GmbH, Feucht, Germany; negatively tested for the presence of pestiviruses), 100 U/mL penicillin and 100 µg/mL streptomycin. All cells were maintained at 37 °C and 5% CO_2. Cell culture-derived LINDA was used for the experiments. After initial cell culture isolation from a clinical case (passage 1, P1), a primary LINDA stock was generated containing a 50% tissue culture infectious dose ($TCID_{50}$) of 1.1×10^7 (P2, GenBank® KY436034) [6]. The virus was titrated in an endpoint dilution assay and supernatant from a single focus was harvested (P3) to ensure freedom from other pathogens. A master stock (P4) was prepared and characterized by RT-PCR and subsequent Sanger sequencing. In direct comparison with the consensus sequence of the original isolate no mutations were detected. All LINDA infection doses used for animal inoculations were recovered from the master stock and thus represent cell culture passage 5 of LINDA. BVDV-1a strain NADL and BVDV-1b strain NCP7 were obtained from E. Dubovi (Cornell University College of Veterinary Medicine, Ithaca, NY) [28]. CSFV 2.3 Alfort-Tübingen [29], BDV-1 X818 [30], BVDV-2 strain 890 [31], and BVDV-3 (unpublished strain isolated from FCS, South American origin) were obtained from the virus collection of the Institute of Virology in Giessen (Justus-Liebig-University, Giessen, Germany). Pestivirus strain Giraffe-1 [32] was a gift from D. J. Paton, Animal Health and Veterinary Laboratory Agency (AHVLA, Weybridge, United Kingdom). BUNGO was obtained from stocks of the Elizabeth Macarthur Agricultural Institute (Department of Primary Industries, Menangle, New South Wales, Australia) [2].

2.2. Virus Infection and Titration

Infections of MDBK and SK-6 cells with various strains of the different pestivirus species were performed with the indicated multiplicity of infection (MOI). Virus stocks for the experiments were generated using 10-cm cell culture dishes infected with a MOI of 0.1. At 72 h post-infection, the cell culture supernatant was harvested, filtered through a 0.45 µm cellulose filter (Sartorius, Göttingen, Germany), aliquoted and stored at −80 °C until use. The $TCID_{50}$ of viral supernatants was determined in three replicates by an end-point dilution assay (EPDA). The virus titer was calculated using the Spearman-Kaerber algorithm [33]. Re-isolation of LINDA was performed using SK-6 cells and fresh sample material.

2.3. RT-PCR Detection

RNA was extracted from serum and tissue samples, saliva, feces, cultured cells or virus cell culture supernatant using the QIAamp Viral RNA Mini Kit and the RNeasy Mini Kit (Qiagen, Hilden, Germany) according to the manufacturer's instructions. RNA was eluted in 60 µL RNase free distilled water and directly used for RT-PCR or stored at –80 °C for subsequent analysis. RT-PCR was carried out using the OneTaq One-Step RT-PCR Kit (NEB, Ipswich, USA) or the One Step RT-PCR Kit (Qiagen) using the oligonucleotides PPF 5′-GTKATHCAATACCCTGARGC-3′ and PPR 5′-GGRTTCCAGGARTACATCA-3′ [6]. PCR amplicons were subjected to gel electrophoresis and purified with the peqGOLD Gel Extraction Kit (Peqlab, Erlangen, Germany), if needed.

2.4. Calibration-Curve-Estimated Copy Number of LINDA Genome Equivalents

For the quantification of virus excretion, viremia and virus loads of organ homogenates, qRT-PCRs were performed on an ABI 7500 cycler (Applied Biosystems, Foster City, USA) using the LINDA and BUNGO specific primers LVqRTfor196 (5′-CACTGGWAAGGATCACCCACT-3′) and LVqRTrev351 (5′-AATYACAACGGATAWTMTTTATACTGG-3′) and the FAM/TAMRA labeled probe, LVqRTprobe322 (5′-Fam-ATAGGATGCCGGCGGATGCCCTGT-TamRa -3′). For the generation of a calibration curve, a T-vector plasmid fragment harboring the cDNA target sequence was produced using EcoRI digest, gel-purified and spectrophotometrically quantified. The copy number of recombinant plasmid fragment was calculated following the formula: N (molecules per µL) = (C

(DNA concentration in µg/µL) / K (fragment size in bp)) × 185.5 × 10^{13}. The factor results from the weight of the genome-equivalent ssDNA molecules (330 daltons per base), the volume projection factors and the Avogadro constant (6.02 × 10^{23} mol^{-1}). In order to obtain a standard curve, a ten-fold dilution series of the DNA control was included in the qRT-PCR setup [34]. Cycling conditions were 50 °C for 10 min, 95 °C for 1 min and 40 cycles of 95 °C for 10 sec, 60 °C for 1 min (amplification and fluorescence detection step). Semi-quantitative copy number estimation was calculated by 7500 System SDS Software (Applied Biosystems) based on the calibration curve. The amplification of the dsDNA standard does not include reverse transcription, which is why minor deviations may occur between RNA samples and dsDNA standard amplification depending on the efficiency of the cDNA synthesis. The qRT-PCR assay presented here is linear up to 10 copies of the dsDNA template per reaction (Ct value of 38). Since 10 copies per reaction were included as the lowest template amount in the standard series, this value represents the limit of quantification and our lower assay cutoff. The genome equivalents of 1.0 µL of purified RNA were converted to copies per 1 mL liquid sample, 1 g tissue sample or copies per swab using appropriate projection factors. The samples from the animal experiment were measured once due to high numbers and limited financial resources. However, the qRT-PCR data were validated by virus isolation experiments carried out in parallel with all samples obtained during the experiment.

2.5. *Virus Neutralization Assay*

All virus neutralization assays (VNA) were prepared in triplicate in 96-well microtiter plates. Viruses used in the VNA were diluted in DMEM without FCS from stock solutions generating a final titer of 100 to 300 $TCID_{50}$ per 0.1 mL. Initial two-fold dilution series of the serum samples were prepared with DMEM without FCS generating a final serum dilution of 1:256 in the last wells. Highly reactive sera were further diluted in five-fold series to a final serum dilution of 1:$10^{5.6}$ in the last wells. 100 µL of each serum dilution were mixed with 100 µL of the respective virus solution containing between 100 to 300 $TCID_{50}$. After the virus was added, the VNA was incubated for 2 h at 37 °C in 5% CO_2. One hundred microliters of this virus/serum mixtures were added to 96-well flat bottom plates containing confluent cell monolayers and incubated for 48 h at 37 °C in 5% CO_2. Viral infections were detected by indirect immunofluorescence using murine monoclonal antibodies (MAbs) as indicated below. Back titration of each virus solution was performed in parallel. Defined positive and negative sera against the respective virus or groups were used as controls. The titers obtained from the VNAs were calculated using the Spearman-Kaerber algorithm and reported as the reciprocal of the serum dilution that inhibited infection of 50% of the cells (neutralization dose 50%, ND_{50}).

2.6. *Indirect Immunofluorescence Assay and Antibodies*

The immunofluorescence assays were performed as previously described [8]. Briefly, the cells were fixed with 4% paraformaldehyde for 20 min at 4 °C, permeabilized with 1% (vol/vol) Triton-X 100 (Merck, Darmstadt, Germany) in PBS and stained with the mouse MAb 6A5 [35] and A18 [36]. The monoclonal antibody 6A5 was used to detect the E2 molecule of BVDV-1, BVDV-2, BVDV-3, BDV, BUNGO, giraffe pestivirus and LINDA infections. CSFV E2 was detected by MAb A18. Goat anti-mouse IgG conjugated with Cy3 (Dianova, Hamburg, Germany) was used as a secondary antibody. A porcine BUNGO antiserum (748-09.10-1) collected from a naturally infected sow, which had produced an abnormal litter, was kindly provided by the Elizabeth Macarthur Agricultural Institute.

2.7. *Animal Experiment*

The animal experiment was approved by the ethics committee of the University of Veterinary Medicine, Vienna and the Federal Ministry of Science, Research and Economy according to the §§ 26ff. of the Austrian Animal Experiments Act from 2012 (Permission code: BMWF-68.205/0130-WF/V/3b/2017, permission date: 07.07.2017). Post-weaning crossbreed piglets (*Sus scrofa domestica*) were used for the study of LINDA pathogenesis and virulence in immunocompetent hosts. The pigs came from the breeding farm Medau of the University of Veterinary Medicine, Vienna. The entire herd of the farm was negative tested for swine pathogenic influenza A viruses (IAV; H3N2, H1N1, H1N2 and H1N1pan) and porcine reproductive and respiratory syndrome viruses (PRRSV). The mother sows of the experimental animals were vaccinated against parvovirus and erysipelas during lactation according to the manufacturer's instructions (Parvoruvac; Merial GmbH, Hallbergmoos, Germany). The piglets themselves were protected on day 21 with a combined vaccination against *Mycoplasma hyopneumoniae* and porcine circovirus-2 (Circoflex and Mycoflex; both from Boehringer Ingelheim Vetmedica GmbH, Ingelheim, Germany). No further special diagnostic tests were carried out as there were neither pathological nor clinical signs of disease in these piglets.

A week prior to the beginning of the trial (study days –7 to –1) 21 weaned piglets in the 13th week of life were housed in a biological safety unit (BSL-2) for adaptation. At the beginning of the animal experiment, six piglets were housed separately in order to later serve as sentinel animals. The remaining piglets were divided into three groups of five animals each and housed in separate units. One group was not infected and served as a negative control, one group was inoculated intramuscularly (i.m.) with 1×10^7 $TCID_{50}$ LINDA and the last group was infected intranasally (i.n.) with 1×10^7 $TCID_{50}$ LINDA (study day 0). One day after infection (study day 1), three sentinel animals were added to each of the infection-groups. A daily clinical score was determined for each individual animal. The general condition, behavior, body temperature, feed intake and weight gain of all animals were assessed and measured after the infection over a period of 28 days. Particular attention was paid to the occurrence of signs of disease. Each animal was assigned a daily clinical score, which included individual values for behavior, feed intake, dyspnea, ocular and nasal discharge, coughing and diarrhea in the range from 0 (physiological) to 3 (severe clinical symptoms) according to an established evaluation scheme of the University Clinic for Swine of the University of Veterinary Medicine, Vienna [37]. While the body temperature of all animals was also monitored daily, the body weight was assessed at the time point of arrival and on study days 0–7, 9, 14, 21 and 28. Blood and fecal samples as well as nasal and oral swabs were taken on the study days 0, 3, 5, 7, 14, 21 and 28/29. Urine samples (spontaneous urine samples or collected via cystocentesis) were obtained on study day 3 from most animals. All animals were euthanized with T61 (5.0 mg/mL tetracaine hydrochloride, 50 mg/mL mebezonium iodide and 200 mg/mL embutramide; 1 mL/10 kg) on study day 28 or 29 under general anesthesia (1.3 mg/kg azaperone and 10 mg/kg ketamine hydrochloride). During necropsy, organ samples were taken for molecular and pathohistological analysis. In particular, the LINDA RNA loads were analyzed in samples taken from the kidney, bladder, cerebellum, cerebrum, spinal cord, dorsal root ganglia, thymus, spleen, tonsils, lymph nodes (Lnn. inguinales, mesenteriales and mandibularis), parotid and sublingual glands, heart, lung, liver and intestinal segments from all animals.

2.8. Pathological Examinations

Tissue samples of the brain and spinal cord, dorsal root ganglia, liver, spleen, kidney, urinary bladder, thymus, spleen, tonsils, inguinal lymph node and mandibular gland were taken from all animals for histological examination. Additional samples of the coeliac ganglion, sciatic nerve, mesenteric and mandibular lymph nodes and parotid gland were included from the experimentally infected animals and sentinels. Five coronary sections of the brain were taken at the levels of telencephalon, diencephalon, cerebellum, mesencephalon and metencephalon. Transversal sections of the spinal cord were taken from cervical, thoracic and lumbar regions. The organ samples were fixed in 10% neutral buffered formalin, embedded in paraffin wax, sectioned at 2 μm and stained with hematoxylin and eosin (HE).

3. Results

3.1. Pathogenicity and Virulence of LINDA in Weaned Piglets

The animal experiment was performed with weaned piglets to investigate the clinical effects of LINDA infection in the immunocompetent host. The animals of all three groups (infected intranasally, intramuscularly or mock), including the sentinel animals, showed normal physiological parameters and good general condition throughout the 28-day infection period. Neither the individually examined clinical values nor the additive clinical score showed major pathological changes in individual animals or significant changes between the different groups (Figure 1). Mild ocular and nasal discharge and cough were observed early after infection (study days 0 to 7) in most animals of both LINDA infected groups. However, these changes were also observed in sentinel animals of these groups that did not show seroconversion to LINDA (n.i. sent., described below). Mild fever (maximum rectal temperature of 40.7 °C) occurred in all i.m. infected animals within the first three days after inoculation. The single sentinel animal that later seroconverted showed an elevated body temperature and mild diarrhea. No differences were found between the groups in necropsy and histological examination. Gross examination revealed alveolar emphysema of the lung in animals of all the groups. Alveolar edema and pulmonary hyperemia were detected in some animals of all groups. Macroscopically, the central nervous system of all animals appeared normal. However, in the histological examination the majority of the animals from all groups ($n = 17$) showed mild, oligofocal, randomly distributed perivascular, mononuclear infiltrations and some glial nodules in the brain and/or spinal cord. A slight follicular hyperplasia of the spleen was observed in one sentinel animal of the i.n. and i.m. group. Stomach lesions such as ulcerations, hyperkeratinization or follicular hyperplasia were found in almost all animals. Mild, predominantly mononuclear, sometimes suppurating, interstitial nephritis or cortical infarction were found in many animals ($n = 13$) including animals from the non-infected control group. An interstitial lymphocytic infiltration of the mandibular gland was evident in one sentinel animal each of intranasally and intramuscularly infected group. An abscess was observed in the snout of an intranasally infected animal. The initial body weight of each animal was defined as 100% and the relative weight gain for each piglet and the average weight gain of the groups were calculated for each study day. The non-infected control group showed a slightly higher weight gain compared to the infected groups and the sentinel animals. However, the differences in weight gain between the experimental groups were neither pronounced nor statistically significant (Figure 2).

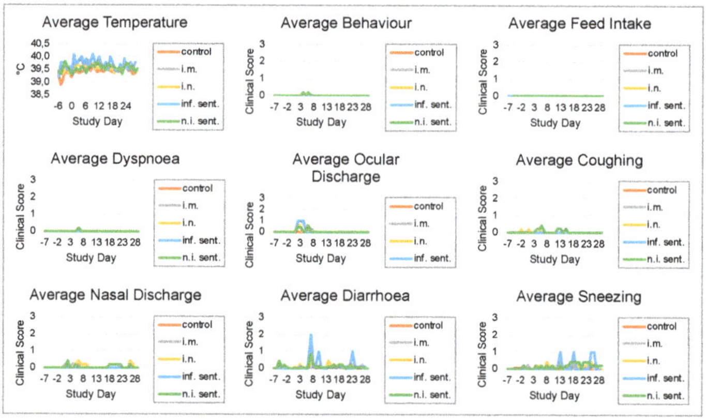

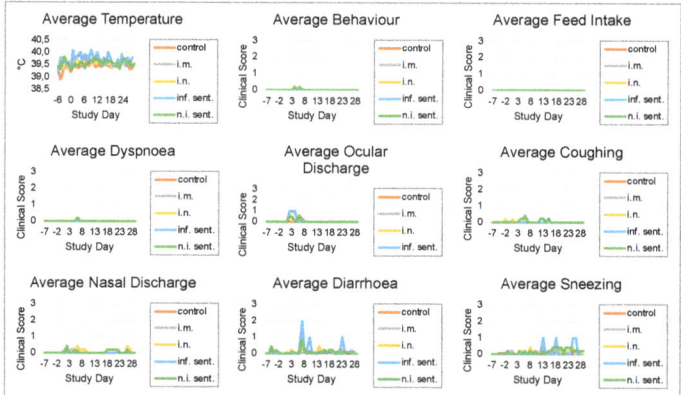

Figure 1. Pathogenicity of lateral-shaking inducing neuro-degenerative agent (LINDA) in immunocompetent piglets. A total of 21 piglets were divided in five groups: five negative control, five intramuscularly (i.m.) infected and five intranasally (i.n.) infected animals as well as three sentinel animals for the i.m. and three sentinel animals for the i.n. group. The animals from the i.m. and i.n. infection groups were inoculated with LINDA (1×10^7 $TCID_{50}$/mL) on study day 0. Body temperature, behavior, feed intake, dyspnea, ocular discharge, coughing, nasal discharge, diarrhea and sneezing were assessed daily and symptoms were classified by a scoring system with scores from 0 (physiological) to 3 (severe clinical symptoms). The mean of gathered parameters was calculated for the control, i.m. infected, i.n. infected, infected sentinel (inf. sent.) and non-infected sentinel (n.i. sent.) animals. No severe LINDA virus associated clinical signs were observed comparing infected and non-infected animals. However, mild fever and other signs of disease were seen in some infected animals, such as the infected sentinel animal within the i.m. group.

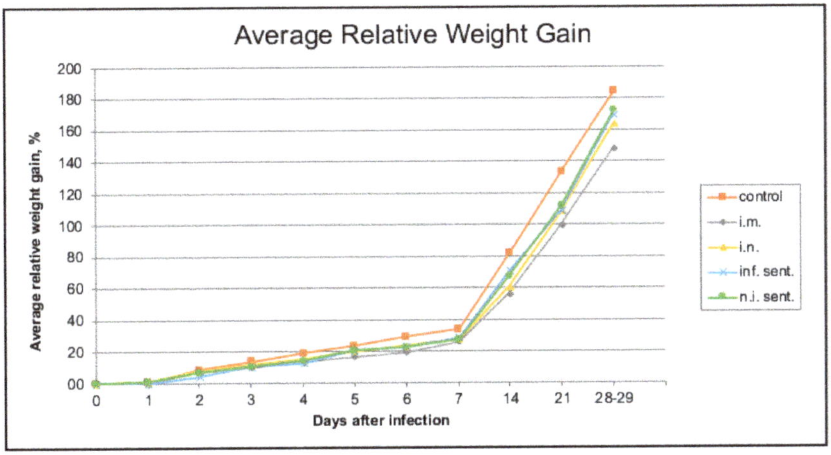

Figure 2. Average relative daily weight gain. Piglets were weighed on study days 0–7, 14, 21 and 28/29. The initial body weight was set as 100% value and relative weight gain was calculated for every individual at the indicated time-points. The mean value was calculated for every trial group. Infected groups show a slightly reduced weight gain compared to the negative control group after 7 days post-infection. However, the absolute differences between the groups were not significant.

3.2. Replication of LINDA in the Immunocompetent Porcine Host

Before the beginning of the experiment, blood samples of all animals were taken and tested for the presence of LINDA RNA and LINDA neutralizing antibodies. Serum samples were taken on study days 0, 3, 7, 14 and 28/29 of the experiment and analyzed in a LINDA-specific qRT-PCR assay as well as in virus isolation studies. No infectious LINDA or viral RNA were detected in the mock infected group at any time of the experiment. Most i.m. infected and all i.n. infected animals did not show a detectable viremia after LINDA inoculation. We observed a low level of viremia in two i.m. infected animals (5.0×10^5 GE/mL for animal 8 i.m. and 5.5×10^5 for animal 9 i.m. on study day 7) and a higher level in one sentinel animal of this group (2.3×10^7 GE/mL on study day 14, Figure 3). LINDA could be isolated from each of these qRT-PCR positive serum samples. Virus shedding was assessed using oral and nasal swabs as well as fecal and urine samples. Most of these samples gave negative results in the LINDA virus-specific qRT-PCR and virus isolation experiments. The oral swabs of one of the viremic piglets from the i.m. group (study day 7) and the oral and nasal swabs from three animals of the i.n. group gave signals below assay cutoff in the qRT-PCR assay (study days 3, 7 and/or 14). The RNA loads in these swabs were very low and cell culture virus isolation was not successful. An additional conventional RT-PCR was performed on the questionable samples as described before [6]. The amplification of LINDA-specific RT-PCR products was verified by nucleotide sequencing (Figure S1). LINDA RNA was not detectable in the fecal samples obtained on study days 3, 7 and 14 from the experimentally infected animals. However, the viremic sentinel animal (number 11) from the i.m. infected group showed substantial LINDA RNA loads in the feces (1.31×10^5 GE/g on study day 14) that also allowed successful virus isolation. All urine samples gave negative results in qRT-PCR and virus isolation. Multiple organs were sampled during necropsy on days 28/29 post-infection. Most organ samples gave negative results for the presence of LINDA RNA in the qRT-PCR. However, LINDA genomes were detectable in several lymphoid organs, such as the inguinal lymph nodes ($n = 4$), spleen ($n = 2$) and tonsils ($n = 4$) of animals of the i.m. infected group reaching values between 4.0×10^3 and 2.3×10^6 GE/g. LINDA RNA was also found in the infected sentinel animal of the i.m. group (animal 11) in the inguinal lymph nodes and spleen. Interestingly, the virus was detectable in the tonsils of all i.n. infected animals ($n = 5$), but only found in the inguinal lymph node of one of these animals ($n = 1$, Figure 3).

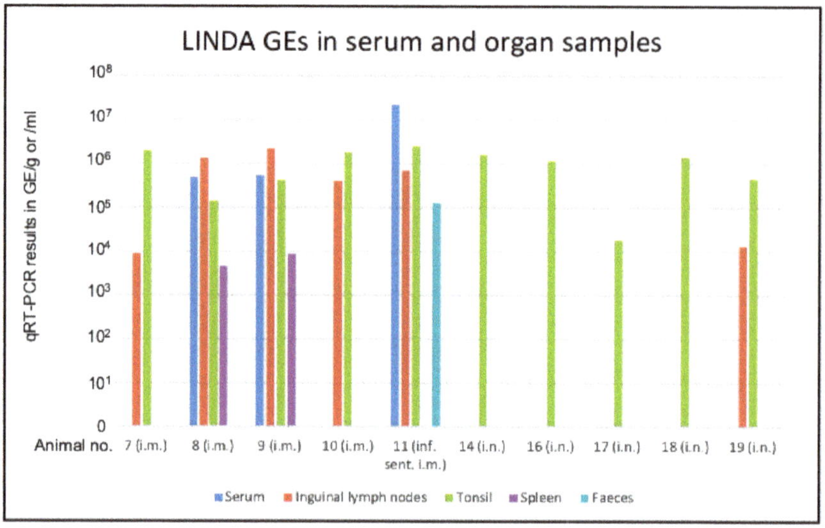

Figure 3. qRT-PCR results of the animal experiment. Multiple samples were analyzed for the presence of LINDA RNA before, during and after the experimental infection of piglets with LINDA. Blood and fecal samples as well as nasal and oral swabs were taken on study days 0, 3, 5, 7, 14, 21 and 28/29. Fecal samples were collected on study days 3, 7 and 14, while urine samples were only taken from most animals on day 3. Kidney, bladder, cerebellum, cerebrum, spinal cord, dorsal root ganglia, thymus, spleen, tonsils, different lymph nodes (Lnn. inguinales, mesenteriales and mandibularis), parotid and sublingual glands, heart, lung, liver and intestinal segments were sampled from all animals during necropsy on days 28/29. Most samples gave negative results for LINDA RNA. Only two i.m. infected animals showed detectable LINDA RNA loads in serum samples taken on study day 7 and one sentinel animal of the i.m. group on day 14. These serum samples also allowed the re-isolation of LINDA. Viral RNA was not detectable in urine samples and only detected in one fecal sample (animal 11, i.m. group, infected sentinel). However, the RNA of LINDA was detected in multiple samples from lymphoid organs demonstrating the presence of LINDA in the experimentally infected animals as well as in the infected sentinel animal until the end of the experiment.

3.3. Humoral Immune Response against LINDA

All serum samples obtained before, during and at the end of the experiment were tested for LINDA-specific antibodies using a LINDA virus neutralization assay (Figure 4). No virus neutralizing activity was measured in the serum samples taken before the start of the experiment ($ND_{50} < 1/2$, below limit of detection). Sera from the mock infected group and sera from the LINDA RNA negative sentinel animals showed no virus neutralizing activity at the end of the trial. A gradual onset of humoral immune responses was observed in all infected animals (i.m. and i.n.) between days 7 and 14 post-infection, reaching peak values of up to 1/8,640 ND_{50}/mL. A comparably strong reactivity of 1/1,028 was also seen in the serum sample of the i.m. infected animal 6, in which no LINDA RNA replication was detected throughout the experiment. Interestingly, the development of the humoral immune response was delayed to study day 21 in the LINDA infected sentinel animal 11.

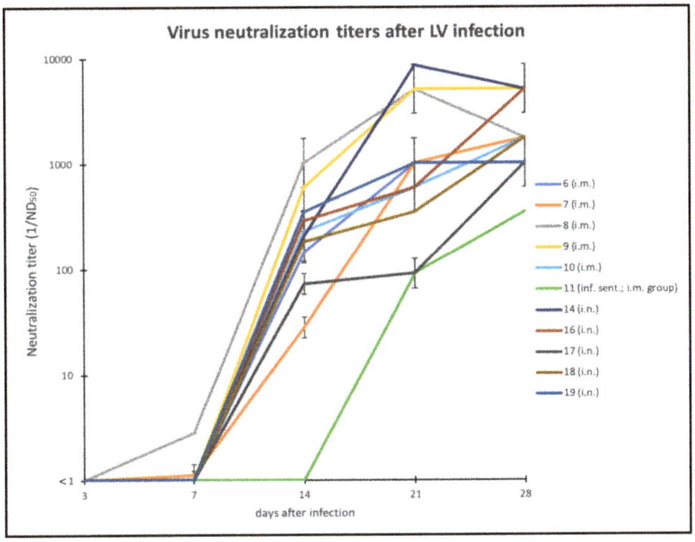

Figure 4. Virus neutralization activities of LINDA immune sera. The serum samples taken on study days 0, 3, 7, 14, 21 and 28/29 were tested in VNAs against LINDA. All VNAs were performed in triplicate and ND_{50} was calculated using the Spearman-Kaerber algorithm. The reciprocal ND_{50} value is presented for each serum sample including error bars for positive and negative standard deviation. All $1/ND_{50}$ values were less than 1 until study day 7, when the first weak neutralizing activities were measured in single animals. Significant neutralizing activities ($ND_{50} > 1{:}10$) were detected in all infected animals starting on day 14. Note the later onset of humoral immune response in the infected sentinel animal from the i.m. group on day 21.

3.4. Cross Neutralization of LINDA-Immune Sera with Other Pestivirus Species

To assess antigenic relations to other pestiviruses, we characterized the LINDA-immune sera from study days 28/29 for their cross-neutralizing activities against multiple pestiviral strains. In particular, we performed VNAs with the pestivirus species BVDV-1a (strain NADL), BVDV-1b (strain NCP7), BVDV-2 (strain 890), CSFV 2.3 (strain Alfort-Tübingen), BDV-1 (strain X818), pestivirus giraffe (strain giraffe 1), BUNGO and an unpublished BVDV-3 strain. We found no neutralizing activity of the LINDA-immune sera ($ND_{50} < 1/2$) against any of these viruses. Additionally, we tested a porcine BUNGO convalescence serum (748-09.10-1), initially obtained for immunodetection of BUNGO infections of cultured cells. This antiserum efficiently neutralized BUNGO (BUNGO ND_{50} 1/3200) as well as LINDA in our VNAs (LINDA ND_{50} 1/1,600) but had no effect on the infection of BVDV-1, BVDV-2 and CSFV (all $ND_{50} < 1/2$).

4. Discussion

The aim of this study was to investigate the pathogenicity of LINDA during post-natal infections and to characterize the humoral immune response against LINDA in order to obtain important basic data for sero-surveillance studies. Therefore, a small-scale animal experiment was set up using LINDA infections in post-weaning piglets. In this experiment, we found no evidence of severe acute disease in weaners caused by LINDA infection. Only mild clinical signs, such as mild fever, nasal discharge and mild changes in fecal consistency, were observed in single animals without significant influence on the growth rates of these piglets. Inflammatory infiltrates and glial nodules occurred in the brain and/or spinal cord of animals from all groups including the control group. Hence, these lesions were not associated with LINDA infection and were consequently interpreted as non-specific experimental background. Other findings of necropsy could be interpreted as random findings or might represent stress-related diseases, such as the typical gastric lesions. However, productive LINDA infections were observed in all animals, which were experimentally inoculated with 1×10^7 $TCID_{50}$, regardless of the infection route used. LINDA virus excretion was detected using qRT-PCR in oronasal fluids of some infected animals and excretion via feces was documented for a single animal. Viremia was hardly detectable in the serum of most animals, while all except one experimentally infected animal (animal 6 i.m.) showed a long-lasting presence of the virus in the tonsils and/or lymphatic organs. Unfortunately, no peripheral mononuclear blood cells (PBMCs) were preserved from the experiment that could have been used for a potentially more sensitive virus detection in the blood [38]. Therefore, future studies, also including pregnant sows, will investigate whether LINDA virus is associated with PBMCs and whether the analysis of isolated PBMCs allows a more sensitive diagnosis. The infection of one sentinel animal in the i.m. group confirmed that infectious loads of virus were secreted in this simulated acute infection scenario. When we consider the positive detection of LINDA in serum, nasal secretions and feces of some infected animals, blood contacts or oronasal uptake of the pathogen may be considered as possible pathways of infection as also demonstrated by the successful artificial i.m. and i.n. infection. Since the minimum infection dose of LINDA for piglets is unknown, it is up to future studies to clarify the natural routes of infection. The clinical data are in accordance with clinical and epidemiological data about other pestiviruses, such as BVDV, BDV or CSFV, showing that mononuclear cells and lymphoid organs are the primary targets of pestivirus infections [39,40]. Most pestiviruses are well-adapted to their host species and the acute infection of immunocompetent animals usually leads to mild to subclinical disease with limited virus replication [41,42]. Severe pestiviral disease, such as abortion, malformation or neurological disorders, is mostly a consequence of intrauterine infections and responsible for the high economic losses following pestiviral infections [43]. This circumstance also explains the differences in the clinical picture between the animal experiment with immunocompetent piglets presented here and the LINDA virus outbreak in a piglet breeding farm from which LINDA virus was originally isolated [6]. Similar results were obtained in infection studies with BUNGO in weaner piglets, where few clinical signs, a short phase of viremia (3 to 10 days post-inoculation) and low levels of virus excretion were observed [17]. We conclude that LINDA is not only the closest genetic relative of Bungowannah virus but also shows similar pathogenicity in pigs. The results of our post-natal experimental infection demonstrate that the clinical picture of BUNGO and LINDA in immunocompetent animals is distinct from symptoms of piglets affected by CSFV strains of high and moderate virulence but might be similar to low virulence CSFV strains [44]. Future studies evaluating LINDA infections of the unborn fetus will be necessary to assess the potential hazard of LINDA [18,45].

The acute infection of piglets with LINDA led to the development of a strong humoral immunity starting at about seven days post-infection (Figure 4). Despite high titers of neutralizing antibodies, viral RNA persisted in the tonsils and lymphoid organs as has also been shown for other pestiviruses. This phenomenon has been described for several pestiviruses and studied in detail for BVDV pointing to a potential risk of virus transmission from convalescent animals [46]. We used the generated LINDA-antisera to evaluate antigenic cross-reactivity between LINDA and other pestiviruses. Unfortunately, we could not include APPV in these tests, because no APPV strain was available that showed the necessary infectivity in the cell culture [19]. Our VNA data clearly demonstrate that antibodies from acute LINDA infections do not provide protection against infections with classical pestiviruses. Therefore, a serological interference with established VNAs for CSFV diagnosis is unlikely. However, further studies are needed to evaluate possible false positive reactions in serological CSFV tests using a larger sample size of LINDA-antisera as well as highly reactive immune sera obtained from sows infected during pregnancy with viremic, persistently infected piglets. Our data support the hypothesis that LINDA forms an independent species (*Pestivirus L*) within the genus *Pestivirus* with a highly divergent antigenic profile. Interestingly, the high titer BUNGO-antiserum showed a strong cross-neutralization activity in VNA against LINDA, while low to absent neutralization profiles of BUNGO-antisera against other pestiviruses were observed in previous studies [47]. This result is puzzling, because it is in conflict with the results obtained with the LINDA-antisera. A possible explanation could be that the species BUNGO and LINDA form an antigenic group within the genus *Pestivirus*, sharing conserved antigenic motifs important for virus neutralization. Again, future studies may answer this question by analyzing a larger sample size of BUNGO- and LINDA-antisera in cross-protection VNAs and, even more importantly, in controlled animal experiments.

5. Patents

The authors B.L., L.S. and T.R. are inventors on a patent on Linda pestivirus (PCT/EP2017/084453; Isolation of a novel pestivirus causing congenital tremor).

Supplementary Materials: The following are available online at http://www.mdpi.com/1999-4915/11/11/975/s1, Figure S1: Detection of LINDA virus by conventional RT-PCR in oro-nasal fluids.

Author Contributions: Conceptualization, K.S., L.S. and B.L.; Data curation, A.K. (Alexandra Kiesler), H.P., S.D, H.-W.C. and B.L.; Formal analysis, A.K. (Alexandra Kiesler), K.S., L.S., K.B. and B.L.; Funding acquisition, T.R. and B.L.; Investigation, A.K. (Alexandra Kiesler), K.S., L.S., K.B., H.P., E.S., S.D., S.H., A.K. (Andrea Klang), C.R., H.-W.C., M.M., P.K., H.W., A.L., T.R. and B.L.; Methodology, K.S., L.S., K.B., H.P., E.S., S.D., S.H., A.K. (Alexandra Kiesler), C.R., H.-W.C., M.M. and T.R.; Project administration, B.L.; Resources, P.K., H.W., A.L., T.R. and B.L.; Supervision, H.W., A.L. and B.L.; Validation, A.K. (Alexandra Kiesler), L.S., H.P. and B.L.; Visualization, A.K. (Alexandra Kiesler) and B.L.; Writing—original draft, A.K. (Alexandra Kiesler), T.R. and B.L.; Writing—review and editing, A.K. (Alexandra Kiesler), L.S., T.R. and B.L.

Funding: This research received no external funding.

Acknowledgments: The authors acknowledge Markus Cihar who was responsible for animal care and supporting during sampling. The authors thank Boehringer Ingelheim Vetmedica GmbH (Binger Strasse 173, 55216 Ingelheim am Rhein, Germany) for their financial support (Contract number: 396806) and Eric Vaughn (Boehringer Ingelheim Animal Health USA Inc., 2412 S Loop Dr, Ames, IA, United States) for valuable scientific discussions.

Conflicts of Interest: The authors declare no conflict of interest.

References

1. Smith, D.B.; Meyers, G.; Bukh, J.; Gould, E.A.; Monath, T.; Scott Muerhoff, A.; Pletnev, A.; Rico-Hesse, R.; Stapleton, J.T.; Simmonds, P.; et al. Proposed revision to the taxonomy of the genus Pestivirus, family Flaviviridae. *J. Gen. Virol* **2017**, *98*, 2106–2112. [CrossRef] [PubMed]
2. Kirkland, P.D.; Frost, M.J.; Finlaison, D.S.; King, K.R.; Ridpath, J.F.; Gu, X. Identification of a novel virus in pigs–Bungowannah virus: A possible new species of pestivirus. *Virus Res.* **2007**, *129*. [CrossRef] [PubMed]
3. Nagai, M.; Aoki, H.; Sakoda, Y.; Kozasa, T.; Tominaga-Teshima, K.; Mine, J.; Abe, Y.; Tamura, T.; Kobayashi, T.; Nishine, K.; et al. Molecular, biological, and antigenic characterization of a Border disease virus isolated from a pig during classical swine fever surveillance in Japan. *J. Vet. Diagn. Invest.* **2014**, *26*, 547–552. [CrossRef] [PubMed]
4. Hause, B.M.; Collin, E.A.; Peddireddi, L.; Yuan, F.; Chen, Z.; Hesse, R.A.; Gauger, P.C.; Clement, T.; Fang, Y.; Anderson, G. Discovery of a novel putative atypical porcine pestivirus in pigs in the USA. *J. Gen. Virol* **2015**, *96*, 2994–2998. [CrossRef]
5. Deng, Y.; Sun, C.Q.; Cao, S.J.; Lin, T.; Yuan, S.S.; Zhang, H.B.; Zhai, S.L.; Huang, L.; Shan, T.L.; Zheng, H.; et al. High prevalence of bovine viral diarrhea virus 1 in Chinese swine herds. *Vet. Microbiol* **2012**, *159*, 490–493. [CrossRef]
6. Lamp, B.; Schwarz, L.; Hogler, S.; Riedel, C.; Sinn, L.; Rebel-Bauder, B.; Weissenbock, H.; Ladinig, A.; Rumenapf, T. Novel Pestivirus Species in Pigs, Austria, 2015. *Emerg Infect. Dis* **2017**, *23*, 1176–1179. [CrossRef]
7. Simmonds, P.; Becher, P.; Bukh, J.; Gould, E.A.; Meyers, G.; Monath, T.; Muerhoff, S.; Pletnev, A.; Rico-Hesse, R.; Smith, D.B.; et al. ICTV Virus Taxonomy Profile: Flaviviridae. *J. Gen. Virol* **2017**, *98*, 2–3. [CrossRef]
8. Lamp, B.; Riedel, C.; Roman-Sosa, G.; Heimann, M.; Jacobi, S.; Becher, P.; Thiel, H.J.; Rumenapf, T. Biosynthesis of classical swine fever virus nonstructural proteins. *J. Virol.* **2011**, *85*, 3607–3620. [CrossRef]
9. Rümenapf, T.; Unger, G.; Strauss, J.H.; Thiel, H.J. Processing of the envelope glycoproteins of pestiviruses. *J. Virol.* **1993**, *67*, 3288–3294.
10. Stark, R.; Meyers, G.; Rümenapf, T.; Thiel, H.J. Processing of pestivirus polyprotein: Cleavage site between autoprotease and nucleocapsid protein of classical swine fever virus. *J. Virol.* **1993**, *67*, 7088–7095.

11. Lamp, B.; Riedel, C.; Wentz, E.; Tortorici, M.A.; Rumenapf, T. Autocatalytic cleavage within classical swine fever virus NS3 leads to a functional separation of protease and helicase. *J. Virol.* **2013**, *87*. [CrossRef] [PubMed]
12. Lackner, T.; Muller, A.; Pankraz, A.; Becher, P.; Thiel, H.J.; Gorbalenya, A.E.; Tautz, N. Temporal modulation of an autoprotease is crucial for replication and pathogenicity of an RNA virus. *J. Virol.* **2004**, *78*, 10765–10775. [CrossRef] [PubMed]
13. Tautz, N.; Elbers, K.; Stoll, D.; Meyers, G.; Thiel, H.J. Serine protease of pestiviruses: Determination of cleavage sites. *J. Virol.* **1997**, *71*, 5415–5422. [PubMed]
14. Inch, C. World Animal Health. *Can. Vet. J.* **2006**, *47*, 790–791.
15. Lohse, L.; Nielsen, J.; Uttenthal, A. Early pathogenesis of classical swine fever virus (CSFV) strains in Danish pigs. *Vet. Microbiol.* **2012**, *159*, 327–336. [CrossRef]
16. Kirkland, P.D.; Read, A.J.; Frost, M.J.; Finlaison, D.S. Bungowannah virus–a probable new species of pestivirus–what have we found in the last 10 years? *Anim. Health. Res. Rev.* **2015**, *16*, 60–63. [CrossRef]
17. Finlaison, D.S.; King, K.R.; Gabor, M.; Kirkland, P.D. An experimental study of Bungowannah virus infection in weaner aged pigs. *Vet. Microbiol.* **2012**, *160*, 245–250. [CrossRef]
18. Finlaison, D.S.; Cook, R.W.; Srivastava, M.; Frost, M.J.; King, K.R.; Kirkland, P.D. Experimental infections of the porcine foetus with Bungowannah virus, a novel pestivirus. *Vet. Microbiol.* **2010**, *144*, 32–40. [CrossRef]
19. Schwarz, L.; Riedel, C.; Högler, S.; Sinn, L.J.; Voglmayr, T.; Wöchtl, B.; Dinhopl, N.; Rebel-Bauder, B.; Weissenböck, H.; Ladinig, A.; et al. Congenital infection with atypical porcine pestivirus (APPV) is associated with disease and viral persistence. *Vet. Res.* **2017**, *48*, 1. [CrossRef]
20. Postel, A.; Hansmann, F.; Baechlein, C.; Fischer, N.; Alawi, M.; Grundhoff, A.; Derking, S.; Tenhundfeld, J.; Pfankuche, V.M.; Herder, V.; et al. Presence of atypical porcine pestivirus (APPV) genomes in newborn piglets correlates with congenital tremor. *Sci. Rep.* **2016**, *6*, 27735. [CrossRef]
21. Pan, S.; Yan, Y.; Shi, K.; Wang, M.; Mou, C.; Chen, Z. Molecular characterization of two novel atypical porcine pestivirus (APPV) strains from piglets with congenital tremor in China. *Transbound. Emerg. Dis.* **2019**, *66*, 35–42. [CrossRef] [PubMed]
22. Beer, M.; Wernike, K.; Drager, C.; Hoper, D.; Pohlmann, A.; Bergermann, C.; Schroder, C.; Klinkhammer, S.; Blome, S.; Hoffmann, B. High prevalence of highly variable atypical porcine pestiviruses found in Germany. *Transboun. Emerg. Dis.* **2016**. [CrossRef] [PubMed]
23. de Groof, A.; Deijs, M.; Guelen, L.; van Grinsven, L.; van Os-Galdos, L.; Vogels, W.; Derks, C.; Cruijsen, T.; Geurts, V.; Vrijenhoek, M.; et al. Atypical Porcine Pestivirus: A Possible Cause of Congenital Tremor Type A-II in Newborn Piglets. *Viruses* **2016**, *8*, 271. [CrossRef] [PubMed]
24. Arruda, B.L.; Arruda, P.H.; Magstadt, D.R.; Schwartz, K.J.; Dohlman, T.; Schleining, J.A.; Patterson, A.R.; Visek, C.A.; Victoria, J.G. Identification of a divergent lineage porcine pestivirus in nursing piglets with congenital tremors and reproduction of disease following experimental inoculation. *PLoS ONE* **2016**, *11*, e0150104. [CrossRef] [PubMed]
25. Cagatay, G.N.; Antos, A.; Meyer, D.; Maistrelli, C.; Keuling, O.; Becher, P.; Postel, A. Frequent infection of wild boar with atypical porcine pestivirus (APPV). *Transbound. Emerg. Dis.* **2018**, *65*, 1087–1093. [CrossRef] [PubMed]
26. Kasza, L.; Shadduck, J.A.; Christofinis, G.J. Establishment, viral susceptibility and biological characteristics of a swine kidney cell line SK-6. *Res. Vet. Sci.* **1972**, *13*, 46–51. [CrossRef]
27. Madin, S.H.; Darby, N.B., Jr. Established kidney cell lines of normal adult bovine and ovine origin. *Proc. Soc. Exp. Biol. Med.* **1958**, *98*, 574–576. [CrossRef]
28. Corapi, W.V.; Donis, R.O.; Dubovi, E.J. Monoclonal antibody analyses of cytopathic and noncytopathic viruses from fatal bovine viral diarrhea virus infections. *J. Virol.* **1988**, *62*, 2823–2827.
29. Meyers, G.; Rumenapf, T.; Thiel, H.J. Molecular cloning and nucleotide sequence of the genome of hog cholera virus. *Virology* **1989**, *171*, 555–567. [CrossRef]
30. Becher, P.; Shannon, A.D.; Tautz, N.; Thiel, H.J. Molecular characterization of border disease virus, a pestivirus from sheep. *Virology* **1994**, *198*, 542–551. [CrossRef]
31. Ridpath, J.F.; Bolin, S.R. The genomic sequence of a virulent bovine viral diarrhea virus (BVDV) from the type 2 genotype: Detection of a large genomic insertion in a noncytopathic BVDV. *Virology* **1995**, *212*, 39–46. [CrossRef] [PubMed]

32. Schmeiser, S.; Mast, J.; Thiel, H.J.; Konig, M. Morphogenesis of pestiviruses: New insights from ultrastructural studies of strain Giraffe-1. *J. Virol.* **2014**, *88*, 2717–2724. [CrossRef] [PubMed]
33. Hierholzer, J.C.; Killington, R.A. 2 - Virus isolation and quantitation. In *Virology Methods Manual*; Mahy, B.W.J., Kangro, H.O., Eds.; Academic Press: London, UK, 1996; Volume 1, pp. 25–46.
34. Sanders, R.; Mason, D.J.; Foy, C.A.; Huggett, J.F. Considerations for accurate gene expression measurement by reverse transcription quantitative PCR when analysing clinical samples. *Anal. Bioanal. Chem.* **2014**, *406*, 6471–6483. [CrossRef] [PubMed]
35. Gilmartin, A.A.; Lamp, B.; Rumenapf, T.; Persson, M.A.; Rey, F.A.; Krey, T. High-level secretion of recombinant monomeric murine and human single-chain Fv antibodies from Drosophila S2 cells. *Protein Eng. Des. Sel.* **2012**, *25*, 59–66. [CrossRef]
36. Weiland, E.; Ahl, R.; Stark, R.; Weiland, F.; Thiel, H.J. A second envelope glycoprotein mediates neutralization of a pestivirus, hog cholera virus. *J. Virol.* **1992**, *66*, 3677–3682.
37. Sinn, L.J.; Zieglowski, L.; Koinig, H.; Lamp, B.; Jansko, B.; Mosslacher, G.; Riedel, C.; Hennig-Pauka, I.; Rumenapf, T. Characterization of two Austrian porcine reproductive and respiratory syndrome virus (PRRSV) field isolates reveals relationship to East Asian strains. *Vet. Res.* **2016**, *47*, 17. [CrossRef] [PubMed]
38. Summerfield, A.; Hofmann, M.A.; McCullough, K.C. Low density blood granulocytic cells induced during classical swine fever are targets for virus infection. *Vet. Immunol. Immunopathol.* **1998**, *63*, 289–301. [CrossRef]
39. Karikalan, M.; Rajukumar, K.; Mishra, N.; Kumar, M.; Kalaiyarasu, S.; Rajesh, K.; Gavade, V.; Behera, S.P.; Dubey, S.C. Distribution pattern of bovine viral diarrhoea virus type 1 genome in lymphoid tissues of experimentally infected sheep. *Vet. Res. Commun.* **2016**, *40*, 55–61. [CrossRef]
40. Ridpath, J.F.; Falkenberg, S.M.; Bauermann, F.V.; VanderLey, B.L.; Do, Y.; Flores, E.F.; Rodman, D.M.; Neill, J.D. Comparison of acute infection of calves exposed to a high-virulence or low-virulence bovine viral diarrhea virus or a HoBi-like virus. *Am. J. Vet. Res.* **2013**, *74*, 438–442. [CrossRef]
41. Lanyon, S.R.; Hill, F.I.; Reichel, M.P.; Brownlie, J. Bovine viral diarrhoea: Pathogenesis and diagnosis. *Vet. J.* **2014**, *199*, 201–209. [CrossRef] [PubMed]
42. Strong, R.; La Rocca, S.A.; Paton, D.; Bensaude, E.; Sandvik, T.; Davis, L.; Turner, J.; Drew, T.; Raue, R.; Vangeel, I.; et al. Viral Dose and Immunosuppression Modulate the Progression of Acute BVDV-1 Infection in Calves: Evidence of Long Term Persistence after Intra-Nasal Infection. *PLoS ONE* **2015**, *10*, e0124689. [CrossRef] [PubMed]
43. Bielefeldt-Ohmann, H.; Tolnay, A.E.; Reisenhauer, C.E.; Hansen, T.R.; Smirnova, N.; Van Campen, H. Transplacental infection with non-cytopathic bovine viral diarrhoea virus types 1b and 2: Viral spread and molecular neuropathology. *J. Comp. Pathol.* **2008**, *138*, 72–85. [CrossRef] [PubMed]
44. Tarradas, J.; de la Torre, M.E.; Rosell, R.; Perez, L.J.; Pujols, J.; Munoz, M.; Munoz, I.; Munoz, S.; Abad, X.; Domingo, M.; et al. The impact of CSFV on the immune response to control infection. *Virus Res.* **2014**, *185*, 82–91. [CrossRef] [PubMed]
45. Passler, T.; Riddell, K.P.; Edmondson, M.A.; Chamorro, M.F.; Neill, J.D.; Brodersen, B.W.; Walz, H.L.; Galik, P.K.; Zhang, Y.; Walz, P.H. Experimental infection of pregnant goats with bovine viral diarrhea virus (BVDV) 1 or 2. *Vet. Res.* **2014**, *45*, 38. [CrossRef] [PubMed]
46. Collins, M.E.; Heaney, J.; Thomas, C.J.; Brownlie, J. Infectivity of pestivirus following persistence of acute infection. *Vet. Microbiol.* **2009**, *138*, 289–296. [CrossRef]
47. Kirkland, P.D.; Frost, M.J.; King, K.R.; Finlaison, D.S.; Hornitzky, C.L.; Gu, X.; Richter, M.; Reimann, I.; Dauber, M.; Schirrmeier, H.; et al. Genetic and antigenic characterization of Bungowannah virus, a novel pestivirus. *Vet. Microbiol.* **2015**, *178*, 252–259. [CrossRef]

viruses

MDPI

Review

Pestivirus K (Atypical Porcine Pestivirus): Update on the Virus, Viral Infection, and the Association with Congenital Tremor in Newborn Piglets

Alais M. Dall Agnol [1,2,3], Alice F. Alfieri [1,2,3] and Amauri A. Alfieri [1,2,3,*]

1 Laboratory of Animal Virology, Department of Veterinary Preventive Medicine, Universidade Estadual de Londrina, Londrina, CEP 86057-970 Paraná, Brazil; alaisagnol@hotmail.com (A.M.D.A.); aalfieri@uel.br (A.F.A.)
2 Multi-User Animal Health Laboratory, Molecular Biology Unit, Department of Veterinary Preventive Medicine, Universidade Estadual de Londrina, CEP 86057-970 Paraná, Brazil
3 Rodovia Celso Garcia Cid Road-Campus Universitário, Londrina, PO Box 10011, CEP 86057-970 Paraná, Brazil
* Correspondence: alfieri@uel.br; Tel.: +55-43-3371-5876; Fax: +55-43-3371-4485

Received: 4 July 2020; Accepted: 23 July 2020; Published: 18 August 2020

Abstract: The atypical porcine pestivirus (APPV) belongs to the species *Pestivirus K* of the genus *Pestivirus* and the family *Flaviviridae*, and it has been associated with congenital tremor (CT) type A-II in newborn piglets. Although APPV was discovered in 2015, evidence shows that APPV has circulated in pig herds for many years, at least since 1986. Due to the frequently reported outbreaks of CT on different continents, the importance of this virus for global pig production is notable. Since 2015, several studies have been conducted to clarify the association between APPV and CT. However, some findings regarding APPV infection and the measures taken to control and prevent the spread of this virus need to be contextualized to understand the infection better. This review attempts to highlight advances in the understanding of APPV associated with type A-II CT, such as etiology, epidemiology, diagnosis, and control and prevention measures, and also describes the pathophysiology of the infection and its consequences for pig production. Further research still needs to be conducted to elucidate the host's immune response to APPV infection, the control and prevention of this infection, and the possible development of vaccines.

Keywords: APPV; pestiviruses; congenital tremor type A-II; persistent infection; pigs

1. Introduction

In 2015, the atypical porcine pestivirus (APPV) was first described and identified by next-generation sequencing (NGS) in pig serum samples [1]. Initially, the virus was not believed to be associated with any clinical manifestations; later, an experimental inoculation study conducted in the United States (US) demonstrated that APPV was associated with the occurrence of congenital tremor (CT) in newborn piglets [2]. At present, the occurrence of APPV associated with CT is frequently reported by several studies on different continents, including the Americas (North and South), Europe, and Asia [3,4].

Since 2015, several studies have been conducted to clarify the association between APPV and CT. However, several findings regarding the infection and the measures taken to control and prevent the spread of the virus need to be contextualized to characterize the infection better. This review aims to address advances in understanding APPV associated with CT type A-II and further describe the pathophysiology of this virus and its consequences for pig production.

2. Congenital Tremor Syndrome

CT is a neurological disorder that affects newborn piglets and is characterized by muscle spasms in the head and body, which can be localized or generalized [5,6]. The intensity of the tremors is variable, and in more severe cases, they include generalized tremors that result in difficulty in standing or walking and, consequently, inability to suckle, resulting in death due to starvation. The clinical manifestation of CT is also called shaker pig syndrome, trembling pigs, or congenital myoclonus [6,7].

CT syndrome is classified into A and B according to the presence or absence of histopathological lesions in the central nervous system (CNS). In CT type A, histopathological changes are found in the brain or spinal cord. When histopathological lesions are not observed, it is classified as CT type B. Based on the etiology of CT type A, this type is subdivided into five subgroups (I-V) [5]. CT type A-I is characterized by cerebellar hypoplasia, dysplasia, hypomyelination of the spinal cord, and its etiology associated with other porcine pestivirus infections in the classical swine fever virus (CSFV) [8].

For a long time, the cause of CT A-II was not defined, although its infectious etiology has always been related to this type of tremor [9]. Since 2016, the etiology of CT A-II has been attributed to the newly described APPV, which has already been demonstrated by experimental infections [2,10]. Other viruses also were detected coinfection with APPV, such as porcine circovirus type 2 [11], astrovirus [12], porcine circovirus type 3 [13], linda virus [14], porcine circovirus-like virus P1 [15], and *Teschovirus A* [16], however, the role played by these three viruses as the primary cause of CT A-II is debatable.

CT type A-III is related to a genetic defect presented only by the Landrace breed, in which a lack of myelin is observed together with a reduction in the number of oligodendrocytes [17]. CT type A-IV is caused by a recessive genetic defect in the Saddleback swine and is characterized by hypomyelination of the brain and spinal cord [18,19]. Type A-V is caused by metrifonate (trichlorfon) intoxication. The occurrence of this type of CT may be associated with the use of metrifonate as an antiparasitic medication, and when administered during pregnancy, CT type A-V can cause cerebellar hypoplasia and cause piglets to be born with ataxia and tremor [20,21]. Finally, CT type B has no known specific cause, and changes in the CNS are not observed [5].

3. Atypical Porcine Pestivirus

3.1. Classification and Etiology

Until 2017, only four "classical" species belonged to the genus *Pestivirus*: bovine viral diarrhea virus 1 (BVDV-1); bovine viral diarrhea virus 2 (BVDV-2); border disease virus (BDV); and CSFV [22]. However, a large number of viruses linked to pestiviruses have been described [1,23–32], and Smith et al. [22] proposed that the study group of the *Flaviviridae* family of the International Committee on Viral Taxonomy (ICTV) revise the taxonomy of the genus *Pestivirus*. Thus, the genus *Pestivirus* was divided into 11 species, *Pestivirus A* (BVDV-1) [33], *Pestivirus B* (BVDV-2) [34], *Pestivirus C* (CSFV) [35], *Pestivirus D* (BDV) [36], *Pestivirus E* (pronghorn antelope pestivirus) [29], *Pestivirus F* (porcine pestivirus) [26], *Pestivirus G* (giraffe pestivirus) [23], *Pestivirus H* (HoBi-like pestivirus) [27], *Pestivirus I* (Aydin-like pestivirus) [24], *Pestivirus J* (rat pestivirus) [25], and *Pestivirus K* (atypical porcine pestivirus) [1], and one unclassified virus, bat pestivirus [30].

APPV is the single species of *Pestivirus K*, to the genus *Pestivirus*, and to the family *Flaviviridae* [37]. The viral particle is spherical, with a diameter of approximately 60 nm [38] and enveloped. Viral genomes are single-stranded, positive-sense RNA, exhibiting a genome size of approximately 11 to 11.6 kb. The genome comprises a 5′-noncoding region (NCR), one single open reading frame region encoding a single polyprotein with 3635 amino acids, and a 3′-NCR region [1,39]. The polyprotein is processed into 12 proteins, C (capsid protein), E^{rns}, E1, E2 (envelope proteins), and nonstructural proteins N^{pro}, p7, NS2, NS3, NS4A, NS4B, NS5A, and NS5B [40].

Phylogenic analyzes of the complete or partial genome of APPV strains detected in different countries and years, and in pigs from commercial farms and wild boars, demonstrate a high degree of genetic variability, subdividing into different clusters [3,41–44].

3.2. Pathogenesis and Clinical-Pathological Manifestation Associated with APPV

Two studies of experimental infection were carried out to associate APPV infection with CT and elucidate the pathogenesis. The first experimental inoculation was performed by Arruda et al. [2] and occurred in the US. Five pregnant sows, a group at 45 days and another at 62 days of gestation, were inoculated in the fetal amniotic vesicle, intranasally, and intravenously with known serum samples positive for APPV. In the period following the inoculation, no inoculated sow showed detectable viremia by qRT-PCR and no observation of clinical signs suggestive of CT. After birth, piglets were monitored for CT, and 57% to 100% of piglets showed signs of tremor. In the second experimental study carried out in the Netherlands by de Groof et al. [10], APPV-positive sera were inoculated intramuscularly in three gilts at 32 days of gestation. At 10 days postinfection, the sera from three gilts were RT-PCR-positive for the virus, and one of the gilts presented a relatively lower viral load than the other gilts. At birth, two of the three litters contained piglets showing clinical signs of CT, ranging from mild to severe signs in almost all piglets. No piglets born from the gilt that had low viremia showed signs of CT.

In both experimental studies, clinical signs were classified according to intensity; the most severe cases were characterized by intense tremors throughout the body, while the mild ones were characterized by muscle fasciculation in the limbs. Another clinical sign frequently observed was the presence of piglets with the splay leg [2,10]. This clinical manifestation is a syndrome characterized by temporary dysfunctionality of the posterior leg muscles, occurring shortly after birth, and resulting in difficulty or inability to stand and walk [45]. The presence of APPV RNA was detected in different organs of all piglets that showed signs of CT, and in some of the piglets not affected by CT [2,10]. These experiments were essential to elucidate the association of APPV with CT. However, Koch's postulate was not fulfilled by the two studies, probably due to the difficulty of viral isolation in cell cultures.

In the various studies that describe natural and experimental infections, the virus can be detected by conventional RT-PCR, qRT-PCR, and immunohistochemical in a wide variety of organs. The highest viral loads are observed primarily in the CNS and lymphoid tissues. In the CNS, the virus is found in greater quantities in the cerebellum and lymphoid tissues, mainly in the inguinal lymph nodes, submandibular lymph nodes, and thymus, suggesting that the cerebellum and lymph nodes are target organs of APPV [46,47]. In other organs, such as the brain, brain stem, spinal cord, heart, liver, spleen, lungs, and kidneys, the viral load is relatively lower [42,46,47].

To date, no significant macroscopic lesions have been described in piglets with CT [48–50]. Histopathological findings in APPV infections primarily include vacuolization of the white matter in the cerebellum [48,50,51]. Studies using Luxol® Fast Blue staining have shown a reduction in the intensity of myelin staining in the white matter of the spinal cord, cerebrum, cerebellum, and sciatic nerve, which implies a decrease in the amount of myelin in these tissues [16,48,50,52,53]. Gliosis has also been observed in the cerebral cortex and, to a lesser extent, in the spinal cord [16,49,50]. Other histopathological lesions observed in affected piglets were neuronal necrosis, neuronophagia with satellitosis, particularly at the cerebral cortex and spinal cord, Wallerian degeneration of the spinal cord, and necrosis of Purkinje cells of the cerebellum [49]. Transmission electron microscopy performed in two CNS tissues revealed not only hypomyelination of the cerebellar white matter but also more severe changes, such as interruption and breakdown of myelin, in animals affected by APPV [50]. These histopathological lesions described in piglets with CT, primary demyelination, and/or hypomyelination, can be considered a neuropathological characteristic associated with APPV infection [49]. The high viral loads of APPV in the CNS, along with the lesions observed in these tissues, explains the intensity of the neurological symptoms caused by APPV [46,47].

3.3. Viral Shedding and Transmission

Piglets that developed CT and recovered were monitored by some studies to assess the possibility of viral excretion and duration. Therefore, de Groof et al. [10] followed five pigs that presented CT and recovered at 8.5 months of age; these animals continued to shed APPV in their feces. Excretion in feces was also suggested by Postel et al. [53] since high viral loads were detected in the duodenum, pancreas, and colon. APPV is found in salivary glands; therefore, the virus is shedding by saliva until at least the sixth month of life [50]. In these same animals, the virus was also found in semen. Another study evaluated semen collection from three commercial boar herds and detected the presence of the viral genome in swab and fluid preputial and in the semen from 15% of boars from a commercial boar stud, suggesting that artificial insemination may be an important form of transmission [54].

Regarding transmission, the horizontal and vertical forms of transmission were described, with the latter being the most important and responsible for the development of CT in newborn piglets from naturally infected sows. Transplacental transmission was demonstrated by experimental inoculation. Pregnant sows were infected with serum samples known to be positive for APPV, and the birth of piglets with CT of different levels of severity was observed [2,10].

In a controlled exposure, using fetal fluid from litters that had exhibited CT, a batch of 91 gilts was submitted to oral exposure to the antigen approximately 54 days before breeding. Signs of CT were observed in 45.0% of the litters and 30.8% of all piglets born [51]. These data suggest probable horizontal transmission, due to oral exposure, and the development of CT cases.

In piglets, horizontal transmission was described after the monitoring of piglets that showed signs of CT and obtained a confirmed diagnosis for APPV. After weaning, naive piglets were mixed with APPV-positive piglets. Due to close contact, infection of susceptible piglets occurred, suggesting that transmission occurs after mixing infected with susceptible piglets. No manifestation of disease was observed in this type of late infection; moreover, these animals were transiently infected [55]. It is likely that this type of transmission is related to environmental contamination and implies the spread of the virus. However, little is known about the forms of transmission and their possible effects; therefore, further investigations are necessary for a satisfactory clarification on this topic.

3.4. Epidemiology and Implications in Pig Production

In the United States, APPV was first identified by NGS in pig serum samples collected in 2014 from five different states [1]. In subsequent years, several studies from different countries reported the circulation of APPV in the US [1,2,51,54,56], Germany [53,55,57–61], Netherlands [10], Sweden [62], Austria [50], England [63], China [38,42,44,46,47,61,64–73], Spain [41,74], South Korea [75], Brazil [16,48,49,76], Great Britain [61], Italy [43,61], Serbia [61], Switzerland [61,77], Taiwan [61], Canada [52], and Hungary [78].

Epidemiological surveys demonstrate a high number of the pigs positive for APPV, depending on the region evaluated and the assay used. Postel and colleagues [61] used qRT-PCR to analyze 1460 serum samples from asymptomatic pigs from Germany, Italy, Serbia, Great Britain, Switzerland, China, and Taiwan and found a prevalence of 8.9% (130/1460) of APPV-positive samples. Additionally, in this same study, the highest (17.5%; 35/200) prevalence of positive samples was in Italy. Michelitsch et al. [59], by indirect immunofluorescence test (IFI), found an antibody prevalence of 16.3% (182/1115) in pig serum samples in Germany. In Switzerland, 1080 sera from pigs for slaughter, which were obtained between 1986 and 2018, were evaluated by qRT-PCR, and the APPV genome was detected in 13% [77].

Although this virus was discovered recently, some studies have shown that APPV has been circulating in pig herds for a long time. A study conducted in Spain detected the presence of the viral genome in pig serum samples collected in 1997 [41]. In Switzerland, another investigation analyzed 1080 pig sera collected between 1986 and 2018 and revealed that 7% (6/87) of the samples collected in slaughter pigs in 1986 were APPV-positive [77]. These findings correspond to reports of the long-term presence of APPV in the world.

The presence of APPV in wild boars has been demonstrated in Germany, Spain, and Italy, and the prevalence of the virus varies according to the country. In Germany, 456 wild boar serum samples were analyzed, and 19% of samples revealed the presence of the viral genome, while 52% of the serum samples had antibodies against the E^{rns} protein [50]. In Spain and Italy, the prevalence observed was lower, 0.23% (1/437), and 0.69% (3/430) of the evaluated sera had the virus genome, respectively [43,74]. Transmission of APPV from wild boars to pigs or vice versa has not been demonstrated to date. However, wild boars are sources of various pathogens that infect domestic pigs, including other pestiviruses, such as CSFV [79], and can contribute to their transmission. Due to the importance of the transmission of viral pathogens, these wild animals are a challenge for pig health and must be considered for effective control plans to be devised.

Regarding the occurrence of CT, the litters from first parity sows are most affected by APPV infection, and this infection rarely occurs in higher birth order [10,48,49]. During an outbreak on a farm in the Netherlands, 48 litters with CT were born from gilts and were monitored, the number of piglets affected within each litter ranged from <10 to 100%. Furthermore, the total mortality of piglets, reached 26%, with 60% of these deaths being attributed to CT [10]. In Brazil, an outbreak of CT observed in piglets born to gilts, lasted for three weeks, and the mortality rate reached 30% [49].

3.5. *APPV Infection and Immunity*

The dynamics of APPV infection can be hypothesized in two ways: persistently infected and transiently infected animals. These forms of infection are known to other pestiviruses, as is the case with CSFV [79] and BVDV in cattle [80]. In APPV infection, the dynamics have not been fully elucidated; however, this phenomenon can be explained by two studies [50,55]. Schwarz et al. [50] monitored the health status of two piglets (one female and one male) aged up to six months; shortly after birth, these two animals showed CT. Specific antibodies to APPV NS3 were detected at birth and at up to eight weeks of age. The tremor symptoms decreased and disappeared completely until 14 weeks; however, both piglets still presented viremia, antibody titer, and shedding of the virus by saliva. At six months the male piglet reached sexual maturity, and a high viral load was detected in saliva and semen; on the other hand, viremia was reduced.

Cagatay et al. [55], using direct and indirect tests, monitored 20 piglets from unaffected and affected litters by CT from birth to slaughter. In the vertically infected and symptomatic piglets, viremia was detected from the first days of life until slaughter. For the presence of antibodies, these piglets showed high levels of antibodies at six days of age, and these antibodies were undetectable at 21 and/or 48 days of age. It is possible that these antibodies came from the sows and disappeared with the drop in passive immunity. On the other hand, piglets infected horizontally after being mixed at weaning with those infected vertically showed viremia at 48 days of age and high titers of specific antibodies to E2 when evaluated at 69 and 161 days of life, suggesting the induction of protective immunity against infection.

Based on piglets infected horizontally, the immune response was higher for the E2 protein, and neutralizing antibody titers correlated with the presence of E2-specific antibodies, while a correlation with E^{rns}-specific antibodies was not observed [58].

Due to the longevity of viral shedding, detectable viremia, together with the disappearance of specific antibodies over time, may suggest that persistent infection (PI) can be attributed to piglets that are intrauterine-infected. Soon after birth, these piglets can show signs of CT, which usually regresses over time [41,50,61]. On the other hand, piglets infected horizontally, through contact with persistently infected animals, develop a transient infection, with viremia detected for several weeks, but over time the piglet develops active immunity against APPV, and the virus becomes undetectable [55]. These studies provide evidence of the dynamics of infections and the immune response; however, the studies examine a small number of animals, meaning that further studies are required for complete elucidation.

3.6. *Diagnostic Methods of APPV*

Currently, a wide variety of diagnostic techniques are available for use in elucidating APPV infection in pigs. Due to practicality, speed, sensitivity, and specificity, molecular tests are the most commonly used. Among these tests, both conventional RT-PCR [16,48,49] and qRT-PCR [2] have been described in viral detection in several studies. Different clinical samples can be subjected to viral detection. In clinically affected animals, the CNS and the lymphoid organs are specimens of choice, since higher viral loads are found in these tissues, especially the cerebellum and lymph nodes [46,47]. Another available technique is NGS [2,10,51], which has been used since the first viral description in 2015 [1], and NGS is still employed as a diagnostic tool, assisting in the detection of the viral genome. In addition, NGS provides data on possible coinfections and primarily obtains larger fragments of the viral genome, which favors the phylogenetic study of circulating strains.

Histological tests performed from tissues fixed in paraffin are of great diagnostic importance, enabling the visualization of lesions caused by APPV. Luxol®® Fast Blue staining helps to observe the demyelination caused by the APPV in the CNS, primarily located in the cerebellum and spinal cord [16,48,50,52,53]. Histopathology, together with an immunohistochemical technique [46], and in situ hybridization [53,60] enables the detection of the viral agent (protein or nucleic acid) at the lesion site.

Other important methods that are used to assist in the diagnosis of infectious diseases are serological tests. The tests described for APPV infection are IFI [59], virus neutralization [55], and indirect enzyme-linked immunosorbent assay (ELISA) to the NS3, E2, and E^{rns} proteins [50,55,58,61]. These tests have many important applications, and they can be used for population-based epidemiological studies and monitoring of infection in the herd; the tests also feature easy execution and low cost.

3.7. Control and Prevention

To date, there are no effective drugs or vaccines available to treat or prevent APPV infection. Zhang et al. [81] constructed a recombinant baculovirus of APPV glycoprotein E2, which induced a robust humoral and cellular immune response in mice. Based on these studies and knowing that the E2 protein is responsible for inducing neutralizing antibodies [55], this vaccine appears to be a promising tool as likely prevention of APPV in pigs. However, further viral challenge studies are needed to demonstrate an effective immune response.

As litters of gilts are most affected by APPV infection [10,49], it is probable that the introduction of naive gilts in the herd is an important issue in the epidemiology of the disease [49,55]. Therefore, preventive measures are necessary to address this issue. The use of acclimatization for replacement gilts is an interesting tool that can be used in diseases that do not have vaccines and/or when available are not completely effective for prevention. Gatto et al. [3] suggest the protocol similar to what is used for the control of enzootic pneumonia caused by *Mycoplasma hyopneumoniae*. However, once horizontal transmission in gilts after oral exposure to fetal fluid antigen positive for APPV has been demonstrated, followed by the induction of CT in the litters of these gilts [51], this measure needs to be considered carefully. There is a need to conduct studies focused on this tool and, therefore, the development of protocols aimed at infection by APPV.

Another important point to be considered when designing control programs is the possibility of PI animals. Sensitive diagnostic methods that are able to detect PI animals, such as qRT-PCR, especially in the case of breeders, are highly important. After identifying these animals, they must be removed from the herd due to viral shedding by semen, feces, and oral fluid, which favors viral transmission. Another issue to be monitored is the semen used in the practice of artificial insemination; one possible measure is to include testing for APPV in the routine tests performed on these boars and their semen.

Despite specific measures to control APPV, biosecurity measures cannot be overlooked. When replacing breeders, animals from herds with a good health strategy must be purchased, and quarantining must be performed before introduction on the farm. Traffic control of people and vehicles on farms must be considered in addition to the execution of adequate cleaning and disinfection programs in the facilities. Finally, the control of wild animals, rodents, and insects contributes to the health of the herd in general.

4. Conclusions and Perspectives

The new porcine pestivirus (APPV) is strongly related to cases of CT type A-II, which until 2016, had an undefined etiology. Despite the recent discovery of APPV, evidence has shown that this virus has been circulating in pig herds for many years, since at least 1986, along with longstanding reports of CT, and the importance of this virus for global pig production is notable. In addition, APPV belongs to the genus *Pestivirus*, presenting important biological characteristics for the epidemiology of the disease, such as viral persistence, which can represent viral maintenance and a constant source of both horizontally and vertically transmitted infection in pig herds. These points are highly important for viral prevention. However, these issues and others regarding this disease still need to be elucidated, and further research should investigate the host's immune response, the control and prevention of APPV infection, and the development of vaccines.

Author Contributions: Conceptualization, A.M.D.A. and A.A.A.; Writing Original Draft Preparation, A.M.D.A.; Writing Review and Editing, A.F.A. and A.A.A. All authors have read and agreed to the published version of the manuscript.

Funding: This work was supported by the National Council of Technological and Scientific Development (CNPq grant number 305.062/2015-8).

Acknowledgments: The authors thank the following Brazilian Institutes for financial support: the National Council of Technological and Scientific Development (CNPq), the Brazilian Federal Agency for Support and Evaluation of Graduate Education (CAPES), the Financing of Studies and Projects (FINEP), and the Araucaria Foundation (FAP/PR). A.A.A. and A.F.A. are recipients of CNPq fellowships. A.M.D.A. is recipients of INCT-Leite/CAPES fellowship (grant number 88887.495081/2020-00).

Conflicts of Interest: The authors declare no conflict of interest.

References

1. Hause, B.M.; Collin, E.A.; Peddireddi, L.; Yuan, F.; Chen, Z.; Hesse, R.A.; Gauger, P.C.; Clement, T.; Fang, Y.; Anderson, G. Discovery of a novel putative atypical porcine pestivirus in pigs in the USA. *J. Gen. Virol.* **2015**, *96*, 2994–2998. [CrossRef]
2. Arruda, B.L.; Arruda, P.H.; Magstadt, D.R.; Schwartz, K.J.; Dohlman, T.; Schleining, J.A.; Patterson, A.R.; Visek, C.A.; Victoria, J.G. Identification of a divergent lineage porcine pestivirus in nursing piglets with congenital tremors and reproduction of disease following experimental inoculation. *PLoS ONE* **2016**, *11*, e0150104. [CrossRef]
3. Gatto, I.R.H.; Sonálio, K.; de Oliveira, L.G. Atypical Porcine Pestivirus (APPV) as a new species of pestivirus in pig production. *Front. Vet. Sci.* **2019**, *6*, 35. [CrossRef]
4. Pan, S.; Mou, C.; Chen, Z. An emerging novel virus: Atypical porcine pestivirus (APPV). *Rev. Med. Virol.* **2019**, *29*, 2018. [CrossRef]
5. Done, J.T. Congenital nervous diseases of pigs: A review. *Lab. Anim.* **1968**, *2*, 207–218. [CrossRef]
6. Stenberg, H.; Jacobson, M.; Malmberg, M. A review of congenital tremor type A-II in piglets. *Anim. Health Res. Rev.* **2020**, 1–5. [CrossRef] [PubMed]
7. Leman, A.D.; Hurtgen, J.P.; Hilley, H.D. Influence of intrauterine events on postnatal survival in the pig. *J. Anim. Sci.* **1979**, *49*, 221–224. [CrossRef] [PubMed]
8. Done, J.T.; Harding, J.D. The relationship of maternal swine fever infection to cerebellar hypoplasia in piglets. *Proc. R. Soc. Med.* **1966**, *59*, 1083–1084. [PubMed]
9. Vandekerckhove, P.; Maenhout, D.; Curvers, P.; Hoorens, J.; Ducatelle, R. Type A2 congenital tremor in piglets. *Zentralbl. Veterinarmed. A* **1989**, *36*, 763–771. [CrossRef] [PubMed]
10. de Groof, A.; Deijs, M.; Guelen, L.; van Grinsven, L.; van Os-Galdos, L.; Vogels, W.; Derks, C.; Cruijsen, T.; Geurts, V.; Vrijenhoek, M.; et al. Atypical Porcine Pestivirus: A possible cause of congenital tremor type A-II in newborn piglets. *Viruses* **2016**, *8*, 271. [CrossRef]
11. Stevenson, G.W.; Kiupel, M.; Mittal, S.K.; Choi, J.; Latimer, K.S.; Kanitz, C.L. Tissue distribution and genetic typing of porcine circoviruses in pigs with naturally occurring congenital tremors. *J. Vet. Diagn. Investig.* **2001**, *13*, 57–62. [CrossRef]

12. Blomström, A.L.; Ley, C.; Jacobson, M. Astrovirus as a possible cause of congenital tremor type AII in piglets? *Acta Vet. Scand.* **2014**, *56*, 82. [CrossRef] [PubMed]
13. Chen, G.H.; Mai, K.J.; Zhou, L.; Wu, R.T.; Tang, X.Y.; Wu, J.L.; He, L.L.; Lan, T.; Xie, Q.M.; Sun, Y.; et al. Detection and genome sequencing of porcine circovirus 3 in neonatal pigs with congenital tremors in South China. *Transbound. Emerg. Dis.* **2017**, *64*, 1650–1654. [CrossRef] [PubMed]
14. Lamp, B.; Schwarz, L.; Högler, S.; Riedel, C.; Sinn, L.; Rebel-Bauder, B.; Weissenböck, H.; Ladinig, A.; Rümenapf, T. Novel pestivirus species in pigs, Austria, 2015. *Emerg. Infect. Dis.* **2017**, *23*, 1176–1179. [CrossRef] [PubMed]
15. Wen, L.; Mao, A.; Jiao, F.; Zhang, D.; Xie, J.; He, K. Evidence of porcine circovirus-like virus P1 in piglets with an unusual congenital tremor. *Transbound. Emerg. Dis.* **2018**, *65*, e501–e504. [CrossRef]
16. Possatti, F.; Headley, S.A.; Leme, R.A.; Dall Agnol, A.M.; Zotti, E.; de Oliveira, T.E.S.; Alfieri, A.F.; Alfieri, A.A. Viruses associated with congenital tremor and high lethality in piglets. *Transbound. Emerg. Dis.* **2018**, *65*, 331–337. [CrossRef]
17. Blakemore, W.F.; Harding, J.D.; Done, J.T. Ultrastructural observations on the spinal cord of a Landrace pig with congenital tremor type AIII. *Res. Vet. Sci.* **1974**, *17*, 174–178. [CrossRef]
18. Patterson, D.S.; Sweasey, D.; Brush, P.J.; Harding, J.D. Neurochemistry of the spinal cord in British Saddleback piglets affected with congenital tremor, type A-IV, a second form of hereditary cerebrospinal hypomyelinogenesis. *J. Neurochem.* **1973**, *21*, 397–406. [CrossRef]
19. Patterson, D.S.; Sweasey, D.; Harding, J.D. Lipid deficiency in the central nervous system of Landrace piglets affected with congenital tremor A3. A form of cerebrospinal hypomyelinogenesis. *J. Neurochem.* **1972**, *19*, 2791–2799. [CrossRef]
20. Bölske, G.; Kronevi, T.; Lindgren, N.O. Congenital tremor in pigs in Sweden. A case report. *Nord. Vet. Med.* **1978**, *30*, 534–537.
21. Knox, B.; Askaa, J.; Basse, A.; Bitsch, V.; Eskildsen, M.; Mandrup, M.; Ottosen, H.E.; Overby, E.; Pedersen, K.B.; Rasmussen, F. Congenital ataxia and tremor with cerebellar hypoplasia in piglets borne by sows treated with Neguvon vet. (metrifonate, trichlorfon) during pregnancy. *Nord. Vet. Med.* **1978**, *30*, 538–545. [PubMed]
22. Smith, D.B.; Meyers, G.; Bukh, J.; Gould, E.A.; Monath, T.; Scott Muerhoff, A.; Pletnev, A.; Rico-Hesse, R.; Stapleton, J.T.; Simmonds, P.; et al. Proposed revision to the taxonomy of the genus Pestivirus, family Flaviviridae. *J. Gen. Virol.* **2017**, *98*, 2106–2112. [CrossRef] [PubMed]
23. Avalos-Ramirez, R.; Orlich, M.; Thiel, H.J.; Becher, P. Evidence for the presence of two novel pestivirus species. *Virology* **2001**, *286*, 456–465. [CrossRef]
24. Becher, P.; Schmeiser, S.; Oguzoglu, T.C.; Postel, A. Complete genome sequence of a novel pestivirus from sheep. *J. Virol.* **2012**, *86*, 11412. [CrossRef] [PubMed]
25. Firth, C.; Bhat, M.; Firth, M.A.; Williams, S.H.; Frye, M.J.; Simmonds, P.; Conte, J.M.; Ng, J.; Garcia, J.; Bhuva, N.P.; et al. Detection of zoonotic pathogens and characterization of novel viruses carried by commensal Rattus norvegicus in New York City. *mBio* **2014**, *5*, e01933-14. [CrossRef] [PubMed]
26. Kirkland, P.D.; Frost, M.J.; Finlaison, D.S.; King, K.R.; Ridpath, J.F.; Gu, X. Identification of a novel virus in pigs—Bungowannah virus: A possible new species of pestivirus. *Virus Res.* **2007**, *129*, 26–34. [CrossRef]
27. Liu, L.; Kampa, J.; Belák, S.; Baule, C. Virus recovery and full-length sequence analysis of atypical bovine pestivirus Th/04_KhonKaen. *Vet. Microbiol.* **2009**, *138*, 62–68. [CrossRef]
28. Neill, J.D.; Ridpath, J.F.; Fischer, N.; Grundhoff, A.; Postel, A.; Becher, P. Complete genome sequence of pronghorn virus, a pestivirus. *Genome Announc.* **2014**, *2*. [CrossRef]
29. Vilcek, S.; Ridpath, J.F.; Van Campen, H.; Cavender, J.L.; Warg, J. Characterization of a novel pestivirus originating from a pronghorn antelope. *Virus Res.* **2005**, *108*, 187–193. [CrossRef]
30. Wu, Z.; Ren, X.; Yang, L.; Hu, Y.; Yang, J.; He, G.; Zhang, J.; Dong, J.; Sun, L.; Du, J.; et al. Virome analysis for identification of novel mammalian viruses in bat species from Chinese provinces. *J. Virol.* **2012**, *86*, 10999–11012. [CrossRef]
31. Sozzi, E.; Lavazza, A.; Gaffuri, A.; Bencetti, F.C.; Prosperi, A.; Lelli, D.; Chiapponi, C.; Moreno, A. Isolation and full-length sequence analysis of a pestivirus from aborted lamb fetuses in Italy. *Viruses* **2019**, *11*, 744. [CrossRef]
32. Hurtado, A.; Aduriz, G.; Gómez, N.; Oporto, B.; Juste, R.A.; Lavin, S.; Lopez-Olvera, J.R.; Marco, I. Molecular identification of a new pestivirus associated with increased mortality in the Pyrenean Chamois (Rupicapra pyrenaica pyrenaica) in Spain. *J. Wildl. Dis.* **2004**, *40*, 796–800. [CrossRef] [PubMed]

33. Colett, M.S.; Larson, R.; Gold, C.; Strick, D.; Anderson, D.K.; Purchio, A.F. Molecular cloning and nucleotide sequence of the pestivirus bovine viral diarrhea virus. *Virology* **1988**, *165*, 191–199. [CrossRef]
34. Ridpath, J.F.; Bolin, S.R. The genomic sequence of a virulent bovine viral diarrhea virus (BVDV) from the type 2 genotype: Detection of a large genomic insertion in a noncytopathic BVDV. *Virology* **1995**, *212*, 39–46. [CrossRef] [PubMed]
35. Ruggli, N.; Moser, C.; Mitchell, D.; Hofmann, M.; Tratschin, J.D. Baculovirus expression and affinity purification of protein E2 of classical swine fever virus strain Alfort/187. *Virus Genes* **1995**, *10*, 115–126. [CrossRef] [PubMed]
36. Becher, P.; Shannon, A.D.; Tautz, N.; Thiel, H.J. Molecular characterization of border disease virus, a pestivirus from sheep. *Virology* **1994**, *198*, 542–551. [CrossRef] [PubMed]
37. International Committee on Taxonomy of Viruses. Virus Taxonomy: 2019 Release. Available online: https://talk.ictvonline.org/taxonomy/ (accessed on 26 May 2020).
38. Liu, J.; Ren, X.; Li, H.; Yu, X.; Zhao, B.; Liu, B.; Ning, Z. Development of the reverse genetics system for emerging atypical porcine pestivirus using in vitro and intracellular transcription systems. *Virus Res.* **2020**, *283*, 197975. [CrossRef] [PubMed]
39. Simmonds, P.; Becher, P.; Bukh, J.; Gould, E.A.; Meyers, G.; Monath, T.; Muerhoff, S.; Pletnev, A.; Rico-Hesse, R.; Smith, D.B.; et al. ICTV virus taxonomy profile: Flaviviridae. *J. Gen. Virol.* **2017**, *98*, 2–3. [CrossRef]
40. Blome, S.; Beer, M.; Wernike, K. New leaves in the growing tree of Pestiviruses. *Adv. Virus Res.* **2017**, *99*, 139–160.
41. Muñoz-González, S.; Canturri, A.; Pérez-Simó, M.; Bohórquez, J.A.; Rosell, R.; Cabezón, O.; Segalés, J.; Domingo, M.; Ganges, L. First report of the novel atypical porcine pestivirus in Spain and a retrospective study. *Transbound. Emerg. Dis.* **2017**, *64*, 1645–1649. [CrossRef]
42. Shen, H.; Liu, X.; Zhang, P.; Wang, L.; Liu, Y.; Zhang, L.; Liang, P.; Song, C. Identification and characterization of atypical porcine pestivirus genomes in newborn piglets with congenital tremor in China. *J. Vet. Sci.* **2018**, *19*, 468–471. [CrossRef] [PubMed]
43. Sozzi, E.; Salogni, C.; Lelli, D.; Barbieri, I.; Moreno, A.; Alborali, G.L.; Lavazza, A. Molecular survey and phylogenetic analysis of Atypical Porcine Pestivirus (APPV) identified in swine and wild boar from Northern Italy. *Viruses* **2019**, *11*, 1142. [CrossRef]
44. Yan, X.L.; Li, Y.Y.; He, L.L.; Wu, J.L.; Tang, X.Y.; Chen, G.H.; Mai, K.J.; Wu, R.T.; Li, Q.N.; Chen, Y.H.; et al. 12 novel atypical porcine pestivirus genomes from neonatal piglets with congenital tremors: A newly emerging branch and high prevalence in China. *Virology* **2019**, *533*, 50–58. [CrossRef] [PubMed]
45. Papatsiros, V.G. The splay leg syndrome in piglets: A review. *American J. Anim. Vet. Sci.* **2012**, *7*, 80–83.
46. Liu, J.; Li, Z.; Ren, X.; Li, H.; Lu, R.; Zhang, Y.; Ning, Z. Viral load and histological distribution of atypical porcine pestivirus in different tissues of naturally infected piglets. *Arch. Virol.* **2019**, *164*, 2519–2523. [CrossRef]
47. Yuan, J.; Han, Z.; Li, J.; Huang, Y.; Yang, J.; Ding, H.; Zhang, J.; Zhu, M.; Zhang, Y.; Liao, J.; et al. Atypical Porcine Pestivirus as a novel type of pestivirus in pigs in China. *Front. Microbiol.* **2017**, *8*, 862. [CrossRef]
48. Mósena, A.C.S.; Weber, M.N.; da Cruz, R.A.S.; Cibulski, S.P.; da Silva, M.S.; Puhl, D.E.; Hammerschmitt, M.E.; Takeuti, K.L.; Driemeier, D.; de Barcellos, D.; et al. Presence of atypical porcine pestivirus (APPV) in Brazilian pigs. *Transbound. Emerg. Dis.* **2018**, *65*, 22–26. [CrossRef]
49. Possatti, F.; de Oliveira, T.E.S.; Leme, R.A.; Zotti, E.; Dall Agnol, A.M.; Alfieri, A.F.; Headley, S.A.; Alfieri, A.A. Pathologic and molecular findings associated with atypical porcine pestivirus infection in newborn piglets. *Vet. Microbiol.* **2018**, *227*, 41–44. [CrossRef]
50. Schwarz, L.; Riedel, C.; Högler, S.; Sinn, L.J.; Voglmayr, T.; Wöchtl, B.; Dinhopl, N.; Rebel-Bauder, B.; Weissenböck, H.; Ladinig, A.; et al. Congenital infection with atypical porcine pestivirus (APPV) is associated with disease and viral persistence. *Vet. Res.* **2017**, *48*, 1. [CrossRef]
51. Sutton, K.M.; Lahmers, K.K.; Harris, S.P.; Wijesena, H.R.; Mote, B.E.; Kachman, S.D.; Borza, T.; Ciobanu, D.C. Detection of atypical porcine pestivirus genome in newborn piglets affected by congenital tremor and high preweaning mortality. *J. Anim. Sci.* **2019**, *97*, 4093–4100. [CrossRef]
52. Dessureault, F.G.; Choinière, M.; Provost, C.; Gagnon, C.A. First report of atypical porcine pestivirus in piglets with congenital tremor in Canada. *Can. Vet. J.* **2018**, *59*, 429–432. [PubMed]

53. Postel, A.; Hansmann, F.; Baechlein, C.; Fischer, N.; Alawi, M.; Grundhoff, A.; Derking, S.; Tenhündfeld, J.; Pfankuche, V.M.; Herder, V.; et al. Presence of atypical porcine pestivirus (APPV) genomes in newborn piglets correlates with congenital tremor. *Sci. Rep.* **2016**, *6*, 27735. [CrossRef] [PubMed]
54. Gatto, I.R.H.; Arruda, P.H.; Visek, C.A.; Victoria, J.G.; Patterson, A.R.; Krull, A.C.; Schwartz, K.J.; de Oliveira, L.G.; Arruda, B.L. Detection of atypical porcine pestivirus in semen from commercial boar studs in the United States. *Transbound. Emerg. Dis.* **2018**, *65*, e339–e343. [CrossRef]
55. Cagatay, G.N.; Meyer, D.; Wendt, M.; Becher, P.; Postel, A. Characterization of the humoral immune response induced after infection with Atypical Porcine Pestivirus (APPV). *Viruses* **2019**, *11*, 880. [CrossRef] [PubMed]
56. Chen, F.; Knutson, T.P.; Braun, E.; Jiang, Y.; Rossow, S.; Marthaler, D.G. Semi-quantitative duplex RT-PCR reveals the low occurrence of Porcine Pegivirus and Atypical Porcine Pestivirus in diagnostic samples from the United States. *Transbound. Emerg. Dis.* **2019**, *66*, 1420–1425. [CrossRef]
57. Beer, M.; Wernike, K.; Dräger, C.; Höper, D.; Pohlmann, A.; Bergermann, C.; Schröder, C.; Klinkhammer, S.; Blome, S.; Hoffmann, B. High prevalence of highly variable atypical porcine pestiviruses found in Germany. *Transbound. Emerg. Dis.* **2017**, *64*, e22–e26. [CrossRef]
58. Cagatay, G.N.; Antos, A.; Meyer, D.; Maistrelli, C.; Keuling, O.; Becher, P.; Postel, A. Frequent infection of wild boar with atypical porcine pestivirus (APPV). *Transbound. Emerg. Dis.* **2018**, *65*, 1087–1093. [CrossRef]
59. Michelitsch, A.; Dalmann, A.; Wernike, K.; Reimann, I.; Beer, M. Seroprevalences of newly discovered porcine pestiviruses in German pig farms. *Vet. Sci.* **2019**, *6*, 86. [CrossRef]
60. Pfankuche, V.M.; Hahn, K.; Bodewes, R.; Hansmann, F.; Habierski, A.; Haverkamp, A.K.; Pfaender, S.; Walter, S.; Baechlein, C.; Postel, A.; et al. Comparison of different In Situ Hybridization techniques for the detection of various RNA and DNA viruses. *Viruses* **2018**, *10*, 384. [CrossRef]
61. Postel, A.; Meyer, D.; Cagatay, G.N.; Feliziani, F.; De Mia, G.M.; Fischer, N.; Grundhoff, A.; Milićević, V.; Deng, M.C.; Chang, C.Y.; et al. High abundance and genetic variability of Atypical Porcine Pestivirus in pigs from Europe and Asia. *Emerg. Infect. Dis.* **2017**, *23*, 2104–2107. [CrossRef]
62. Blomström, A.L.; Fossum, C.; Wallgren, P.; Berg, M. Viral metagenomic analysis displays the co-infection situation in healthy and PMWS affected pigs. *PLoS ONE* **2016**, *11*, e0166863. [CrossRef]
63. Williamson, S. Congenital tremor associated with atypical porcine pestivirus. *Vet. Rec.* **2017**, *180*, 42–43.
64. Zhang, K.; Wu, K.; Liu, J.; Ge, S.; Xiao, Y.; Shang, Y.; Ning, Z. Identification of atypical porcine pestivirus infection in swine herds in China. *Transbound. Emerg. Dis.* **2017**, *64*, 1020–1023. [CrossRef]
65. Guo, Z.; Wang, L.; Qiao, S.; Deng, R.; Zhang, G. Genetic characterization and recombination analysis of atypical porcine pestivirus. *Infect. Genet. Evol.* **2020**, *81*, 104259. [CrossRef]
66. Pan, S.; Yan, Y.; Shi, K.; Wang, M.; Mou, C.; Chen, Z. Molecular characterization of two novel atypical porcine pestivirus (APPV) strains from piglets with congenital tremor in China. *Transbound. Emerg. Dis.* **2019**, *66*, 35–42. [CrossRef]
67. Wang, X.; Xie, Y.; He, D.; Yan, H. Near-complete genome sequence of a newly emerging subgenotype of Atypical porcine pestivirus. *Microbiol. Resour. Announc.* **2020**, *9*. [CrossRef]
68. Wu, Z.; Liu, B.; Du, J.; Zhang, J.; Lu, L.; Zhu, G.; Han, Y.; Su, H.; Yang, L.; Zhang, S.; et al. Discovery of diverse rodent and bat pestiviruses with distinct genomic and phylogenetic characteristics in several Chinese provinces. *Front. Microbiol.* **2018**, *9*, 2562. [CrossRef]
69. Xie, Y.; Wang, X.; Su, D.; Feng, J.; Wei, L.; Cai, W.; Li, J.; Lin, S.; Yan, H.; He, D. Detection and genetic characterization of Atypical Porcine Pestivirus in Piglets with congenital tremors in Southern China. *Front. Microbiol.* **2019**, *10*, 1406. [CrossRef]
70. Yin, Y.; Shi, K.; Sun, W.; Mo, S. Complete genome sequence of an Atypical Porcine Pestivirus strain, GX01-2018, from Guangxi Province, China. *Microbiol. Resour. Announc.* **2019**, *8*. [CrossRef]
71. Zhang, H.; Wen, W.; Hao, G.; Hu, Y.; Chen, H.; Qian, P.; Li, X. Phylogenetic and genomic characterization of a novel atypical porcine pestivirus in China. *Transbound. Emerg. Dis.* **2018**, *65*, e202–e204. [CrossRef]
72. Zhang, X.; Dai, R.; Li, Q.; Zhou, Q.; Luo, Y.; Lin, L.; Bi, Y.; Chen, F. Detection of three novel atypical porcine pestivirus strains in newborn piglets with congenital tremor in southern China. *Infect. Genet. Evol.* **2019**, *68*, 54–57. [CrossRef] [PubMed]
73. Zhou, K.; Yue, H.; Tang, C.; Ruan, W.; Zhou, Q.; Zhang, B. Prevalence and genome characteristics of atypical porcine pestivirus in southwest China. *J. Gen. Virol.* **2019**, *100*, 84–88. [CrossRef]

74. Colom-Cadena, A.; Ganges, L.; Muñoz-González, S.; Castillo-Contreras, R.; Bohórquez, J.A.; Rosell, R.; Segalés, J.; Marco, I.; Cabezon, O. Atypical porcine pestivirus in wild boar (Sus scrofa), Spain. *Vet. Rec.* **2018**, *183*, 569. [CrossRef] [PubMed]
75. Kim, S.; Jeong, C.; Yoon, S.; Lee, K.; Yang, M.; Kim, B.; Lee, S.; Kang, S.; Kim, W. Detection of atypical porcine pestivirus (APPV) from a case of congenital tremor in Korea. *Korean J. Vet. Serv.* **2017**, *40*, 209–213.
76. Gatto, I.R.H.; Harmon, K.; Bradner, L.; Silva, P.; Linhares, D.C.L.; Arruda, P.H.; de Oliveira, L.G.; Arruda, B.L. Detection of atypical porcine pestivirus in Brazil in the central nervous system of suckling piglets with congenital tremor. *Transbound. Emerg. Dis.* **2018**, *65*, 375–380. [CrossRef] [PubMed]
77. Kaufmann, C.; Stalder, H.; Sidler, X.; Renzullo, S.; Gurtner, C.; Grahofer, A.; Schweizer, M. Long-term circulation of Atypical Porcine Pestivirus (APPV) within Switzerland. *Viruses* **2019**, *11*, 653. [CrossRef]
78. Dénes, L.; Biksi, I.; Albert, M.; Szeredi, L.; Knapp, D.G.; Szilasi, A.; Bálint, Á.; Balka, G. Detection and phylogenetic characterization of atypical porcine pestivirus strains in Hungary. *Transbound. Emerg. Dis.* **2018**, *65*, 2039–2042. [CrossRef]
79. Moennig, V. The control of classical swine fever in wild boar. *Front. Microbiol.* **2015**, *6*, 1211. [CrossRef]
80. Ridpath, J. Preventive strategy for BVDV infection in North America. *Jpn. J. Vet. Res.* **2012**, *60*, S41–S49.
81. Zhang, H.; Wen, W.; Hao, G.; Chen, H.; Qian, P.; Li, X. A subunit vaccine based on E2 protein of atypical porcine pestivirus induces Th2-type immune response in mice. *Viruses* **2018**, *10*, 673. [CrossRef]

viruses

Article

Molecular Survey and Phylogenetic Analysis of Atypical Porcine Pestivirus (APPV) Identified in Swine and Wild Boar from Northern Italy

Enrica Sozzi *, Cristian Salogni, Davide Lelli, Ilaria Barbieri, Ana Moreno, Giovanni Loris Alborali and Antonio Lavazza

Istituto Zooprofilattico Sperimentale della Lombardia e dell'Emilia Romagna "Bruno Ubertini" (IZSLER), Via Antonio Bianchi 7/9, 25124 Brescia, Italy; cristian.salogni@izsler.it (C.S.); davide.lelli@izsler.it (D.L.); ilaria.barbieri@izsler.it (I.B.); anamaria.morenomartin@izsler.it (A.M.); giovanni.alborali@izsler.it (G.L.A.); antonio.lavazza@izsler.it (A.L.)

* Correspondence: enrica.sozzi@izsler.it; Tel.: +39-030-2290361

Received: 14 November 2019; Accepted: 8 December 2019; Published: 10 December 2019

Abstract: Atypical porcine pestivirus (APPV) is a newly recognized member of the *Flaviviridae* family. This novel porcine pestivirus was first described in 2015 in the USA, where it has been associated with congenital tremor type A-II in new-born piglets. APPV is widely distributed in domestic pigs in Europe and Asia. In this study, a virological survey was performed in Northern Italy to investigate the presence of APPV using molecular methods. Testing of 360 abortion samples from pig herds revealed two APPV strains from distinct provinces in the Lombardy region and testing of 430 wild boar blood samples revealed three strains, one from Lombardy and two from Emilia Romagna. The nucleotide sequencing of a fragment of the nonstructural protein 3-coding region revealed a high similarity to the previously detected European strains (Spanish, German, and Italian) of APPV.

Keywords: pestivirus; pig; APPV; phylogenetic analysis; Italy

1. Introduction

Pestiviruses are highly variable single-stranded RNA genome viruses, belonging to the *Flaviviridae* family. Actually, based on molecular and epidemiological evidence, the genus Pestivirus includes eleven species, indicated with progressive letters from A to K. Thus, the "classical" species are A (Bovine viral diarrhea virus 1), B (Bovine viral diarrhea virus 2), C (Classical swine fever virus), and D (Border disease virus), whereas the new species are E to K [1]. In addition, other pestiviruses were described, likely as three additional species, respectively, in bats (bat-derived pestivirus) [2,3], sheep and goats (Tunisian sheep pestiviruses) [4], and pigs (Linda pestivirus) [5]. In the swine species only, in addition to Pestivirus C, three pestiviruses have been identified to date: (1) Pestivirus F (Bungowannah virus) reported only in Australia, as a cause of reproductive disorders, fetal death, and sudden death in piglets, [6]; (2) The Linda virus described in association with congenital tremors (CTs) in piglets in Austria and thereafter only occasionally reported, leaving its geographical spread and clinical relevance in pigs undefined [5]; and (3) Pestivirus K, commonly known as atypical porcine pestivirus (APPV), which is the most relevant due to the frequency of identification, clinical findings, and economic importance. In fact, it has been identified several times in North America [7,8], South America [9,10], Europe [11–13], and Asia [14,15], and it should be considered as stably present for a long time in domestic and wild pig populations [8,10,15,16]. This is true also for Italy, since in a previous survey, at least four APPV isolates were found between 2015–2017 and a quite high seroprevalence was detected in pig sera [15].

Although APPV has been repeatedly identified in asymptomatic animals, there is clear evidence that it is associated with CT syndrome type A-II (CT A-II) in newborns [7]. Clinically healthy pigs and wild boars may have an epidemiological role as vehicles of APPV, but, considering the different frequency of detection in wild boars, which was quite high in Germany and Serbia [17] and very low in Spain [18], the epidemiology of APPV may vary considerably from country to country with increases in livestock and wild populations, animal breeding, and world trade [18]. While one study described the economic losses caused by a 10% drop in the number of weaned piglets per sow [19], the full economic consequences of APPV outbreaks remain to be determined.

In this study, the presence of the APPV genome in pig fetuses and wild boars from both the Lombardy and Emilia Romagna regions of Northern Italy was determined, and the genetic characterizations of the identified strains are described.

2. Materials and Methods

2.1. Pigs

From 2016 to 2018, 360 fetuses of pigs from pig farms in the Lombardy region were examined at the IZSLER Diagnostic Laboratory in Brescia. All of the samples examined originated from field cases of spontaneous abortions in pig farms and sent to the general diagnostic laboratory of IZSLER to determine the presence of any infectious agent. In none of these cases was the presence of clinical signs referable to CT syndrome specifically reported. During necropsy, samples of organs (brain, lung, spleen, liver, and kidney) were taken from each aborted fetus, then collected into a single farm-specific pool and homogenized (10% w/v) in minimum essential medium (MEM; Gibco, Life Technologies, Paisley, UK) supplemented with an antibiotic (1000 U/mL penicillin, 1 mg/mL streptomycin; Gibco, Life Technologies, Paisley, UK) and anti-mycotic (2.5 µg/mL amphotericin B; Gibco, Life Technologies, Paisley, UK). After centrifugation, the supernatant was analyzed to identify any agents that cause abortions in swine. For bacteriological agents such as Brucella spp., Listeria spp., and Mycoplasma spp., the screening and pathogen identification were conducted according to Office International des Epizooties (OIE) standardized protocols [20]. The presence of Chlamydophila spp. was investigated by real-time Polymerase Chain Reaction (PCR) directly in biological samples [21] and typing by the PCR-restriction fragment length polymorphism (RFLP) assay, targeting the 16S ribosomal gene [22]. For Mycoplasma spp., the PCR method described by van Kuppeveld et al. [23] was used. Virological analyses for detecting the more common pig pathogens were conducted using a panel of PCR methods including porcine reproductive and respiratory syndrome virus (PRRSV) (AgPath-ID™ NA and EU PRRSV Multiplex© Applied Biosystems), porcine circovirus type 2 (PCV-2) [24], porcine parvovirus (PPV) [25], and porcine circovirus type 3 (PCV-3) [26]. The presence of pestiviruses was determined by using a pan-pestivirus real-time RT-PCR [27], and, considering its limited capacity in detecting APPV, by a APPV-specific real-time RT-PCR [19]. In addition, all the tested samples were inoculated on cell cultures (primary embryonic swine kidney cells, swine alveolar macrophages, and monkey kidney cell line MARC-145), which allow the isolation of a broad range of swine viruses. The inoculated cell monolayers were observed daily for 5–7 d for the appearance of a cytopathic effect and then sub-cultured to the second passage, at which time they were independently tested using an "in-house" sandwich enzyme-linked immunosorbent assay (ELISAs) for the presence of pestivirus [28] and PRRSV antigens [29]. Only those samples that were positive for APPV at the initial screening test using the APPV-specific real-time RT-PCR were further subcultured to improve the chance to isolate the APPV until the fifth passage, at which time, even in the absence of a cytopathic effect, they were assessed again with the APPV-specific real-time RT-PCR.

2.2. Wild Boars

In total, 430 blood samples of wild boars, killed during the 2017–2018 hunting season, were collected in the framework of the Lombardy and Emilia Romagna wildlife monitoring plans for classical swine fever (CSF) and Aujeszky disease and transferred to IZSLER for examination. Serum samples were tested for antibodies against swine vesicular disease virus [30,31], encephalomyocarditis virus [32], glycoprotein E of Aujeszky's disease virus [33], pestivirus (A–D) [34], swine influenza virus type A, subtypes H1N1, H1N2, and H3N2 [35,36], and finally for Brucella spp. (Svanovir Brucella–Ab C-ELISA©). Serological analyses were conducted with the methods currently in use at the IZSLER.

2.3. *Identification and Genomic Characterization of Atypical Porcine Pestivirus (APPV)*

For the investigation of the APPV genome, all samples, both the homogenates from fetal organs and the wild boar sera, were screened using the NS5B gene-specific real-time RT-PCR method [19]. Samples that tested positive were characterized by Sanger sequencing using RT-PCR that amplified a fragment of the NS3 region. The nucleotide sequences were aligned using the ClustalW method and compared with sequences present in GenBank [37] using MEGA6 software [38]. The maximum likelihood phylogenetic tree was constructed using IQ-tree software [39] by applying the TIM2+F+G4 model identified using ModelFinder selection [40].

3. Results

3.1. *Pigs*

The examination of the NS5B gene by real-time RT-PCR in the homogenized samples of pig fetuses identified two (0.6%) positive samples in two distinct farrow-to-finish farms, one from the province of Brescia in a pool containing the organs of three fetuses and the other from the province of Mantua in a pool containing the organs of two fetuses (Figure 1).

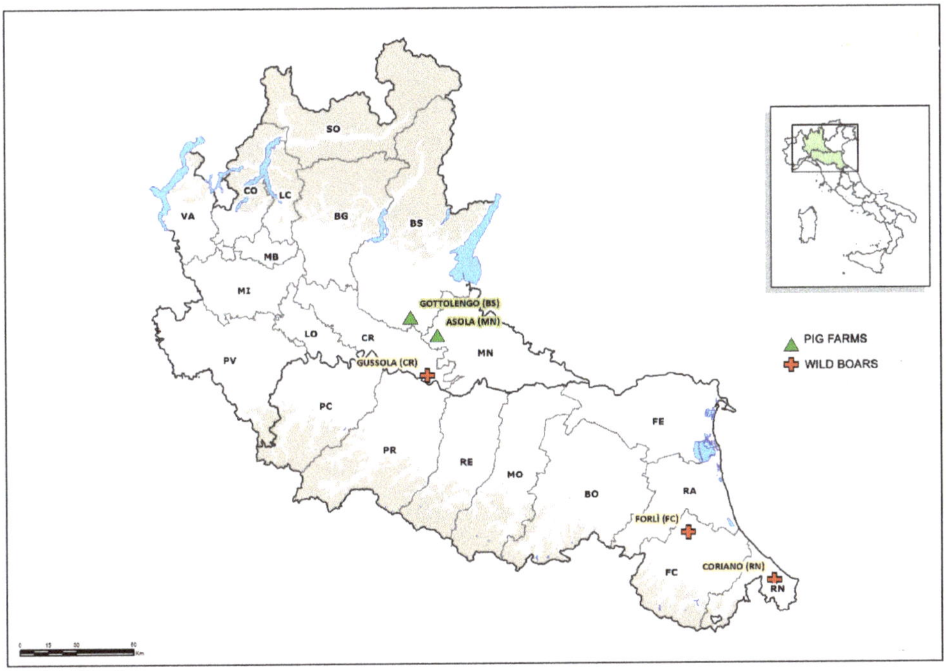

Figure 1. Geographical distribution of the pig farms and wild boar hunting sites where atypical porcine pestivirus (APPV) was identified.

At necropsy, none of these fetuses showed internal macroscopic lesions, and bacteriological investigations consistently produced negative results. Moreover, the molecular examinations for PRRSV, PCV-2, PPV, and pan-pestivirus were all negative, and only the real-time PCR for PCV-3 produced a positive result from a pool of fetuses that originated from the province of Brescia.

The two sequences obtained, APPV_Italy_SW_BS341729_2017 and APPV_Italy_SW_MN212160_2016, had a nucleotide similarity of 92.5% between them and clustered with APPVs previously identified in Europe. In particular, the phylogenetic tree constructed using the NS3 region (Figure 2) revealed that the two identified strains belong to distinct groups: (1) the APPV_Italy_SW_MN212160_2016 strain is related to both the 98/Sp06 strain identified in Spain in 2006 [16] (99.2% identity) and the German strain Bavaria S5/9 identified in 2015 [11] (96.8% identity); and (2) APPV_Italy_SW_BS341729_2017 forms a separate and closely related clade with two sequences identified in 2015 in pigs from Italy, Italy-164 and Italy-181 [15], with a nucleotide similarity of 95.5% and 99.2%, respectively.

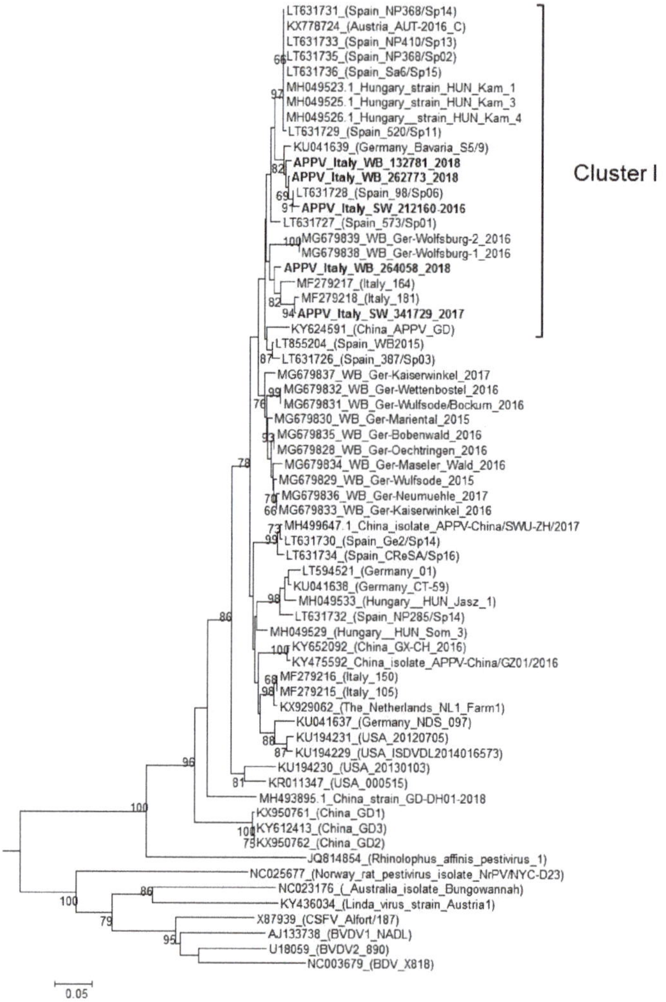

Figure 2. Phylogenetic tree based on a 645-nt fragment of the nonstructural protein 3-encoding region of the atypical porcine pestivirus (APPV) genome present in GenBank. A phylogenetic analysis using the maximum likelihood method including 1.000 bootstrap iterations was performed. Only bootstrap values ≥60 are indicated. Sequences in bold were generated in this study.

Despite the positive results of the real-time RT-PCR test, virological examinations of the organs of swine fetuses by inoculating cell cultures up to the fifth passage produced consistent negative results. Consequently, we did not succeed in isolating any APPV strains.

3.2. Wild Boars

All of the wild boar samples examined in this study including those that tested APPV-positive, originated from hunting activity, and thus, they were considered healthy based on their behavior before shooting and the absence of lesions on carcasses examined at slaughter. Out of the 430 blood samples examined, three (0.69%), originating from the Cremona Province in Lombardy (APPV_Italy_WB_CR264058_2018) and from two neighboring provinces, Rimini and Forlì in Emilia Romagna (APPV_Italy_WB_RN262773_2018 and APPV_Italy_WB_FC132781_2018, respectively) (Figure 1), screened positive for APPV. In agreement with the genomic characterization of the strains identified in pigs, the sequences of those identified in wild boars could be divided into two distinct clusters. The first included the two strains detected in Emilia Romagna (APPV_Italy_WB _RN262773_2018 and APPV_Italy_WB _FC132781_2018) and the sequence identified in the pig farm from Mantua Province in Lombardy (APPV_Italy_SW_MN212160_2016), which had identities of 95.6% and 96%, respectively; plus the previously characterized strains BavariaS5/9 and Spain 98/Sp06, with which the WB strains had a higher nucleotide identity of 98–99.2%.

The second cluster included the APPV_Italy_WB_CR264058_2018, the APPV Italy_SW_ BS341729_2017 (96.4% identity), the Italian pig strains, Italy-164 and Italy-181, which had nucleotide of 96.4% and 96%, respectively, and the strains identified in 2016 in Lower Saxony, Germany (94.3% identity).

For the collateral serological examinations, all APPV positive samples were negative for all serological tests employed except for: (a) one serum (APPV_Italy_WB_CR264058_2018) that tested positive for anti-Brucella spp. antibodies and (b) one (APPV_Italy_WB_FC132781_2018) that tested positive for glycoprotein E antibodies to the Aujeszky's disease virus.

The nucleotide sequences of the five APPV isolates were deposited in GenBank (NCBI) with the accession numbers MN736974–MN736978.

4. Discussion

The examination of 360 swine fetuses and 430 hunted wild boar blood samples identified APPVs in the area, two strains from the former and three strains from the latter. Based on the phylogenetic analysis performed, all five of these strains clustered with APPV strains previously identified in Spain and Germany. Although all of the sequences obtained can be grouped into the putative Cluster I described by Muñoz-González et al. [16], which should include viruses that have a common origin, the strains identified here could be further divided in two distinct sub-clusters.

Based on the absence of lesions on both fetuses and wild boars, we were keen to exclude a definite pathogenic role for APPV in these specific cases. However, it remains to be clarified whether the PCV-3 plus APPV (APPV_Italy_SW_BS341729_2017) co-infection, which was detected in the pool of fetuses originating from Brescia Province, may have had a synergistic effect in infected piglets, resulting in a clinical form characterized by abortion and natimortality. The very low prevalence of APPV found in the examined pig samples (0.6%) is largely different from the sole previous study on APPV in Italy [15], which reported a higher prevalence of viral detection (17.5%). This could be likely due to the sample types and selection, since we included in the survey only aborted pig fetuses with the aim of trying to establish if any correlation existed between the clinical case and the detection of APPV. A more specific monitoring program focused on the two APPV positive farms, and in general, to have systematically reported and analyzed cases of CT are anticipated in order to better define the prevalence of APPV in pig farms in Italy and clarify its effects.

Wild boars are susceptible to APPV infection, but their role in the epidemiology of the virus remains unknown. The low prevalence of APPV in the examined wild boar population is well correlated with the numbers reported by Colom-Cadena et al. [18] in Spain, but is in contrast to the high prevalence found in wild boars from northern Germany [17], where APPV seems to be endemic among wild boars in many areas. Considering the small number of strains isolated from the large geographical territory represented by the two regions (Lombardy and Emilia Romagna) and the lack of any particular geographic distribution of the clustered strains, it is impossible to draw epidemiological interpretations on APPV diffusion or on transmission between domestic pigs and wild boars.

5. Conclusions

APPV appears to be well established in the domestic swine populations of different countries in Europe, America, and Asia. Indeed, although identifying APPV has only become possible during the last few years due to the progressive refinement of diagnostic techniques, retrospective studies have indicated its circulation for many years. Based on the available genomic characterization data in the literature and the prevalence in domestic pigs, APPV exhibits a high genetic diversity among viral strains detected in different countries and tends to form independent clusters according to geographic locations. This study confirms the presence and distribution of APPV in populations of domestic and wild pigs present in the Lombardy and Emilia Romagna regions of Italy, which are under the jurisdiction of IZSLER. The high homology levels with strains identified in Germany and Spain reinforce the hypothesis that Italian strains have a European origin, and they confirm the likely determining role in the spread of infection being the commercial trade in pigs among different countries. The detection of new pestiviruses indicates the need to monitor for their presence and distribution using a systematic surveillance and diagnostic approach. In fact, the accurate and constant characterization of circulating strains is necessary to update the serological and virological tests, which in turn may be used to collect more detailed epidemiological information regarding APPV such as routes of entry and dissemination, and genetic evolution.

Author Contributions: C.S. and G.L.A. performed necropsy and routine laboratory examinations; E.S., I.B., D.L., and A.M. performed the laboratory work and analyzed the data; E.S., A.M., D.L., and A.L. interpreted the results and designed the figures; E.S. and A.L. wrote the manuscript. All the authors critically analyzed, revised, and approved the manuscript.

Funding: This study was supported by grants from the project PRC 2015019 of the Italian Ministry of Health.

Acknowledgments: The authors would like to thank Loredana Zingarello and Debora Campagna for their excellent technical assistance.

Conflicts of Interest: The authors declare no conflicts of interest.

References

1. Smith, D.B.; Meyers, G.; Bukh, J.; Gould, E.A.; Monath, T.; Scott Muerhoff, A.; Pletnev, A.; Rico-Hesse, R.; Stapleton, J.T.; Simmonds, P.; et al. Proposed revision to the taxonomy of the genus Pestivirus, family Flaviviridae. *J. Gen. Virol.* **2017**, *98*, 2106–2112. [CrossRef] [PubMed]
2. Wu, Z.Q.; Ren, X.W.; Yang, L.; Hu, Y.F.; Yang, J.; He, G.M.; Zhang, J.P.; Dong, J.; Sun, L.L.; Du, J.; et al. Virome analysis for identification of novel mammalian viruses in bat species from Chinese provinces. *J. Virol.* **2012**, *86*, 10999–11012. [CrossRef] [PubMed]
3. Wu, Z.Q.; Liu, B.; Du, J.; Zhang, J.P.; Lu, L.; Zhu, G.J.; Han, Y.L.; Su, H.X.; Yang, L.; Zhang, S.Y.; et al. Discovery of diverse rodent and bat pestiviruses with distinct genomic and phylogenetic characteristics in several Chinese provinces. *Front. Microbiol.* **2018**, *9*, 2562. [CrossRef] [PubMed]
4. Liu, L.L.; Xia, H.; Wahlberg, N.; Belák, S.; Baule, C. Phylogeny, classification and evolutionary insights into pestiviruses. *Virol.* **2009**, *385*, 351–357. [CrossRef] [PubMed]
5. Lamp, B.; Schwarz, L.; Högler, S.; Riedel, C.; Sinn, L.; Rebel-Bauder, B.; Weissenböck, H.; Ladinig, A.; Rümenapf, T. Novel pestivirus species in pigs, Austria, 2015. *Emerg. Infect. Dis.* **2017**, *23*, 1176–1179. [CrossRef]
6. Kirkland, P.D.; Frost, M.J.; Finlaison, D.S.; King, K.R.; Ridpath, J.F.; Gu, X. Identification of a novel virus in pigs—Bungowannah virus: A possible new species of pestivirus. *Virus Res.* **2007**, *129*, 26–34. [CrossRef]

7. Arruda, B.L.; Arruda, P.H.; Magstadt, D.R.; Schwartz, K.J.; Dohlman, T.; Schleining, J.A.; Patterson, A.R.; Visek, C.A.; Victoria, J.G. Identification of a Divergent Lineage Porcine Pestivirus in Nursing Piglets with Congenital Tremors and Reproduction of Disease following Experimental Inoculation. *PLoS ONE* **2016**, *11*, e0150104. [CrossRef]
8. Hause, B.M.; Collin, E.A.; Peddireddi, L.; Yuan, F.; Chen, Z.; Hesse, R.A.; Gauger, P.C.; Clement, T.; Fang, Y.; Anderson, G. Discovery of a novel putative atypical porcine pestivirus in pigs in the USA. *J. Gen. Virol.* **2015**, *96*, 2994–2998. [CrossRef]
9. Gatto, I.R.H.; Harmon, K.; Bradner, L.; Silva, P.; Linhares, D.C.L.; Arruda, P.H.; de Oliveira, L.G.; Arruda, B.L. Detection of atypical porcine pestivirus in Brazil in the central nervous system of suckling piglets with congenital tremor. *Transbound Emerg. Dis.* **2018**, *65*, 375–380. [CrossRef]
10. Mósena, A.C.S.; Weber, M.N.; da Cruz, R.A.S.; Cibulski, S.P.; da Silva, M.S.; Puhl, D.E.; Hammerschmitt, M.E.; Takeuti, K.L.; Driemeier, D.; de Barcellos, D.E.S.N.; et al. Presence of atypical porcine pestivirus (APPV) in Brazilian pigs. *Transbound Emerg. Dis.* **2018**, *65*, 22–26. [CrossRef]
11. Postel, A.; Hansmann, F.; Baechlein, C.; Fischer, N.; Alawi, M.; Grundhoff, A.; Derking, S.; Tenhündfeld, J.; Pfankuche, V.M.; Herder, V.; et al. Presence of atypical porcine pestivirus (APPV) genomes in newborn piglets correlates with congenital tremor. *Sci. Rep.* **2016**, *6*, 27735. [CrossRef] [PubMed]
12. Beer, M.; Wernike, K.; Dräger, C.; Höper, D.; Pohlmann, A.; Bergermann, C.; Schröder, C.; Klinkhammer, S.; Blome, S.; Hoffmann, B. High Prevalence of Highly Variable Atypical Porcine Pestiviruses Found in Germany. *Transbound Emerg. Dis.* **2017**, *64*, e22–e26. [CrossRef] [PubMed]
13. de Groof, A.; Deijs, M.; Guelen, L.; van Grinsven, L.; van Os-Galdos, L.; Vogels, W.; Derks, C.; Cruijsen, T.; Geurts, V.; Vrijenhoek, M.; et al. Atypical Porcine Pestivirus: A Possible Cause of Congenital Tremor Type A-II in Newborn Piglets. *Viruses* **2016**, *8*, 271. [CrossRef] [PubMed]
14. Yuan, J.; Han, Z.; Li, J.; Huang, Y.; Yang, J.; Ding, H.; Zhang, J.; Zhu, M.; Zhang, Y.; Liao, J.; et al. Atypical Porcine Pestivirus as a Novel Type of Pestivirus in Pigs in China. *Front. Microbiol.* **2017**, *8*, 862. [CrossRef] [PubMed]
15. Postel, A.; Meyer, D.; Cagatay, G.N.; Feliziani, F.; De Mia, G.M.; Fischer, N.; Grundhoff, A.; Milićević, V.; Deng, M.C.; Chang, C.Y.; et al. High Abundance and Genetic Variability of Atypical Porcine Pestivirus in Pigs from Europe and Asia. *Emerg. Infect. Dis.* **2017**, *23*, 2104–2107. [CrossRef] [PubMed]
16. Muñoz-González, S.; Canturri, A.; Pérez-Simó, M.; Bohórquez, J.A.; Rosell, R.; Cabezón, O.; Segalés, J.; Domingo, M.; Ganges, L. First report of the novel atypical porcine pestivirus in Spain and a retrospective study. *Transbound Emerg. Dis.* **2017**, *64*, 1645–1649. [CrossRef] [PubMed]
17. Cagatay, G.N.; Antos, A.; Meyer, D.; Maistrelli, C.; Keuling, O.; Becher, P.; Postel, A. Frequent infection of wild boar with atypical porcine pestivirus (APPV). *Transbound Emerg. Dis.* **2018**, *65*, 1087–1093. [CrossRef]
18. Colom-Cadena, A.; Ganges, L.A.; Muñoz-González, S.; Castillo-Contreras, R.; Alejandro Bohórquez, J.; Rosell, R.; Segalés, J.; Marco, I.; Cabezon, O. Atypical porcine pestivirus in wild boar (Sus scrofa), Spain. *Vet. Rec.* **2018**, *183*, 569.
19. Schwarz, L.; Riedel, C.; Högler, S.; Sinn, L.J.; Voglmayr, T.; Wöchtl, B.; Dinhopl, N.; Rebel-Bauder, B.; Weissenböck, H.; Ladinig, A.; et al. Congenital infection with atypical porcine pestivirus (APPV) is associated with disease and viral persistence. *Vet. Res.* **2017**, *48*, 1. [CrossRef]
20. OIE Manual for Terrestrial Animals 2016 cap 2.1.4 par B1, B1.1, B1.2, B1.3: Brucellosis [Brucella abortus, B. melitensis, B. suis (Infection with B. abortus, B. melitensis, B. suis)]—Diagnostic Techniques—Identification of the Agent—Staining Methods—Collection of Samples and Culture—Identification and Typing. Available online: https://www.oie.int/fileadmin/Home/eng/Health_standards/tahm/3.01.04_BRUCELLOSIS.pdf (accessed on 15 October 2019).
21. Ehricht, R.; Slickers, P.; Goellner, S.; Hotzel, H.; Sachse, K. Optimized DNA microarray assay allows detection and genotyping of single PCR-amplifiable target copies. *Mol. Cell Probes.* **2006**, *20*, 60–66. [CrossRef]
22. Vicari, N.; Santoni, R.; Vigo, P.G.; Magnino, S. A PCR-RFLP assay targeting the 16S ribosomal gene for the diagnosis of animal chlamydioses. In Proceedings of the 5th Meeting of the European Society for Chlamydia Research, Budapest, Hungary, 1–4 September 2004.
23. Van Kuppeveld, F.J.; van der Logt, J.T.; Angulo, A.F.; van Zoest, M.J.; Quint, W.G.; Niesters, H.G.; Galama, J.M.; Melchers, W.J. Genus- and species-specific identification of mycoplasmas by 16S rRNA amplification. *Appl. Environ. Microbiol.* **1992**, *58*, 2606–2615, Erratum in: *Appl. Environ. Microbiol.* **1993**, *59*, 655.

24. Olvera, A.; Sibila, M.; Calsamiglia, M.; Segalés, J.; Domingo, M. Comparison of porcine circovirus type 2 load in serum quantified by a real time PCR in postweaning multisystemic wasting syndrome and porcine dermatitis and nephropathy syndrome naturally affected pigs. *J. Virol. Methods* **2004**, *117*, 75–80. [CrossRef] [PubMed]
25. Kim, J.; Choi, C.; Han, D.U.; Chae, C. Simultaneous detection of porcine circovirus type 2 and porcine parvovirus in pigs with PMWS by multiplex PCR. *Vet. Rec.* **2001**, *149*, 304–305. [CrossRef] [PubMed]
26. Palinski, R.; Pineyro, P.; Shang, P.; Yuan, F.; Guo, R.; Fang, Y.; Byers, E.; Hause, B.M. A novel porcine circovirus distantly related to known circoviruses is associated with porcine dermatitis and nephropathy syndrome and reproductive failure. *J. Virol.* **2016**, *91*, 16. [CrossRef] [PubMed]
27. OIE Manual for Terrestrial Animals 2018 cap 3.4.7 par B1.2: Bovine Viral Diarrhea—Diagnostic Techniques—Detection of the Agent—Nucleic Acid Detection. Available online: https://www.oie.int/fileadmin/Home/eng/Health_standards/tahm/3.04.07_BVD.pdf (accessed on 15 October 2019).
28. Brocchi, E.; Cordioli, P.; Berlinzani, A.; Gamba, D.; De Simone, F. Development of a panel of anti-pestivirus monoclonal antibodies useful for virus identification and antibody assessment. In Proceedings of the second Symposium on Ruminant Pestiviruses, Annecy, France, 1–3 October 1992; Edwards, S., Ed.; Fondation Marcel Mérieux: Lyon, France, 1993; pp. 215–218.
29. Grazioli, S.; Pezzoni, G.; Cordioli, P.; Brocchi, E. Validation of a competitive ELISA for serodiagnosis of PRRS based on recombinant n-protein and monoclonal antibody. In Proceedings of the 8th International Congress of Veterinary Virology, Budapest, Hungary, 24–27 August 2009; Benkõ, M., Harrach, B., Eds.; Hungarian Academy of Sciences: Budapest, Hungary, 2009; p. 199.
30. Heckert, R.A.; Brocchi, E.; Berlinzani, A.; Mackay, D.K. An international comparative analysis of a competitive ELISA for the detection of antibodies to swine vesicular disease virus. *J. Vet. Diagn. Invest.* **1998**, *10*, 295–297. [CrossRef]
31. OIE Manual for Terrestrial Animals 2018 cap 3.8.8 par B2.2: Swine Vesicular Disease—Diagnostic Techniques—Serological Tests—Enzyme-Linked Immunosorbent Assay. Available online: https://www.oie.int/fileadmin/Home/eng/Health_standards/tahm/3.08.08_SVD.pdf (accessed on 15 October 2019).
32. Brocchi, E.; Carra, E.; Koenen, F.; De Simone, F. *Sviluppo di metodi ELISA basati sull'impiego di anticorpi monoclonali per la dimostrazione del virus della encefalomiocardite (EMCV) e dei relativi anticorpi*; La Selezione Veterinaria—Supplemento; IZSLER: Brescia, Italy, 2000; pp. 207–215.
33. OIE Manual for Terrestrial Animals 2018 cap 3.1.2 par B2.2: Aujeszky's Disease (Infection with Aujeszky's Disease Virus)—Diagnostic Techniques—Serological Tests—Enzyme-Linked Immunosorbent Assay (a Prescribed Test for International Trade). Available online: https://www.oie.int/fileadmin/Home/eng/Health_standards/tahm/3.01.02_AUJESZKYS.pdf (accessed on 15 October 2019).
34. OIE Manual for Terrestrial Animals 2019 cap 3.8.3 par B2.3: Classical Swine Fever (Infection with Classical Swine Fever Virus)—Diagnostic Techniques—Serological Tests—Enzyme-Linked Immunosorbent Assay (a Prescribed Test for International Trade). Available online: https://www.oie.int/fileadmin/Home/eng/Health_standards/tahm/3.08.03_CSF.pdf (accessed on 15 October 2019).
35. Kendall, A.P.; Pereira, M.S.; Skehel, J.J. *Concepts and Procedures for Laboratory-Based Influ-Enza Surveillance*; World Health Organization: Geneva, Switzerland, 1982; (Copies available from the WHO Collaborating Centre for Surveillance, Epidemiology and Control of Influenza, CDC, Atlanta, GA).
36. OIE Manual for Terrestrial Animals 2018 cap 3.8.7 par B2.1: Influenza A Virus of Swine–Diagnostic Techniques—Serological Tests—Haemagglutination Inhibition Test. Available online: https://www.oie.int/fileadmin/Home/eng/Health_standards/tahm/3.08.07_INF_A_SWINE.pdf (accessed on 15 October 2019).
37. GenBank® (NCBI). Available online: www.ncbi.nlm.nih.gov (accessed on 22 November 2019).
38. Tamura, K.; Stecher, G.; Peterson, D.; Filipski, A.; Kumar, S. MEGA6: Molecular evolutionary genetics analysis version 6.0. *Mol. Biol. Evol.* **2013**, *30*, 2725–2729. [CrossRef]
39. Nguyen, L.T.; Schmidt, H.A.; von Haeseler, A.; Minh, B.Q. IQ-TREE: A fast and effective stochastic algorithm for estimating maximum-likelihood phylogenies. *Mol. Biol. Evol.* **2015**, *32*, 268–274. [CrossRef]
40. Kalyaanamoorthy, S.; Minh, B.Q.; Wong, T.K.F.; von Haeseler, A.; Jermiin, L.S. ModelFinder: Fast model selection for accurate phylogenetic estimates. *Nat. Methods* **2017**, *14*, 587–589. [CrossRef]

viruses

MDPI

Article

Prevalence and Genetic Diversity of Atypical Porcine Pestivirus (APPV) Detected in South Korean Wild Boars

SeEun Choe [1], Gyu-Nam Park [1], Ra Mi Cha [1], Bang-Hun Hyun [1], Bong-Kyun Park [1,2] and Dong-Jun An [1,*]

1 Virus Disease Division, Animal and Plant Quarantine Agency, Gimchen, Gyeongbuk-do 39660, Korea; ivvi59@korea.kr (S.C.); changep0418@gmail.com (G.-N.P.); rami.cha01@korea.kr (R.M.C.); hyunbh@korea.kr (B.-H.H.); parkx026@korea.kr (B.-K.P.)

2 College of Veterinary Medicine, Seoul University, Gwanak-ro, Gwanak-gu, Seoul 08826, Korea

* Correspondence: andj67@korea.kr; Tel.: +82-54-912-0795

Received: 30 May 2020; Accepted: 22 June 2020; Published: 24 June 2020

Abstract: Atypical porcine pestivirus (APPV), currently classified as *pestivirus K*, causes congenital tremor (CT) type A-II in piglets. Eighteen APPV strains were identified from 2297 South Korean wild boars captured in 2019. Phylogenetic analysis of the structural protein E2 and nonstructural proteins NS3 and Npro classified the APPV viruses, including reference strains, into Clades I, II and III. Clade I was divided into four subclades; however, the strains belonging to the four subclades differed slightly, depending on the tree analysis, the NS3, E2, and Npro genes. The maximum-likelihood method was assigned to South Korean wild boar APPV strains to various subclades within the three trees: subclades I.1 and I.2 in the E2 tree, subclade I.1 in the Npro tree, and subclades I.1 and I.4 in the NS3 ML tree. In conclusion, APPV among South Korean wild boars belonging to Clade I may be circulating at a higher level than among the South Korean domestic pig populations.

Keywords: APPV; wild boar; ML tree; Clade; pestivirus

1. Introduction

Pestiviruses are highly variable single-stranded RNA genome viruses belonging to the family *Flaviviridae*. The genus *Pestivirus* includes animal pathogens that are of worldwide socioeconomic significance—these include bovine viral diarrhea virus (BVDV, *pestivirus A–B*), classical swine fever virus (CSFV, *pestivirus C*), and border disease virus (BDV, *pestivirus D*) [1]. Other *Pestiviruses* include *Pestivirus E* (pronghorn pestivirus), *Pestivirus F* (Bungowannah virus), *Pestivirus G* (giraffe pestivirus), *Pestivirus H* (Hobi-like pestivirus), *Pestivirus I* (Aydin-like pestivirus), and *Pestivirus J* (rat pestivirus) [1].

A novel genetically distinct strain of *pestivirus K*, named atypical porcine pestivirus (APPV), was first identified in the USA in 2015 [2]. It was also identified in Germany, the Netherlands, Austria, Spain, China, Hungary, Brazil, and Sweden in 2016–18 [3–8]. In South Korea, the first case of APPV, which causes congenital tremor (CT) type A-II in suckling piglets, was identified in 2017 [9]. Although APPV has been found in animals with no clinical signs, there is clear evidence that it is associated with CT type A-II in newborns [10]. The presence of APPV in domestic pigs was confirmed repeatedly in several countries in Europe, North and South America, and Asia [2,11], and in wild boars from countries in Europe [10,12,13]. Clinically, healthy pigs and wild boars may have an epidemiological role as vehicles for APPV.

Pestivirus genomes can be classified into genogroups and sub-genogroups [14,15]. Phylogenetic analysis of the E2 and Npro proteins divided APPV into at least four (A–D) [5,16] or five (A–E) different genogroups [17].

In this study, we describe the detection and genetic characterization of APPV in wild boars from South Korea.

2. Material and Methods

2.1. Wild Boars

To satisfy the OIE requirements for the surveillance of wild boars and feral pigs in Classical Swine Fever-free countries, wild boars have been hunted (in co-operation with the Korean Pork Producers Association and the South Korea government) since 2010. Immediately after hunting, blood samples were collected from wild boars captured in nine provinces of South Korea and transported to the APQA. Blood samples from 2297 wild boars (1126 males, 1045 females, and 126 unknown), captured in 2019, were screened for APPV.

2.2. Reverse Transcription-Polymerase Chain Reaction (RT-PCR) of APPV

RT-PCR was performed to detect APPV [12,16]. Briefly, total RNA was extracted from 100 μL of whole blood using the QIAamp viral RNA mini kit (Qiagen, Cat. No. 52904. Hilden, Germany). Extracted RNA was reverse-transcribed using SuperScript III (Invitrogen, Cat. No. 18080093. Carlsbad, CA, USA). For APPV screening and sequencing, PCR was performed using primers designed to target areas of the conserved NS3-encoding region, as described previously [12]. Primers targeting the E2 and Npro genes were used to amplify the complete nucleotide sequences, as described previously [16]. The amplification products were purified using the QIAquick Gel Extraction Kit (Qiagen, Cat. No. 28704. Hilden, Germany) and used directly for sequencing (Cosmogentech Co., Seoul, Korea). PCR and serum neutralization tests were used to detect viral antigens and antibodies specific for CSFV and BVDV in APPV-positive samples, as described previously [18].

2.3. Phylogenetic Analysis of APPV

Multiple nucleotide sequence alignment was carried out by the Clustal X alignment program [19] using APPV sequences available in GenBank as references, and BLAST software (NCBI, Bethesda, Rockville, MD, USA). Outgroup strains comprised *pestiviruses A–H*. The partial sequences of NS3, derived from 18 APPVs detected in South Korean wild boars, were compared with 86 reference sequences (including eight outgroup strains) from Asia, North America, and Europe. The complete E2 sequence of four APPVs and the complete Npro sequence data of five APPVs detected in South Korean wild boars were compared with 70 (including ten outgroup strains) and 69 reference sequences (including ten outgroup strains), respectively. Nucleotide sequences of the NS3, E2, and Npro regions were analyzed phylogenetically using the maximum-likelihood (ML) method, with the Tamura–Nei model and bootstrap analysis (n = 1000) within MEGA 7.0 software (State College, PA, USA) with default parameters [20]. The ML tree was based on rates among sites (Gamma distributed with invariant sites (G+I)) and the ML heuristic method (Nearest-neighbor-interchange (NNI)). The partial NS3 sequences of 18 APPV strains (accession numbers: MT501737–MT501754), the complete E2 sequences of four APPV strains (accession numbers: MT501733–MT501736), and the complete Npro sequences of five APPV strains (accession numbers: MT501555–MT501759) detected in South Korean wild boars were deposited in GenBank.

2.4. Ethical Approval

The authors confirm that the work complies with the ethical policies of the journal. The work was approved by the Institutional Animal Care and Use Committee of the Animal and Plant Quarantine Agency (APQA) (Approval Number: 2019-448).

3. Results

3.1. Geographic Prevalence of APPV

Eighteen APPV strains were identified in 2297 blood samples collected from wild boars in 2019, suggesting that the prevalence of APPV is 0.78%. Of the APPV-positive wild boars, 15 were male (15/18, 83.3%), two were female (2/18, 11.1%), and one was of unknown sex (1/18, 5.6%). APPV strains were detected in wild boars from six provinces and of various weights (seven <30 kg; seven 30–60 kg; four >60 kg). Among the 18 APPVs detected in South Korean wild boars, five were detected in Gyeongnam (GN, 5/292; 1.71%), four in Gangwon (GW, 4/609; 0.66%), three each in Gyeonggi (GG, 3/452; 0.66%) and Chungnam (CN, 3/288; 0.35%), two in Chungbuk (CB, 2/204; 0.98%), and one in Gyeongkuk (GB, 1/275; 0.36%), as shown in Figure 1. All APPV-positive samples were negative for anti-CSFV and BVDV antibodies and antigens.

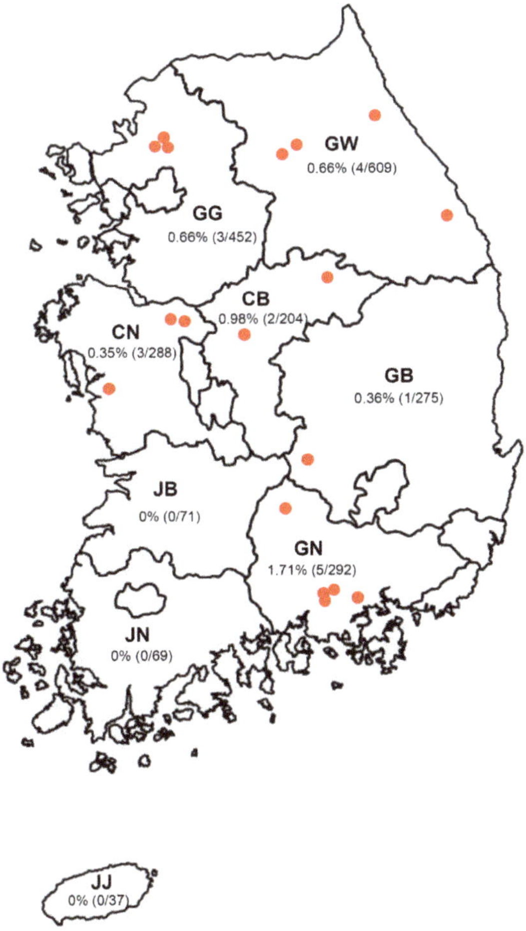

Figure 1. Locations in which APPV-positive wild boars were captured. Locations in which wild boars were captured are marked by a red dot. GW: Gangwon; GG: Gyeonggi; GN: Gyeongnam; GB, Gyeongbuk; JN: Jennam; JB: Jenbuk; CN: Chungnam; CB: Chungbuk; JJ: Jeju.

3.2. ML Trees Based on NS3 Sequences

The NS3 sequences (767 nucleotides (nt)) of the 18 APPVs detected from wild boars were 87.7–99.9% identical at the nt level and 96.5–100% identical at the amino acid (aa) level. ML tree analysis of NS3, E2, and Npro sequences revealed that *Pestivirus* strains were clearly divided into two groups: *Pestivirus K* (APPVs) and Other *Pestiviruses (A–H)*, as shown in Figures 2–4). All APPVs were classified into three large Clades (I, II, and III) and four smaller subclades (I.1, I.2, I.3, and I.4), as shown in Figures 2–4). ML analysis of the NS3 sequences of the 18 APPVs from South Korean wild boars were included in Clade I (16 in subclade I.1 and two in subclade I.4), as shown in Figure 2. The nt sequence identity between the South Korean APPVs in subclades I.1 and I.4 was 86.8–88.7%; however, identity at the aa sequence level was 95.7–97.6%.

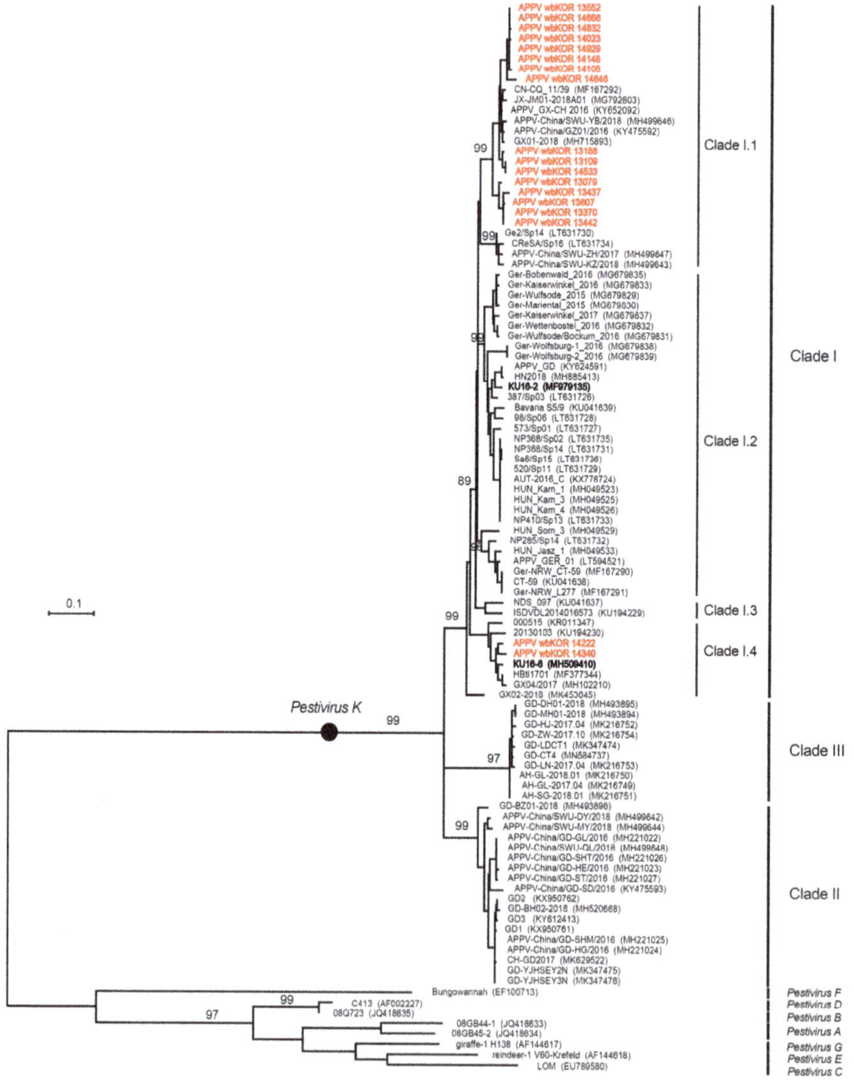

Figure 2. Phylogenetic tree of South Korean wild boar APPVs, based on NS3 sequences. The phylogenetic tree was constructed using the ML method (based on the Tamura–Nei model), with bootstrap analysis

(n = 1000), in MEGA 7.0 software. The 767 nt NS3 sequences of 18 APPVs from South Korean wild boars were compared with 86 reference sequences (including eight outgroup strains: *pestiviruses A–G*) from Asia, North America, and Europe. The Log likelihood (Log L) is −10,191.77, and only bootstrap values ≥70 are indicated on the nodes. South Korean wild boar APPV strains and South Korean domestic pig APPV strains are denoted by red bold and black bold letters. The scale bar indicates the number of nucleotide substitutions per site.

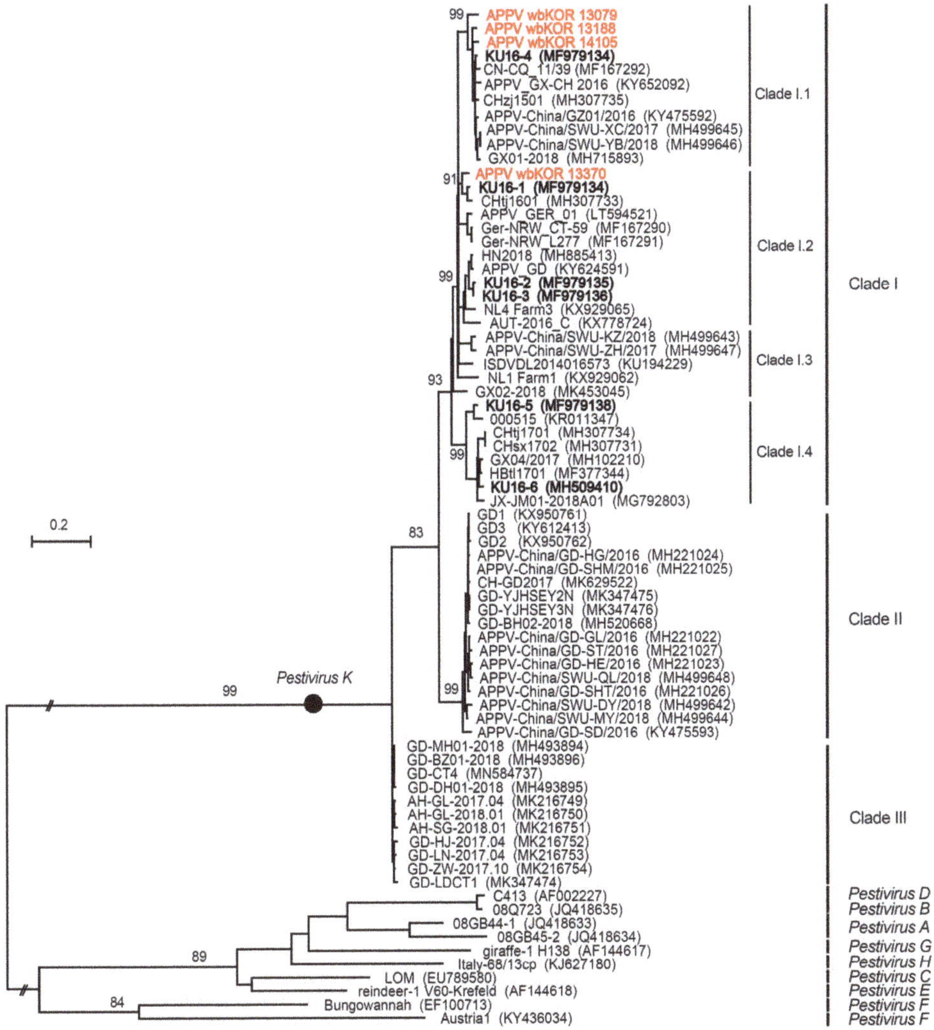

Figure 3. ML tree of South Korean wild boar APPVs, based on E2 sequences. The phylogenetic tree (Log L, −11,438.12) was constructed using the ML method (based on the Tamura–Nei model), with bootstrap analysis (n = 1000). The complete E2 sequences of four APPVs detected in South Korean wild boars were compared with 70 reference sequences (including ten outgroup strains: *pestiviruses A–H*).

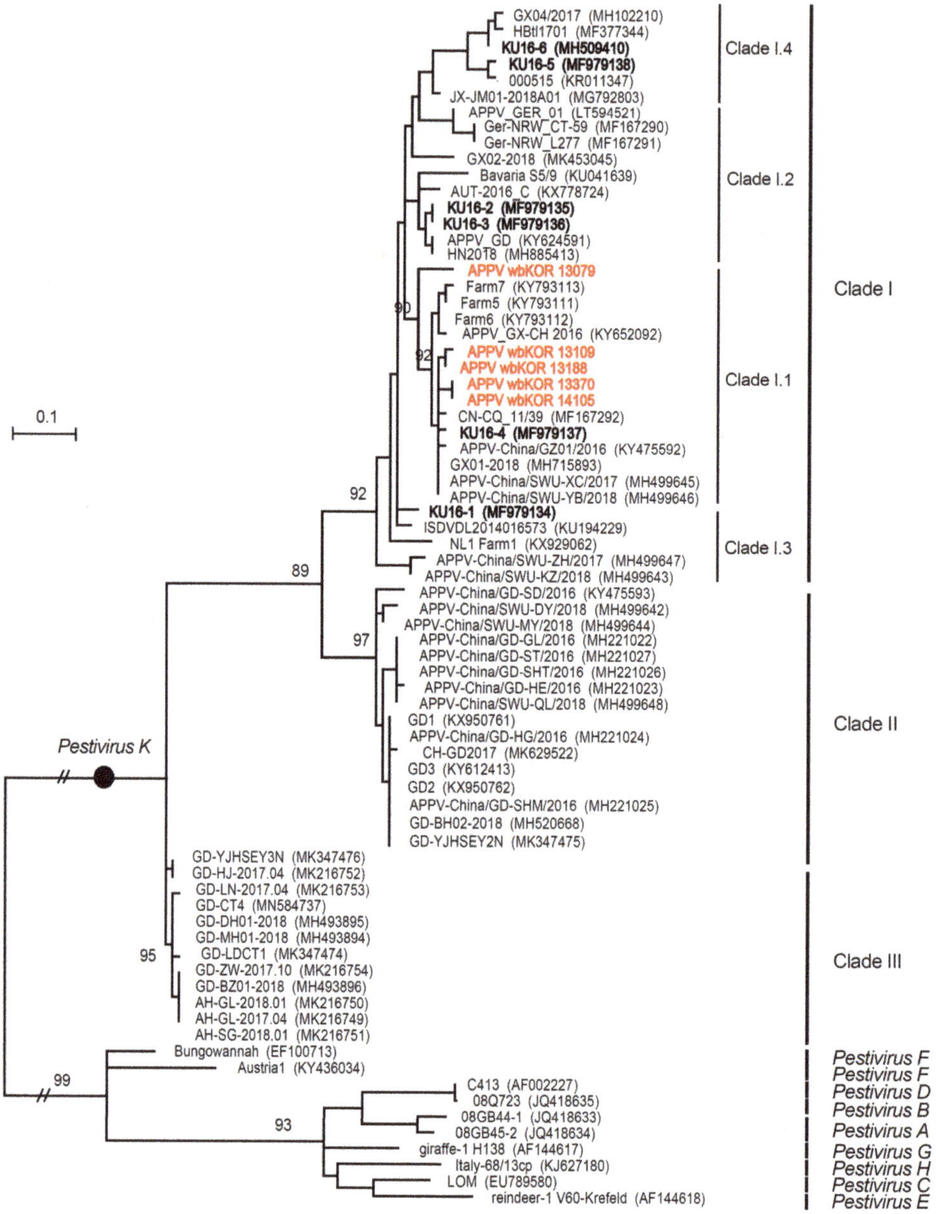

Figure 4. ML tree of South Korean wild boar APPV, based on Npro sequences. The phylogenetic tree (Log L: −5113.82) was constructed using the ML method (based on the Tamura–Nei model), with bootstrap analysis ($n = 1000$). The complete Npro sequences of five APPV detected in South Korean wild boars were compared with 69 reference sequences (including ten outgroup strains: *pestiviruses A–H*).

The 767-nt NS3 sequences of the 16 South Korean wild boar APPVs (Subclade I.1) and 8 APPVs isolated from German wild boars (Subclade I.2), as shown in Figure 2, were 90.9–92.0% identical at the nt level and 97.6–99.2% identical at the aa level. Moreover, NS3 sequences from the two South Korean wild boar APPVs (subclade I.4) and of the eight German wild boar APPVs (subclade I.2) were 88.1–88.9% identical at the nt level and 96.1–97.3% identical at the aa level.

3.3. ML Trees for the E2 Sequences

The E2 sequences (723 nt) from four South Korean wild boar APPVs (wbKOR 13079, -13188, -14105, and -13370) were 91.3–94.3% identical at the nt level and 95.4–97.1% identical at the aa level. ML analysis of the E2 sequences categorized them as Clade I (three in subclade I.1 and one in subclade I.2), as shown in Figure 3. By contrast, six APPVs from South Korean domestic pigs belonged to three subclades (one to subclade I.1, three to subclade I.2, and two to subclade I.4), as shown in Figure 3. The four South Korean wild boar APPVs and six South Korean domestic pig APPVs were 87.4–94.3% identical at the nt level and 94.2–98.8% identical at the aa level. The E2 sequences from the four South Korean wild boar APPVs and the 17 other APPVs in Clade II were 84.4–85.1% identical at the nt level and 94.2–96.3% identical at the aa level, whereas the four South Korean wild boar APPVs and 11 other APPVs in Clade III were 81.5–82.7% identical at the nt level and 90.0–92.5% identical at the aa level.

3.4. ML Trees for the Npro Sequences

The Npro sequences (540 nt) of five APPVs from South Korean wild boars (wbKOR 13079, -13109, -13188, -14105, and -13370) were 92.6–98.7% identical at the nt level and 93.9–99.4% identical at the aa level. ML analysis of the Npro sequences categorized them as Clade I (all five belonged to subclade I.1), as shown in Figure 4. The six APPVs from South Korean domestic pigs were distributed among all subclades (one in subclade I.1, two in subclade I.2, one in subclade I.3, and one in subclade I.4), as shown in Figure 4. The nt sequence identity between the five South Korean wild boar APPVs and the six South Korean domestic pig APPVs was 85.0–95.4%, whereas the aa sequences were 89.4–95.0% identical. The Npro sequences of the five South Korean wild boar APPVs and the 16 APPVs in Clade II were 79.3–80.5% identical at the nt level and 80.6–83.9% identical at the aa level, whereas the sequences from the five South Korean wild boar APPVs and the 12 APPVs in Clade III were 77.0–80.4% identical at the nt level and 80.6–81.1% identical at the aa level.

4. Discussion

Some retrospective studies suggest that APPVs were circulating widely for decades before the recent reports describing their detection [21,22]. An early study identified an APPV strain from samples obtained from piglets with CT; qRT-PCR of samples from a pig herd in the USA revealed an APPV prevalence of 6% [2]. Another study reported APPVs in CT-affected piglets from many countries, which further supports the potential relationship between APPV and CT type A-II [3]. In South Korea, APPV sequences from the domestic pig samples collected in 2016 had been submitted to GenBank (accession numbers: MF979135, MH509410), and the first identification of APPV was reported in suckling piglets with CT in 2017 [9]. Moreover, an APPV detection rate of 2.4–22% was reported in apparently healthy pigs in the USA and Germany [6,12,23]. A recent study of serum samples from apparently healthy pigs revealed that the prevalence of APPV in Europe was 2.3–17.5%, and that in China it was 5–11% [11]. All of these results demonstrate that APPV is highly prevalent in both apparently healthy pigs and CT-affected pigs, suggesting that the virus may have spread worldwide.

Wild boars are susceptible to APPV infection, although the role of this species in the epidemiology of the virus is unknown [13]. A very recent finding is that the prevalence of APPV in serum samples from wild boars in Europe is 0.23–19%, suggesting that wild boars may be a reservoir for APPV [10,12,13]. All wild boars sampled in this study, including APPV-positive wild boars, were considered healthy (based on their behavior) before hunting. The prevalence of APPV among South Korean wild boars is low, as in Spain (0.23%) and Italy (0.69%) [10,13]; however, this is in contrast to the high prevalence detected in northern Germany (19%) [12]. Furthermore, serological investigations in wild boars revealed an antibody detection rate of 52% in northern Germany, where APPV seems to be endemic among wild boars in many areas [12].

Phylogenetic analysis revealed that APPV sequences (complete or partial polyprotein) exhibit high genetic diversity between strains detected in different countries and species (domestic pigs or wild boars), and that they form independent clusters according to geographic location [24].

Recently, phylogenetic analysis in China revealed a high level of genetic variation among APPVs: three Clades (I–III), with four subgroups (1–4) in Clade I [25]. Sequence analysis of the APPV NS3, E2, and Npro genes, followed by the construction of ML trees, classified them into three Clades—APPVs isolated from China were assigned to all Clades, whereas APPVs isolated from Asia, North America, and European countries belonged only to Clade I [25,26]. APPV strains isolated from South Korean wild boars and South Korean domestic pigs belonged to Clade I, without a preference for any particular subclade. This means that APPVs may have high the potential for spreading between wild boars and domestic pigs. In the three ML trees (NS3, E2, and Npro) constructed for this study, Clade I contained four subclades; however, the APPV strains contained within these subclades were not clearly distinguished by the three ML trees. A previous study suggests that recombination events occur between Clades (Clades II and III) or within a Clade (Clade I) [25]. Therefore, further research is needed to determine whether the difference in subclade of South Korean APPV strains in three ML trees are because of the recombination events within Clade I.

5. Conclusions

Here, we present the first report of APPV detected in South Korean wild boars. We found that the overall prevalence of APPV in South Korean wild boars was low (0.78%). South Korean wild boars harbor genetically diverse APPV strains belonging to Clade I. Wild boars may be an important virus reservoir for APPV. More epidemiological information will help to establish effective control measures and to eradicate the virus from affected pig herds in the future.

Author Contributions: Conceptualization, S.C. and B.-K.P.; methodology, G.-N.P. and R.M.C.; writing—review and editing, B.-H.H. and D.-J.A. All authors have read and agreed to the published version of the manuscript.

Funding: This project was supported by grants (Project Code No. B-1543083-2019-21-02) from the Animal and Plant Quarantine Agency, Republic of Korea.

Conflicts of Interest: The authors declare no conflicts of interest.

References

1. Smith, D.B.; Meyers, G.; Bukj, J.; Gould, E.A.; Monath, T.; Scott Muerhoff, A.; Pletnev, A.; Rico-Hesse, R.; Stapleton, J.T.; Simmonds, P.; et al. Proposed revision to the taxonomy of the genus Pestivirus, family Flaviviridae. *J. Gen. Virol.* **2017**, *98*, 2016–2112. [CrossRef]
2. Hause, B.; Collin, E.A.; Peddireddi, L.; Yuan, F.; Chen, Z.; Hesse, R.A.; Anderson, G. Discovery of a novel putative atypical porcine pestivirus in pigs in the United States. *J. Gen. Virol.* **2015**, *96*, 2994–2998. [CrossRef] [PubMed]
3. Yuan, J.; Han, A.; Li, J.; Huang, Y.; Yang, J.; Ding, H.; Zhang, J.; Zhu, M.; Zhang, Y.; Liao, J.; et al. Atypical porcine pestivirus as a novel type of pestivirus in pigs in China. *Front. Microbiol.* **2017**, *8*, 862. [CrossRef] [PubMed]
4. Beer, M.; Wernlike, K.; Drager, D.; Hoper, D.; Pohlmann, A.; Bergermann, C.; Hoffmann, B. High prevalence of highly variable atypical porcine pestiviruses found in Germany. *Transboud. Emerg. Dis.* **2016**, *64*, e22–e26. [CrossRef]
5. De Groof, A.; Deijs, M.; Guelen, K.; van Grinsven, L.; van Os-Galdos, L.; Vogels, W.; van der Hoek, L. Atypical porcine pestivirus: A possible cause of congenital tremor type A-II in newborn piglets. *Viruses* **2016**, *8*, 271. [CrossRef]

6. Postel, A.; Hansmann, F.; Baechlein, C.; Fischer, N.; Alawi, M.; Grundhoff, A.; Becher, P. Presence of atypical porcine pestivirus (APPV) genome in newborn piglets correlates with congenital tremor. *Sci. Rep.* **2016**, *6*, 27735. [CrossRef] [PubMed]
7. Schwarz, L.; Riedel, C.; Hogler, S.; Sinn, L.J.; Voglmayr, T.; Wochtl, B.; Lamp, B. Congenital infection with atypical porcine pestivirus (APPV) is associated with disease and viral persistence. *Vet. Res.* **2017**, *48*, 1–14. [CrossRef]
8. Gatto, I.R.H.; Harmon, K.; Bradner, L.; Silva, P.; Kinhares, D.C.L.; Arruda, P.H.; de Oliverira, L.G.; Arruda, B.L. Detection of atypical porcine pestivirus in Brazil in the central nervous system of suckling piglets with congenital tremor. *Transboud Emerg. Dis.* **2018**, *65*, 375–380. [CrossRef]
9. Kim, S.; Jeong, C.; Yoon, S.; Lee, K.; Yang, M.; Kim, B.; Lee, S.Y.; Kang, S.J.; Kim, W.I. Detection of atypical porcine pestivirus (APPV) from a case of congenital tremor in Korea. *Korean. J. Vet. Sev.* **2017**, *40*, 209–213.
10. Enrica, S.; Saloni, C.; Lelli, D.; Barbieri, I.; Moreno, A.; Loris, G.; Lavazza, A. Molecular Survey and Phylogenetic Analysis of Atypical Porcine Pestivirus (APPV) Identification in Swine and Wild Boar from Northern Italy. *Viruses.* **2019**, *11*, 1142.
11. Postel, A.; Meyer, D.; Cagatay, G.N.; Feliziani, F.; De Mia, G.M.; Fischer, N.; Grundho_, A.; Mili´cevi´c, V.; Deng, M.C.; Chang, C.Y.; et al. High Abundance and Genetic Variability of Atypical Porcine Pestivirus in Pigs from Europe and Asia. *Emerg. Infect. Dis.* **2017**, *23*, 2104–2107. [CrossRef] [PubMed]
12. Cagatay, G.N.; Antos, A.; Meyer, D.; Maistrelli, C.; Keuling, O.; Becher, P.; Postel, A. Frequent infection of wild boar with atypical porcine pestivirus (APPV). *Transboud Emerg. Dis.* **2018**, *65*, 1087–1093. [CrossRef] [PubMed]
13. Colom-Cadena, A.; Ganges, L.A.; Munoz-Gonsalez, S.; Castillo-Contreras, R.; Alejandro Bochorquez, J.; Rosell, R.; Seglaes, J.; Marco, I.; Cabezon, O. Atypical porcine pestivirus in wild boar (Sus Scrofa), Spain. *Vet. Rec.* **2018**, *183*, 569. [CrossRef] [PubMed]
14. Weber, M.N.; Streck, A.F.; Silveira, S.; Mosena, A.C.S.; da Silva, M.S.; Canal, C.W. Homomolgours recombination in pestiviruses: Identification of three putative novel events between different subtypes/genogroups. *Infect. Genet. Evol.* **2015**, *30*, 2109–2224. [CrossRef] [PubMed]
15. Greiser-Wilke, I.; Depner, K.; Fritzemeier, J.; Haas, L.; Moenning, V. Application of a computer program for genetic typing of classical swine fever virus isolates from Germany. *J. Virol. Methods.* **1998**, *75*, 141–150. [CrossRef]
16. Zhang, H.; Wen, W.; Hao, G.; Hu, Y.; Chen, H.; Qian, P.; Li, X. Phylogenetic and genomic characterization of a novel atypical porcine pestivirus in China. *Transboud Emerg. Dis.* **2018**, *65*, e202–e204. [CrossRef] [PubMed]
17. Zhou, K.; Yue, H.; Tang, C.; Ruan, W.; Zhou, Q.; Zhang, B. Prevalence and genome characteristics of atypical porcine pestivirus in Southwest China. *J. Gen. Virol.* **2019**, *100*, 84–88. [CrossRef] [PubMed]
18. Choe, S.; Kim, J.H.; Kim, K.S.; Song, S.; Kang, W.C.; Kim, H.J.; Park, G.N.; Cha, R.M.; Cho, I.S.; Hyun, B.H.; et al. Impact of a live attenuated classical swine fever virus introduced onto Jeju Island, a CSF-free area. *Pathogens.* **2019**, *8*, 251. [CrossRef]
19. Thompson, J.D.; Gibson, T.J.; Plewniak, F.; Jeanmougin, F.; Higgins, D.G. The CLUSTAL_X windows interface: Flexible strategies for multiple sequence alignment aided by quality analysis tools. *Nucleic Acids Res.* **1997**, *25*, 4876–4882. [CrossRef]
20. Kumar, S.; Stecher, G.; Tamura, K. MEGA7: Molecular Evolutionary Genetics Analysis Version 7.0 for Bigger Datasets. *Mol. Biol. Evol.* **2016**, *33*, 1870–1874. [CrossRef]
21. Muñoz-González, S.; Canturri, A.; Pérez-Simó, M.; Bohórquez, J.A.; Rosell, R.; Cabezón, O.; Segalés, J.; Domingo, M.; Ganges, L. First report of the novel atypical porcine pestivirus in Spain and a retrospective study. *Transbound Emerg. Dis.* **2017**, *64*, 1645–1649. [CrossRef] [PubMed]
22. Kaufmann, C.; Stalder, H.; Sidler, X.; Renzullo, S.; Gurtner, C.; Grahofer, A.; Schweizer, M. Long-Term Circulation of Atypical Porcine Pestivirus (APPV) within Switzerland. *Viruses.* **2019**, *11*, 653. [CrossRef] [PubMed]
23. Saitou, N.; Nei, M. The neighbor-joining method: A new method for reconstructing phylogenetic trees. *Mol. Biol. Evol.* **1987**, *4*, 406–426. [PubMed]
24. Gatto, I.; Sonalio, K.; Oliveira, L. Atypical porcine pestivirus (APPV) as a New Species of Pestivirus in Pig Production. *Front. Vet. Sci.* **2019**, *6*, 35. [CrossRef] [PubMed]

25. Guo, Z.; Wang, L.; Qaio, S.; Deng, R.; Zhang, G. Genetic characterization and recombination analysis of atypical porcine pestivirus. *Infect. Genet. Evol.* **2020**, *81*, 104259. [CrossRef] [PubMed]
26. Yan, X.L.; Li, Y.Y.; He, L.L.; Wu, J.L.; Tang, X.Y.; Chen, G.H.; Mai, K.J.; Wu, R.T.; Li, Q.N.; Chen, Y.H.; et al. 12 novel atypical porcine pestivirus genomes from neonatal piglets with congenital tremors: A newly emerging branch and high prevalence in China. *Virology.* **2019**, *533*, 50–58. [CrossRef]

viruses

MDPI

Article

Atypical Porcine Pestivirus Circulation and Molecular Evolution within an Affected Swine Herd

Alba Folgueiras-González [1,2], Robin van den Braak [1], Bartjan Simmelink [1], Martin Deijs [2], Lia van der Hoek [2] and Ad de Groof [1,*]

1 Department Discovery & Technology, MSD Animal Health, Wim de Körverstraat 35, P.O. Box 31, 5830AA Boxmeer, The Netherlands; alba.folgueiras.gonzalez@merck.com (A.F.-G.); robin.braak.van.den@merck.com (R.v.d.B.); bartjan.simmelink@merck.com (B.S.)

2 Laboratory of Experimental Virology, Department of Medical Microbiology, Amsterdam UMC Location AMC, University of Amsterdam, Meibergdreef 9, 1105AZ Amsterdam, The Netherlands; m.deijs@amsterdamumc.nl (M.D.); c.m.vanderhoek@amsterdamumc.nl (L.v.d.H.)

* Correspondence: ad.de.groof@merck.com

Received: 27 May 2020; Accepted: 23 September 2020; Published: 25 September 2020

Abstract: Atypical porcine pestivirus (APPV) is a single-stranded RNA virus from the family Flaviviridae, which is linked to congenital tremor (CT) type A-II in newborn piglets. Here, we retrospectively investigated the molecular evolution of APPV on an affected herd between 2013 and 2019. Monitoring was done at regular intervals, and the same genotype of APPV was found during the entire study period, suggesting no introductions from outside the farm. The nucleotide substitutions over time did not show substantial amino acid variation in the structural glycoproteins. Furthermore, the evolution of the virus showed mainly purifying selection, and no positive selection. The limited pressure on the virus to change at immune-dominant regions suggested that the immune pressure at the farm might be low. In conclusion, farms can have circulation of APPV for years, and massive testing and removal of infected animals are not sufficient to clear the virus from affected farms.

Keywords: pestivirus; atypical porcine pestivirus (APPV); viral persistence; congenital tremor; swine; asymptomatic; genomic sequence; phylogenetic analysis; purifying selection

1. Introduction

The genus Pestivirus, belonging to the Flaviviridae family, includes single-stranded, positive-sense RNA viruses of veterinary importance, causing economically relevant diseases in livestock animals, but also affecting wildlife species [1]. While for decades the International Committee on Taxonomy of Viruses (ICTV) only recognized the four so-called classical pestiviruses, namely bovine viral diarrhea virus types 1 and 2 (BVDV-1 and BVDV-2), classical swine fever virus (CSFV) and border disease virus (BDV), the discovery of seven novel atypical pestiviruses in the last few years led to the proposal of a new species-independent classification from Pestivirus A to Pestivirus K. Thus, atypical porcine pestivirus (APPV) is now classified as Pestivirus K [2]. Moreover, new pestivirus genomes in bats [3] and piglets (LINDA virus [4]) have been discovered lately, but not yet classified.

Congenital tremor (CT) in newborn piglets, also known as myoclonia congenita, "shaker pigs" or "dancing pigs", was first reported in 1922 [5]. Its characteristic clinical signs involve tremors of the head and limbs which worsen on stress situations, but are almost gone during sleep. Even though most piglets are clinically healthy at weaning time, the earlier shivering hinders their ability to feed normally from their mother, thus increasing the risk for inadequate colostrum intake, growth retardation or even death by starvation [6]. CT is divided based on its pathology into type A, which displays morphological lesions on the central nervous system, and type B, which does not show any visible morphological lesion [7].

Regarding the distinct causes of disease, type A was further differentiated into five subtypes. Type A-I is associated with one of the classical pestiviruses, CSFV, and characterized by cerebellar hypoplasia in affected pigs [8]. Types A-III and A-IV are both related to different genetic defects only present in the Landrace breed and Saddleback breed, respectively [9,10]. Type-V, also defined by cerebellar hypoplasia, is caused by food poisoning by trichlorfon in pregnant sows [11]. Meanwhile, during the decades, the causing agent of CT type A-II remained unknown, although an infectious agent was suspected [12].

APPV was first characterized in 2015 by Hause et al. [13] through metagenomic sequencing performed on a sample positive for porcine reproductive and respiratory syndrome virus (PRRSV). After its discovery, two independent studies proved APPV's link to congenital tremor in newborn piglets via the inoculation of pregnant sows with infectious animal material [14,15]. However, Koch's postulates have not been fulfilled yet due to the difficulties in establishing an appropriate substrate for in vitro culture and retrieval of pure virus stocks [16]. In 2017, Lamb et al. [4] reported a novel pestivirus, named LINDA virus (lateral-shaking inducing neurodegenerative agent), in an Austrian cohort of CT piglets. Although the clinical manifestations and pathology were similar to the ones of APPV, further phylogenetic analysis showed its proximity to Bungowannah virus, a pestivirus never detected outside Australia.

A research study by Postel et al. [17] analyzed 1460 samples of apparently healthy pigs from different countries. They showed that the prevalence of APPV genotypes was around 10% and up to 60% of the tested animals were seropositive. Kaufmann et al. [18] showed that APPV has been circulating in Switzerland at least since 1986, which was the earliest detection point worldwide. To date, APPV has been reported on four continents and fourteen different countries [19]. Mortality is high in CT-affected litters, from around 10% to as high as 30% in some affected farms [14–16].

APPV is a highly variable enveloped virus which contains a linear, non-segmented, positive-sense RNA strand of around 11–12 kb. A single open-reading frame (ORF) encodes a single polyprotein of 3635 amino acids on average, which is assumed to be cleaved co- and post-translationally into four structural (the core protein C and the envelope glycoproteins Erns, E1 and E2) and eight non-structural proteins (Npro, p7, NS2, NS3, NS4A, NS4B, NS5A and NS5B) [13]. Several novel APPV sequences have been described since the first report in 2015, further dividing the phylogenetic tree and proposing distinct divisions into three, four or five different genotypes and into numerous particular subclades [20]. To date, 58 unique full-genome sequences are available in NCBI GenBank (access date: 1 April 2020). The APPV full-genome sequences have 80–99% identity, showing high variability even within the same country, with difficulty to infer the origin, dissemination and common ancestor of several strains of the virus [19,21].

High variability in RNA viruses, caused by the lack of proofreading repair mechanisms in RNA polymerases, provides them with a selective advantage, promoting rapid adaptation to novel hosts and environments [22]. On the other hand, excessive error rates can also lead to lethal mutagenesis and threaten the viability of the viral populations [23]. Therefore, hypervariable regions that are thought to contribute to immune escape and adaptation, as well as hyperconserved elements that are fundamental for virus replication and maintenance in the population are main research topics for molecular evolution studies [24].

In the current study, we retrospectively investigated the molecular evolution and functional implication of nucleotide and amino acid substitutions in the APPV genome in a farm in the North Brabant province of The Netherlands over a period of seven years. Sampling was done consistently over time in the controlled environment of a closed-herd farm, where the full farming process is carried out within the farm, and the breeding stock is replaced by gilts from the farm itself. Therefore, by eliminating the input of animals from outside, the risk of novel viral introductions is minimized. In 2013, a first CT outbreak with positive cases of APPV was reported in the farm. Two years later, in 2015, a second APPV outbreak occurred in the same farm with clinically affected, trembling piglets that tested positive. Since then, serum samples were taken from CT piglets at the time of the outbreaks, as well as from apparently healthy pigs as part of regular farm monitoring for APPV and other porcine viruses. Pigs that tested positive were removed from the breeding population to stop the spread of the virus via persistently infected animals. However, several re-emerging peaks of viremic animals were been detected in the farm during the monitoring period, suggesting the continued circulation of APPV in clinically healthy animals.

To assess the possible biological significance of the selective pressures that shape the evolution of APPV within the farm, we looked for evidence of purifying and diversifying selection in the APPV genome. Although viral surface glycoproteins like E2 are well-known targets of positive selection, and variations in these regions can help in immune escape, we also took a full-genome study approach to get a more general, helicopter view and investigate other regions that might have been underestimated until now. The correlation of clinical CT outbreaks with fixation events of non-synonymous, but also synonymous, substitutions can reveal the link to pathogenicity and infectivity of the virus. Here, we reported, to our knowledge, the first longitudinal study on the evolution of the full-length APPV genome within a controlled farm environment. We compared molecular evolution and variability in the genome sequences found on the farm with full-genome sequences from different locations around the world.

2. Materials and Methods

2.1. Sample Collection

In 2013, the farm under study increased the population of breeding sows from 280 to 570 sows and relocated to new facilities. Piglets stayed in the farrowing unit after weaning (i.e., litters were not mixed). Piglets were vaccinated with a porcine reproductive and respiratory syndrome virus (PRRSV) live vaccine at 2 weeks of age, and with a combined porcine circovirus type 2 (PCV2) and *Mycoplasma hyopneumoniae* (*M. hyo*) vaccine at 3 weeks of age. PRRSV was regularly diagnosed in the farm and therefore a vaccination strategy in sows and piglets was set up. There were no other pathogens diagnosed in regular screenings.

Serum and fecal samples were obtained from gilts and congenitally trembling piglets during the CT outbreaks in 2013 and 2015. Between January 2016 and April 2016, serum was collected from all gilts selected for breeding (196 animals) at the age of first insemination for APPV monitoring and removal of positive animals.

To monitor the presence of APPV in serum over time, five litters with CT born from gilts during the 2015/2016 outbreak were repeatedly sampled until the age of 18 weeks. Moreover, 6 gilts (4 APPV-positive at 5 and 18 weeks of age, and 2 negative) and 2 boars (1 APPV-positive at 5 and 18 weeks of age, and 1 negative), were followed for APPV presence in serum and fecal shed until the age of 10 months. Two APPV-positive gilts were co-housed with one negative gilt, and the boars were housed next to each other with direct contact.

From April 2016 to 2019, clinically healthy 10-week-old gilts in the sow breeding line maintained on the farm were regularly monitored for the presence of APPV in serum. Samples from animals with congenital tremor symptoms were also taken for analysis. In total, 1498 samples were taken during this period.

Blood was collected using the Vacuette 5/8 mL Sep Clot Activator (Greiner-Bio One, Kremsmünster, Austria) and serum was obtained by centrifugation for 10 min at 3200× g at 4 °C. Fecal samples were collected using Sigma Virocult swabs and vials (MWE, Corsham, UK), vortexed, transferred to 1.5 mL Eppendorf tubes and centrifuged for 10 min at 10,000× g at 4 °C.

RNA was extracted from 200 µL samples by the automated MagNA Pure 96 system (Roche Applied Science, Manheim, Germany) using the protocol 'Viral NA plasma external lysis SV3.1'.

2.2. *Quantitative Reverse Transcription*

A universal, quantitative, reverse transcription PCR (qRT-PCR) was used to quantitatively detect APPV in samples using primers in the 5′-untranslated region (UTR) of the genome (APPV-PAN2-F3-B: CGYGCCCAAAGAGAAATCGG and APPV-PAN2-R3-B: CCGGCACTCTATCAAGCAGT) [14]. One-step qPCR reactions were performed in a final volume of 50 μL containing 1 μL SuperScript III RT/Platinum Taq Mix (Thermo Fisher Scientific, Waltham, MA, USA), 25 μL 2× SYBR Green Reaction Mix (Thermo Fisher Scientific, Waltham, MA, USA), 17 μL water, 1 μL forward primer (10 μM), 1 μL reverse primer (10 μM) and 5 μL of the RNA isolate. Thermocycling was performed in the CFX96 Touch real-time PCR detection system (Bio-Rad Laboratories, Hercules, CA, USA) starting with an RT reaction for 3 min at 55 °C, a pre-denaturation step for 5 min at 95 °C and 40 cycles of 15 s at 95 °C (denaturation) and 30 s at 60 °C (annealing and elongation). The specificity of SYBR Green qPCR was validated by melting curve analysis between 65 °C and 95 °C with an increasing gradient of 0.5 °C per 5 s. Results were analyzed with the CFX Manager software (Bio-Rad Laboratories, Hercules, CA, USA).

To obtain a standard curve for the quantitative analysis, a recombinant bacterial plasmid containing the 5′-UTR PCR target was made (GenScript, Piscataway, NJ, USA). The copy number of the recombinant plasmid was calculated, and eight dilution series (10^8–10^1 copies/μL) were included in duplicate in RT-qPCR to calculate the number of virus copies per μL.

2.3. *Viral Genome Amplification and Sanger Sequencing*

A starting sequence of 1073 bp obtained from a previous study using Illumina Sequencing (Amsterdam UMC, Amsterdam, The Netherlands) was used as reference [14]. This short read was mapped to the 58 full-genome sequences available in NCBI GenBank (access date: 1 April 2020). The three most similar full-genome sequences (GenBank accession numbers: KY624591.1, MH885413.1 and KX778724.1) were aligned by CLUSTALW using Geneious Prime v2019.0.4 (http://www.geneious.com) (Geneious, Auckland, New Zealand), and the consensus sequence was extracted for primer design.

The cDNA from the extracted RNA of the clinical samples (Section 2.1) was synthesized using the QuantiTect reverse transcription kit (Qiagen, Hilden, Germany) following the manufacturer's manual. The entire viral genome from the 2013 sample was amplified using an initial series of twelve overlapping PCRs. Gaps and faulty PCRs were solved following a genome walking strategy (amplification and sequencing primers available in Supplementary Table S1). PCR reactions were performed in a final volume of 25 μL containing 12.5 μL 2× Phusion High-Fidelity PCR Master Mix with HF Buffer (New England BioLabs, Ipswich, MA, USA), 0.75 μL DMSO 100%, 6.75 μL water, 1.25 μL FW 10 μM primer, 1.25 μL REV 10 μM primer and 2.5 μL of template cDNA. Thermocycling was performed in the CFX96 Touch real-time PCR detection system (Bio-Rad Laboratories, Hercules, CA, USA) using an initial denaturation step at 98 °C for 5 min, 40 cycles of 30 s at 98 °C (denaturation), 30 s at the annealing temperature optimal for the primer set and 1 min and 30 s at 72 °C (elongation), followed by 7 min at 72 °C. The PCR products were stored at 4 °C until further analysis. Five microliters of the PCR products were analyzed by agarose gel electrophoresis (1.4% *w*/*v* agarose) to check the fragment size and the specificity of the amplification. PCR products were purified using the QiaQuick PCR purification kit (Qiagen, Hilden, Germany) following the manufacturer's manual.

Sequencing PCR reactions were performed using the BigDye Terminator v3.1 cycle sequencing kit (Applied Biosystems, Carlsbad, CA, USA) in a final volume of 20 μL containing 4 μL BigDye Terminator ready reaction mix, 3 μL 2× Phusion High-Fidelity PCR Master Mix with HF Buffer (New England BioLabs, Ipswich, MA, USA), 0.5 μL DMSO 100%, 2.5 μL WFI, 2.5 μL FW or REV 10 μM primer and 7.5 μL of the purified PCR product. Thermocycling was performed in 30 cycles of 10 s at 95 °C (denaturation), 10 s at the annealing temperature optimal for the primer set and 2 min at 60 °C (elongation). The PCR products were stored at 4 °C for further analysis. The cycle sequencing products were purified using the DyeEx 2.0 Spin kit (Qiagen, Hilden, Germany) following the manufacturer's manual. Capillary electrophoresis was performed using the 3500 Genetic Analyzer (Applied Biosystems,

Carlsbad, CA, USA) and data were analyzed using the Sequencher 5.4.6 (Gene Codes, Ann Arbor, MI, USA) and Geneious Prime v2019.0.4 (Biomatters Ltd., Auckland, New Zealand) software.

Based on the obtained sequence, Sanger sequencing was performed following the same methodology on the other five samples (years 2015, 2016, 2017, 2018 and 2019) by GenScript (Piscataway, NJ, USA).

2.4. Determination of E2 Sequences

The amplification of E2-coding sequences was performed using a two-step RT-PCR protocol as described in Section 2.3. The amplification and sequencing PCR reactions of 857 bp and 953 bp fragments were performed using two primer pairs flanking the E2-coding region (E2-F1: 5′-TGGTGCCTATTGTTGTCAGG-3′, E2-R1: 5′-AGTTCTTCCTTGACGGCTAG-3′, E2-F2: 5′-GCCCTGGTGAACATAGTCAC-3′ and E2-R2: 5′-TCCTTGACGGCTAGCATTATG-3′). Capillary electrophoresis was performed using the 3500 Genetic Analyzer (Applied Biosystems, Carlsbad, CA, USA) and trimmed sequences (702 bp) were analyzed using Geneious Prime v2019.0.4 (Biomatters Ltd., Auckland, New Zealand).

2.5. Submission of Sequences

The sequences of the APPV genomes were deposited in GenBank under the accession numbers MT512531–MT512537. The partial E2-coding sequences were deposited in GenBank under the accession numbers MW011356–MW011406.

2.6. Nucleotide and Amino Acid Analysis of Variants

The obtained partial E2-coding sequences were aligned using the CLUSTALW translation alignment in Geneious Prime v2019.0.4 (Biomatters Ltd., Auckland, New Zealand) with BLOSUM cost matrix, a gap open cost of 5 and a gap extended cost of 4.

The fifty-eight full-genome APPV sequences from NCBI GenBank database (access date: 1 April 2020) were aligned along with the sequences obtained in the current study [25]. Multiple sequence alignment was performed using the CLUSTALW translation alignment in Geneious Prime v2019.0.4 (Biomatters Ltd., Auckland, New Zealand) with BLOSUM cost matrix, a gap open cost of 10 and a gap extended cost of 6.66. For the farm dataset, we called single nucleotide substitutions on the aligned genomes using Geneious Prime v2019.0.4 and differentiated them into synonymous and non-synonymous. Ambiguous nucleotide and amino acid calls, which are considered by the software as variants, were treated as missing data.

The prediction of *O*-glycosylation and *N*-glycosylation motifs in the Erns, E1 and E2 glycoproteins were performed with DictyOGlyc1.1, NetOGlyc4.0 and NetNGlyc1.0 prediction algorithms via the webserver from the Technical University of Denmark Department of Bio and Health Informatics (DTU Bioinformatics, http://www.cbs.dtu.dk/services/) [26,27].

2.7. Phylogenetic Analysis and Estimation of Evolutionary Rates within the Farm

The Recombination Detection Program version 4 (RDP4) was used to screen for recombination events on the multiple sequence alignment of the six APPV genomes found in the farm, using RDP, GENECONV, BootScan, Maxchi, Chimaera, Siscan and 3Seq methods. Recombination was considered when p-value was <0.0001 and the recombinant score was >0.6 [28].

The full-genome APPV sequences available in NCBI and the six APPV in-farm sequences obtained in this study were aligned using the CLUSTALW translation alignment in Geneious Prime v2019.0.4 (Biomatters Ltd., Auckland, New Zealand) with BLOSUM cost matrix, a gap open cost of 10 and a gap extended cost of 6.66. The phylogenetic tree was created with the MEGA X software using the neighbor-joining method with the Kimura two-parameter substitution model [29,30]. Complete deletion was done in case of gaps or missing data. The analysis was performed for 500 bootstrap replicates.

An estimation of the evolutionary rates was calculated for the obtained E2-coding sequences, as well as for the six full-genome sequences using the TempEst v1.5.3 software [31].

2.8. Selection Pressure Analysis

Selection pressure analyses were done for the multiple sequence alignment of the six farm sequences as well as for the multiple sequence alignment including the NCBI full-genome APPV sequences. Selection pressure analyses were performed on the full ORF, using mixed-effects model of evolution (MEME) and fixed-effects likelihood (FEL). All these algorithms were implemented on the HyPhy (Hypothesis Testing using Phylogenies) open-source software package and can be accessed through the Datamonkey webserver (https://www.datamonkey.org/) [32].

MEME uses a mixed-effect maximum likelihood approach. The algorithm estimates a synonymous α parameter (d_S) and a two-category mixture of non-synonymous (d_N) β parameters for each site. MEME infers two ω (d_N/d_S) classes and uses a likelihood ratio test to compare between the models and check for episodic diversifying selection. A significance threshold of $p < 0.1$ was used on the analysis [33].

The fixed-effects likelihood (FEL) algorithm was used in order to detect negatively selected sites on the alignment. FEL uses a maximum likelihood approach to infer non-synonymous (d_N) and synonymous (d_S) substitution rates, also assuming a constant selection pressure for each site on the alignment. All branches were tested for selection. A model with synonymous rate variation, where the d_S parameter in the codon model is allowed to vary across sites, was used for the analysis. A significance threshold of $p < 0.1$ was used on the analysis [24].

3. Results

3.1. 2013–2016: Congenital Tremor Outbreaks, Follow Up Studies and Eradication Strategy Design

In 2013, the farm increased the internal population of breeding sows, aimed to breed gilts for the replacement of production sows (sows breeding line) from 280 to 570 sows and, at the same time, relocated to new facilities with state-of-the-art climate control and housing conditions. Piglets stayed in the farrowing unit after weaning, thus litters were not mixed. In the same year, a first large-scale CT outbreak was reported on the farm and a second large-scale outbreak occurred at the end of 2015/early 2016. Both outbreaks were related with positive cases of APPV and various other clinical observations (e.g., reduced vitality and mortality of the piglets, and increased return to estrus percentage). In the time between 2013 and 2015, no major abnormalities were observed, although incidentally few piglets with tremors were seen, but with no impact on production.

During the 2015–2016 outbreak, CT prevalence varied from 5% in litters from fourth parity sows, up to 55% in litters from first and second parity sows. In the latter case, the mortality in the farrowing unit ranged from 25% in litters from second parity sows to 69% in the ones from first parity sows. The related effects on the production data are shown in Supplementary Table S2.

After the second large-scale outbreak, it was decided to monitor all gilts selected for breeding for the presence of APPV, both purebred line gilts and production gilts. Serum of those animals was analyzed at the age of first insemination with the aim to remove positive gilts from the breeding population. A universal, quantitative, reverse transcription PCR (qRT-PCR) was used to quantitatively detect APPV on pig serum samples. This strategy was applied between January and April 2016, during which time 196 gilts were tested and 15% of them tested positive for APPV (Supplementary Table S3).

During parallel monitoring of pigs born with CT and positive for APPV in serum, we observed that a significant percentage of pigs turned PCR-negative for APPV in serum around the age of 18 weeks. The piglets born with CT were considered as persistent carriers of the virus and, at the time of weaning, the virus was still present in the piglets that still showed recognizable, but less severe, tremors. A follow-up of these piglets showed that the percentage of APPV-positive PCR scores in serum was reduced to 45% at 18 weeks of age, with viral copy numbers in serum also being greatly reduced (Supplementary Table S4).

To gain further insight into the dynamics of APPV viremia, 6 gilts (4 APPV-positive at 5 and 18 weeks of age, and 2 negative), and 2 boars (1 APPV-positive at 5 and 18 weeks of age, and 1 negative), were followed for APPV presence in serum until the age of 10 months. Two APPV-positive gilts were co-housed with one negative gilt, in separate cages, and the boars were housed next to each other with direct contact. No APPV was detected in any of the 8 animals at the age of 24 weeks and at any time point thereafter, with the exception of 1 boar testing PCR-positive at the age of 32 weeks (Supplementary Table S5). Fecal shed from the same animals was also monitored until the age of 10 months via qPCR. Results showed, in the age range between 24 and 44 weeks, intermittent shedding in the feces of the persistently infected (PI) gilts and temporary presence in the feces of horizontally infected gilts (Supplementary Table S6).

3.2. 2016–2019: Re-Emergence of Congenital Tremor and Monthly Quantitative Detection of APPV

Since April 2016, when regular sampling from clinically healthy 10-week-old gilts in the sow breeding line maintained on the farm started, serum samples were received monthly for APPV monitoring purposes. Until December 2019, a total of 1505 serum samples from 10–16-week-old pigs were analyzed.

Figure 1 shows the percentage of APPV-positive animals among the tested set for each month since the screening started in April 2016 until the last analyzed data from December 2019. In this period, two peaks of CT symptoms in newborn piglets were detected in the farm in April 2016 and May 2017 (blue columns in Figure 1). In order to diminish horizontal transmission of the virus in the sow breeding population, infected APPV pigs detected during the monitoring were removed from the population as these were likely persistently infected.

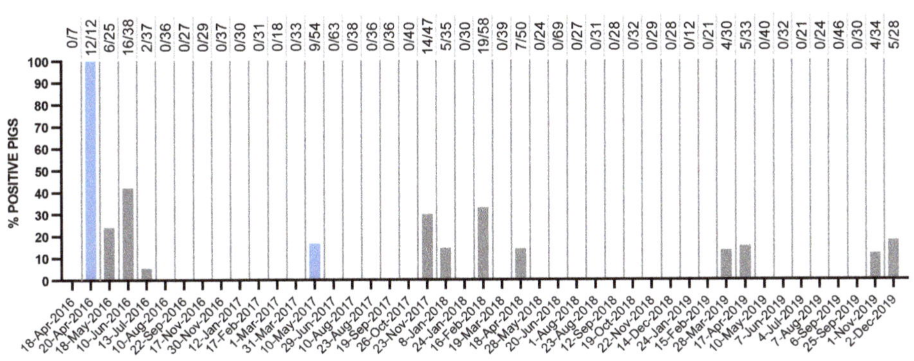

Figure 1. Percentage of atypical porcine pestivirus (APPV)-positive pigs detected during monthly screening in the farm by RT-qPCR analysis. The x-axis shows the sampling date from April 2016 to December 2019. The percentage of positive samples is shown in the y-axis. Blue-colored columns show that positive PCR results were accompanied by clinical CT symptoms on those specific months—i.e., April 2016 and May 2017. The number of PCR-positive animals in relation to the total number tested each specific month is shown above the columns.

3.3. Characterization of APPV Sequences

Forty-eight E2-coding sequences were determined from the APPV-positive samples collected during monthly monitoring in the farm at different time points between 2016 and 2019, along with three sequences obtained previously in 2013 and 2015 during the large-scale CT outbreaks (Supplementary Table S7). The E2-coding sequences showed a maximum pairwise genetic distance of only 0.71%. The low genetic diversity observed within the farm supported the hypothesis that only one viral strain was circulating in the herd with no viral introductions from the outside.

The amino acid sequences were also compared according to the collection date (Supplementary Figure S1). Samples obtained from different animals at the same time point showed a completely identical genome, with the exception of the highly mutated codon D752, varying to either Gly, Ser or Asn residues (Section 3.5). Besides, a more thorough approach was taken for the samples obtained from affected animals at the time of the CT outbreaks, with 8 out of 12 sequences obtained in April 2016 and 8 out of 9 in May 2017. No characteristic differences were seen at those time points that could lead to an explanation of the distinct symptoms in relation to the nucleotide or amino acid sequences.

3.4. Phylogenetic Analysis and Estimation of Evolutionary Rates of APPV within the Farm

A phylogenetic analysis of the full-length APPV sequences available worldwide and obtained from the NCBI database and six APPV sequences obtained in the present study (Section 3.5) confirmed the division of APPV into three different clades, two of them containing only sequences from China (Figure 2). APPV sequences from the current study clustered in one subtree together with two German sequences (NC_030653 and KU041639, from 2015), one Austrian sequence (KX778724, from 2016), two Chinese sequences (MH885413, from 2018, and KY624591, from 2016) and one South Korean sequence (MF979135, from 2016). The other full-APPV genome from The Netherlands available in NCBI was clustered within the same clade, but into a different subtree, together with eight sequences from Switzerland and two from the US (Figure 2).

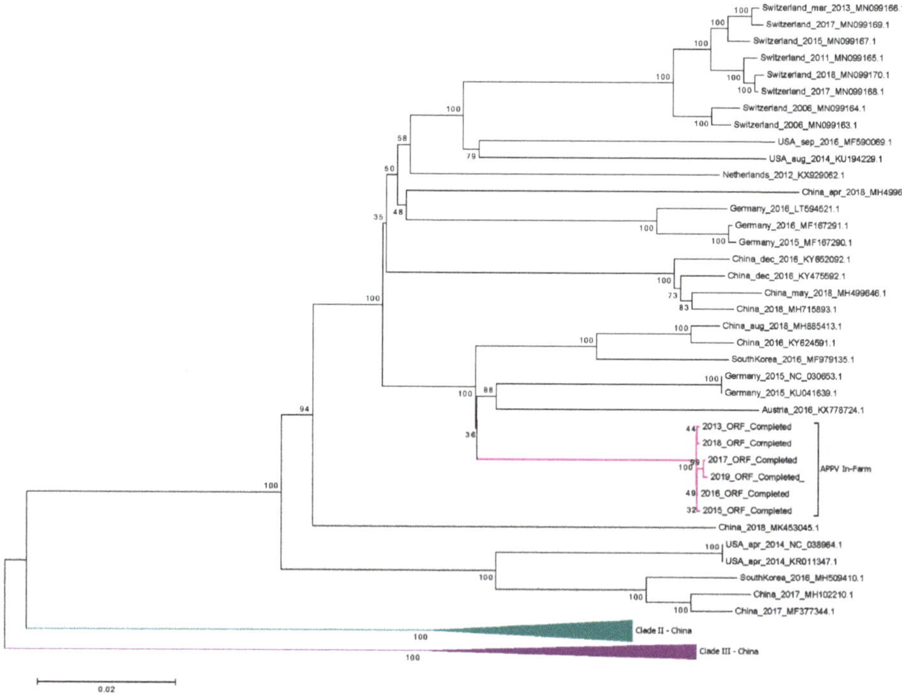

Figure 2. Phylogenetic analysis of the complete coding region alignment of atypical porcine pestivirus (APPV) sequences worldwide obtained from NCBI database and six APPV sequences retrieved in the present longitudinal study on the Dutch swine farm between 2013 and 2019. The neighbor-joining method with the Kimura two-parameter substitution model was used. Gaps and missing data were

subjected to complete deletion. Bootstrap values are provided at the root of the clusters. The scale bar is a measure of the proportion of divergence. Run for 500 bootstrap replicates. Clade I shown as expanded tree. Clades II and III, containing only sequences from China, are collapsed and shown in green and purple in the lower part of the tree. APPV branches obtained in the current study are shown in pink.

3.5. Genome and Protein Variations of APPV within the Farm

The samples with the highest numbers of APPV genome copies were selected from each of the years, and six full-length APPV genomic sequences from samples collected in the farm on 2013 and each year between 2015 and 2019, were determined. The 2013 and 2015 APPV genome sequences were obtained from serum samples from CT-affected piglets at the time of the large-scale outbreaks. The length of the complete APPV in-farm coding sequence (excluding the 5′-UTR region) was 10,908 nt, in line with the previously reported length of the coding region [14]. Multiple sequence alignment was performed on these six sequences using the CLUSTALW translation alignment in Geneious Prime v2019.0.4. The sequences from the different years showed a high similarity at the nucleotide level, between 99.74% and 99.90%. The number of nucleotide differences ranged from only 10 nucleotide substitutions between 2017 and 2019 genomes to 26 nucleotide substitutions between 2013 and 2017 genomes. On the amino acid level, the differences ranged from only 1 amino acid substitution between 2017 and 2019 to 7 amino acid substitutions between 2013 and 2017 and between 2016 and 2017 (Table 1).

Table 1. Distance matrix from the multiple sequence alignment of six in-farm APPV sequences.

	2013	2015	2016	2017	2018	2019
2013		3 aa	7 aa	6 aa	3 aa	5 aa
2015	18 nt		4 aa	3 aa	2 aa	3 aa
2016	17 nt	15 nt		7 aa	6 aa	6 aa
2017	26 nt	18 nt	23 nt		4 aa	1 aa
2018	16 nt	13 nt	15 nt	19 nt		5 aa
2019	25 nt	21 nt	21 nt	10 nt	20 nt	

The nucleotide substitutions found in the alignment were classified regarding their protein location and differentiating synonymous substitutions and non-synonymous substitutions. All substitutions are shown in Supplementary Table S8. Forty-five nucleotide substitutions were found along the APPV genome during the monitoring years in the farm from which thirty-four positions corresponded to synonymous substitutions. Eleven non-synonymous substitutions occurred on the APPV genome in the farm between 2013 and 2019. Among those, one amino acid change, V563A, was in the structural glycoprotein gene E1, while two amino acid changes, I726V and D752G, were in the E2 structural glycoprotein encoding region. The other eight non-synonymous substitutions occurred in the genes of non-structural proteins: two in Npro (S48P and H152L), one in NS2 (R1122K), three in NS5A (K2437R, G2550E and A2824T) and two in NS5B (D2979N and N3166T) (Supplementary Table S8).

Some amino acid changes in the strains have not been described before. The E1 non-synonymous substitution V563A only occurred within the farm, while the genotype of this site worldwide was always conserved as a Val residue. The synonymous substitution found on the same coding gene, V567, was also exclusive, when compared with other sequences in Clade I. Interestingly, site T608 in the in-farm sequences possessed a unique genotype compared with the consensus I608, and the V608 variant present as well in Clades II and III. Moreover, site I790 also showed a unique genotype, found only in one sequence from Switzerland from 2011 (accession number MN099165), while a Val residue is encoded on that position in the sequences worldwide. The glycosylation status of the pestivirus glycoproteins plays an important role in virulence. In this line, the substitution on E2 glycoprotein D752 to a Gly, Ser or Asn residue (Supplementary Table S8, Supplementary Figure S1) modified the 752NDT754 N-glycosylation motif predicted by NetNGlyc1.0 server (http://www.cbs.dtu.dk/services/). In the non-structural proteins, a unique motif was found in the N-terminal protease. The site 126KPAPASR132 was unique within the APPV sequences retrieved from the current farm study, including up to four modifications on its amino acid chain, in comparison with other geographically distinct APPV full genome sequences.

An estimation of the evolutionary rates within the farm was calculated for the full ORF of the APPV genome, as well as for the E2 glycoprotein-the coding sequence using TempEst v1.5.3 software [31]. An evolutionary rate of 3.224×10^{-4} substitutions/site/year (correlation coefficient = 0.9009; $R^2 = 0.8117$) was estimated for the full ORF of APPV over a period of six years of evolution, from 2013 to 2019, within the studied farm. The envelope glycoprotein E2, considered in all pestiviruses as the main antigen eliciting immune response, had an estimated evolutionary rate of 9.347×10^{-4} substitutions/site/year (correlation coefficient = 0.7706; $R^2 = 0.5938$). Both evolutionary rates were in line with the ones reported in other pestivirus species. The evolution of the CSFV E2 gene was estimated between 1.73×10^{-3} and 5.76×10^{-4} substitutions/site/year [34,35], while the full genome estimation was 1.03×10^{-4} substitutions/site/year [36]. For BVDV, evolutionary rates of 1.40×10^{-4} substitutions/site/year for the full genome and 1.26×10^{-3} substitutions/site/year for the E1-E2 coding region were reported [37]. A recombination analysis was done using the RDP4 software in order to find breakpoints within the multiple sequence alignment that might have undergone recombination, and to screen them for evidence of phylogenetic incongruence previous to phylogenetic tree building [28]. No recombination events were found at the farm level.

3.6. Selection Patterns within a Farm: Six-Year Evolution Study on an APPV in-Farm Variant

Negative, also called purifying, selection consists of the evolutionary pressure hindering the fixation of non-beneficial or deleterious protein variations in the population. On the other hand, positive or diversifying selection is the evolutionary pressure that promotes the fixation of beneficial protein variations in the population. This last one can be considered pervasive, when it is constant through time, or episodic, when it is sporadic and affecting only some lineages. Using the algorithms implemented in the HyPhy software packages via the Datamonkey webserver [32], we were able to identify a number of codons potentially subjected to selection pressures on our APPV genome between 2013 and 2019.

No positively selected codons, neither episodic nor pervasive, were found by any of the methods. Thus, positive selection cannot be considered as an evolution pattern within the farm during the six years of the study. On the other hand, FEL (fixed-effects likelihood) found evidence of pervasive negative/purifying selection at 9 sites among the 3635 codons, with a p-value threshold of 0.1. The nine codons subjected to purifying selection were found to be in the N-terminal protease (codon 132), the Erns glycoprotein (codon 418), the NS2 protein (codons 1054 and 1198), the NS3 protein (codon 1814), the NS5A protein (codons 2676 and 2848) and the NS5B protein (codons 2905 and 3243). FEL algorithm indicated, under the global MG94xREV model, a total non-synonymous/synonymous rate ratio for the in-farm alignment equal to 0.172, indicating the prevalence of negative selection within the APPV genome sequence.

4. Discussion

In 2013, a first large-scale CT outbreak was reported in a farm in The Netherlands, with a second large-scale outbreak at the end of 2015/early 2016, both related to APPV infections. After the second outbreak, all gilts selected for breeding were monitored by serum PCR screening at the age of first insemination for the presence of APPV with the aim of removing infected animals from the breeding population. However, there were some practical limitations to this approach: (1) The removal of a high percentage of the selected gilts at the age of first insemination severely disrupted the normal farm sow replacement strategy. (2) Serological tests were not available at the time and thus, the difference between persistent carriers and gilts with horizontal infections could not be established. (3) Parallel monitoring of persistently infected pigs from the second outbreak revealed that, in contrast to what it may have been expected based on experience with other pestiviruses, APPV disappeared from the serum in adult pigs. The data published by Schwartz et al. [16] showed stable presence of APPV in serum until 14 weeks of age, but our observations on a larger number of animals showed that a significant percentage of pigs turned PCR-negative for APPV around the age of 18 weeks.

Fecal shed was also monitored to potentially identify persistently APPV-infected pigs. In the age range between 24 and 44 weeks, monitoring APPV infection is difficult due to the absence of the virus in serum and intermittent shedding in the feces of PI gilts. Horizontally infected gilts show temporary presence in feces and serum. Therefore, a gilt-monitoring strategy was technically complicated and could lead to false-negative test results and consequently, the maintenance of persistently infected animals in the breeding population. It is of note that APPV-positive fecal shed has not proven to be infectious.

Based on these considerations, we reasoned that the early removal of PI animals in the purebred gilt breeding population, by performing qPCR analysis on serum samples before the age of 16 weeks, was the best strategy to control APPV at the farm level. Hence, a monthly monitoring program for APPV infections in 10-week-old selected breeding gilts of the sow breeding line was set up in the "closed" CT-affected farm with the purpose of reducing the risk of horizontal APPV infections to pregnant gilts and subsequent vertical infection of newborn piglets. It was assumed that these persistent carriers of APPV, often born without showing clinical symptoms after vertical transmission of the virus and with subsequent use in the farm sow replacement breeding program, contributed significantly to the risk of infections of pregnant gilts and low-parity sows in group housing. The results of our analysis showed that even with stringent removal of APPV PCR-positive, assumed persistently infected replacement gilts, the virus cannot be completely eliminated from the farm. The genotype remained unchanged throughout the years, as demonstrated by the low genetic diversity observed in the partial E2-coding sequences from the samples obtained during monthly monitoring, making it unlikely that new introductions via sperm were involved in the continuing APPV infections in the farm. Re-emerging peaks of viremic animals, often without symptoms of the disease but incidentally with the birth of CT piglets during these six years within this closed-herd farm, even with infected breeding animals removed from the population, showed that the virus was difficult to control let alone eliminate. Besides, the E2-coding sequences obtained from the majority of affected animals at the time of the CT outbreaks did not show any characteristic difference at the nucleotide nor amino acid level when compared to the asymptomatic carriers that could lead to an explanation for the clinical manifestation.

The hypervariable regions, which contribute to virus escape from the immune system, as well as the conserved elements involved in crucial virus replication processes and maintenance of the viral population, are obvious targets for drug and vaccine design [24]. Moreover, it can be hypothesized that certain substitutions in the viral coding sequence may be involved in the cyclic peaks of viremia and/or correlate with the pathogenicity of the virus. In the present study, forty-five nucleotide substitutions were found along the APPV genome during the monitoring years in the farm, eleven of them causing an amino acid change. Among those, two occurred on the E2 protein-coding gene, usually recognized as the main antigen able to elicit neutralizing antibodies in infected animals [38,39]. Recent studies on subunit E2 vaccines from APPV have shown strong immune responses in mice [40]. Moreover, E1–E2 heterodimers have been considered in other pestivirus species as key for viral infectivity [38]. Although the results obtained after the analysis of the APPV sequences on the followed farm did not point at any positively selected sites, two of the non-synonymous substitutions found on the heterodimer-coding genes were almost unique to the farm under study. More interestingly, the substitution D752G in E2 glycoprotein, which showed high variability throughout the fifty-one partial E2 sequences analyzed, modified the predicted 752NDT754 *N*-glycosylation motif. Generally, E2 glycosylation sites are highly conserved due to their primary role in viral entry and infection and their removal has shown viral attenuation in other pestivirus species [41]. Further research needs to be done to decipher the functionality of this highly variable motif.

The unique motif 126KPAPASR132 was found in the N-terminal protease in all APPV sequences retrieved from the farm. This motif included up to four modifications on its amino acid chain, in comparison with other geographically distinct APPV full-genome sequences. Although the exact functionality of this site is not known, Npro plays an important role in viral evasion of the innate immune response in other pestivirus species. Npro activity decreases the levels of the interferon regulatory factor 3 (IRF3) via proteasomal degradation, inhibiting its downstream signaling and thus suppressing type-I interferon responses in infected animals [42]. Previous research studies have related the functionality of this protein with the enhancement of co-infections with bovine respiratory syncytial virus (BRSV) in cattle [43]. APPV in persistently infected animals is commonly found together with other viruses and, it may therefore intensify those secondary infections. More importantly, experimental Erns and Npro mutations in the BVDV genome have failed to induce persistent infections in cattle [44].

The other 34 substitutions found along the APPV genome during the six years evolution in the farm, but not leading to amino acid changes, are not less important as the codon usage bias has been related to virus translation efficiency [45], RNA structures critical for replication and packing [46] and enhanced virulence [47]. Future studies on the codon usage and effects on RNA structure using reverse genetics may therefore shed light on the actual effect of synonymous substitutions. We are aware of the fact that, with only six full-genome sequences included in the current study, our results were based on a limited amount of sequence data. However, given the dynamics of the virus at the farm, we had maximized our analysis possibilities. Pigs with high APPV loads in serum, which are needed for a successful sequencing strategy, were only occasionally observed and, in addition, pigs from the same litter were expected to have the same viral genotype, as demonstrated by the partial E2 sequences obtained from a large number of samples, and further sequence analysis would not add meaningful data. Moreover, the limited genome variation shown within the farm suggested that more sequences obtained during the same years would not radically change the observations and conclusions drawn in the present study. Nonetheless, we acknowledge that an increase in the number of APPV full-length sequences available, especially within longitudinal studies, will be key for the scientific community to further understand the evolutionary dynamics and genomic features of this pestivirus.

With regard to the evolutionary pressures of APPV, the present study showed a general genome-wide purifying pressure, especially strong on the non-structural proteins. These results indicated the importance for the virus to maintain the functionality of the non-structural viral proteins, avoiding the fixation of detrimental amino acid substitutions that might hinder the virus ability to evade the host immune system and cause persistent infections. In Flaviviridae, non-structural proteins have been determined as key players in viral escape from the host immune system. Hepatitis C virus (HCV) NS3 protein inhibits tumor necrosis factor alpha (TFN-α) stimulated NF-κB activation to evade the host innate immunity [48]; NS4A proteins of dengue virus 1 (DENV-1) inhibit the interferon- β (IFN-β) signaling pathway mediated by the retinoic acid-inducible gene I (RIG-I) and TANK-binding kinase-1 (TBK1) proteins [49]; and NS5 protein of several flaviviruses is also involved in the inhibition of IFN signaling by degrading the signal transducer and activator of transcription 1 (STAT1) or STAT2 [50,51]. Even though the signaling pathways and target sites of the majority of the non-structural proteins remain unknown for several virus species, these examples acknowledge their potential function in the host innate immune evasion. The strong purifying selection we found could be explained by a lack of pressure from the host immune system, as the persistently infected animals may potentially be immunotolerant, although experimental evidence is needed to ultimately confirm this hypothesis. Regular use of ELISA systems, to monitor whether immunity is raised to the virus, would bring light to this issue [39].

Monitoring studies in APPV-infected farms during the last years showed that APPV caused persistent infections in piglets when they were infected in-utero, occasionally developing congenital tremors, while horizontal infections within the herd were transient and without visible clinical signs [39]. As aforementioned, this is similar to BVDV infections in cattle, which only causes persistent infections in the calves when the virus is transmitted in-utero before 120 days of gestation. While some persistently infected calves are born with congenital malformations, others are clinically healthy with BVDV infections going unnoticed until the onset of clinical signs in the new parities [52]. A widely accepted hypothesis, supported by several research studies, is that non-cytopathic BVDV strains fail to induce type-I interferon responses in the infected fetus developing persistent infections and immunotolerant calves [52,53]. These persistently infected animals shed the virus continuously during their lifetimes, provoking transient horizontal infections within the herds, but more importantly infecting pregnant cows leading to the birth of persistently infected calves [54].

Our study results of APPV circulation on the affected sow farm showed similarities with the ways that BVDV persists within the cattle populations. Persistently APPV-infected animals remained undetected in a population that was not continuously monitored by PCR because they did not always show congenital tremors. Although our control strategy did not result in full elimination of the virus, at least it resulted in no further large-scale outbreaks since early 2016. Our approach was based on the hypothesis that persistently infected gilts are the main contributors to the spread of the virus and thus, removing those indirectly leads to the prevention of infections of pregnant sows. Nevertheless, the combination with a direct approach based on the elimination of infected pregnant animals could further improve the eradication strategy, as experimental studies have shown that virus infection during early stages of gestation leads to viremia and transplacental transfer.

Research on APPV vaccines is still limited. High strain heterogenicity and persistent infections have been a major bottleneck in the case of BVDV vaccination [55], and for APPV the same problems will be encountered when vaccines are designed and tested. Importantly, effective controlling strategies in closed farm situations should start with intensive testing of all the animals in the herd, followed by removal of positive carriers from the population. Testing should not only be done via RT-qPCR in serum, but also in fecal swabs, as it has been observed that the virus disappears from blood while it can still (intermittently) be present in feces [14]. The strategy presented in this study was not adequate to fully eliminate the virus; APPV kept circulating in the sow population after removal of persistent carriers in the sow breeding herd, suggesting that all gilts should be monitored. In addition, serological data could reveal if the virus is still circulating within the herd via horizontal infections. Vaccination of seronegative gilts and sows is an option that can be considered if removal of carriers alone does not sufficiently control APPV infections.

Supplementary Materials: The following are available online at http://www.mdpi.com/1999-4915/12/10/1080/s1, Table S1: List of amplification and sequencing primers used in the study; Table S2: Details on the production data of the studied farm during the 2015–2016 CT outbreak; Table S3: Serum APPV monitoring of gilts selected for breeding at the age of first insemination between January and April in 2016; Table S4: Follow-up of five litters with CT born from gilts for the presence of APPV in serum over time (2015/2016 outbreak); Table S5: Follow-up of boars and gilts until the age of 10 months for the presence of APPV in serum (2015/2016 outbreak); Table S6: Follow-up of boars and gilts until the age of 10 months for the presence of APPV in fecal shed (2015/2016 outbreak); Table S7: APPV E2 sequencing sample list and corresponding APPV viral loads; Table S8: Nucleotide substitutions in the APPV in-farm multiple sequence alignment; Figure S1: Multiple sequence alignment of the partial APPV E2 amino acid sequences chronologically ordered.

Author Contributions: Conceptualization, A.F.-G., A.d.G. and L.v.d.H.; methodology, A.F.-G., L.v.d.H., M.D., R.v.d.B. and B.S.; software, A.F.-G., R.v.d.B. and B.S.; data analysis, A.F.-G.; writing—original draft preparation, A.F.-G.; writing—review and editing, A.d.G., L.v.d.H. and R.v.d.B.; supervision, A.d.G. and L.v.d.H. All authors have read and agreed to the published version of the manuscript.

Funding: This research received funding from the European Union's Horizon 2020 research and innovation program, under the Marie Skłodowska-Curie Actions grant agreement no. 721367 (HONOURs).

Acknowledgments: The authors wish to acknowledge the input of the veterinarian Jolanda Rooijendijk, from DAC de Peelhorst, and the farmers who sent in samples of diseased animals. We thank Ignacio Postigo-Hidalgo for useful discussions on bioinformatics.

Conflicts of Interest: The authors of this manuscript have the following competing interests: A.F.G., R.v.d.B., B.S. and A.d.G. are employed at MSD Animal Health, a commercial company. M.D., A.d.G. and L.v.d.H. are inventors on a patent application on APPV.

References

1. Blome, S.; Beer, M.; Wernike, K. New Leaves in the Growing Tree of Pestiviruses. *Adv. Virus Res.* **2017**, *99*, 139–160. [CrossRef] [PubMed]
2. Smith, D.B.; Meyers, G.; Bukh, J.; Gould, E.A.; Monath, T.; Scott Muerhoff, A.; Pletnev, A.; Rico-Hesse, R.; Stapleton, J.T.; Simmonds, P.; et al. Proposed revision to the taxonomy of the genus Pestivirus, family Flaviviridae. *J. Gen. Virol.* **2017**, *98*, 2106–2112. [CrossRef] [PubMed]
3. Wu, Z.; Liu, B.; Du, J.; Zhang, J.; Lu, L.; Zhu, G.; Han, Y.; Su, H.; Yang, L.; Zhang, S.; et al. Discovery of Diverse Rodent and Bat Pestiviruses With Distinct Genomic and Phylogenetic Characteristics in Several Chinese Provinces. *Front. Microbiol.* **2018**, *9*, 2562. [CrossRef]
4. Lamp, B.; Schwarz, L.; Hogler, S.; Riedel, C.; Sinn, L.; Rebel-Bauder, B.; Weissenbock, H.; Ladinig, A.; Rumenapf, T. Novel Pestivirus Species in Pigs, Austria, 2015. *Emerg. Infect. Dis.* **2017**, *23*, 1176–1179. [CrossRef]
5. Kinsley, A.T. Dancing pigs. *Vet. Med.* **1922**, *17*, 123.
6. Done, J.T.; Woolley, J.; Upcott, D.H.; Hebert, C.N. Porcine congenital tremor type AII: Spinal cord morphometry. *Br. Vet. J.* **1986**, *142*, 145–150. [CrossRef]
7. Done, J.T. Congenital Nervous Diseaseas of Pigs: A Review. *Lab. Anim.* **1968**, *2*, 207–217. [CrossRef]
8. Bradley, R.; Done, J.T.; Hebert, C.N.; Overby, E.; Askaa, J.; Basse, A.; Bloch, B. Congenital tremor type AI: Light and electron microscopical observations on the spinal cords of affected piglets. *J. Comp. Pathol.* **1983**, *93*, 43–59. [CrossRef]
9. Blakemore, W.F.; Harding, J.D.; Done, J.T. Ultrastructural observations on the spinal cord of a Landrace pig with congenital tremor type AIII. *Res. Vet. Sci.* **1974**, *17*, 174–178. [CrossRef]
10. Patterson, D.S.; Sweasey, D.; Brush, P.J.; Harding, J.D. Neurochemistry of the spinal cord in British Saddleback piglets affected with congenital tremor, type A-IV, a second form of hereditary cerebrospinal hypomyelinogenesis. *J. Neurochem.* **1973**, *21*, 397–406. [CrossRef]
11. Knox, B.; Askaa, J.; Basse, A.; Bitsch, V.; Eskildsen, M.; Mandrup, M.; Ottosen, H.E.; Overby, E.; Pedersen, K.B.; Rasmussen, F. Congenital ataxia and tremor with cerebellar hypoplasia in piglets borne by sows treated with Neguvon vet. (metrifonate, trichlorfon) during pregnancy. *Nord. Vet. Med.* **1978**, *30*, 538–545. [PubMed]
12. Patterson, D.S.; Done, J.T.; Foulkes, J.A.; Sweasey, D. Neurochemistry of the spinal cord in congenital tremor of piglets (type AII), a spinal dysmyelimogenesis of infectious origin. *J. Neurochem.* **1976**, *26*, 481–485. [CrossRef] [PubMed]
13. Hause, B.M.; Collin, E.A.; Peddireddi, L.; Yuan, F.; Chen, Z.; Hesse, R.A.; Gauger, P.C.; Clement, T.; Fang, Y.; Anderson, G. Discovery of a novel putative atypical porcine pestivirus in pigs in the USA. *J. Gen. Virol.* **2015**, *96*, 2994–2998. [CrossRef] [PubMed]
14. de Groof, A.; Deijs, M.; Guelen, L.; van Grinsven, L.; van Os-Galdos, L.; Vogels, W.; Derks, C.; Cruijsen, T.; Geurts, V.; Vrijenhoek, M.; et al. Atypical Porcine Pestivirus: A Possible Cause of Congenital Tremor Type A-II in Newborn Piglets. *Viruses* **2016**, *8*, 271. [CrossRef]
15. Arruda, B.L.; Arruda, P.H.; Magstadt, D.R.; Schwartz, K.J.; Dohlman, T.; Schleining, J.A.; Patterson, A.R.; Visek, C.A.; Victoria, J.G. Identification of a Divergent Lineage Porcine Pestivirus in Nursing Piglets with Congenital Tremors and Reproduction of Disease following Experimental Inoculation. *PLoS ONE* **2016**, *11*, e0150104. [CrossRef]
16. Schwarz, L.; Riedel, C.; Hogler, S.; Sinn, L.J.; Voglmayr, T.; Wochtl, B.; Dinhopl, N.; Rebel-Bauder, B.; Weissenbock, H.; Ladinig, A.; et al. Congenital infection with atypical porcine pestivirus (APPV) is associated with disease and viral persistence. *Vet. Res.* **2017**, *48*, 1–14. [CrossRef]
17. Postel, A.; Meyer, D.; Cagatay, G.N.; Feliziani, F.; De Mia, G.M.; Fischer, N.; Grundhoff, A.; Milicevic, V.; Deng, M.C.; Chang, C.Y.; et al. High Abundance and Genetic Variability of Atypical Porcine Pestivirus in Pigs from Europe and Asia. *Emerg. Infect. Dis.* **2017**, *23*, 2104–2107. [CrossRef]
18. Kaufmann, C.; Stalder, H.; Sidler, X.; Renzullo, S.; Gurtner, C.; Grahofer, A.; Schweizer, M. Long-Term Circulation of Atypical Porcine Pestivirus (APPV) within Switzerland. *Viruses* **2019**, *11*, 653. [CrossRef]

19. Yan, X.L.; Li, Y.Y.; He, L.L.; Wu, J.L.; Tang, X.Y.; Chen, G.H.; Mai, K.J.; Wu, R.T.; Li, Q.N.; Chen, Y.H.; et al. 12 novel atypical porcine pestivirus genomes from neonatal piglets with congenital tremors: A newly emerging branch and high prevalence in China. *Virology* **2019**, *533*, 50–58. [CrossRef]
20. Xie, Y.; Wang, X.; Su, D.; Feng, J.; Wei, L.; Cai, W.; Li, J.; Lin, S.; Yan, H.; He, D. Detection and Genetic Characterization of Atypical Porcine Pestivirus in Piglets With Congenital Tremors in Southern China. *Front. Microbiol.* **2019**, *10*, 1406. [CrossRef]
21. Beer, M.; Wernike, K.; Drager, C.; Hoper, D.; Pohlmann, A.; Bergermann, C.; Schroder, C.; Klinkhammer, S.; Blome, S.; Hoffmann, B. High Prevalence of Highly Variable Atypical Porcine Pestiviruses Found in Germany. *Transbound. Emerg. Dis.* **2017**, *64*, e22–e26. [CrossRef] [PubMed]
22. Steinhauer, D.A.; Domingo, E.; Holland, J.J. Lack of evidence for proofreading mechanisms associated with an RNA virus polymerase. *Gene* **1992**, *122*, 281–288. [CrossRef]
23. Bull, J.J.; Sanjuan, R.; Wilke, C.O. Theory of lethal mutagenesis for viruses. *J. Virol.* **2007**, *81*, 2930–2939. [CrossRef] [PubMed]
24. Kosakovsky Pond, S.L.; Frost, S.D. Not so different after all: A comparison of methods for detecting amino acid sites under selection. *Mol. Biol. Evol.* **2005**, *22*, 1208–1222. [CrossRef] [PubMed]
25. Sayers, E.W.; Cavanaugh, M.; Clark, K.; Ostell, J.; Pruitt, K.D.; Karsch-Mizrachi, I. GenBank. *Nucleic Acids Res.* **2019**, *47*, D94–D99. [CrossRef] [PubMed]
26. Gupta, R.; Jung, E.; Gooley, A.A.; Williams, K.L.; Brunak, S.; Hansen, J. Scanning the available Dictyostelium discoideum proteome for O-linked GlcNAc glycosylation sites using neural networks. *Glycobiology* **1999**, *9*, 1009–1022. [CrossRef]
27. Steentoft, C.; Vakhrushev, S.Y.; Joshi, H.J.; Kong, Y.; Vester-Christensen, M.B.; Schjoldager, K.T.; Lavrsen, K.; Dabelsteen, S.; Pedersen, N.B.; Marcos-Silva, L.; et al. Precision mapping of the human O-GalNAc glycoproteome through SimpleCell technology. *EMBO J.* **2013**, *32*, 1478–1488. [CrossRef]
28. Martin, D.P.; Murrell, B.; Golden, M.; Khoosal, A.; Muhire, B. RDP4: Detection and analysis of recombination patterns in virus genomes. *Virus Evol.* **2015**, *1*, vev003. [CrossRef]
29. Kimura, M. A simple method for estimating evolutionary rates of base substitutions through comparative studies of nucleotide sequences. *J. Mol. Evol.* **1980**, *16*, 111–120. [CrossRef]
30. Kumar, S.; Stecher, G.; Li, M.; Knyaz, C.; Tamura, K. MEGA X: Molecular Evolutionary Genetics Analysis across Computing Platforms. *Mol. Biol. Evol.* **2018**, *35*, 1547–1549. [CrossRef]
31. Rambaut, A.; Lam, T.T.; Max Carvalho, L.; Pybus, O.G. Exploring the temporal structure of heterochronous sequences using TempEst (formerly Path-O-Gen). *Virus Evol.* **2016**, *2*, vew007. [CrossRef] [PubMed]
32. Delport, W.; Poon, A.F.; Frost, S.D.; Kosakovsky Pond, S.L. Datamonkey 2010: A suite of phylogenetic analysis tools for evolutionary biology. *Bioinformatics* **2010**, *26*, 2455–2457. [CrossRef] [PubMed]
33. Murrell, B.; Wertheim, J.O.; Moola, S.; Weighill, T.; Scheffler, K.; Kosakovsky Pond, S.L. Detecting individual sites subject to episodic diversifying selection. *PLoS Genet.* **2012**, *8*, e1002764. [CrossRef] [PubMed]
34. Choe, S.; Le, V.P.; Shin, J.; Kim, J.H.; Kim, K.S.; Song, S.; Cha, R.M.; Park, G.N.; Nguyen, T.L.; Hyun, B.H.; et al. Pathogenicity and Genetic Characterization of Vietnamese Classical Swine Fever Virus: 2014–2018. *Pathogens* **2020**, *9*, 169. [CrossRef]
35. Yoo, S.J.; Kwon, T.; Kang, K.; Kim, H.; Kang, S.C.; Richt, J.A.; Lyoo, Y.S. Genetic evolution of classical swine fever virus under immune environments conditioned by genotype 1-based modified live virus vaccine. *Transbound. Emerg. Dis.* **2018**, *65*, 735–745. [CrossRef] [PubMed]
36. Kwon, T.; Yoon, S.H.; Kim, K.W.; Caetano-Anolles, K.; Cho, S.; Kim, H. Time-calibrated phylogenomics of the classical swine fever viruses: Genome-wide bayesian coalescent approach. *PLoS ONE* **2015**, *10*, e0121578. [CrossRef]
37. Chernick, A.; van der Meer, F. Evolution of Bovine viral diarrhea virus in Canada from 1997 to 2013. *Virology* **2017**, *509*, 232–238. [CrossRef]
38. Ronecker, S.; Zimmer, G.; Herrler, G.; Greiser-Wilke, I.; Grummer, B. Formation of bovine viral diarrhea virus E1-E2 heterodimers is essential for virus entry and depends on charged residues in the transmembrane domains. *J. Gen. Virol.* **2008**, *89*, 2114–2121. [CrossRef]
39. Cagatay, G.N.; Meyer, D.; Wendt, M.; Becher, P.; Postel, A. Characterization of the Humoral Immune Response Induced after Infection with Atypical Porcine Pestivirus (APPV). *Viruses* **2019**, *11*, 880. [CrossRef]
40. Zhang, H.; Wen, W.; Hao, G.; Chen, H.; Qian, P.; Li, X. A Subunit Vaccine Based on E2 Protein of Atypical Porcine Pestivirus Induces Th2-type Immune Response in Mice. *Viruses* **2018**, *10*, 673. [CrossRef]

41. Risatti, G.R.; Holinka, L.G.; Fernandez Sainz, I.; Carrillo, C.; Lu, Z.; Borca, M.V. N-linked glycosylation status of classical swine fever virus strain Brescia E2 glycoprotein influences virulence in swine. *J. Virol.* **2007**, *81*, 924–933. [CrossRef] [PubMed]
42. Hilton, L.; Moganeradj, K.; Zhang, G.; Chen, Y.H.; Randall, R.E.; McCauley, J.W.; Goodbourn, S. The NPro product of bovine viral diarrhea virus inhibits DNA binding by interferon regulatory factor 3 and targets it for proteasomal degradation. *J. Virol.* **2006**, *80*, 11723–11732. [CrossRef] [PubMed]
43. Alkheraif, A.A.; Topliff, C.L.; Reddy, J.; Massilamany, C.; Donis, R.O.; Meyers, G.; Eskridge, K.M.; Kelling, C.L. Type 2 BVDV N(pro) suppresses IFN-1 pathway signaling in bovine cells and augments BRSV replication. *Virology* **2017**, *507*, 123–134. [CrossRef] [PubMed]
44. Meyers, G.; Ege, A.; Fetzer, C.; von Freyburg, M.; Elbers, K.; Carr, V.; Prentice, H.; Charleston, B.; Schurmann, E.M. Bovine viral diarrhea virus: Prevention of persistent fetal infection by a combination of two mutations affecting Erns RNase and Npro protease. *J. Virol.* **2007**, *81*, 3327–3338. [CrossRef]
45. Coleman, J.R.; Papamichail, D.; Skiena, S.; Futcher, B.; Wimmer, E.; Mueller, S. Virus attenuation by genome-scale changes in codon pair bias. *Science* **2008**, *320*, 1784–1787. [CrossRef]
46. Simmonds, P.; Smith, D.B. Structural constraints on RNA virus evolution. *J. Virol.* **1999**, *73*, 5787–5794. [CrossRef]
47. Lauring, A.S.; Acevedo, A.; Cooper, S.B.; Andino, R. Codon usage determines the mutational robustness, evolutionary capacity, and virulence of an RNA virus. *Cell Host Microbe* **2012**, *12*, 623–632. [CrossRef]
48. Chen, Y.; He, L.; Peng, Y.; Shi, X.; Chen, J.; Zhong, J.; Chen, X.; Cheng, G.; Deng, H. The hepatitis C virus protein NS3 suppresses TNF-alpha-stimulated activation of NF-kappaB by targeting LUBAC. *Sci. Signal.* **2015**, *8*, ra118. [CrossRef]
49. Dalrymple, N.; Cimica, V.; Mackow, E. Dengue Virus NS Proteins Inhibit RIG-I/MAVS Signaling by Blocking TBK1/IRF3 Phosphorylation: Dengue Virus Serotype 1 NS4A Is a Unique Interferon-Regulating Virulence Determinant. *mBio* **2015**, *6*, e00553-15. [CrossRef]
50. Kumthip, K.; Chusri, P.; Jilg, N.; Zhao, L.; Fusco, D.N.; Zhao, H.; Goto, K.; Cheng, D.; Schaefer, E.A.; Zhang, L.; et al. Hepatitis C virus NS5A disrupts STAT1 phosphorylation and suppresses type I interferon signaling. *J. Virol.* **2012**, *86*, 8581–8591. [CrossRef]
51. Grant, A.; Ponia, S.S.; Tripathi, S.; Balasubramaniam, V.; Miorin, L.; Sourisseau, M.; Schwarz, M.C.; Sanchez-Seco, M.P.; Evans, M.J.; Best, S.M.; et al. Zika Virus Targets Human STAT2 to Inhibit Type I Interferon Signaling. *Cell Host Microbe* **2016**, *19*, 882–890. [CrossRef] [PubMed]
52. Charleston, B.; Fray, M.D.; Baigent, S.; Carr, B.V.; Morrison, W.I. Establishment of persistent infection with non-cytopathic bovine viral diarrhoea virus in cattle is associated with a failure to induce type I interferon. *J. Gen. Virol.* **2001**, *82*, 1893–1897. [CrossRef] [PubMed]
53. Peterhans, E.; Jungi, T.W.; Schweizer, M. BVDV and innate immunity. *Biologicals* **2003**, *31*, 107–112. [CrossRef]
54. Brock, K.V. The persistence of bovine viral diarrhea virus. *Biologicals* **2003**, *31*, 133–135. [CrossRef]
55. Griebel, P.J. BVDV vaccination in North America: Risks versus benefits. *Anim. Health Res. Rev.* **2015**, *16*, 27–32. [CrossRef]

viruses

MDPI

Article

The Outcome of Porcine Foetal Infection with Bungowannah Virus Is Dependent on the Stage of Gestation at Which Infection Occurs. Part 1: Serology and Virology

Deborah S. Finlaison * and Peter D. Kirkland

Virology Laboratory, Elizabeth Macarthur Agricultural Institute, New South Wales Department of Primary Industries, Menangle, NSW 2568, Australia; peter.kirkland@dpi.nsw.gov.au
* Correspondence: deborah.finlaison@dpi.nsw.gov.au

Received: 29 May 2020; Accepted: 24 June 2020; Published: 26 June 2020

Abstract: Bungowannah virus is a novel porcine pestivirus identified in a disease outbreak in Australia in 2003. The aim of this study was to determine the outcome of infection of the pregnant pig with this virus. Twenty-four pregnant pigs were infected at days 35, 55, 75 or 90 of gestation. Blood, tonsillar and rectal swabs were collected from each pig at birth and then weekly until euthanasia or death. Tissues were sampled at necropsy. Viral load was measured by real-time reverse-transcription polymerase chain reaction (qRT-PCR) and antibody levels in serum by peroxidase-linked immunoassay. Bungowannah virus was detected in the serum and excretions of all infected pigs at birth regardless of the stage of gestation at which infection occurred. Persistent infections occurred following infection prior to the development of foetal immunocompetence. Unexpectedly some animals infected at day 55 of gestation later cleared the virus and seroconverted. Viraemia and viral shedding resolved quickest following infection in late gestation.

Keywords: Bungowannah virus; foetus; pestivirus; porcine; real-time PCR; serology; virology

1. Introduction

Bungowannah virus is a novel pestivirus identified from an outbreak of disease in a piggery in New South Wales, Australia, in June 2003 [1]. It is genetically distinct from the other recognised pestiviruses of pigs, classical swine fever virus (CSFV) and atypical porcine pestivirus (APPV) [2,3] with its closest genetic relationship to the recently identified Linda virus [4]. The disease was referred to as the porcine myocarditis syndrome, or PMC, because histological changes in affected animals consist almost exclusively of a multifocal non-suppurative myocarditis, with myonecrosis in some cases. The outbreak initially presented as sudden death in 2- to 3-week-old weaning age pigs, but soon after the onset there was a marked increase in the birth of stillborn foetuses and a slight increase in the occurrence of mummified pigs. Cumulative losses in some weeks exceeded 50% of pigs born, and it is estimated that as many as 50,000 pigs died in the initial outbreak. Due to the reproductive effects and disease occurring almost exclusively in the first 2–3 weeks of life it was presumed to be predominantly the consequence of in utero infection. This hypothesis was supported by the detection of elevated serum IgG levels in up to 50% of stillborn pigs and by the absence of disease in pigs soon after weaning or in sows farrowing affected litters [1].

The pestiviruses are well recognised reproductive pathogens where the outcome of infection is dependent on a number of factors including the pathogenicity of the infecting strain, the stage of gestation that infection occurs in relation to organogenesis and development of immune competence, where infection prior to foetal immunocompetence may result in a persistent infection due to immunotolerance [5–14]. Persistently infected (PI) animals remain serologically negative and demonstrate cell-mediated unresponsiveness to the infecting strain, shed virus throughout their lives, and are usually epidemiologically more important in ongoing virus transmission than acute, transiently infected animals [5,13–22]. It has been shown experimentally that pigs infected post-natally with Bungowannah virus develop transient infections that resolve over a 10-day period and transmit the virus inefficiently [23]. PI animals are usually the reservoirs of pestiviruses in nature but no PI pigs surviving past 6–8 weeks old have been identified in the affected piggery. Therefore, to better understand how virus transmission is maintained in an infected population it is important to know if long-term infections can occur and to clarify the significance and sources of viral shedding following in utero infections.

This study examined the virological and serological characteristics of in utero infection of the porcine foetus with Bungowannah virus at different stages of gestation. The primary objectives were to determine:

1. If the porcine foetus becomes infected in utero following intra-nasal exposure of the sow;
2. The concentrations of Bungowannah virus RNA in serum, and shed in oropharyngeal secretions and faeces, and whether this is affected by the stage of pregnancy at which the sow is infected;
3. If the pig foetus mounts a humoral immune response following in utero infection;
4. If persistent infections with Bungowannah virus occur and whether there is a critical stage of gestation at which infection results in this outcome;
5. The optimal tissue samples for the detection of Bungowannah virus;
6. If PI pigs can readily transmit infection to naïve pigs.

Infection of the porcine foetus with Bungowannah virus was successful, and the virological and serological characteristics following in utero infection at different stages of gestation are reported. The clinical signs and gross pathology are described in an accompanying manuscript [24].

2. Materials and Methods

2.1. Londitudinal Study Design

Twenty-four pregnant pigs (22 gilts and two parity-1 sows) with known joining dates were obtained from a piggery known to be free of Bungowannah virus, and sampled and shown to be seronegative on the day of inoculation. Pregnancy was confirmed by ultrasound examination prior to selection for the study. Gilts from this piggery were routinely vaccinated against parvovirus, leptospirosis, erysipelas and *E. coli* with commercial vaccines at selection and again 4 weeks later. Both gilts and sows were also vaccinated against leptospirosis, erysipelas and E. coli at 13 weeks of pregnancy. The animals were moved to facilities at the Elizabeth Macarthur Agricultural Institute 3–5 days before they were due to be infected.

The pigs were challenged intranasally at approximately day 35 (34–36), 55 (55–58), 75 (72–76) or 90 (90–92) of gestation (n = 6 per group referred to as D35, D55, D75 and D90) [24]. These time-points were selected as they were similar to those used in a previous study where foetuses were directly inoculated [25] and they span the gestational age at which the pig foetus is considered to become immunocompetent (70 days). Due to animal accommodation availability and the logistics of undertaking a study of this size, the four treatment groups were challenged and managed separately. The four batch farrowings occurred over a 5-month period. To facilitate intranasal challenge, the pregnant pigs were sedated approximately 30 minutes prior with Azaperone (40 mg/mL—up to 2 mL/20 kg).

All pregnancies were allowed to proceed to 113 days of gestation when all pigs were induced to farrow on day 114 to optimise collection of blood from pigs prior to suckling. Farrowing was induced with 500 μg cloprostenol given intramuscularly on the morning of day 113 and, if required, followed 24 h later with Oxytocin (10 i.u.) intramuscularly.

Pigs from infected litters were weaned when 21–25 days old. As the presence of Bungowannah virus could still be detected at weaning in several challenge groups, as many pigs as the secure containment facilities could hold were weaned and retained until at least 5 to 8 weeks of age (D35, n = 11; D55, n = 22; D75, n = 23; D90, n = 20). Selected pigs from D35 and D55 were kept in the study for a longer period of time to allow ongoing monitoring of virological and serological parameters and clinical signs [24]. Throughout the study, pigs that were moribund, not feeding or did not appear to be viable were euthanased by an intravenous overdose of pentobarbitone sodium.

Pigs from the three litters that did not become infected with Bungowannah virus (on the basis that the virus could not be detected in serum or body fluid for any of the pigs in the litter in a real-time reverse-transcription polymerase chain reaction (qRT-PCR) assay and negative precolostral serology results) were kept as control pigs until they were 14 to 21 days old, at which time they were euthanased.

Individual pig identifications (IDs) have been retained in the text where relevant to facilitate comparison between virological, serological, clinical signs, gross [24] and histopathology findings described in related manuscripts. The first number relates to the litter ID and the second the animal ID within the litter (generally in order of birth) e.g., 8-01 indicates the first piglet to be born in litter 8.

The animal studies were approved by the Animal Ethics Committee of the Elizabeth Macarthur Agricultural Institute, AEC Reference No. M09/02 (6 March 2009).

2.2. Inoculum

The pregnant pigs were challenged with approximately 5.3–5.8 $\log_{10}$ $TCID_{50}$ of Bungowannah virus in 5 mL of phosphate buffered gelatin saline (PBGS; pH 7.3, 2.5 mL per nostril), with the exception of Litter 11 where the sow only received approximately 3.8 $\log_{10}$ $TCID_{50}$ of virus. This inoculum was prepared as previously described [23] and confirmed to be free of porcine parvovirus, porcine circovirus type 2 and other pestiviruses by PCR [26,27]. Testing for porcine reproductive and respiratory syndrome virus was not performed because Australia is free of this virus.

2.3. Sample Collection—Pregnant Animals

A vaginal swab was collected from each sow on the day of farrowing. In addition, a piece of placenta was collected and swabbed. Vaginal swabs were then collected every 2–3 days until 14 days post-farrowing and then once to twice weekly until 21–28 days.

All swabs collected during the course of the study were placed in 2 mL of PBGS and stored at 4 °C prior to testing.

2.4. Sample Collection—Piglets

A clotted blood sample, and oropharyngeal and rectal swabs were collected from pigs shortly after birth and where possible before they suckled. While this was achieved in most instances, some animals farrowed earlier than 114 days or commenced farrowing earlier than 24 h after receiving the cloprostenol injection and had suckled. For pigs that were stillborn, body cavity fluid (in preferential order of pericardial>pleural>abdominal fluid) was collected rather than serum.

Thereafter, oropharyngeal and rectal swabs and clotted blood samples were collected weekly up to 8 weeks of age. After this time-point, any remaining pigs were sampled every 10–14 days up to 3 months of age. Three animals from the D55 group that were seronegative at birth were followed for an extended period (10-1 to 189 days, and 8-01 and 8-05 to 329 days of age). Urine was collected opportunistically from weaned pigs.

All pigs were subjected to a detailed necropsy and a wide range of samples were collected for virology, serology and histopathology. Samples for qRT-PCR were collected by firmly rubbing a swab across the freshly cut surface of a section of heart, lung, thymus, spleen, small intestine, inguinal lymph node and brain, and from the surface of the tonsil from pigs of infected litters; from pigs of uninfected litters, swabs were collected directly from the surface of the tonsil and from a cut section of lung and spleen.

2.5. Transmission Study Design

Two 5-and-a-half week-old pigs were obtained from the same piggery as the pregnant animals. After an acclimatisation period of 2 days in isolation, the two naïve pigs were housed for 28 days in the same room and pen with two pigs from D55 (8-01 and 10-01) considered PI with Bungowannah virus. The PI pigs were selected for the study based on low pre-suckle IgG levels (448 and 135 µg/mL), and high quantities of Bungowannah virus RNA in serum (6.9 and 5.5 $\log_{10}$ copies/mL serum) and on oropharyngeal swabs (6.1 and 6.8 $\log_{10}$ copies/swab) at 6 weeks of age compared with cohorts. Retrospective testing demonstrated both were seronegative for Bungowannah virus antibodies at birth, when tested in the peroxidase-linked immunoassay.

2.6. Sample Collection—Transmission Study

Serum, oropharyngeal and rectal swabs were collected from the introduced pigs on day 0, and then from the introduced and PI pigs on days 1, 8 (serum) or 9 (oropharyngeal and rectal swabs), 15, 22 and 28 days post-introduction.

2.7. RNA Extraction and Real-Time Reverse-Transcription Polymerase Chain Reaction (qRT-PCR)

Total nucleic acid was extracted from 50 uL of sera and PBGS containing swabs using the MagMax™-96 Viral RNA Isolation Kit (Ambion, Austin, Texas Cat. No. 1836). A Kingfisher®96 magnetic particle handling system (Thermo Fisher Scientific, Waltham, MA, USA) was used for the extraction process and the purified nucleic acid was eluted in 50 µL of buffer.

qRT-PCR was used for the detection and quantification of Bungowannah virus RNA. Five microlitres of nucleic acid was amplified using the AgPath-ID™ One-Step RT-PCR kit (Applied Biosystems, Foster City, CA, USA, Cat No. 4387424) with primers and probe as previously published [28]. The reaction was run in an Applied Biosystems 7500 Fast or 7900HT Real-Time PCR System (Foster City, CA, USA) both run in standard mode. A 10-fold dilution series of Bungowannah virus RNA standards ranging from 10^7–10^2 RNA copies/5 µL [25] was included in each assay and the quantity of Bungowannah virus RNA in a sample determined from the standard curve. For quantification purposes 2.3 and 2.6 $\log_{10}$ copies/mL were considered the sensitivity of the assay for serum and swabs/tissues respectively although the accuracy of quantification was considered likely to be variable below 4.3 or 4.6 $\log_{10}$ copies/mL. Additionally, means were calculated against a baseline of 2.3 or 2.6 $\log_{10}$ copies/mL for negative samples depending on the sample type.

2.8. Serology

A peroxidase-linked immunoassay was used for the detection of Bungowannah virus antibodies [28]. Where an animal recorded an antibody titre ≥5120, a titre of 5120 (12.3 $\log_2$) was used for calculation of means; where no antibody was detected, a titre of 8 (3 $\log_2$) was assigned for calculation of means.

2.9. Analysis of Results

To facilitate analysis of data, Bungowannah qRT-PCR results for serum, oropharyngeal and rectal swabs for each challenge time-point were grouped into time intervals based on the age of the pigs (0–1, 2–9, 10–17, 18–24, 25–31, 32–38, 39–45, 46–52, 53–60, 61–70 and 71–80 days). The same process was applied for serology results although the first time intervals were 0 and 1–9 days. Pigs that where known to have suckled prior to Day 0 sampling were classified as having been sampled on day 1 for the purpose of this study. Where an animal had been sampled twice within one of these intervals, the mean result for that animal was used when calculating the time-point mean. In addition, the results for animals in the D35 and D55 challenge groups were subdivided based on the presence or absence of antibodies to Bungowannah virus in samples collected prior to suckling.

3. Results

3.1. Longitudinal Study

3.1.1. Pregnant Pigs

All pregnant pigs became infected with Bungowannah virus (as confirmed by seroconversion), with transplacental infection detected for 20 of the 23 (87%) litters. Transplacental infection did not occur for one pregnant pig in each of the D35, D55 and D75 challenge groups [24]. One pig from D55 was found to be not pregnant. For each sow that produced an infected litter, virus was detected in high quantities on vaginal swabs (mean 7.2 $\log_{10}$ copies/swab; range 5.8–8.2 $\log_{10}$ copies/swab) on the day of farrowing, decreasing rapidly by 5–7 days post-farrowing (mean 3.3 $\log_{10}$ copies/swab; range <2.6–5.4 $\log_{10}$ copies/swab) (Figure 1). By day 12–14 post-farrowing the mean amount of virus detected was 2.9 $\log_{10}$ copies/swab (range <2.6–4.3 $\log_{10}$ copies/swab). Viral RNA was not detected on any vaginal swabs collected after 23 days post-farrowing (n = 15). The mean quantity of Bungowannah virus RNA detected on placental swabs at day 0 (7.0 $\log_{10}$ copies/swab; range 5.7–8.1 $\log_{10}$ copies/swab) from the infected litters was comparable to that detected on vaginal swabs. Bungowannah virus RNA was not detected on vaginal or placental swabs collected from sows farrowing uninfected litters.

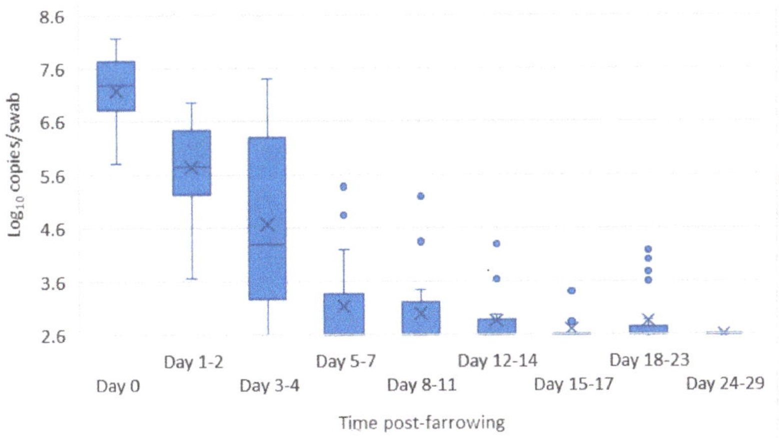

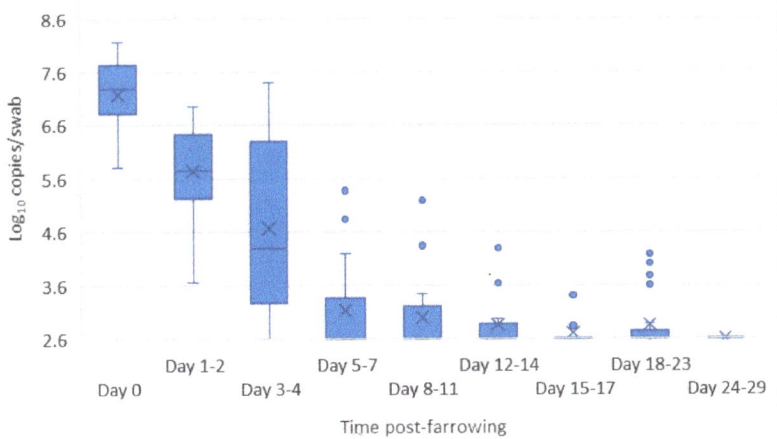

Figure 1. Box and whisker plot summarising the quantity of Bungowannah virus detected in vaginal swabs between 0–29 days post-farrowing following in utero infection (x = mean; — = median; the whiskers extend up from the top of the box to the largest data element that is less than or equal to 1.5 times the interquartile range (IQR) and down from the bottom of the box to the smallest data element that is larger than 1.5 times the IQR. Values outside this range are considered outliers and are represented by dots).

3.1.2. Piglets—qRT-PCR

Bungowannah virus RNA was detected in the serum, body cavity fluid or internal tissues of 225/226 pigs in the 20 infected litters at birth regardless of the stage of gestation that the dam had been infected. It was concluded that one pig (11-15) had not become infected in utero based on the absence of viraemia at birth and because it became seronegative at 9 weeks old indicating the absence of a humoral immune response to infection. This pig was born after it was believed farrowing had concluded and had an antibody titre of ≥10,240 when sampled the next day, presumably due to ingestion of colostrum. No viraemia was detected during the 10 weeks it was followed in the study.

All pigs from three litters (n = 42) were uninfected at birth (based on the absence of viraemia and seronegative serology at birth). The pigs in these three litters all developed high antibody titres after ingestion of colostrum. They did not become infected with Bungowannah virus (based on the absence of a viraemia) despite being in the same room as pigs from infected litters where no direct contact was possible. Occasionally low quantities of Bungowannah virus were detected on oropharyngeal (n = 4; mean 3.5 $\log_{10}$ copies/swab; range 3.3–3.9 $\log_{10}$ copies/swab) and rectal swabs (n = 7; mean 3.2 $\log_{10}$ copies/swab; range 2.7–4.9 $\log_{10}$ copies/swab), presumably due to environmental contamination as a result of sharing a room with an infected litter. Viral RNA was never detected on oropharyngeal swabs from the animal in which 4.9 $\log_{10}$ copies were detected on the rectal swab, or on tissue samples at necropsy. Additionally, all tissue samples collected from these pigs were negative for Bungowannah virus RNA.

Results for the D35 and D55 groups have been subdivided based on whether Bungowannah virus antibodies were detected at birth (Ab +ve) or not (Ab −ve). Samples were collected from all weaned/surviving pigs in the D35, D75 and D90 groups until 75, 56 and 35 days of age respectively. In some cases on the day of euthanasia a direct tonsillar swab was collected rather than an oropharyngeal swab. This result is captured under the results for tissues. In the D55 group, samples were collected through until 120 days in the Ab +ve group (n = 4) and to 329 days in the Ab −ve group (n = 2; 8-01 and 8-05). The number of animals from which serum or body fluid was collected at each time-point for Bungowannah virus PCR is summarised in Table S1.

- Serum: the mean quantity of Bungowannah virus RNA detected in serum at birth exceeded 6.4 $\log_{10}$ copies/mL for all groups of pigs, with the highest levels recorded in those animals where no antibody was detected at birth and where the sow had been infected prior to foetal immunocompetence (D35 (Ab −ve) 7.8 $\log_{10}$ copies/mL; D55 (Ab −ve) 8.4 $\log_{10}$ copies/mL) (Figure 2). For those animals seropositive at birth, the infection gradually cleared over time, although low-level viraemias were still detected at 75 days of age in some animals. At around 28 days of age when maternal antibodies started to wane (Figure 3) the viral load in serum started to rise for the pigs born seronegative in the D35 and D55 groups (Figure 2). By day 120 viral RNA could no longer be detected in the serum of the four remaining D55 (Ab +ve) pigs (data not shown).
- Oropharyngeal swabs: the mean quantity of viral RNA detected on oropharyngeal swabs at birth ranged from 6.5 (D90) to 7.2 $\log_{10}$ copies/swab (D55 Ab −ve & Ab +ve) (Figure 4). Through to 75 days of age the amount of RNA detected on oropharyngeal swabs remained elevated for the D35 (Ab −ve) and D55 (Ab −ve) groups (Figure 4). In contrast, for all other groups, the amount of RNA decreased over time, reducing most rapidly for the D75 and D90 animals. By day 120 viral RNA could no longer be detected on oropharyngeal swabs of the three D55 (Ab +ve) pigs sampled (data not shown).
- Rectal swabs: the mean level of viral RNA detected on rectal swabs at birth ranged from 6.3 (D90) to 7.1 $\log_{10}$ copies/swab (D55 Ab −ve) (Figure 5). As observed for oropharyngeal swabs, the amount of RNA detected on rectal swabs remained elevated for the D35 (Ab −ve) and D55 (Ab −ve) animals through to day 75. In contrast, for all other groups where the pigs were seropositive at birth, the amount of RNA decreased over time, reducing most rapidly for the D75 and D90 animals. By day 120 viral RNA could not be detected on rectal swabs from the four remaining D55 (Ab +ve) pigs (data not shown). Three animals from the D55 (Ab −ve) group were followed for >6 months (8-01, 8-05 and 10-01). Pig 10-01 appeared to seroconvert between 42 to 49 days old, with a marked reduction in viral load in serum and on rectal swabs from around 84 days of age, and oropharyngeal swabs on day 98 (Figure 6A). Virus was not detected in the serum of 10-01 at time of euthanasia on day 189, but low levels of viral RNA were still detected on oropharyngeal and rectal swabs. Pigs 8-01 and 8-05 seroconverted between 127 and 190 days of age (Figure 6B,C) after which the viraemia was cleared. The amount of virus detected on oropharyngeal and rectal swabs decreased more rapidly after seroconversion for 8-01 compared with 8-05 (Figure 6B,C).

- Mummified foetuses: the mean quantity of viral RNA detected on swabs of the internal organs of mummified foetuses was 5.6 log_{10} copies/swab for D35 (n = 13; range 4.9–6.9 log_{10} copies/swab), 5.6 log_{10} copies/swab for D75 (n = 2; range 5.5–5.7 log_{10} copies/swab) and 3.0 log_{10} copies/swab for D90 (n = 1).
- Tissues: at birth and in the first 10 days of life, virus was readily detected in all tissues collected from infected pigs in groups D35, D55 and D75 (Table 1). For group D90, virus was not always detected in tissues sampled at birth but was most likely to be detected in tonsillar swabs and from lymph nodes and heart. For all groups, in the first 10 days of life, the greatest quantity of viral RNA was detected from the tonsils (range of means 5.8 to 7.2 log_{10} copies/swab) and the brain (range of means 4.8 to 7.0 log_{10} copies/swab). Subsequently, the amount of Bungowannah viral RNA detected in tissues decreased over time and the proportion of tissues in which viral RNA was detected decreased in those groups where the pigs were seropositive at birth. This decrease occurred most rapidly for the D75 and D90 animals (Table 1). There were only two D35 (Ab +ve) animals identified and by the end of the study (day 75) Bungowannah viral RNA was either no longer detected in tissues or was at low levels, with the highest levels in lymph nodes followed by tonsil (Table 1). In contrast, for D35 (Ab −ve) virus was detected at all time-points for all animals and remained elevated in all tissues with highest quantities detected from tonsil, and in lymph nodes and brain. For the D55 (Ab −ve) group the amount of RNA detected in tissues was generally higher for the first two time-points compared with D55 (Ab +ve) although the number of animals sampled was low. An extensive range of tissues was collected from the presumptively PI pigs (8-01 and 8-05) that were followed until 11 months of age (Table 2). The quantity of Bungowannah virus RNA detected in epididymal semen (9.8 log_{10} copies/mL) is the highest amount of viral RNA detected in any sample collected throughout the course of the study.
- Urine: urine was collected directly from the bladder of six D35 (Ab −ve) animals at necropsy at between 57–77 days of age in with a mean 6.2 log_{10} copies/mL (range 5.1–6.9) of viral RNA detected. Virus was also detected in the urine of four D55 (Ab +ve) pigs between 80–90 days of age (mean 4.4 log_{10} copies/mL) but by day 117 viral RNA was no longer detectable in three. In contrast, Bungowannah virus RNA was only detected in the urine of 2/12 animals from D90 in low quantities when euthanased between 19 and 22 days of age (2.7 log_{10} copies/mL).

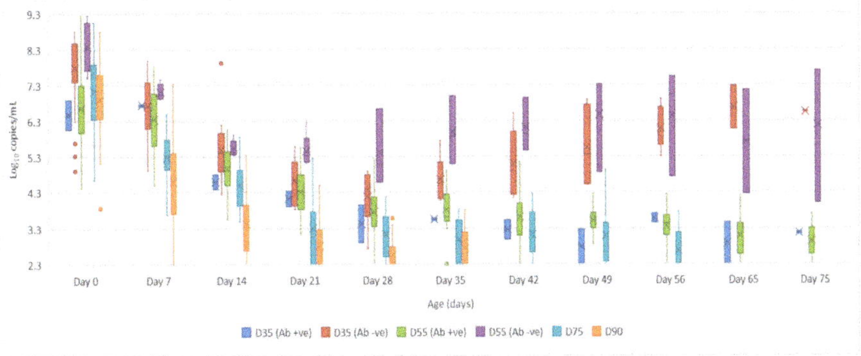

Figure 2. Box and whisker plot summarising the quantity of Bungowannah virus detected in serum between 0 to 75 days of age following in utero infection at one of four stages of gestation (x = mean; — = median; the whiskers extend up from the top of the box to the largest data element that is less than or equal to 1.5 times the interquartile range (IQR) and down from the bottom of the box to the smallest data element that is larger than 1.5 times the IQR. Values outside this range are considered outliers and are represented by dots).

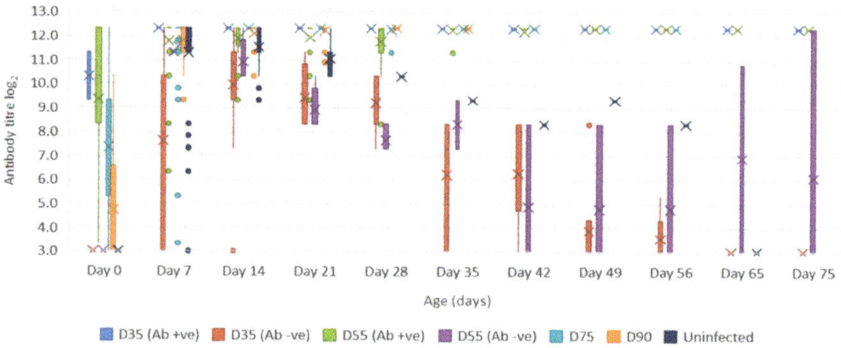

Figure 3. Box and whisker plot summarising the antibody titre against Bungowannah virus in serum as measured by peroxidase-linked immune assay between 0 to 75 days of age following in utero infection at one of four stages of gestation (x = mean; — = median; the whiskers extend up from the top of the box to the largest data element that is less than or equal to 1.5 times the interquartile range (IQR) and down from the bottom of the box to the smallest data element that is larger than 1.5 times the IQR. Values outside this range are considered outliers and are represented by dots).

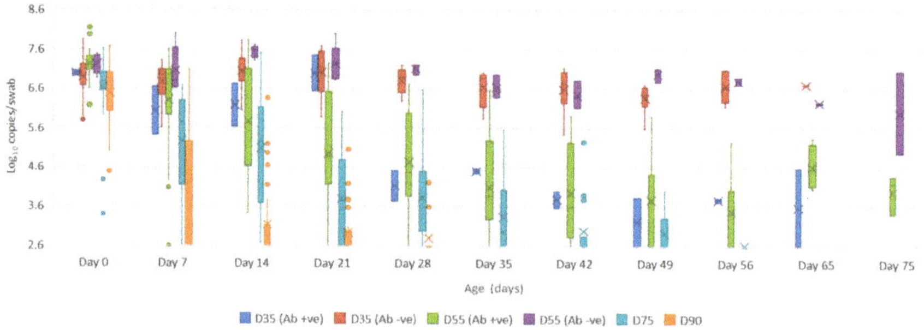

Figure 4. Box and whisker plot summarising the quantity of Bungowannah virus detected on oropharyngeal swabs between 0 to 75 days of age following in utero infection at one of four stages of gestation (x = mean; — = median; the whiskers extend up from the top of the box to the largest data element that is less than or equal to 1.5 times the interquartile range (IQR) and down from the bottom of the box to the smallest data element that is larger than 1.5 times the IQR. Values outside this range are considered outliers and are represented by dots).

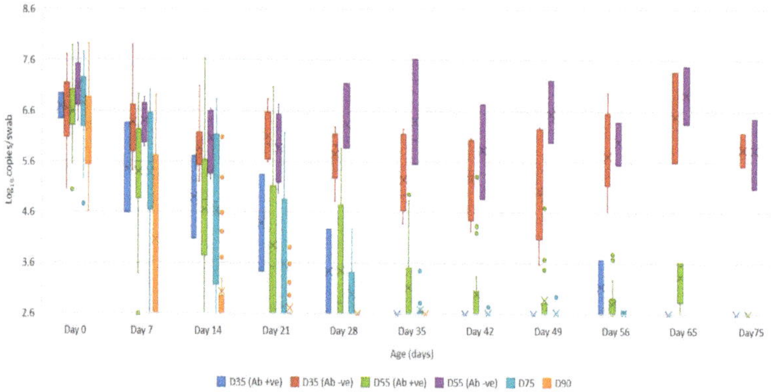

Figure 5. Box and whisker plot summarising the quantity of Bungowannah virus detected on rectal

swabs between 0 to 75 days of age following in utero infection at one of four stages of gestation (x = mean; — = median; the whiskers extend up from the top of the box to the largest data element that is less than or equal to 1.5 times the interquartile range (IQR) and down from the bottom of the box to the smallest data element that is larger than 1.5 times the IQR. Values outside this range are considered outliers and are represented by dots).

3.1.3. Piglets—Serology

The antibody titres of surviving pigs in each of the challenge groups between age 0 to 75 days are presented in Figure 3. Pigs that were known to have fed and were seropositive were excluded from the day 0 antibody titre calculations. The results can be broadly divided into four groups:

- Animals are detected with antibody prior to suckling (or are presumably in the process of seroconverting – D90). The percentage of seropositive (titres ≥10) pigs for each challenge group seropositive was 4% (D35), 82% (D55), 98% (D75) and 50% (D90). The highest antibody titres were detected in those litters infected earliest in gestation: D35 (mean = 1280; n = 2), D55 (mean = 654; n = 32), D75 (mean = 164; n = 45) and D90 (mean = 27; n = 50). After ingestion of maternal antibodies, the antibody titre rises and stays elevated for the remainder of the study (D35 (Ab +ve); D55 (Ab +ve); D75; D90).
- Animals are seronegative at birth; antibody levels increase with ingestion of maternal antibody and then gradually wane with the animal becoming seronegative again at around 40–60 days of age (D35 (Ab −ve)).
- Animals are seronegative at birth (D55 (Ab −ve); n = 7). These animals have the same serological profile as the D35 group after the ingestion of colostrum. At a variable time after losing maternal antibody these animals seroconvert (10-01 at between 42–49 days of age; 8-01 and 8-05 between 127 and 190 days of age; Figure 6). Note that the seroconversion of 10-01 has resulted in the large box plot for D55 (Ab −ve) from day 42 onwards, as the other two animals in the group (8-01 and 8-05) were seronegative at these time-points.
- Uninfected (seronegative) pigs born to an infected dam; antibody levels increase with ingestion of maternal antibody and then gradually wane with the animal becoming seronegative again at around 60–70 days of age.

3.2. Transmission Study

Bungowannah viral RNA was detected on oropharyngeal swabs collected from the introduced pigs at 24 h after their entry to the room/pen containing the presumptively PI animals. Peak viraemia and viral shedding were detected at 8 (serum) or 9 (oropharyngeal and rectal swabs) days post-introduction. Both introduced pigs seroconverted between days 8 and 15 with a titre of 5120 recorded at day 15. Figure 7 illustrates the differences in levels of viraemia and shedding observed between the PI and transiently infected pigs in this study. The differences in viral load detected on oropharyngeal swabs were >2 $\log_{10}$ copies per swab higher for the PI pigs compared to the transiently infected in-contact pigs at peak shedding on day 9. This difference was more than 10-fold higher for the rectal swabs. At the time of peak viraemia in the transiently infected pigs the viral load was equal to that of pig 10-01 but still >2 $\log_{10}$ copies/mL less than that detected in 8-01.

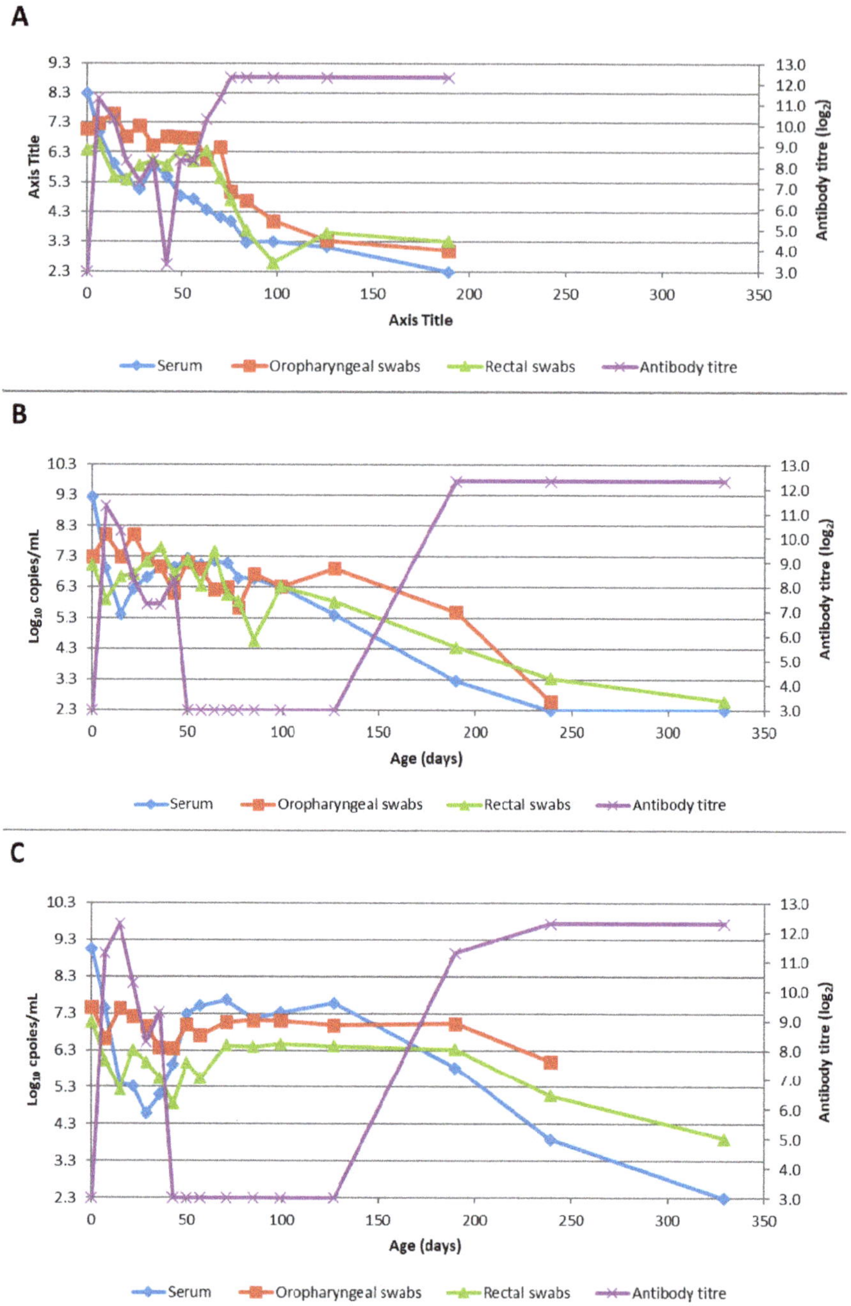

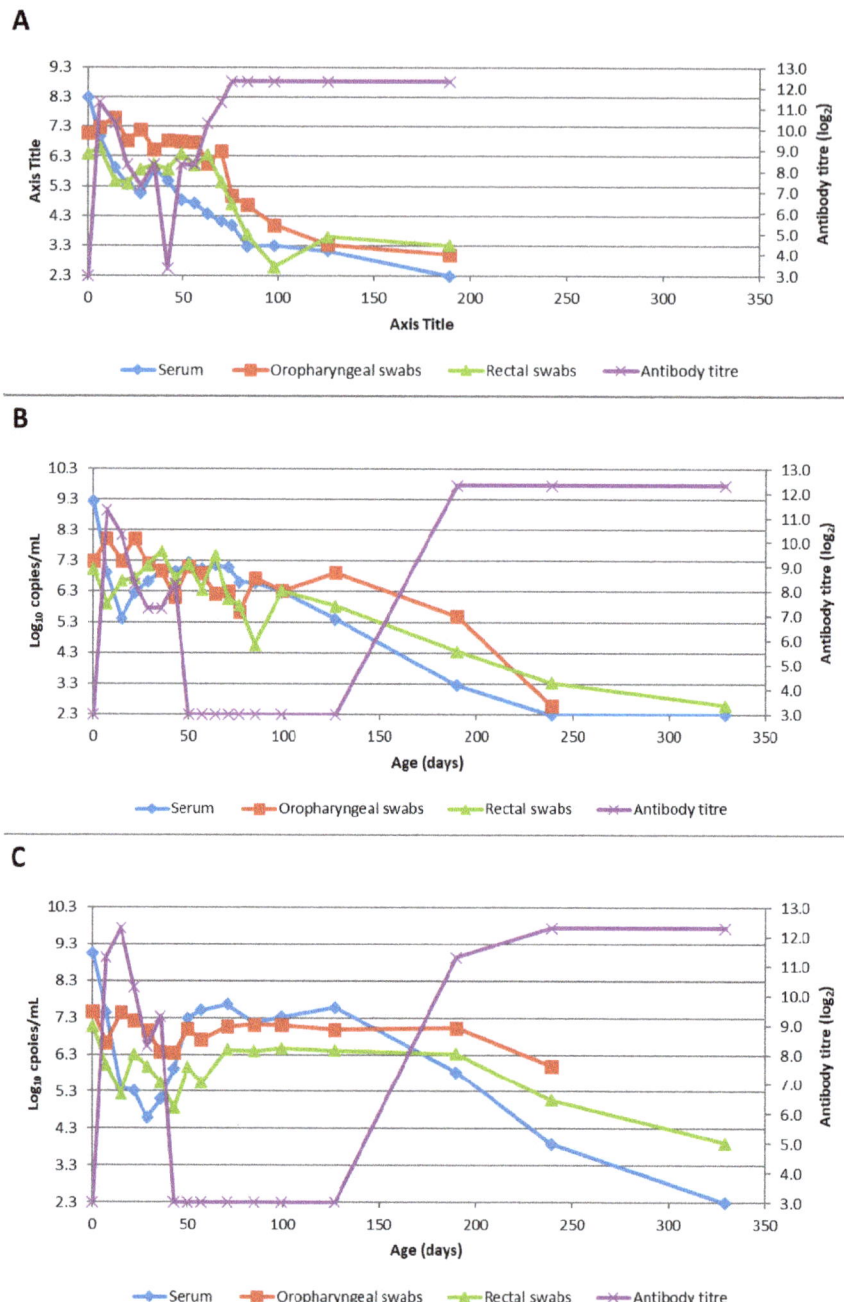

Figure 6. Quantity of Bungowannah virus detected in serum, oropharyngeal and rectal swabs, and antibody titre against Bungowannah virus detected in: (**A**) 10-01 from birth to 189 days of age; (**B**) 8-01 (D55 Ab −ve) from birth to 329 days of age; (**C**) 8-05 (D55 Ab −ve) from birth to 329 days of age.

Table 1. Quantity of Bungowannah virus RNA detected in selected tissues at four time periods after birth in challenge groups D35, D55, D75 and D90.

		D35 (Ab +ve)			D35 (Ab −ve)			D55 (Ab +ve)			D55 (Ab −ve)			D75			D90		
Tissue	Age	No. +ve (%) [a]	Mean [b]	Range [b]	No. +ve (%)	Mean	Range	No. +ve (%)	Mean	Range	No. +ve (%)	Mean	Range	No. +ve (%)	Mean	Range	No. +ve (%)	Mean	Range
Tonsil	0–10				28 (100%)	7.1	6.2–8.3	10 (100%)	7.2	6.3–7.8	2 (100%)	7.2	7.0–7.5	30 (100%)	6.7	5.0–7.8	24 (96%)	5.8	ND–7.7
	11–30				4 (100%)	6.7	5.2–7.6	5 (71%)	4.4	ND–7.1	2 (100%)	6.5	6.4–6.6	3 (75%)	4.0	ND–6.4	1 (5%)	2.6	ND–2.8
	31–60				5 (100%)	6.9	6.4–7.4							8 (44%)	2.9	ND–4.3	1 (8%)	2.6	ND–2.8
	>60	1 (50%)	3.3	ND–4.0	3 (100%)	7.0	6.5–7.4	10 (48%)	3.2	ND–6.8	0 (0%)	<2.6	NA						
Lymph Node	0–10				31 (100%)	5.3	4.4–6.3	10 (100%)	5.4	4.5–6.4	2 (100%)	4.5	3.9–5.2	34 (100%)	5.3	3.9–6.5	26 (100%)	4.7	3.2–6.4
	11–30				4 (100%)	4.5	3.4–5.7	7 (100%)	5.1	2.8–6.0	2 (100%)	5.5	5.0–6.1	4 (100%)	4.8	3.7–5.4	12 (55%)	3.6	ND–5.6
	31–60				5 (100%)	5.3	4.5–6.6							12 (71%)	3.9	ND–6.3	10 (77%)	3.9	ND–5.4
	>60	2 (100%)	4.2	4.0–4.5	3 (100%)	6.0	5.7–6.2	17 (77%)	4.5	ND–6.5	1 (100%)	6.5	NA						
Spleen	0–10				30 (100%)	5.9	4.5–7.3	10 (100%)	5.2	4.7–5.8	2 (100%)	4.9	4.7–5.0	34 (100%)	5.4	3.6–6.9	18 (69%)	3.6	ND–5.2
	11–30				4 (100%)	4.5	3.4–5.9	7 (100%)	3.8	2.9–4.7	2 (100%)	4.3	3.8–4.8	4 (100%)	3.5	2.8–4.6	1 (5%)	2.6	ND–2.8
	31–60				5 (100%)	4.6	3.8–5.8							9 (50%)	3.1	ND–4.4	0 (0%)	<2.6	NA
	>60	1 (50%)	2.8	ND–3.1	3 (100%)	5.8	5.3–6.4	14 (67%)	3.3	ND–4.9	1 (100%)	3.9	NA						
Thymus	0–10				29 (100%)	5.6	4.1–6.4	10 (100%)	5.2	4.3–5.8	2 (100%)	4.9	4.8–5.1	32 (100%)	5.4	3.3–6.8	19 (79%)	4.1	ND–5.7
	11–30				4 (100%)	4.3	3.0–5.1	7 (100%)	3.8	2.9–4.9	2 (100%)	4.5	4.5–4.6	1 (25%)	2.9	ND–3.7	2 (10%)	2.6	ND–3.2
	31–60				5 (100%)	4.6	4.0–5.4							7 (39%)	2.8	ND–3.9	1 (8%)	2.6	ND–3.0
	>60	0 (0%)	<2.6	ND	3 (100%)	5.2	4.7–5.5	9 (41%)	3.0	ND–4.5	1 (100%)	3.0	NA						
Lung	0–10				31 (100%)	6.4	5.1–7.6	10 (100%)	6.0	5.2–6.8	2 (100%)	6.2	6.2–6.2	33 (100%)	6.5	4.4–8.2	21 (81%)	4.2	ND–6.2
	11–30				4 (100%)	4.8	4.2–5.6	7 (100%)	3.8	2.7–5.5	2 (100%)	4.7	4.5–4.9	4 (100%)	3.9	3.6–4.2	1 (5%)	2.6	ND–2.9
	31–60				5 (100%)	4.8	4.3–5.5							11 (61%)	3.4	ND–6.8	2 (15%)	2.7	ND–3.1
	>60	0 (0%)	<2.6	ND	3 (100%)	5.0	4.4–6.1	12 (57%)	3.2	ND–5.5	1 (100%)	4.1	NA						
Heart	0–10				31 (100%)	6.1	4.3–7.2	10 (100%)	6.0	4.9–7.0	2 (100%)	5.6	5.5–5.8	34 (100%)	6.0	4.2–7.9	26 (100%)	5.4	3.9–6.9
	11–30				4 (100%)	4.3	3.7–5.2	7 (100%)	3.7	2.6–4.8	2 (100%)	4.2	4.2–4.2	4 (100%)	3.4	2.9–4.2	8 (36%)	2.9	ND–4.1
	31–60				5 (100%)	4.2	3.4–5.2							4 (22%)	2.8	ND–4.3	4 (31%)	2.7	ND–3.1
	>60	0 (0%)	<2.6	ND	3 (100%)	4.3	3.8–5.1	7 (32%)	2.9	ND–4.9	0 (0%)	<2.6	NA						
Intestine	0–10				31 (100%)	5.4	3.4–7.4	10 (100%)	4.3	3.3–5.7	2 (100%)	4.6	4.4–4.8	34 (100%)	5.0	2.6–6.8	17 (68%)	3.7	ND–5.5
	11–30				4 (80%)	4.4	ND–5.4	3 (43%)	3.0	ND–4.5	1 (50%)	3.8	2.6–5.1	3 (75%)	3.5	ND–4.3	2 (10%)	2.7	ND–3.8
	31–60				5 (100%)	6.2	4.2–7.1							3 (17%)	2.7	ND–3.7	4 (31%)	2.8	ND–3.3
	>60	0 (0%)	<2.6	ND	3 (100%)	5.7	5.2–6.1	10 (45%)	3.1	ND–6.2	1 (100%)	4.5	NA						
Brain	0–10				32 (100%)	7.0	6.1–7.8	10 (100%)	6.5	5.4–7.7	2 (100%)	6.8	6.5–7.2	30 (100%)	6.3	5.2–7.4	23 (92%)	4.8	ND–6.2
	11–30				4 (100%)	6.1	5.8–6.5	7 (100%)	5.3	4.4–6.1	2 (100%)	5.9	5.8–6.0	4 (100%)	4.3	3.8–5.0	4 (18%)	2.9	ND–4.8
	31–60				5 (100%)	6.0	5.5–6.5							14 (78%)	3.5	ND–4.5	8 (62%)	3.1	ND–3.9
	>60	1 (50%)	3.1	ND–3.7	3 (100%)	6.4	5.8–7.2	17 (77%)	4.0	ND–5.9	1 (100%)	3.3	NA						

[a] Percentage positive of total animals sampled at that time–point; [b] Log_{10} copies/swab; ND = Not detected.

Table 2. Quantity of Bungowannah virus RNA detected in selected tissues and fluids from pigs 8-01 and 8-05 (D55 Ab −ve) at 11 months of age.

Animal ID	Sample	Viral Load [a]	Animal ID	Sample	Viral Load [a]
08-01	Tonsil	4.74	08-05	Tonsil	7.46
	Lymph node	4.82		Lymph node	7.31
	Spleen	4.36		Spleen	4.93
	Thymus	ND		Thymus	ND
	Brain	ND		Brain	4.18
	Heart	4.25		Heart	3.78
	Lung	ND		Lung	ND
	Intestine	4.78		Intestine	6.34
	Urine	6.9		Urine	5.8
	Kidney	ND		Kidney	3.7
	Epididymal semen	9.8		Ovarian follicular fluid	ND
	Seminal fluid	3.4		Cervix	ND
	Bulbourethral gland	5.8		Ovary	2.8
	Epididymis (head)	6.2		Uterus	4.3
	Epididymis (tail)	8.0		Vagina	5.2
	Prostate	4.1			
	Seminal vesicle	ND			
	Testis	6.3			

[a] Log_{10} copies/mL or swab; ND = Not detected.

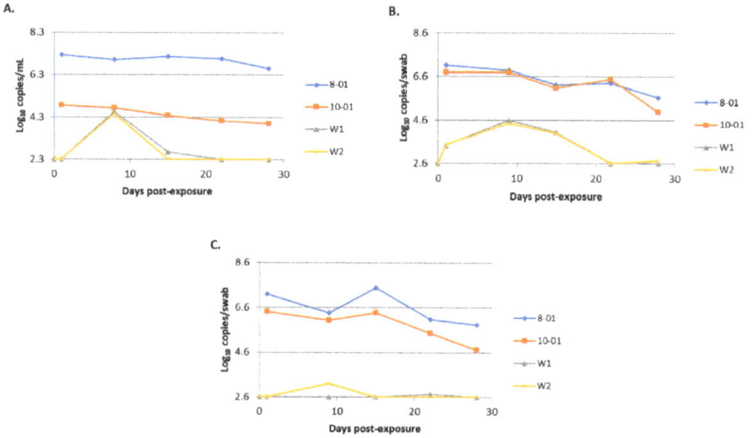

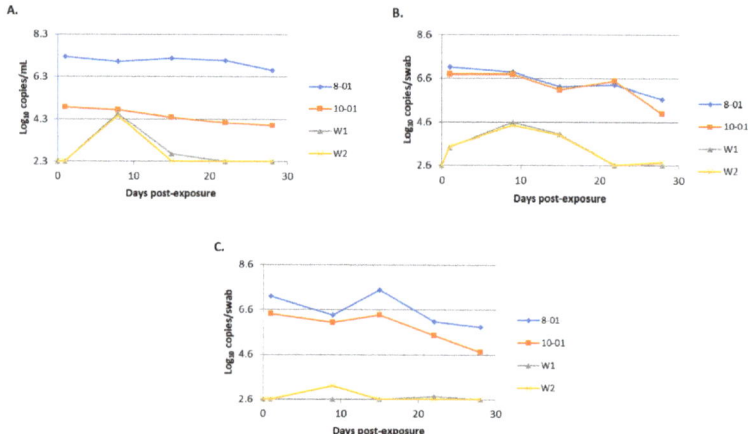

Figure 7. Comparison of the quantity of virus detected in serum (**A**), oropharyngeal swabs (**B**) and rectal swabs (**C**) for persistently infected (PI) animals (8-01 and 10-01) and transiently infected in-contact animals (W1 and W2).

4. Discussion

This study demonstrates that the porcine foetus can become infected with Bungowannah virus following intranasal challenge of the dam and characterises the virological and serological responses following in utero infection of the porcine foetus at different stages of gestation. Persistent infections, as described for other pestiviruses, were observed, as was a chronic infection state where animals that had been presumed to be PI animals later seroconverted and cleared the viral infection. Transmission of Bungowannah virus infection from chronically infected pigs to naïve pigs was readily achieved, with evidence of virus transfer within 24 h.

Preliminary investigations to identify the causative agent of the porcine myocarditis syndrome utilised a direct foetal inoculation route [25]. In the current study, the goal was to determine the outcomes using a natural route of infection at different stages of gestation. Transplacental infection was successfully achieved following intra-nasal exposure of the dam in 87% (20/23) of the litters. In addition, this outcome was not affected by the stage of gestation of the sow at the time of challenge.

Regardless of the stage of gestation that infection occurred, Bungowannah virus was detected in the serum, body fluid and excretions of infected pigs at birth and this was unrelated to the presence of precolostral Bungowannah virus-specific antibody in these animals. Generally, there was minimal difference in virus shedding between the pigs in the different challenge groups at birth, although the mean and median were lowest for the D90 group. The highest viral loads in serum or body fluid were recorded in the D35 (Ab −ve) and D55 (Ab −ve) groups. Over the course of the study it was noted that viral shedding reduced more quickly the later in gestation that the dam was challenged for those pigs which were seropositive at birth or in the D90 group. For those pigs seronegative at birth (D35 and D55 groups), viral shedding remained elevated throughout the course of the study. The exception was those animals that seroconverted after 6–7 weeks (10-01) and between 18 and 28 weeks (8-01 and 8-05) after which viral shedding gradually decreased, although the reduction varied between pigs. In contrast, the viral load in serum declined for all challenge groups after the ingestion of colostrum, but, as maternal antibodies waned in those animals that were seronegative at birth (D35 and D55), the viral load in serum increased as has been described for CSFV [13–15]. Despite animals 8-01, 8-05 and 10-01 seroconverting, it still took approximately 3 months for the viraemia to clear after this event.

It is interesting to compare our results with experimental studies of classical swine fever virus in the porcine foetus which were conducted 40–50 years ago with the less sensitive technologies of virus isolation and antigen detection by immunofluorescence [13,14,21,29]. While test sensitivity may affect the direct comparison of the proportion of animals infected with CSFV at birth compared to Bungowannah virus, the findings are similar. Virus could be detected in a proportion of all pigs born following infection with CSFV both pre- and post-immunocompetence as we have observed. Van Oirschot [13] also described congenitally infected pigs that cleared their CSFV infection by 2 weeks of age. The proportion of pigs that recovered increased the later in gestation that infection took place and was also associated with a low viral load in serum at birth. While precolostral CSFV neutralising antibodies were only detected in one animal (sow infected at 90 days of gestation), these virology findings are similar to those observed for congenital Bungowannah virus infections.

The pig foetus is able to mount a humoral immune response to Bungowannah virus following in utero infection. The highest antibody titres at birth were observed in those pigs that seroconverted and were the progeny of sows challenged at D35 and D55. Only one pig at D70 had no detectable antibody and it was stillborn. Antibody titres in the D90 group were generally low at birth presumably due to the short interval between infection and birth. However, the detection of viraemia indicates that all pigs in the D90 litters were infected prior to birth. Pigs infected post-natally generally seroconvert from 12–14 days post-infection [23]. Therefore, based on the D90 data, it is considered that transplacental and subsequent foetal infection had occurred within 8–10 days of intra-nasal exposure of the sow.

Unfortunately, farrowings were not monitored throughout the night and in some cases pigs were able to feed prior to precolostral blood samples being collected. While not optimal, we were still able to identify pigs in the D35 and D55 groups that did and did not mount a humoral immune response in utero and demonstrate that the time of gestation at which infection occurs has an influence on the ability of the porcine foetus to mount a humoral immune response to Bungowannah virus. The porcine foetus becomes competent to respond to infection with CSFV between days 70–90 of gestation and the serological findings of this study would suggest that the timing is similar, although likely closer to 70 days for Bungowannah virus [13,14,21]. The failure of some pigs in the D35 and D55 litters to seroconvert is indicative of immunotolerance, which occurs following infection prior to foetal immunocompetence and is recognised with other in utero pestivirus infections [5,13,14,21]. In the D55 group it was in Litters 8 and 10, that were challenged at 56 and 55 days of gestation, respectively, that the birth of litters with a mix of both seronegative and seropositive pigs resulted. A previous study [25], suggested that direct in utero transmission of Bungowannah virus to adjacent foetuses is probable and this has also been suggested for CSFV [22]. We speculate that the mixed outcomes of seronegative and seropositive pigs at birth in the D55 and to a lesser extent the D35 group are a consequence of the timing of foetal infection following transplacental transmission, and timing of any subsequent infections that result from direct in utero transmission.

While the current study did not measure neutralising antibody, the findings have similarities to those of Frey et al. [21] who did not detect neutralising antibody against CSFV in progeny at term when pregnant gilts were infected with CFSV at 65–67 days of gestation. In contrast, 3/4 (75%) of foetuses from gilts infected at 85 days of gestation had neutralising antibodies at term; additionally, antibodies were only detected in 2/6 (33%) of infected pigs when infection occurred after 94 days of gestation. As these foetuses were collected via hysterectomy at term it is not possible to determine if the failure to detect a neutralising antibody response is due to insufficient time between infection and sampling or the result of immunotolerance. Interestingly, Van Oirschot [13] with one exception did not detect precolostral antibodies in any pigs following challenge of the dam with CSFV at 40, 65 or 90 days gestation although the ability of some pigs (highest percentage at 90 days) to clear their infection by 2 weeks of age indicates they were not immunotolerant. Further comparison of antibody titres as measured by PLA and the virus neutralisation test may clarify the nature of the humoral immune response that develops in the period immediately following the development of immunocompetence compared with later in gestation.

The results also suggest that the cell-mediated immune response to Bungowannah virus may be suppressed if the foetus is infected before 90 days gestation or that the antibody response at this time is non-neutralising. In those foetuses from dams infected at 75 days of gestation or earlier that mounted an antibody response, even after ingestion of colostrum, the time taken for these animals to clear the infection was generally delayed compared to those from dams infected at 90 days gestation.

The identification of viraemic pigs that were seronegative following infection of the sow with Bungowannah virus at 35 or 55 days of gestation suggests that persistent infections may result. Persistent pestivirus infections occur following in utero infection of the foetus prior to it becoming immunocompetent and have been described following infection prior to day 70 to 90 of gestation for CSFV, bovine viral diarrhoea virus (BVDV) and Border disease-like viruses in the pig [13,14,30,31]. Despite two pigs in D35 not becoming PI, infection at this stage of gestation results in a high probability of occurrence. By day 55 of gestation the likelihood of maternal infection resulting in persistent infections is declining and may result in a mixed outcome for foetuses within that litter. Based on the results of this study and direct foetal inoculation [25], infection of the pregnant pig after approximately day 60 of gestation (approximately day 70 for foetal infection) seems unlikely to result in persistent infections.

At birth, PI animals could only be differentiated in the laboratory by their seronegative status. While the mean viral load in serum at birth was higher than for those animals born seropositive, individual animal variation was sufficient to prevent the titre of viraemia being a distinguishing feature. Furthermore, it was not until maternal antibodies began to wane from approximately 4–5 weeks of age in the PI pigs that the viral load in serum was higher than the non-PI animals. In contrast, the viral load on oropharyngeal and rectal swabs from PI animals remained elevated throughout the study period but the viral load on swabs for those animals seropositive at birth declined from 3 weeks of age. PI pigs from the D35 group grew poorly and died or were euthanased by 75 days of age [24]. While those from the D55 group were stunted [24], they had survived longer.

Three pigs, from the D55 group (8-01, 8-05 and 10-01) were monitored over a 6 to 11 month period to assess their virological and serological status over time and for any abnormal clinical signs [24]. Unexpectedly, these animals that appeared to be PI seroconverted at varying times during this observation period. This phenomenon has been described rarely in pigs infected with BVDV or border disease virus (BDV) [30,31]. Like these previous studies with BVDV and BDV, the Bungowannah virus viraemia ceased after seroconversion but virus could still be detected in many tissues for a further 5–6 months and at high levels for the female (8-05). For the male (8-01), the highest viral load was 9.8 $\log_{10}$ copies/mL in epididymal semen and was generally above 6.0 $\log_{10}$ copies/mL for other testicular samples (Table 2). Such high virus levels have also been observed in a PI bull [32] and in the pig 8-01 presumably remained high following seroconversion due to the immunologically privileged status of the testes. In contrast, for the female (8-05) the highest viral loads were detected in lymphoid tissues (tonsil and lymph node) (Table 2). It is also of interest to note the high quantity of Bungowannah virus detected in the seminal fluid of the male in the absence of spermatozoa [24]. This has also been described in a boar PI with BVDV [32]. The mechanism for the failure of virus to be cleared from tissues at the same time as the resolution of the viraemia is not clear but may be related to clearance from infected cells being dependent on cell-mediated immunity rather than just the presence of antibody and its inability to access immunologically privileged sites.

Chronic infections resulting from in utero infections with pestivirus have not been reported in ruminants with one exception of BDV in a sheep [17]. It seems probable that the mechanism for development of the chronic infections observed in pigs are related to differences in the time required for maturation of the immune system of ruminants and pigs given that the porcine foetus becomes immunocompetent at a proportionately later stage of gestation compared to ruminants.

While the two chronically infected pigs followed to 11 months of age were quite stunted and would not have been selected as breeding animals [24], they have the potential to be more important in the ongoing transmission of Bungowannah virus than the PI pigs that die early in life. To this end, we were able to show that Bungowannah virus is readily transmitted by these animals to naïve pigs in close proximity.

At least 100–1000 times more Bungowannah virus RNA was detected on oropharyngeal and faecal swabs from pigs at birth in this study compared with peak viral shedding of transiently infected animals [23]. The difference in quantity of virus shed by persistently/chronically infected animals in oropharyngeal swabs compared with peak shedding by transiently infected animals remained approximately 100 times higher until seroconversion was detected in the chronically infected animals. In addition, while the virus load in faeces is negligible in the transiently infected pig, faeces along with urine appear to be an important route of shedding in the PI pig (100–1000 times greater quantity), emphasising the role of PI animals in the epidemiology of pestivirus infections. Placental and vaginal secretions also appear to be an important source of virus for environmental contamination and ongoing transmission and has previously been recognised for BVDV [33]. High quantities of viral RNA could be detected in the placental and vaginal secretions of recently farrowed animals that carried infected foetuses. Although the quantity detected decreased rapidly over the first 5–7 days after farrowing, these reproductive materials could provide a significant source of virus in a population of pregnant animals to sustain transmission cycles even in the absence of a PI animal. This study indicates that all recently farrowed Bungowannah virus infected litters are also potentially adding large quantities of virus into the environment. Further studies are required to determine if these animals are shedding an infectious virus and, therefore, a source of virus for ongoing transmission, although given the quantities of Bungowannah virus RNA detected on oropharyngeal swabs we believe this is highly likely to be the case.

Finally, examination of the virus loads in tissue samples indicates that Bungowannah virus is widely disseminated in the porcine foetus at birth and provides an insight into sample selection for diagnostic purposes from live and dead animals. While the virus could usually be detected in a wide range of samples collected at birth, the optimal tissues to sample for virus detection were serum/body fluid, oropharyngeal swabs, tonsils and lungs. In older animals, virus can also be readily detected in a wide range of samples from PI animals, although serum, oropharyngeal swabs and lymph nodes are preferred for the greatest chance of detection of infection in animals greater than 10 days of age that may not be PI. The high quantity of virus detected in tonsillar samples suggests that the tonsils are the primary source of Bungowannah virus found in oropharyngeal secretions.

5. Conclusions

The results of this study provide a unique insight into the biology of Bungowannah virus infections in the porcine foetus and the subsequent pre- and post-weaning period following infection at different stages of gestation. While the birth of PI animals was expected based on our knowledge of pestivirus infections in other species, the birth of chronically infected pigs that later went on to seroconvert was unexpected. This study provides further information about how the porcine foetus responds to in utero pestivirus infections.

Supplementary Materials: Supplementary materials can be found at http://www.mdpi.com/1999-4915/12/6/691/s1. Summary of the number of animals from which serum or body fluid was collected at each time-point for Bungowannah virus PCR (Table S1).

Author Contributions: Conceptualization, D.S.F. and P.D.K.; methodology, D.S.F. and P.D.K.; investigation, D.S.F.; resources, D.S.F. and P.D.K.; data curation, D.S.F.; writing—original draft preparation, D.S.F.; writing—review and editing, D.S.F. and P.D.K.; visualization, D.S.F. and P.D.K.; supervision, P.D.K.; project administration, D.S.F. and P.D.K.; funding acquisition, P.D.K. All authors have read and agreed to the published version of the manuscript.

Funding: This research was funded by Australian Pork Limited (Project No. 2188), the Australian Biosecurity Cooperative Research Centre for Emerging Infectious Disease and NSW Department of Primary Industries.

Acknowledgments: We gratefully acknowledge the staff of the Virology Laboratory at the Elizabeth Macarthur Agricultural Institute (EMAI) for technical assistance (particularly Katherine King), veterinary assistance (Andrew Read), and assistance with sample collection, and both farm and laboratory staff, especially Glenda Macnamara, for their care of the animals used in this study.

Conflicts of Interest: The authors declare no conflict of interest. The funders had no role in the design of the study; in the collection, analyses, or interpretation of data; in the writing of the manuscript; or in the decision to publish the results.

References

1. McOrist, S.; Thornton, E.; Peake, A.; Walker, R.; Robson, S.; Finlaison, D.; Kirkland, P.; Reece, R.; Ross, A.; Walker, K.; et al. An infectious myocarditis syndrome affecting late-term and neonatal piglets. *Aust. Vet. J.* **2004**, *82*, 509–511. [CrossRef]
2. Kirkland, P.D.; Frost, M.J.; Finlaison, D.S.; King, K.R.; Ridpath, J.F.; Gu, X. Identification of a novel virus in pigs-Bungowannah virus: A possible new species of pestivirus. *Virus Res.* **2007**, *129*, 26–34. [CrossRef]
3. Hause, B.M.; Collin, E.A.; Peddireddi, L.; Yuan, F.; Chen, Z.; Hesse, R.A.; Gauger, P.C.; Clement, T.; Fang, Y.; Anderson, G. Discovery of a novel putative atypical porcine pestivirus in pigs in the USA. *J. Gen. Virol.* **2015**, *96*, 2994–2998. [CrossRef]
4. Lamp, B.; Schwarz, L.; Hogler, S.; Riedel, C.; Sinn, L.; Rebel-Bauder, B.; Weissenbock, H.; Ladinig, A.; Rumenapf, T. Novel pestivirus species in pigs, Austria, 2015. *Emerg. Infect. Dis.* **2017**, *23*, 1176–1179. [CrossRef]
5. Moennig, V.; Liess, B. Pathogenesis of intrauterine infections with bovine viral diarrhea virus. *Vet. Clin. North Am. Food Anim. Pract.* **1995**, *11*, 477–487. [CrossRef]
6. Roeder, P.L.; Jeffrey, M.; Cranwell, M.P. Pestivirus fetopathogenicity in cattle: Changing sequelae with fetal maturation. *Vet. Rec.* **1986**, *118*, 44–48. [CrossRef]
7. Baker, J.C. The Clinical Manifestations of Bovine Viral Diarrhea Infection. *Vet. Clin. N. Am.-Food A.* **1995**, *11*, 425–445. [CrossRef]
8. McGowan, M.R.; Kirkland, P.D. Early reproductive loss due to bovine pestivirus infection. *Br. Vet. J.* **1995**, *151*, 263–270. [CrossRef]
9. Garcia-Perez, A.L.; Minguijon, E.; Estevez, L.; Barandika, J.F.; Aduriz, G.; Juste, R.A.; Hurtado, A. Clinical and laboratorial findings in pregnant ewes and their progeny infected with Border disease virus (BDV-4 genotype). *Res. Vet. Sci.* **2009**, *86*, 345–352. [CrossRef] [PubMed]
10. Arruda, B.L.; Arruda, P.H.; Magstadt, D.R.; Schwartz, K.J.; Dohlman, T.; Schleining, J.A.; Patterson, A.R.; Visek, C.A.; Victoria, J.G. Identification of a divergent lineage porcine pestivirus in nursing piglets with congenital tremors and reproduction of disease following experimental inoculation. *PLoS ONE* **2016**, *11*, e0150104. [CrossRef]
11. Nettleton, P.F. Border disease. In *Infectious Diseases of Livestock*, 2nd ed.; Coetzer, J.A.W., Tustin, R.C., Eds.; Oxford University Press Southern Africa: Cape Town, South Africa, 2004; Volume 2, pp. 970–974.
12. Trautwein, G.; Hewicker, M.; Liess, B.; Orban, S.; Grunert, E. Studies on transplacental transmissibility of a bovine virus diarrhoea (BVD) vaccine virus in cattle III. Occurrence of central nervous system malformations in calves born from vaccinated cows. *J. Vet. Med. B.* **1986**, *33*, 260–268. [CrossRef] [PubMed]
13. Van Oirschot, J.T. Experimental production of congenital persistent swine fever infections: I. Clinical, pathological and virological observations. *Vet. Microbiol.* **1979**, *4*, 117–132. [CrossRef]

14. Meyer, H.; Liess, B.; Frey, H.R.; Hermanns, W.; Trautwein, G. Experimental transplacental transmission of hog cholera virus in pigs. IV. Virological and serological studies in newborn piglets. *Zentralbl Veterinarmed B* **1981**, *28*, 659–668. [CrossRef]
15. Van Oirschot, J.T.; Terpstra, C. A congenital persistent swine fever infection. I. Clinical and virological observations. *Vet. Microbiol.* **1977**, *2*, 121–132. [CrossRef]
16. Van Oirschot, J.T. A congenital persistent swine fever infection. II. Immune response to swine fever virus and unrelated antigens. *Vet. Microbiol.* **1977**, *2*, 133–142. [CrossRef]
17. Nettleton, P.F.; Gilmour, J.S.; Herring, J.A.; Sinclair, J.A. The production and survival of lambs persistently infected with border disease virus. *Comp. Immunol. Microbiol. Infect. Dis.* **1992**, *15*, 179–188. [CrossRef]
18. Barlow, R.M.; Vantsis, J.T.; Gardiner, A.C.; Rennie, J.C.; Herring, J.A.; Scott, F.M.M. Mechanisms of natural transmission of Border disease. *J. Comp. Pathol.* **1980**, *90*, 57–65. [CrossRef]
19. Plateau, E.; Vannier, P.; Tillon, J.P. Atypical hog cholera infection: Viral isolation and clinical study of in utero transmission. *Am. J. Vet. Res.* **1980**, *41*, 2012–2015.
20. Moennig, V.; Floegel-Niesmann, G.; Greiser-Wilke, I. Clinical signs and epidemiology of classical swine fever: A review of new knowledge. *Vet. J.* **2003**, *165*, 11–20. [CrossRef]
21. Frey, H.R.; Liess, B.; Richter-Reichhelm, H.B.; von Benten, K.; Trautwein, G. Experimental transplacental transmission of hog cholera virus in pigs. I. Virological and serological studies. *Zentralbl Veterinarmed B* **1980**, *27*, 154–164. [CrossRef] [PubMed]
22. Van Oirschot, J.T. Congenital infections with nonarbo togaviruses. *Vet. Microbiol.* **1983**, *8*, 321–361. [CrossRef]
23. Finlaison, D.S.; King, K.R.; Gabor, M.; Kirkland, P.D. An experimental study of Bungowannah virus infection in weaner aged pigs. *Vet. Microbiol.* **2012**, *160*, 245–250. [CrossRef]
24. Finlaison, D.S.; Kirkland, P.D. The outcome of porcine foetal infection with Bungowannah virus is dependent on the stage of gestation at which infection occurs. Part 2: Clinical signs and gross pathology. *Viruses* **2020**. submitted for publication.
25. Finlaison, D.S.; Cook, R.W.; Srivastava, M.; Frost, M.J.; King, K.R.; Kirkland, P.D. Experimental infections of the porcine foetus with Bungowannah virus, a novel pestivirus. *Vet Microbiol* **2010**, *144*, 32–40. [CrossRef] [PubMed]
26. Kim, J.; Choi, C.; Han, D.U.; Chae, C. Simultaneous detection of porcine circovirus type 2 and porcine parvovirus in pigs with PMWS by multiplex PCR. *Vet. Rec.* **2001**, *149*, 304–305. [CrossRef]
27. Hoffmann, B.; Depner, K.; Schirrmeier, H.; Beer, M. A universal heterologous internal control system for duplex real-time RT-PCR assays used in a detection system for pestiviruses. *J. Virol. Methods* **2006**, *136*, 200–209. [CrossRef]
28. Finlaison, D.S.; King, K.R.; Frost, M.J.; Kirkland, P.D. Field and laboratory evidence that Bungowannah virus, a recently recognised pestivirus, is the causative agent of the porcine myocarditis syndrome (PMC). *Vet. Microbiol.* **2009**, *136*, 259–265. [CrossRef]
29. Cowart, W.O.; Morehouse, L. Effects of attenuated hog cholera virus in pregnant swine at various stages of gestation. *J. Am. Vet. Med. Assoc.* **1967**, *151*, 1788–1794.
30. Paton, D.J.; Done, S.H. Congenital infection of pigs with ruminant-type pestiviruses. *J. Comp. Pathol.* **1994**, *111*, 151–163. [CrossRef]
31. Terpstra, C.; Wensvoort, G. A congenital persistent infection of bovine virus diarrhoea virus in pigs: Clinical, virological and immunological observations. *Vet. Q.* **1997**, *19*, 97–101. [CrossRef]
32. Kirkland, P.D.; Richards, S.G.; Rothwell, J.T.; Stanley, D.F. Replication of bovine viral diarrhoea virus in the bovine reproductive tract and excretion of virus in semen during acute and chronic infections. *Vet. Rec.* **1991**, *128*, 587–590. [CrossRef] [PubMed]
33. Kirkland, P.D.; McGowan, M.R.; Mackintosh, S.G.; Moyle, A. Insemination of cattle with semen from a bull transiently infected with pestivirus. *Vet. Rec.* **1997**, *140*, 124–127. [CrossRef] [PubMed]

viruses

MDPI

Article

The Outcome of Porcine Foetal Infection with Bungowannah Virus Is Dependent on the Stage of Gestation at Which Infection Occurs. Part 2: Clinical Signs and Gross Pathology

Deborah S. Finlaison * and Peter D. Kirkland

Virology Laboratory, New South Wales Department of Primary Industries, Elizabeth Macarthur Agricultural Institute, Menangle, NSW 2568, Australia; peter.kirkland@dpi.nsw.gov.au
* Correspondence: deborah.finlaison@dpi.nsw.gov.au

Received: 6 July 2020; Accepted: 3 August 2020; Published: 10 August 2020

Abstract: Bungowannah virus is a novel pestivirus identified from a disease outbreak in a piggery in Australia in June 2003. The aim of this study was to determine whether infection of pregnant pigs with Bungowannah virus induces the clinical signs and gross pathology observed during the initial outbreak and how this correlates with the time of infection. Twenty-four pregnant pigs were infected at one of four stages of gestation (approximately 35, 55, 75 or 90 days). The number of progeny born alive, stillborn or mummified, and signs of disease were recorded. Some surviving piglets were euthanased at weaning and others at ages up to 11 months. All piglets were subjected to a detailed necropsy. The greatest effects were observed following infection at 35 or 90 days of gestation. Infection at 35 days resulted in a significant reduction in the number of pigs born alive and an increased number of mummified foetuses (18%) and preweaning mortalities (70%). Preweaning losses were higher following infection at 90 days of gestation (29%) and were associated with sudden death and cardiorespiratory signs. Stunting occurred in chronically and persistently infected animals. This study reproduced the clinical signs and gross pathology of the porcine myocarditis syndrome and characterised the association between the time of infection and the clinical outcome.

Keywords: Bungowannah virus; foetus; pestivirus; porcine

1. Introduction

Bungowannah virus is a novel pestivirus identified from an outbreak of a disease in a piggery in New South Wales, Australia, in June 2003 [1]. It is genetically distinct from the other recognised pestiviruses of pigs, namely, classical swine fever virus (CSFV) and atypical porcine pestivirus (APPV) [2,3], with its closest genetic relationship to the recently identified Linda virus [4]. The disease was referred to as the porcine myocarditis syndrome, or PMC, because histological changes in the first affected animals consisted almost exclusively of multifocal, non-suppurative myocarditis, with myonecrosis observed in some cases. The outbreak initially presented as sudden death in 2–3-week-old weaning-age pigs, but soon after the onset, there was a marked increase in the birth of stillborn foetuses and a slight increase in the occurrence of mummified pigs [1]. Cumulative losses in some weeks exceeded 50% of pigs born, and it is estimated that as many as 50,000 pigs died in the initial outbreak. Due to the reproductive effects and disease becoming apparent almost exclusively in the first 2–3 weeks of life, it was presumed to be predominantly the consequence of in utero infection. This hypothesis was supported by the detection of elevated serum immunoglobulin G levels in up to 50% of stillborn pigs, and by the absence of disease in pigs after the immediate postweaning period or in sows farrowing affected litters [1].

The pestiviruses are well-recognised reproductive pathogens with the outcome of infection being dependent on a number of factors, including the stage of gestation that the infection occurs in relation to organogenesis and the development of immune competence [5–12]. CSFV has historically been the main pestivirus recognised to infect pigs (although infections with bovine viral diarrhoea virus (BVDV) and border disease virus (BDV) have been identified), and low virulence strains of CSFV may cause in utero infection and reproductive losses in the absence of disease in the rest of the herd [13–18].

Subsequent to the discovery of Bungowannah virus, two more pestiviral infections of pigs have been described: APPV [3] and Linda virus [4]. Both viruses have been associated with congenital tremors in piglets [4,10,19].

While there is strong epidemiological, virological [20] and preliminary experimental [21] evidence that Bungowannah virus is the causative agent of the PMC syndrome, the disease has not been reproduced experimentally. In an accompanying report [22], we described the virological and serological characteristics following the successful infection of the porcine foetus at four different stages of gestation. Regardless of the stage of gestation at which infection occurred, Bungowannah virus was detected in the serum/body fluid and excretions of infected pigs at birth and this was unrelated to the presence or absence of a precolostral Bungowannah virus-specific antibody in these animals. Persistent infections, as described for other pestiviruses, were observed following infection of the dam at 35 days of gestation, as was a chronic infection state where animals that had been presumed to be persistently infected (PI) following infection of the dam at 55 days later seroconverted. Viraemia, virus detection in tissues and viral shedding cleared more rapidly the later in gestation that infection occurred.

The aim of the current study was to determine whether infection of pregnant pigs with Bungowannah virus at different stages of gestation induces the clinical signs and gross pathology of PMC in the progeny that were observed during the initial outbreak. An in-depth description of the histopathological findings will be reported separately.

2. Materials and Methods

2.1. Study Design

Twenty-four pregnant pigs (22 gilts and two parity-1 sows) with known joining dates were obtained from a piggery known to be free of Bungowannah virus. Full details of the management of these pigs are described in a companion paper [22]. Pregnancy was confirmed using an ultrasound examination prior to selection for the study. The pigs were challenged intranasally at approximately 35 (34–36), 55 (55–58), 75 (72–76) or 90 (90–92) days of gestation (n = 6 per group—referred to as D35, D55, D75 and D90; Table 1). These time-points were selected as they were similar to those used in a previous study where foetuses were directly inoculated [21] and they span the gestational age at which the pig foetus is considered to become immunocompetent (70 days). All pregnancies were allowed to proceed to 113 days of gestation; then, all pigs were induced to farrow on day 114 to optimise the collection of blood from piglets prior to suckling.

Table 1. Summary of challenge details and reproductive outcomes following the inoculation of pregnant pigs with Bungowannah virus.

Group [1]	Sow Number (Litter ID)	Day of Gestation at Challenge	Day of Farrowing	Born Alive	Stillborn/AutolySing (%)	Mummified (%)	Total Litter Size	Preweaning Losses (%)	Number Weaned	Comments
D35	4558 (15)	34	112	8	1 (7%)	5 (36%)	14	8 (100%)	0	
D35	4536 (17)	34	114	7	6 (46%)	0 (0%)	13	2 (29%)	5	
D35	4540 (12)	35	111	7	3 (25%)	2 (17%)	12	7 (100%)	0	
D35	4596 (16)	35	114	4	4 (40%)	2 (20%)	10	3 (75%)	1	
D35	4570 (14)	36	114	9	0 (0%)	2 (18%)	11	4 (44%)	5	
Mean				**7.0**	**24%**	**18%**	**12**	**70%**	**2.2**	
D55	4261 (10)	55	114	11	1 (8%)	0 (0%)	12	2 (18%)	9	
D55	325 (11)	55	114	13	2 (13%)	0 (0%)	15	3 (23%)	10	Parity 1; pig 11-15 not infected
D55	4409 (8)	56	114	6	2 (25%)	0 (0%)	8	0 (0%)	6	
D55	4268 (7)	58	114	11	0 (0%)	0 (0%)	11	2 (18%)	6	3 piglets savaged
Mean				**10.3**	**12%**	**0%**	**11.5**	**15%**	**7.8**	
D75	4294 (3)	73	114	11	0	0	11	3 (27%)	8	
D75	4300 (4)	73	114	7	1	2	10	0 (0%)	7	
D75	4486 (5)	73	114	10	1	0	11	0 (0%)	0 [2]	10 piglets savaged/poor milk production
D75	4279 (1)	76	110	11	3	0	14	0 (0%)	0 [2]	11 piglets savaged
D75	4281 (2)	76	114	7	5	0	12	0 (0%)	7	
Mean				**9.2**	**16%**	**4%**	**11.6**	**5%**	**7.3**	
D90	739 (18)	90	111	6	3	1	10	1 (17%)	1	4 piglets savaged
D90	735 (22)	90	113	9	2	0	11	5 (56%)	4	
D90	738 (23)	90	114	10	1	0	11	0 (0%)	10	
D90	736 (20)	91	112	9	3	0	12	2 (22%)	7	
D90	2622 (21)	91	114	10	0	0	10	4 (40%)	6	
D90	2623 (19)	92	113	8	0	0	8	3 (38%)	5	
Mean				**8.7**	**14%**	**2%**	**10.3**	**29%**	**5.5**	
D35	35 (13)	36	114	13	3	1	17	1 (8%)	12	NI [3]; Parity 1
D55	4260 (9)	56	114	11	2	2	15	4 (38%)	6	NI; 1 death due to accident
D75	4292 (6)	72	114	10	0	0	10	1 (10%)	9	NI
Mean				**11.3**	**10%**	**6%**	**14.0**	**18%**	**9**	

[1] One sow from the D55 group did not farrow and has been excluded from the table. [2] Excluded from mean. [3] Litter not infected.

At birth, all pigs were weighed, their crown–rump lengths recorded, individually identified with ear tags and had their litter identity recorded. They were given an iron supplement intramuscularly between 2 and 4 days of age.

Farrowing outcomes, including the number of progeny born alive, stillborn or mummified, were recorded for each litter. Signs of disease observed throughout the study were documented.

Pigs from infected litters were weaned when they were 21–25 days old. As the presence of Bungowannah virus could still be detected at weaning in several challenge groups, as many pigs as the secure containment facilities could hold were weaned and retained until at least 8 weeks old. Selected pigs from D35 and D55 were kept in the study for a longer period to monitor clinical signs and virological and serological parameters [22]. Throughout the study, pigs that were moribund, not feeding or did not appear to be viable were euthanased.

Pigs from the three litters that did not become infected with Bungowannah virus (on the basis that the virus could not be detected in serum or body fluid for any of the pigs in the litter in a real-time reverse-transcription polymerase chain reaction (qRT-PCR) assay and negative precolostral serology results) were kept as control pigs until 14 to 21 days old, at which time they were euthanased. Because of the logistical considerations associated with the management of breeding sows in a secure containment facility, no mock-inoculated treatment group was included.

Individual pig IDs have been retained in the text where relevant to facilitate comparison between histopathology and virology and serology findings described in related manuscripts [22]. The first number relates to the litter ID and the second relates to the animal ID within the litter (generally in order of birth), e.g., 8-01 indicates the first pig to be born in litter 8.

All piglets were subjected to a detailed necropsy.

The animal studies were approved by the Animal Ethics Committee of the Elizabeth Macarthur Agricultural Institute, AEC Reference No. M09/02 (6 March 2009).

2.2. Inoculum

The pregnant pigs were challenged with a total of 5.3–5.8 $\log_{10}$ $TCID_{50}$ of Bungowannah virus in 5 mL of phosphate-buffered gelatin saline (2.5 mL per nostril), with the exception of litter 11, where the sow only received approximately 3.8 $\log_{10}$ $TCID_{50}$ of the virus. This inoculum was prepared as previously described [22,23].

2.3. Statistics

The confounding effect of treatment due to the separate cohorts was considered. The proportions of live, stillborn, mummified (of total born), weaned and lost/died pigs were analysed using a generalised linear model (GLM) as quasi-binomial (allowing for additional variation between sows). Each model included the time of challenge as a fixed effect. For the crown–rump length and birth weight data, a mixed model was fitted, with the time of challenge as a fixed effect, as well as random sow effects.

All calculations were performed with the R statistical software [24] using the glm function to fit the GLM model or the ASReml-R library [25] to fit the mixed models for length and weight.

3. Results

All 24 pigs challenged became infected. One pig from D55 failed to farrow and was not pregnant. As previously reported, the foetuses of 20/23 (87%) of the infected dams became infected with Bungowannah virus, as determined by the detection of Bungowannah virus nucleic acid using qRT-PCR in serum or body fluid at birth [22]. Within the 20 infected litters, 226 pigs (including stillborn and mummified foetuses) were born, of which, 225 had been infected in utero [22]. As all infected pigs were viraemic at birth, where individual animals are mentioned, reference is only made to their precolostral antibody level at birth. Three litters (n = 42 pigs) were uninfected at birth (as determined by negative qRT-PCR and precolostral serology results) and remained uninfected for the duration they remained in the study (14–21 days).

There were no significant differences in the crown–rump length ($p = 0.594$) and birth weight ($p = 0.180$) of pigs in the four treatment groups and uninfected litters.

3.1. Incidence of Stillbirths, Preweaning Deaths and Mummified Pigs

The day of gestation that the pregnant animals were challenged and their reproductive outcomes, including preweaning losses are summarised in Table 1. The greatest effects of infection were observed in the litters of dams infected at 35 or 90 days of gestation. While the use of cohorts confounded results, significant differences were associated with the time of challenge for the proportion born alive (of total) ($p = 0.029$) and proportion mummified ($p = 0.008$), where these values are lower and higher for the D35 group, respectively. For live-born pigs, there were significant differences between challenge groups in the proportion of preweaning losses ($p = 0.004$). The preweaning losses were significantly higher for D35 (there was also a numerically higher mean for D90 compared with D55 and D75; however, this was not statistically significant).

3.2. Clinical Signs

Table 2 summarises the reproductive effects, the likelihood of development of persistent infections and the most common clinical signs and necropsy findings observed in the progeny of sows following in utero infection with Bungowannah virus at different stages of gestation.

Table 2. Summary of the reproductive effects, and most common clinical signs and necropsy findings following in utero infection with Bungowannah virus.

Day of Gestation at Challenge	Reproductive Effects [1]	Persistent Infections? [2]	Clinical Signs		Necropsy Findings	
			Preweaning	Postweaning	Preweaning	Postweaning
35 days	Born alive ↓↓↓ Stillbirths ↑↑ Mummified foetuses ↑↑↑ Preweaning deaths ↑↑↑	Yes (>90%)	Purpura, subcutaneous oedema, absence of facial whiskers, weak-born pigs that did not suckle, neurological signs—abnormal behaviour	Stunting, pallor due to anaemia, conjunctivitis, subcutaneous facial skin mass, leukopaenia, thrombocytopaenia, purplish discolouration of the snout and ears, ecchymoses in the skin of the hind-quarter	Purpura, subcutaneous oedema, white foci in the cerebrum, orange/mottled liver, fibrinous adhesions in pleural and abdominal cavities (older animals)	Blood clots and serosanguinous fluid in abdomen/thoracic cavities, petechiae, red/haemorrhagic lymph nodes, ulcerated tonsils, facial mass
55 days	Stillbirths ↑	At birth, 18% were seronegative to Bungowannah virus and apparently persistently infected. The 3 seronegative animals retained postweaning later seroconverted	Melaena, pallor, ataxia	Stunting of apparently persistently infected pigs, occasional diarrhoea	Occasional fibrinous adhesions in thoracic and abdominal cavities	No significant findings
75 days	Stillbirths ↑	No	No disease observed	No disease observed	No significant findings	No significant findings
90 days	Stillbirths ↑ Preweaning deaths ↑↑	No	Sudden death, tachypnoea and dyspnoea	No disease observed	Cardiomegally, liver enlargement, fibrin tags in the thoracic and/or abdominal cavities, blood clots in abdominal cavity, pericardial/thoracic effusions, red/haemorrhagic mesenteric and/or inguinal lymph nodes	No significant findings

[1] ↑—mild increase, ↑↑—moderate increase, ↑↑↑—marked increase, ↓↓↓—marked decrease. [2] [22].

3.2.1. Group D35

The exposure of the pregnant animals at around 35 days of gestation resulted in a high percentage of infected pigs that were seronegative (96%) and presumably persistently infected at birth [22]. Clinically, a high proportion of stillborn (24%) and mummified pigs (18%) and a very high level of preweaning mortalities (70%) resulted (Table 1). The live-born pigs in these litters were often very weak, had difficulties moving from the rear of the sow after birth and some showed a limited ability to suckle. Many died soon after birth or were overlain, presumably as a consequence of their weakness or were euthanased due to their inability to feed. Additional abnormalities observed at birth in both stillborn and live-born pigs included purpura extending over the skin ($n = 15$) (Figure 1a,b) and/or subcutaneous oedema, particularly extending over the head, neck and ventral abdomen ($n = 9$) (Figure 1c). Pigs born with subcutaneous oedema had poor viability. All nine died soon after birth, were weak and were overlain or were euthanased on humane grounds within 3 days of birth. Mild subcutaneous oedema was noted in an additional two pigs at necropsy at 3 and 11 days of age. Of the four pigs with purpura that survived to be weaned, the purpura resolved with age and three of the four appeared to grow normally until weaning, although two developed conjunctivitis at approximately 3 weeks of age. They subsequently died or were euthanased when 26, 34, 35 and 57 days old. Several pigs exhibited neurological signs at birth, including walking into walls (14-04) and screaming, due to an inability to find the udder despite being able to suckle (most of litter 12). One pig (14-04) also developed haemorrhagic diarrhoea in the first week of life that resolved following antibiotic treatment. Three pigs developed diarrhoea in the first week of life, two of which were cross-fostered from their original mother due to milk supply issues, presumably due to poor udder stimulation from weak piglets and were ultimately euthanased (12-02, 12-05) by 2 weeks of age, and a third that had diarrhoea for 3–4 days that resolved following antibiotic treatment and then died suddenly 6 days later (16-02). One 2-week-old pig was euthanased as it failed to grow.

From the five infected litters, eleven pigs survived to weaning and were monitored until approximately 11 weeks old. Of these 11 pigs, nine were considered PI with Bungowannah virus based on ongoing viraemia and shedding of the virus throughout the study until death or euthanasia and negative precolostral serology at birth (14-01, 14-02, 14-04, 14-07, 14-09, 16-07, 17-01, 17-03, 17-04) [22]. The PI pigs grew poorly compared to their two cohorts (17-02, 17-06), which were not PI, based on their precolostral seropositive status at birth [22]. The PI pigs became stunted from soon after weaning (Figure 1d) and exhibited generalised skin pallor from 7 weeks of age. The first pig (17-03) was euthanased within 4 days of weaning after the sudden onset of recumbency; mouth breathing with a small amount of purulent nasal discharge; and purplish skin discolouration of the snout, ears, ventral abdomen and hind quarters (Figure 1e,f). Conjunctivitis was an ongoing problem in two pigs (17-01, 17-04). In addition, three pigs developed firm subcutaneous masses under their eyes between 5 and 9 weeks of age, which forced the lower eyelid dorsally, leaving them unable to open the affected eye (16-07, 17-01, 17-04) (Figure 1g); another developed an erosive lip lesion that resolved following antibiotic treatment at 6 weeks old (14-09). Investigation of the pallor diagnosed anaemia and only the PI pigs were affected. When sampled at 8 weeks old, the packed cell volume of the five affected animals ranged from 8% to 21% (reference interval for weaner pigs 26–41%; [26]). The low levels of circulating reticulocytes (0–0.4%) (corrected reticulocyte percentages <1%) and the normal mean corpuscular volume indicated that the anaemias were non-regenerative and normocytic. In addition, three animals were leukopaenic [26] and three were thrombocytopaenic. While one PI pig survived to 75 days of age (14-01), the remaining PI pigs that did not succumb to other problems were ultimately euthanased due to severe anaemia.

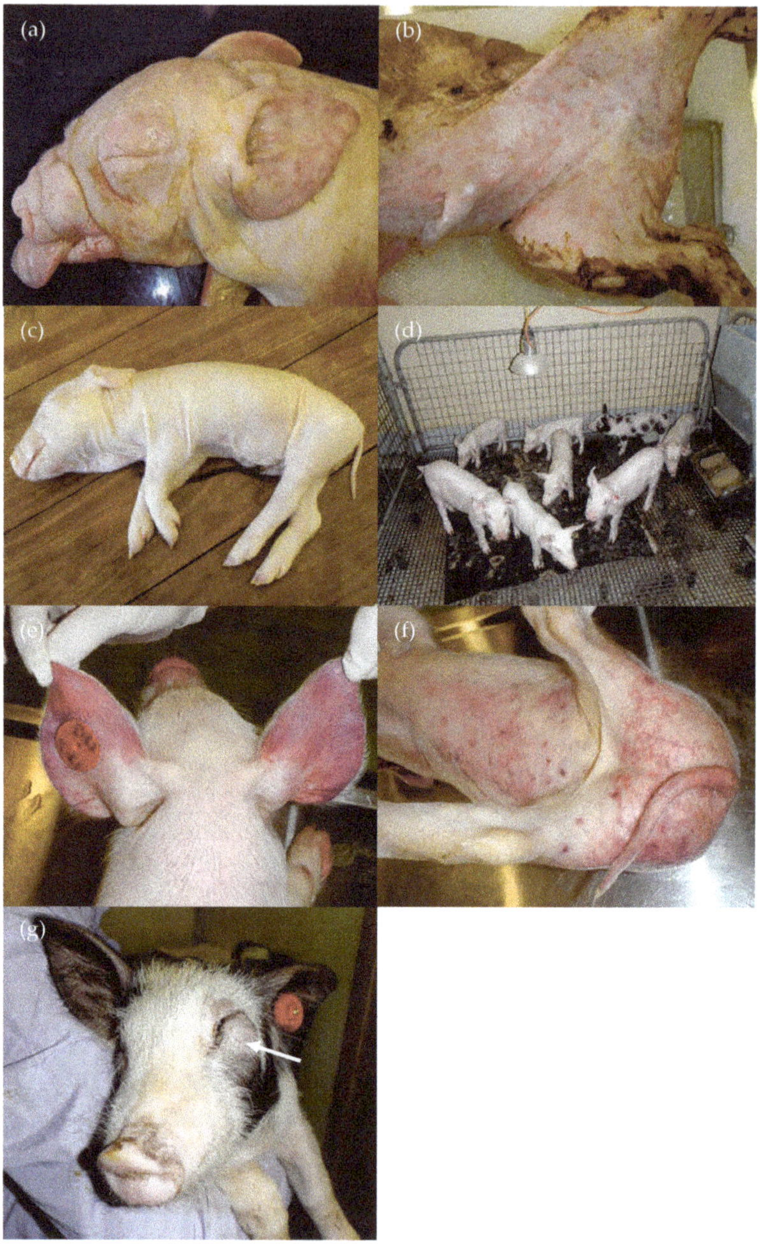

Figure 1. Clinical signs observed in D35 pigs: (**a**) purpura, subcutaneous oedema and absence of facial whiskers (12-09, stillborn); (**b**) purpura (17-11, stillborn); (**c**) subcutaneous oedema—most evident ventrally (12-10, 1 day old); (**d**) remaining pigs at 7 weeks of age with stunting of 6 smaller pigs compared with the 2 larger, non-PI pigs; (**e**) 26-day-old pig with purplish discolouration of the ears and snout (17-03); (**f**) same pig as (e) with purplish discolouration ventrally and on hind-quarters; and (**g**) 16-07 (62 days old) with mass under left eye, resulting in closure of the eyelids (arrow).

3.2.2. Group D55

There were no clear reproductive effects observed following infection at approximately 55 days gestation (Table 1), with 82% of the infected pigs being seropositive at birth. Two of the stillborn pigs that were seronegative [22] exhibited mild subcutaneous oedema, although the significance is unclear as they were also moderately autolysed. In the immediate postnatal period (first 2 days of life), deaths were attributed to savaging (n = 3), overlay (n = 3) and runts (n = 1). Commencing at about one week after birth, in two litters, four pigs had dark faeces on a rectal swab that was consistent with melaena, suggesting upper gastrointestinal tract bleeding (7-10, 7-11, 10-04, 10-12) and one pig passed bloody liquid at the time of sampling (10-10) (the precolostral antibody titres for 10-04 and 10-10 were 10 and 80 respectively; the three other pigs had suckled prior to the initial sampling). Melaena was detected again on a rectal swab at 14 days of age in one of these pigs (7-11). This pig exhibited pallor, grew poorly compared to its littermates and was euthanased when it failed to gain weight and was unable to compete for food. Another pig (10-12) also grew poorly after melaena was detected at 6 days of age and was euthanased rather than weaned. The third pig (10-04) with melaena grew normally until 13 days old but then grew little in the following week, and at 20 days old, it exhibited pallor and mild jaundice. It was weaned at 28 days, and at this time, it was still growing poorly and exhibiting mild hind limb ataxia. At 19 days old, one pig (7-06; precolostral antibody titre of 1280) was pyrexic (40.7 °C), mildly ataxic with a slight tremor and showed retarded growth compared to its littermates. Despite antibiotic treatment, it was still pyrexic (40.2 °C) 2 days later and continued to exhibit neurological signs, including poor balance with a wide base stance; there was no obvious head tilt or nystagmus. It was noticed that one pig from litter 8 (8-01; seronegative at birth) appeared weaker than the other pigs in its litter in the first couple of days of life and was sometimes slow to move to the teat to feed.

In litters 7 (n = 4), 8 (n = 3, including 8-01 and 8-05) and 10 (n = 5, including 10-01), the pigs with the highest viral loads in serum and oropharyngeal secretions at 2 weeks of age, as well as all pigs surviving to weaning age from litter 11 (n = 10), were weaned to clarify whether the ongoing viraemia was due to a persistent infection, given that the dams were infected prior to the expected timing of foetal immunocompetence [22]. All weaned pigs were raised until at least 10 weeks old. Disease was not observed during this period, with the exception of occasional diarrhoea, and at 6.5 and 8 weeks of age, 10-04 had two episodes of pyrexia and lethargy, which responded well to antibiotic treatment on both occasions. The most dramatic clinical difference was the variation in size of the littermates from litter 8 (Figure 2) and litter 10 when 11 weeks old. Across both of these litters, five pigs were live born and seronegative for Bungowannah virus antibody [22]. The two small pigs (Figure 2; 8-01 and 8-05) appeared to be PI on the basis of qRT-PCR results and the absence of antibodies against Bungowannah virus at birth and were markedly stunted compared to their non-PI littermates (seropositive at birth). When 6 to 7 months old, Bungowannah virus RNA could no longer be detected in the serum of these stunted animals; they concurrently developed high antibody titres and their growth rate improved markedly and they are described here as chronically rather than persistently infected [22].

One of the stunted pigs was reared until 6 months of age (10-01), and another two until 11 months of age (8-01 and 8-05). At approximately 3.5 months old, 8-01 and 10-01 experienced an episode of diarrhoea due to a *Salmonella* infection, but otherwise, no disease was observed. At 189 days of age, pig 10-01 weighed 105 kg; at 11 months of age, 8-01 and 8-05 weighed 183 and 153 kg respectively.

Figure 2. Littermates from litter 8 in D55 at 77 days old—note the stunting of two chronically infected littermates (8-01 and 8-05) compared with the larger non-persistently or chronically infected animal (8-03).

3.2.3. Group D75

Losses in the preweaning period were the result of savaging (n = 18), overlay (n = 2), splay leg (n = 1), runt (n = 1) and poor mothering/milk production (n = 2) and occurred in the first week of life (Table 1). Pigs surviving to weaning did not exhibit any signs of disease and 98% of the infected pigs were seropositive at birth [22].

After weaning, several pigs lost condition compared to others in the group and one developed a skin infection. Otherwise, no disease was observed. No postweaning mortalities occurred and 23 pigs in this group were reared up to 8 weeks of age.

3.2.4. Group D90

Preweaning losses (29%) in this group were higher compared to the D55 and D75 groups (Table 1) and were clinically different from the D35 group. Only 50% of the infected pigs were seropositive at birth, presumably because of the short interval between infection and birth [22]. While some losses due to savaging (n = 4) and overlay (n = 1) were recorded, some pigs also exhibited tachypnoea and dyspnoea prior to death or euthanasia. Sudden death was recorded for seven apparently healthy pigs between 2 and 5 days of age without any prior signs (19-03, 19-07, 20-04, 22-03, 22-05, 22-06 and 22-08).

Six pigs developed a notable increase in respiratory rate and effort (19-02, 20-06, 21-02, 21-03, 21-09 and 23-11). The earliest onset of respiratory signs was at 4 days and the latest at 7 days old. Three pigs were euthanased due to the severity of their clinical signs. Two pigs, one of which was thinner than its littermates, died 5 days after the detection of an increased respiratory rate. Finally, one pig continued to exhibit an increased respiratory rate when stressed and was thinner than its littermates through to weaning and subsequently to 5 weeks age.

Twenty pigs from this group were weaned and reared up to 5 weeks of age. No disease was evident in these animals.

3.2.5. Uninfected Litters

No disease was observed with the exception of diarrhoea in one pig. The only preweaning losses were due to overlay (n = 2), splay leg (n = 3), runt (n = 1) and accident (n = 1) (Table 1). While there was no statistical difference in birth weight between the infected and uninfected litters, the mean weight for the uninfected litters at approximately 7 and 14 days of age was 0.5–0.8 kg and 0.6–1.1 kg greater, respectively.

3.3. Necropsy Findings

Table 2 summarises the most common gross pathology findings observed following in utero infection with Bungowannah virus at different stages of gestation. The viral load in tissues at necropsy was dependent on the stage of gestation that infection occurred in, whether the pig was able to mount an immune response in utero and the age of the pig at necropsy [22].

3.3.1. Group D35

Subcutaneous oedema was observed in nine pigs and was most commonly observed ventrally and around the head and neck, giving rise to palpebral oedema and the impression of a Roman nose (Figure 1a,c). Petechiae were observed on the hearts of two stillborn pigs. Purpura (pigs 0 to 3 days old) were most easily identified on the ears, head, ventral abdomen and sometimes on the dorsal body (Figure 1a,b). Petechiae were observed on the pleural surface of the lungs of an overlain pig. White pin-point foci were observed throughout the brain of one stillborn pig (12-09) and a region of presumptive necrosis was observed in the left cerebral hemisphere of another (17-11). Other observations at necropsy in the preweaning period included haemorrhages on the tonsils (n = 2) and tongue (n = 1), mottled orange liver (n = 3) and the absence of facial whiskers (n = 1). Red mottled lungs were commonly observed and presumptively attributed to hypostasis. Thoracic and abdominal effusions in stillborn pigs were associated with autolysis. Fibrinous adhesions were observed in the thorax and or abdomen of two pigs that died at 12 and 15 days (both had previously been treated for diarrhoea).

A range of lesions was observed in the PI pigs that survived past weaning. These animals remained viraemic throughout their life, shed high levels of the virus and had high quantities of the virus in all tissues examined at their necropsies [22]. All became severely stunted, with weights ranging at time of euthanasia from 4.0 kg at 34 days to 13.2 kg at 77 days. Two animals (14-04, 14-09) showed evidence of a bleeding disorder, including the presence of clots in the abdominal cavity and serosanguinous fluid/effusion in the thoracic and/or abdominal cavities; blood streaks were observed in the stomach contents of one of these two pigs. Other changes that indicated a bleeding disorder or vasculitis in the PI pigs included red/haemorrhagic inguinal, mesenteric and/or lumbar lymph nodes in four animals; petechiae were occasionally observed on a number of organs, including the heart (n = 3), kidney (n = 2) (Figure 3a), liver (n = 1), serosal surface of the intestines/colon (n = 2), peritoneal surface (n = 1), cerebellum (n = 1) and oropharynx (n = 1). Other abnormalities observed included ulcerated tonsils (16-07) (Figure 3b), increased pericardial fluid (n = 4; up to 20 mL), cobblestone pattern to the liver (n = 1), mild pulmonary interlobular oedema (n = 4) and fibrin tags on intestinal serosa (n = 1). The facial mass observed in three pigs (Figure 1g) was firm with an intimate connection between the skin and the subcutis. There was no evidence of wounds in the skin or oral cavity connecting with the mass. Mild subcutaneous oedema was detected in the neck of one pig.

No abnormalities were detected in the two pigs that were not PI (17-02 and 17-06) [22]. At 75 days of age, their mean weight was 30.75 kg. Only low quantities of Bungowannah virus were detectable in these two animals at necropsy [22].

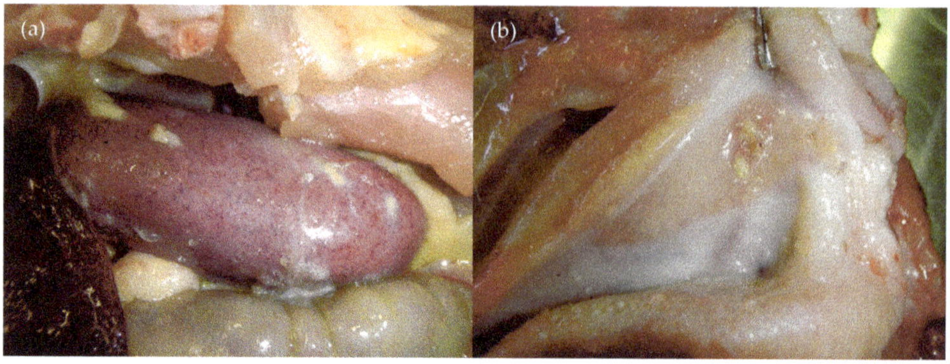

Figure 3. Necropsy findings from D35: (**a**) petechiation of a kidney (17-03—same pig as Figure 1e,f) and (**b**) ulcerated tonsil (16-07).

3.3.2. Group D55

Serosanginous pleural and abdominal fluid, as well as mild subcutaneous oedema, was noted in two stillborn pigs that exhibited autolytic changes. Pericardial petechiation was observed in one stillborn pig. Mild interlobular pulmonary oedema and congestion was observed in the lungs of six animals (ages 2–22 days) and was considered to be due to hypostasis. One animal had fibrinous pleural and abdominal adhesions, and another, fibrinous adhesions between the spleen and the liver; both suffered a preweaning death. Other lesions included a cobblestone pattern on the liver, kidney enlargement, oedematous spiral colon and dark blood clots in the colon.

The two chronically infected pigs (8-01 and 8-05) [22] were reared to 11 months of age. Few abnormalities were observed at necropsy. The testes of 8-01 were soft on palpation, fluid collected from the epididymis was of low opacity and sperm were not observed via a microscopic examination, suggesting infertility. The highest viral load detected throughout this study was in the epididymal semen of this animal (9.8 $\log_{10}$ copies/mL) [22]. Pig 8-05 was female and penned with 8-01 from 4 months old. From 8 months old, she was observed in oestrus approximately every 3 weeks and did not become pregnant. This was confirmed at necropsy, and an examination of her ovaries showed many follicles and multiple corpora lutea, indicating reproductive maturity.

3.3.3. Group D75

No significant gross pathology was observed for this group. Occasionally petechiae in the heart wall ($n = 2$; stillborn pigs) and consolidated lung lobules ($n = 6$; presumptively due to hypostasis) were observed. Serosanginous pleural and abdominal fluid was noted in two stillborn pigs that exhibited autolytic changes. Bungowannah virus was cleared relatively rapidly from this group, with only low viral loads generally detected after 21–28 days of age [22].

3.3.4. Group D90

Serosanguinous thoracic or abdominal effusions were observed in stillborn pigs with autolytic changes ($n = 3$). Occasional petechiation of the pericardium was also observed ($n = 1$). Two stillborn pigs had fibrin tags in the peritoneal cavity and equivocally enlarged hearts (18-07, 18-08), while another had red tonsils.

Of the preweaning mortalities, four of the five pigs observed to have increased respiratory rates exhibited cardiomegaly (20-06, 21-02, 21-03, 21-09) based on subjective observation of an increased base to apex height and increased sternal contact (Figure 4a), and two 5 mm white foci were observed in the superficial myocardium of 20-6. The fifth pig had a marked haemorrhagic pericardial effusion (>50 mL) and blood clots in the abdominal cavity (19-02) (Figure 4b). Thoracic and/or abdominal effusions ($n = 2$), fibrin tags in the thoracic and/or abdominal cavity ($n = 2$), liver enlargement ($n = 4$), reddened inguinal and mesenteric lymph nodes ($n = 3$) and oedema of the spiral colon ($n = 3$) were also observed amongst these five cases. The lungs of the most severely affected pigs did not deflate and a large amount of fluid/froth drained from the lungs after death. Viral RNA was detected in the hearts of all pigs sampled in the first 10 days of life [22]. The pigs that died suddenly (19-03, 19-07, 20-04, 22-03, 22-05, 22-06 and 22-08) appeared to be behaving normally and eating well prior to death. Bruising to suggest they had been overlain was not observed. A range of changes was observed, including subcutaneous oedema of the ventral neck ($n = 2$), pericardial effusion ($n = 1$), fibrin tags in thoracic and/or abdominal cavities ($n = 2$), blood clots in the abdominal cavity ($n = 2$), petechiae and haemorrhages on the kidney ($n = 1$), red/haemorrhagic mesenteric and/or inguinal lymph nodes ($n = 3$), haemorrhages on the spiral colon and caecum ($n = 2$) and oedema of the spiral colon ($n = 1$).

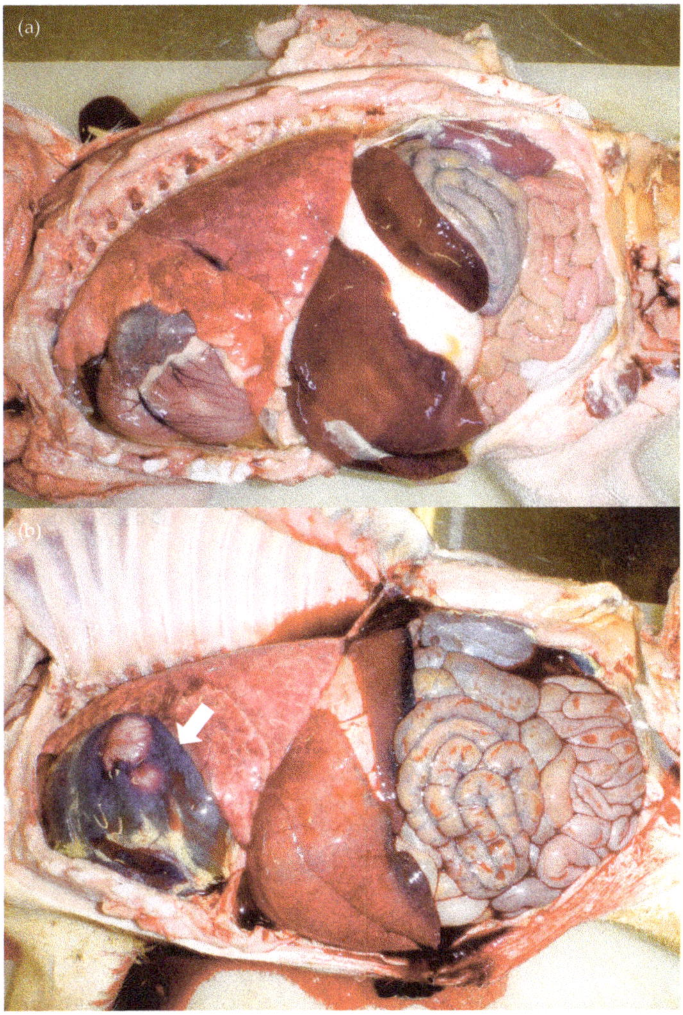

Figure 4. Necropsy findings from D90: (**a**) cardiac and hepatic enlargement (20-06; 7 days old) and (**b**) free blood in the unopened pericardial sac (arrow) and abdomen (19-02; 12 days old).

From weaning to termination of the monitoring at 5 weeks of age, few gross lesions were observed. Bungowannah virus was cleared most rapidly from this group with virus shedding and the viral loads in tissues were only at low levels from 11 days of age or often not detected at all [22]. Interlobular pulmonary oedema was observed in two animals and fine fibrin tags were observed in the peritoneal cavity of five animals. The pig that exhibited an increased respiratory rate after exercise and survived to 5 weeks old (23-11) also showed mild interlobular pulmonary oedema and a smaller than expected thymus, but did not have definite cardiac enlargement. However, the ventricular walls appeared thinner than normal, suggesting possible cardiac dilation.

3.3.5. Uninfected Litters

No significant abnormalities were observed at necropsy in pigs from this group. The ages of pigs examined ranged from stillborn to 21 days old. The most frequent changes observed were reddening of the lungs (presumptively due to hypostasis), with some interlobular oedema in four pigs. Other changes observed included petechial haemorrhages on the pericardium, tonsil and skull of an overlain pig, a moderate increase in the amount of thoracic fluid of stillborn pigs, circular skin deficits over the left carpus of two stillborn pigs in litter 9 and moderately enlarged inguinal lymph nodes in a splay-legged pig.

4. Discussion

In this study, the clinical signs and gross pathological lesions observed during the outbreak of the porcine myocarditis syndrome were reproduced in the progeny of pregnant pigs experimentally infected with Bungowannah virus. The disease and lesions associated with in utero Bungowannah virus infection at different gestational ages were further characterised. Infection of the foetus at 35 or 90 days gestation resulted in the most severe clinical and pathological effects.

Losses in the field due to PMC were characterised by increased stillbirths, preweaning losses and to a lesser extent, mummified foetuses. At the peak of the outbreak on the farm, on the most severely affected production unit (a gilt unit), preweaning losses approached 50%, stillbirths 40% and mummified foetuses 10–15%, although losses did vary across production units [1,27]. The results of the current study indicated that Bungowannah virus is capable of causing these effects and, depending on the presence of sows at critical stages of pregnancy, could cause significant reproductive losses if introduced into a naïve piggery. The two most critical time-points for exposure of the pregnant pig in relation to disease appears to be around 35 and 90 days of gestation. Our observations following infection at these stages of gestation are consistent with events observed during the early stages of the PMC outbreak, which suggests that the introduction of the virus occurred at a single time-point. The first losses observed were sudden death at 2–3 weeks of age, which is consistent with the outcomes of experimental infection at around 90 days gestation. These initial losses were followed approximately 4 weeks later by a marked increase in the number of stillborn pigs, which would indicate infection at around 40–50 days gestation if 90 days is considered the critical time-point for sudden death at 2–3 weeks of age [20,27].

The description of porcine myocarditis syndrome arose from the fact that early in the field outbreak, affected animals often had an enlarged/dilated heart with evidence of congestive heart failure and non-suppurative myocarditis [1]. Based on the results of this study, it appears that the cardiac enlargement previously reported occurs principally following infection in late gestation (approximately 90 days) and results in a moderate increase in preweaning losses. This was the only challenge group where the previously described [1] multifocal, non-suppurative myocarditis with myonecrosis was observed [28]. Based on these findings, it is clear that the outcome of infection with Bungowannah virus and the associated pathological changes are more extensive than the original field description "porcine myocarditis syndrome" would suggest. Clinically, similar to the field cases, sudden death or increased respiratory rate and effort were observed. Unlike the field situation, cyanosis of the snout and ears and excessive vocalisation before death were not observed during the course of this study.

The cardiorespiratory signs and sudden deaths observed have not been reported for CSFV, APPV or Linda viruses, the other pestiviruses primarily infecting pigs. Postnatal deaths are uncommon following infection after 90 days gestation with low virulence strains of CSFV [13,29].

The reproductive effects of in utero infection with Bungowannah virus were most severe following infection around 35 days gestation and are presumably a consequence of the foetus being unable to mount an immune response to the virus at this stage of gestation [22]. These effects included a marked increase in preweaning mortalities (particularly in the first few days of life) and mummified foetuses, as well as a moderate to marked increase in stillbirths, all of which suggests that these effects in the field were the result of infection around 35 days of gestation, reducing by day 55. These clinical findings are similar to those observed following in utero infections with low virulence strains of CSFV that, prior to 70 days gestation, are characterised by high prenatal mortality due to stillborn and mummified foetuses, and increased postnatal mortality with weak-born pigs, although congenital deformities were not observed for Bungowannah virus [13,29–31]. The increased levels of stillbirths following exposure of the dam at 55, 75 or 90 days gestation during this study, when compared with industry targets, suggests that Bungowannah virus may have also caused a mild to moderate increase in the stillbirth rate at these time-points, although a greater number of pregnant animals would need to be studied to confirm this finding.

Subcutaneous oedema that spread principally over the head and thorax was observed in a proportion of stillborn pigs during the PMC outbreak [1]. Experimentally, this clinical sign was only observed following Bungowannah virus infection in early gestation (approximately 35–55 days). While purpura was not reported in the field, it was observed in 31% of live and stillborn pigs born to dams infected at around 35 days gestation in the current study. Purpura/petechiae/haemorrhages and subcutaneous oedema have also been described following in utero infection with CSFV in foetuses infected before 70 days gestation [13,29–31].

Other clinical similarities between infection with Bungowannah virus at 35 days gestation and in utero CSFV infection prior to 70 days include weak-born pigs that may exhibit or develop poor sucking ability, alopecia, ascites/hydrothorax, petechiae/haemorrhages in organs other than the skin, pallor, diarrhoea, conjunctivitis, anorexia, lethargy or cyanosis of the snout and ears, with some of these changes being observed postweaning [13,29–31]. In contrast, the clinical signs following experimental Bungowannah virus infection at 55 days more closely parallel what is observed when in utero infection with CSFV occurs between 70 and 90 days gestation, which is associated with fewer prenatal deaths [29–31], although some melaena and diarrhoea was observed in the preweaning period. It is not clear whether the diarrhoea and melaena were increased compared to D75 and D90 pigs due to in utero Bungowannah virus infection or is unrelated.

Following the initial postnatal losses in the D35 group, the next major clinical effect noted following Bungowannah virus infection at either 35 or 55 days of gestation was variability in pig size within a litter postweaning. This disparity in size was noted in those pigs that were born seronegative to Bungowannah virus and, on the basis of PCR results, were considered to be persistently infected [22]. Persistent infections with CSFV are recognised following in utero infection before approximately 70 days gestation and with decreasing frequency up to 90 days gestation [13,29,30]. Clinically, CSFV PI pigs that survive past weaning may present similarly to chronic cases of classical swine fever (CSF), where after an initial, albeit slow recovery, they eventually relapse and die. Initially, they may be indistinguishable from uninfected pigs, but at a variable time (usually weeks to months) after weaning, they often develop severe runting and growth retardation, or "late-onset CSF", characterised by increasing anorexia and lethargy, pyrexia, conjunctivitis, dermatitis, locomotion disturbances and intermittent diarrhoea [13,15,30,31]. These clinical signs were generally consistent with those observed in pigs infected with Bungowannah virus at D35 that were considered PI. With the exception of a slightly slower growth rate, these pigs often appeared normal while still suckling, but at a variable time after weaning, developed severe growth retardation. Pallor (due to anaemia) and conjunctivitis were the most frequently observed clinical signs. Ataxia and locomotion disturbances were not observed. In contrast, those pigs that were infected at D55 and were born viraemic, seronegative and survived past weaning, while clearly stunted compared to their littermates, were generally disease-free. Whether this would be the case in the field, where there is greater exposure to pathogens, is not clear.

Two of the pigs (8-01 and 8-05) in the D55 group that were initially considered PI were raised until 11 months of age. Unexpectedly, these two animals seroconverted at approximately 6 months of age and cleared their viraemia, at which point, their growth rate increased [22]. After seroconversion, the gilt started cycling at around 8 months of age, which is older than expected and was presumably delayed due to her stunted growth. These animals are similar to the chronically infected pigs described when pigs were experimentally infected in utero with BDV or BDV-like viruses, or BVDV [14,16]. Similar to our study they also noted an improved growth rate after seroconversion. The reason for this improvement is not clear but it is suggested that the effect on growth is the result of a continuing process and not of an irreversible lesion acquired in early development [14].

Stunting is also observed in cattle PI with BVDV, although some animals may appear normal for several years before succumbing to a disease predominantly induced as a result of the persistent BVDV infection. While not observed during the course of this study, it is important to determine whether pigs persistently or "chronically" infected with Bungowannah virus can remain clinically normal for prolonged periods, as these animals are an ongoing source of infection in the herd and their selection as breeding animals could result in disastrous consequences if they were to be introduced into a naïve herd.

Congenital tremors were not observed in this study or in the field following Bungowannah virus infection. Aside from the absence of congenital tremors, the outcomes of infection of Bungowannah virus and CSFV after foetal infection in early gestation are remarkably similar and presumably have a similar pathogenesis. To date, Bungowannah virus has only been identified in Australia, but as its origin remains unknown, it should be considered in the differential diagnosis in reproductive investigations, especially if CFSV is suspected but has been excluded by laboratory testing.

It was not possible to weigh live animals once they reached 5 kg; therefore, all weights postweaning were obtained at the time of necropsy. In retrospect, weekly weights would have been useful to compare animal weights within litters and in relation to infection and serological status at birth. Any comments on stunting in this manuscript are unfortunately subjective rather than objective and relative to that challenge group cohort/littermates only. Additionally, while qRT-PCR results were typically available within a couple of days after sample collection, the serology (peroxidase-linked assay) was batch-tested at a convenient time [22]. The availability of these results may have been useful when selecting animals to follow postweaning, particularly for the D55 group, to better understand the impact of these "chronic" infections on growth rates. As each treatment group was reared individually and at separate times, direct comparisons regarding weight are difficult; in addition, direct comparisons against commercial growth targets are not appropriate.

5. Conclusions

This study demonstrates that in utero infection with Bungowannah virus was able to produce the clinical signs and gross pathology observed during the field outbreak and has the potential to be a significant reproductive pathogen of the pig, as well as impacting on postnatal mortality, principally in the first few weeks of life. The outcome of infection was dependent on the stage of gestation at which infection occurs. Maternal infection at around 35 days gestation resulted in high pre- and postnatal mortalities, which reduced when infected at about day 55. Reproductive effects and clinical signs were similar to those observed following in utero infection with low virulence strains of CSFV. Increased preweaning mortalities, cardiorespiratory clinical signs and gross lesions in the heart of the progeny of sows infected at around 90 days gestation suggests that infection in late gestation was required for the development of cardiac pathology. Persistent infections resulted in the development of clinical signs and lesions similar to "late-onset" CSFV.

Author Contributions: Conceptualization, D.S.F. and P.D.K.; methodology, D.S.F. and P.D.K.; investigation, D.S.F.; resources, D.S.F. and P.D.K.; data curation, D.S.F.; writing—original draft preparation, D.S.F.; writing—review and editing, D.S.F. and P.D.K.; visualization, D.S.F. and P.D.K.; supervision, P.D.K.; project administration, D.S.F. and P.D.K.; funding acquisition, P.D.K. All authors have read and agreed to the published version of the manuscript.

Funding: This research was funded by Australian Pork Limited (Project No. 2188), the Australian Biosecurity Cooperative Research Centre for Emerging Infectious Disease and NSW Department of Primary Industries.

Acknowledgments: We gratefully acknowledge the staff of the Virology Laboratory at the Elizabeth Macarthur Agricultural Institute (EMAI) for technical assistance and assistance with sample collection, and both farm and laboratory staff, especially Glenda Macnamara, for their care of the animals used in this study. We thank Damien Collins for assistance with statistical analysis, and Sarah Gestier for constructive feedback on the gross pathology descriptions in the manuscript.

Conflicts of Interest: The authors declare no conflict of interest. The funders had no role in the design of the study; in the collection, analyses or interpretation of data; in the writing of the manuscript; or in the decision to publish the results.

References

1. McOrist, S.; Thornton, E.; Peake, A.; Walker, R.; Robson, S.; Finlaison, D.; Kirkland, P.; Reece, R.; Ross, A.; Walker, K.; et al. An infectious myocarditis syndrome affecting late-term and neonatal piglets. *Aust. Vet. J.* **2004**, *82*, 509–511. [CrossRef]
2. Kirkland, P.D.; Frost, M.J.; Finlaison, D.S.; King, K.R.; Ridpath, J.F.; Gu, X. Identification of a novel virus in pigs-Bungowannah virus: A possible new species of pestivirus. *Virus Res.* **2007**, *129*, 26–34. [CrossRef] [PubMed]
3. Hause, B.M.; Collin, E.A.; Peddireddi, L.; Yuan, F.; Chen, Z.; Hesse, R.A.; Gauger, P.C.; Clement, T.; Fang, Y.; Anderson, G. Discovery of a novel putative atypical porcine pestivirus in pigs in the USA. *J. Gen. Virol.* **2015**, *96*, 2994–2998. [CrossRef] [PubMed]
4. Lamp, B.; Schwarz, L.; Hogler, S.; Riedel, C.; Sinn, L.; Rebel-Bauder, B.; Weissenbock, H.; Ladinig, A.; Rumenapf, T. Novel Pestivirus Species in Pigs, Austria, 2015. *Emerg. Infect. Dis.* **2017**, *23*, 1176–1179. [CrossRef] [PubMed]
5. Moennig, V.; Liess, B. Pathogenesis of intrauterine infections with bovine viral diarrhea virus. *Vet. Clin. N. Am. Food Anim. Pract.* **1995**, *11*, 477–487. [CrossRef]
6. Roeder, P.L.; Jeffrey, M.; Cranwell, M.P. Pestivirus fetopathogenicity in cattle: Changing sequelae with fetal maturation. *Vet. Rec.* **1986**, *118*, 44–48. [CrossRef]
7. Baker, J.C. The Clinical Manifestations of Bovine Viral Diarrhea Infection. *Vet. Clin. N. Am. Food A* **1995**, *11*, 425–445. [CrossRef]
8. McGowan, M.R.; Kirkland, P.D. Early reproductive loss due to bovine pestivirus infection. *Br. Vet. J.* **1995**, *151*, 263–270. [CrossRef]
9. Garcia-Perez, A.L.; Minguijon, E.; Estevez, L.; Barandika, J.F.; Aduriz, G.; Juste, R.A.; Hurtado, A. Clinical and laboratorial findings in pregnant ewes and their progeny infected with Border disease virus (BDV-4 genotype). *Res. Vet. Sci.* **2009**, *86*, 345–352. [CrossRef]
10. Arruda, B.L.; Arruda, P.H.; Magstadt, D.R.; Schwartz, K.J.; Dohlman, T.; Schleining, J.A.; Patterson, A.R.; Visek, C.A.; Victoria, J.G. Identification of a Divergent Lineage Porcine Pestivirus in Nursing Piglets with Congenital Tremors and Reproduction of Disease following Experimental Inoculation. *PLoS ONE* **2016**, *11*, e0150104. [CrossRef]
11. Trautwein, G.; Hewicker, M.; Liess, B.; Orban, S.; Grunert, E. Studies on Transplacental Transmissibility of a Bovine Virus Diarrhoea (BVD) Vaccine Virus in Cattle III. Occurrence of Central Nervous System Malformations in Calves Born from Vaccinated Cows. *J. Vet. Med. Ser. B* **1986**, *33*, 260–268. [CrossRef] [PubMed]
12. Nettleton, P.F. Border disease. In *Infectious Diseases of Livestock*, 2nd ed.; Coetzer, J.A.W., Tustin, R.C., Eds.; Oxford University Press Southern Africa: Cape Town, South Africa, 2004; Volume 2, pp. 970–974.
13. Van Oirschot, J.T. Experimental production of congenital persistent swine fever infections: I. Clinical, pathological and virological observations. *Vet. Microbiol.* **1979**, *4*, 117–132. [CrossRef]
14. Paton, D.J.; Done, S.H. Congenital infection of pigs with ruminant-type pestiviruses. *J. Comp. Pathol.* **1994**, *111*, 151–163. [CrossRef]
15. Van Oirschot, J.T.; Terpstra, C. A congenital persistent swine fever infection. I. Clinical and virological observations. *Vet. Microbiol.* **1977**, *2*, 121–132. [CrossRef]

16. Terpstra, C.; Wensvoort, G. A congenital persistent infection of bovine virus diarrhoea virus in pigs: Clinical, virological and immunological observations. *Vet. Q.* **1997**, *19*, 97–101. [CrossRef]
17. Young, G.A.; Kitchell, R.L.; Luedke, A.J.; Sautter, J.H. The effect of viral and other infections of the dam on fetal development in swine. I. Modified live hog cholera viruses; immunological, virological, and gross pathological studies. *J. Am. Vet. Med. Assoc.* **1955**, *126*, 165–171.
18. Leforban, Y.; Vannier, P.; Cariolet, R. Pathogenicity of border disease and bovine viral diarrhoea for pig: Experimental study on the vertical and horizontal transmission of the viruses. In Proceedings of the 11th International Pig Veterinary Society Congress, Lausanne, Switzerland, 1–5 July 1990; p. 204.
19. Postel, A.; Hansmann, F.; Baechlein, C.; Fischer, N.; Alawi, M.; Grundhoff, A.; Derking, S.; Tenhundfeld, J.; Pfankuche, V.M.; Herder, V.; et al. Presence of atypical porcine pestivirus (APPV) genomes in newborn piglets correlates with congenital tremor. *Sci. Rep.* **2016**, *6*, 27735. [CrossRef]
20. Finlaison, D.S.; King, K.R.; Frost, M.J.; Kirkland, P.D. Field and laboratory evidence that Bungowannah virus, a recently recognised pestivirus, is the causative agent of the porcine myocarditis syndrome (PMC). *Vet. Microbiol.* **2009**, *136*, 259–265. [CrossRef]
21. Finlaison, D.S.; Cook, R.W.; Srivastava, M.; Frost, M.J.; King, K.R.; Kirkland, P.D. Experimental infections of the porcine foetus with Bungowannah virus, a novel pestivirus. *Vet. Microbiol.* **2010**, *144*, 32–40. [CrossRef]
22. Finlaison, D.S.; Kirkland, P.D. The outcome of porcine foetal infection with Bungowannah virus is dependent on the stage of gestation at which infection occurs. Part 1: Serology and virology. *Viruses* **2020**, *12*, 691. [CrossRef]
23. Finlaison, D.S.; King, K.R.; Gabor, M.; Kirkland, P.D. An experimental study of Bungowannah virus infection in weaner aged pigs. *Vet. Microbiol.* **2012**, *160*, 245–250. [CrossRef] [PubMed]
24. R Core Team. *R: A Language and Environment for Statistical Computing*; R Foundation for Statisical Computing: Vienna, Austria, 2019; Available online: http://www.R-project.org (accessed on 1 April 2019).
25. Butler, D.G.; Cullis, B.R.; Gilmour, A.R.; Gogel, B.J. *ASReml-R Reference Manual, release 3 ed.*; Queensland Department of Primary Industries and Fisheries: Brisbane, Australia, 2009.
26. Friendship, R.M.; Lumsden, J.H.; McMillan, I.; Wilson, M.R. Hematology and biochemistry reference values for Ontario swine. *Can. J. Comp. Med.* **1984**, *48*, 390–393. [PubMed]
27. Finlaison, D.S. Studies of the Porcine Myocarditis Syndrome. Ph.D. Thesis, University of Sydney, Sydney, Australia, 2010.
28. Finlaison, D.S.; Gestier, S.; Kirkland, P.D. The outcome of porcine foetal infection with Bungowannah virus is dependent on the stage of gestation at which infection occurs—Histopathology. in preparation.
29. Frey, H.R.; Liess, B.; Richter-Reichhelm, H.B.; von Benten, K.; Trautwein, G. Experimental transplacental transmission of hog cholera virus in pigs. I. Virological and serological studies. *Zentralbl. Veterinarmed. B* **1980**, *27*, 154–164. [CrossRef] [PubMed]
30. Meyer, H.; Liess, B.; Frey, H.R.; Hermanns, W.; Trautwein, G. Experimental transplacental transmission of hog cholera virus in pigs. IV. Virological and serological studies in newborn piglets. *Zentralbl. Veterinarmed. B* **1981**, *28*, 659–668. [CrossRef]
31. Hermanns, W.; Trautwein, G.; Meyer, H.; Liess, B. Experimental transplacental transmission of hog cholera virus in pigs. V. Immunopathological findings in newborn pigs. *Zentralbl. Veterinarmed. B* **1981**, *28*, 669–683. [CrossRef]

viruses

MDPI

Article

Infection of Ruminants, Including Pregnant Cattle, with Bungowannah Virus

Andrew J. Read, Deborah S. Finlaison and Peter D. Kirkland *

Virology Laboratory, Elizabeth Macarthur Agriculture Institute, Woodbridge Road, Menangle, New South Wales 2568, Australia; andrew.j.read@dpi.nsw.gov.au (A.J.R.); deborah.finlaison@dpi.nsw.gov.au (D.S.F.)

* Correspondence: peter.kirkland@dpi.nsw.gov.au; Tel.: +61-2-4640-6331

Received: 2 June 2020; Accepted: 22 June 2020; Published: 26 June 2020

Abstract: Bungowannah virus is a pestivirus known to cause reproductive losses in pigs. The virus has not been found in other species, nor is it known if it has the capacity to cause disease in other animals. Eight sheep, eight calves and seven pregnant cows were experimentally infected with Bungowannah virus. It was found that sheep and calves could be infected. Furthermore, it was shown that the virus is able to cross the bovine placenta and cause infection of the foetus. These findings demonstrate the potential for species other than pigs to become infected with Bungowannah virus and the need to prevent them from becoming infected.

Keywords: Bungowannah virus; pestivirus F; ruminant infection

---

1. Introduction

In June 2003 a syndrome of sudden death in sucker pigs, followed by a marked increase in stillborn foetuses and pre-weaning losses, occurred on a large farm in NSW, Australia [1]. Myocarditis and myonecrosis were also observed in affected pigs. A novel pestivirus, known as Bungowannah virus, was subsequently identified [2]. A series of field and laboratory studies have provided strong evidence that Bungowannah virus is the aetiological agent in this disease [2–8]. Bungowannah virus contains all of the genomic and structural elements of classically described pestiviruses, yet phylogenetic analysis demonstrates that it is genetically remote from any of the other pestivirus species [9,10]. Bungowannah virus is the only extant isolate of *pestivirus F* species [10].

Pestiviruses were initially classified according to their host specificity. Whilst this classification was originally appropriate for classical swine fever virus (CSFV), it was soon shown that bovine viral diarrhea virus (BVDV) and border disease virus (BDV) could naturally infect a variety of ruminants, pigs and other mammals. Recently, CSFV has also been shown to naturally infect cattle [11]. In contrast, Bungowannah virus has only ever been detected in pigs. The origin of this virus is not known, nor what threat it may pose to other species. Bungowannah virus has been shown to replicate in ovine and bovine cells in vitro [12] and so the possibility that it may infect ruminants has been raised. This paper documents the outcome of experimental infections of sheep and cattle with Bungowannah virus. Patterns of virus shedding and pathology are described.

2. Materials and Methods

A series of inoculation experiments were conducted in both sheep and cattle. Cattle were either directly inoculated using intranasal instillation or by co-housing with pigs that were chronically infected with Bungowannah virus. Sheep were either directly inoculated using intranasal instillation or subcutaneous injection or by co-housing with pigs that were chronically infected with Bungowannah virus. The specific details are as follows:

2.1. Virus Amplification

The inoculum used for each of the direct inoculation experiments was derived from pooled pig foetal tissues that were passaged once in PK-15 cells (RIE5–1, Collection of Cell Lines in Veterinary Medicine, Friedrich-Loeffler-Institut, Insel Riems, Germany). The titre of infectious virus was also determined by titration in PK-15 cells using standard methods.

2.2. Viral Transport Medium

Swabs were collected into 3 mL of sterile phosphate buffered saline (137 mM NaCl, 8 mM Na_2HPO_4, 2.7 mM KCl and 1.5 mM KH_2PO_4, pH 7.4) containing 0.5% gelatin (*w/v*), 5000 IU penicillin/mL, 95,000 IU streptomycin, 50 μg/mL amphotericin B and 0.1% (*w/v*) phenol red (PBGS).

2.3. Bungowannah Virus Real-Time Polymerase Chain Reaction (qRT-PCR)

Bungowannah virus RNA was identified from samples using a real-time, reverse transcription PCR (qRT-PCR). The method has been previously described [3]. The fluorescence threshold was set manually at 0.05 and the background was automatically adjusted. qRT-PCR results were expressed as cycle threshold (Ct) values and classified as negative if no amplification was observed after the 45 cycles. For quantification, a 10-fold dilution series of Bungowannah virus RNA standards ranging from 10^7 to 10^2 RNA copies/5 μL [6] was included in the assay and the quantity of Bungowannah virus RNA in a sample was determined from the standard curve.

2.4. Bungowannah Virus Neutralisation Test

Antibody titres against Bungowannah virus were measured by virus neutralisation test (VNT). The VNT was performed as described previously [5]. Selected serum samples were tested in the VNT in a two-fold dilution series commencing at 1/4.

2.5. Infection of Sheep

Sheep used in these trials were obtained from a flock that was free of infection with ruminant pestiviruses and had not been vaccinated against pestiviruses. All sheep were tested for anti-pestivirus antibodies using a bovine viral diarrhea virus agarose gel immunodiffusion assay [13] and were found to be negative.

2.5.1. Direct Inoculation

Six 3-month-old Merino lambs were infected intranasally with 2 mL of cell culture amplified Bungowannah virus (5.6 $\log_{10}$ $TCID_{50}$/mL). Two other sheep were inoculated with the same dose subcutaneously while another two other sheep were held as uninfected controls. The inoculated sheep were held in two 11 m^2 rooms (four intranasally infected sheep in one room, the remaining four infected sheep in the other room). The two uninfected sheep were held in a similar 11 m^2 room and were not challenged.

Conjunctival, nasal, oral and rectal swabs, along with serum samples, were collected from all sheep prior to exposure to Bungowannah virus and daily for 14 days. Blood samples were subsequently collected approximately weekly until 6 weeks post-exposure. Clinical signs, including rectal temperatures, were also recorded daily for the first 14 days. The swabs and sera were tested for the presence of Bungowannah virus using real-time PCR (qRT-PCR). Serum samples were tested for the presence of antibodies against Bungowannah virus using a VNT.

2.5.2. Exposure to Chronically Infected Pigs

Four 3-month-old Merino lambs were held in a small room (16 m^2) with three pigs that were chronically infected with Bungowannah virus. The pigs (08-01, 08-05 and 10-01) had been infected in utero and were shown to be shedding Bungowannah virus (oropharyngeal secretions—6.9, 7.0 and

3.4 $\log_{10}$ copies/swab, respectively) 7 days prior to the trial [7]. The sheep and the pigs were co-housed for 48 h. During this time there were two periods of six hours of direct physical contact between the sheep and pigs. During the remainder of the time the pigs were separated from the sheep by a mesh partition which allowed for limited direct contact. The clinical examination and sampling were conducted as described above.

2.6. Infection of Calves

Calves used in the following trials were obtained from a herd that was free of infection with ruminant pestiviruses and had not been vaccinated against pestiviruses. All calves were tested for anti-pestivirus antibodies using a bovine viral diarrhea virus agarose gel immunodiffusion assay [13] and were found to be negative.

2.6.1. Direct Inoculation

Eight 10-week-old Holstein–Friesian calves were infected intranasally with 2 mL of cell culture-amplified Bungowannah virus (5.6 $\log_{10}$ $TCID_{50}$/mL). Two additional calves were held as uninfected controls and did not receive a challenge. The conditions under which they were held, the clinical examination and sampling were conducted as described above for the sheep.

2.6.2. Exposure to Chronically Infected Pigs

Four 5-week-old Holstein–Friesian calves were held with two pigs (08-01 and 08-05; oropharyngeal secretions 5.5 and 7.1 $\log_{10}$ copies/swab respectively, 6 days prior to trial) [7] chronically infected with Bungowannah virus. The conditions under which they were held, the clinical examination and sampling were conducted as described above for the sheep.

2.7. Infection of Pregnant Cows

Five pregnant Holstein-Friesian and two Illawarra-Shorthorn cows were chosen for the trial. They were obtained from a herd known to be free of infection with ruminant pestiviruses and had not been vaccinated against pestiviruses. All cows were tested for anti-pestivirus antibodies using a bovine viral diarrhea virus agarose gel immunodiffusion assay [13] and found to be negative. They were infected by intranasal instillation of 2 mL of the cell culture amplified Bungowannah virus (4.5 $\log_{10}$ TCID/mL). Cows were between 53 and 65 days of pregnancy at the time of inoculation. Nasal and conjunctival swabs and serum samples were collected daily for 14 days following inoculation. The swabs and sera were tested for the presence of Bungowannah virus using qRT-PCR. Serum samples were then collected monthly until calving and were tested for antibodies against Bungowannah virus using a VNT. Pregnancy was monitored by ultrasound examination on a monthly basis. Cows were induced to calve between 255 and 276 days of pregnancy. Serum samples were collected from each calf after birth and prior to suckling. Conjunctival, nasal, oral and rectal swabs and serum samples were collected from calves every 48–72 h for 14 days. Vaginal swabs were also collected from the cows post-partum. A post-mortem examination of the calves was performed between 2 and 4 weeks of age. A wide range of tissues including brain, myocardium and skin (and testicle from a male) were tested for the presence of Bungowannah virus by qRT-PCR. Samples for qRT-PCR were collected by firmly rubbing a swab across the freshly cut surface of a section of the tissue and placed into 3 mL of PBGS. Skin biopsies were stored in 3 mL PBGS. All samples were stored at 4 °C until tested by qRT-PCR and virus isolation.

2.8. Animal Ethics Approval

The trials described in this paper were approved by the Elizabeth Macarthur Agricultural Institute Animal Ethics Committee. The specific approvals were AEC M09/17 "Studies of the biology of Bungowannah virus infections (PMC) in sheep and cattle" (7 December 2009) and AEC M10/16 "Effects of Bungowannah virus infection in pregnant cattle" (22 December 2010).

3. Results

3.1. Infection of Sheep

3.1.1. Direct Inoculation

Bungowannah virus RNA was detected between days four and 11 in the nasal swabs of five of the six sheep exposed by intranasal instillation. Ct values ranged between 27.9 and 39.9 (Table 1). Low levels of Bungowannah virus RNA were sporadically detected between days three and 13 of infection in serum samples from four of the six intranasally inoculated sheep. Ct values ranged between 36.5 and 38.1. Two of these sheep had Bungowannah virus RNA detected in serum on three occasions (Table 1).

Table 1. Bungowannah virus qRT-PCR results from sheep intranasally and subcutaneously inoculated.

		Days Post Inoculation														
Sheep	**Sample**	**0**	**1**	**2**	**3**	**4**	**5**	**6**	**7**	**8**	**9**	**10**	**11**	**12**	**13**	**14**
1N	Nasal	-	-	-	-	36.8	-	-	33.3	-	37.6	34.3	34.1	-	-	-
	Serum	-	-	-	-	-	-	37.1	-	37.8	-	-	36.5	-	-	-
2N	Nasal	-	-	-	-	36.0	36.6	32.6	-	34.5	32.2	37.4	37.5	-	-	-
	Serum	-	-	-	-	-	-	-	-	37.7	-	-	-	-	-	-
3N	Nasal	-	-	-	-	30.0	32.3	27.9	29.1	29.6	34.2	39.9	-	-	-	-
	Serum	-	-	-	37.2	-	-	-	-	-	-	-	-	-	-	-
4N	Nasal	-	-	-	-	29.4	-	34.2	31.1	-	30.5	-	38.4	-	-	-
	Serum	-	-	-	-	-	-	-	38.1	-	36.6	-	-	-	37.1	-
5N	Nasal	-	-	-	-	-	33.2	-	-	-	-	-	-	-	-	-
	Serum	-	-	-	-	-	-	-	-	-	-	-	-	-	-	-
6N	Nasal	-	-	-	-	-	-	-	-	-	-	-	-	-	-	-
	Serum	-	-	-	-	-	-	-	-	-	-	-	-	-	-	-
7S	Nasal	-	-	-	-	-	-	-	-	-	-	-	-	-	-	-
	Serum	-	-	-	-	-	-	-	-	-	-	-	-	-	-	-
8S	Nasal	-	-	-	-	-	-	-	-	-	-	-	-	-	-	-
	Serum	-	-	-	-	-	-	-	-	-	-	-	-	-	-	-
9C	Nasal	-	-	-	-	-	-	-	-	-	-	-	-	-	-	-
	Serum	-	-	-	-	-	-	-	-	-	-	-	-	-	-	-
10C	Nasal	-	-	-	-	-	-	-	-	-	-	-	-	-	-	-
	Serum	-	-	-	-	-	-	-	-	-	-	-	-	-	-	-

A dash indicates that Bungowannah virus RNA was not detected.

Bungowannah viral RNA was not detected in oral, conjunctival or rectal swabs of any of the inoculated or control sheep, nor in serum or nasal swabs for the control sheep or those inoculated subcutaneously. None of the infected sheep developed clinical signs during the course of the infection. The rectal temperatures remained within normal limits. All eight sheep directly inoculated (either intranasally or subcutaneously) developed Bungowannah virus-specific antibodies (Table 2).

Table 2. Bungowannah VN titres from sheep intranasally and subcutaneously inoculated.

		Days Post Inoculation						
Sheep	**Route of Infection**	**0**	**9**	**15**	**21**	**29**	**36**	**43**
1N	Intranasal	-	-	256	128	256	256	256
2N	Intranasal	-	-	-	32	32	32	64
3N	Intranasal	-	-	64	64	128	128	128
4N	Intranasal	-	-	-	64	128	64	256
5N	Intranasal	-	-	-	64	256	512	256
6N	Intranasal	-	-	-	32	128	128	64
7S	Subcutaneous	-	-	-	128	256	256	256
8S	Subcutaneous	-	-	64	256	256	256	256
9C	Uninfected	-	-	-	-	-	-	-
10C	Uninfected	-	-	-	-	-	-	-

A dash indicates that neutralising anti-Bungowannah virus antibodies were not detected.

3.1.2. Exposure to Chronically Infected Pigs

Viral RNA was detected only in the nasal swabs of three of the four sheep during the period of co-mingling with the chronically infected pigs. The Ct values ranged from 36.1 to 36.4. Bungowannah virus RNA was not detected in serum or any other swabs during this time and was not detected in serum or any swabs after the pigs were removed from the room. None of the sheep developed Bungowannah virus-specific antibodies.

3.2. Infection of Calves

3.2.1. Direct Inoculation

Bungowannah virus RNA was detected in nasal swabs of all eight calves on at least three occasions between days two and 11 (Table 3). Viral RNA was detected sporadically in serum samples from six of these eight calves between days five and 10 of infection and intermittently in the oral swabs from four calves between days four and eight. Bungowannah virus was not detected in any rectal swabs. All eight directly inoculated calves developed antibodies directed against Bungowannah virus by 14 days post-inoculation (Table 4). A very mild nasal discharge was observed in five of the challenged calves. The rectal temperatures remained within normal limits.

Table 3. Bungowannah virus qRT-PCR results from calves intranasally inoculated.

		Days Post Inoculation														
Calf	**Sample**	**0**	**1**	**2**	**3**	**4**	**5**	**6**	**7**	**8**	**9**	**10**	**11**	**12**	**13**	**14**
1N	Nasal	-	-	-	-	-	-	27.5	34.5	32.5	-	-	-	-	-	-
	Oral	-	-	-	-	-	-	37.6	37.7	-	-	-	-	-	-	-
	Serum	-	-	-	-	-	-	-	-	-	-	-	-	-	-	-
2N	Nasal	-	-	34.5	30.4	31.8	31.5	28.1	28.4	-	-	-	-	-	-	-
	Oral	-	-	-	-	-	-	-	-	-	-	-	-	-	-	-
	Serum	-	-	-	-	-	-	-	35.9	-	-	-	-	-	-	-
3N	Nasal	-	-	-	-	-	33.9	30.5	31.2	-	37.9	-	-	-	-	-
	Oral	-	-	-	-	-	-	-	-	-	-	-	-	-	-	-
	Serum	-	-	-	-	-	-	-	37.7	-	-	-	-	-	-	-
4N	Nasal	-	-	37.0	32.1	-	-	33.0	30.0	32.8	31.2	-	-	-	-	-
	Oral	-	-	-	-	-	-	-	-	-	-	-	-	-	-	-
	Serum	-	-	-	-	-	35.7	-	-	37.5	35.7	-	-	-	-	-
5N	Nasal	-	-	33.9	30.2	31.4	27.6	27.1	28.4	29.3	35.2	-	-	-	-	-
	Oral	-	-	-	-	-	-	36.7	-	-	-	-	-	-	-	-
	Serum	-	-	-	-	-	-	-	-	-	-	-	-	-	-	-

Table 3. *Cont.*

Calf	Sample	Days Post Inoculation														
		0	1	2	3	4	5	6	7	8	9	10	11	12	13	14
6N	Nasal	-	-	36.2	34.1	-	31.0	29.7	30.7	27.6	33.6	-	37.4	-	-	-
	Oral	-	-	-	-	-	-	-	-	-	-	-	-	-	-	-
	Serum	-	-	-	-	-	36.3	-	36.3	36.0	37.6	-	-	-	-	-
7N	Nasal	-	-		33.1	31.4	32.5	30.1	29.2	29.7	-	-	-	-	-	-
	Oral	-	-	-	-	36.6		36.6	-	-	-	-	-	-	-	-
	Serum	-	-	-	-	-	-	37.3	36.4		36.5	37.7	-	-	-	-
8N	Nasal	-	-	-	-	31.7	31.6	30.3	31.4	33.4	-	-	-	-	-	-
	Oral	-	-	-	-	-	-	-	-	37.5	-	-	-	-	-	-
	Serum	-	-	-	-	-	37.3	-	-	-	-	-	-	-	-	-
9C	Nasal	-	-	-	-	-	-	-	-	-	-	-	-	-	-	-
	Oral	-	-	-	-	-	-	-	-	-	-	-	-	-	-	-
	Serum	-	-	-	-	-	-	-	-	-	-	-	-	-	-	-
10C	Nasal	-	-	-	-	-	-	-	-	-	-	-	-	-	-	-
	Oral	-	-	-	-	-	-	-	-	-	-	-	-	-	-	-
	Serum	-	-	-	-	-	-	-	-	-	-	-	-	-	-	-

A dash indicates that Bungowannah virus RNA was not detected. Calves 9C and 10C were uninfected.

Table 4. Bungowannah VN titres from calves intranasally inoculated.

Calf	Route of Infection	Days Post Inoculation							
		0	7	14	21	28	35	42	49
1N	Intranasal	-	-	16	512	1024	2048	2048	4096
2N	Intranasal	-	-	64	256	1024	2048	2048	2048
3N	Intranasal	-	-	32	256	1024	4096	8192	4096
4N	Intranasal	-	-	64	2048	2048	16,384	8192	2048
5N	Intranasal	-	-	32	128	512	2048	1024	4096
6N	Intranasal	-	-	16	1024	256	2048	8192	4096
7N	Intranasal	-	-	128	1024	4096	16,384	32,768	32,768
8N	Intranasal	-	-	16	128	1024	2048	4096	8192
9C	Uninfected	-	-	-	-	-	-	-	-
10C	Uninfected	-	-	-	-	-	-	-	-

A dash indicates that neutralising anti-Bungowannah virus antibodies were not detected. Calves 9C and 10C were uninfected.

3.2.2. Exposure to Chronically Infected Pigs

Viral RNA was detected in the nasal swabs of two of the four calves during the 48 h that they were housed in contact with chronically infected pigs. The virus was detected in the oral swab of another calf on day two and the nasal swab of this same calf on day eight. This calf was the only one of the four to develop antibodies against Bungowannah virus. Neutralising antibodies were first detected 16 days post-exposure. The antibody titre in this calf peaked at 2048 between 63 and 84 days post-inoculation. Bungowannah viral RNA was not detected in any other samples collected. Calves remained healthy throughout the study.

3.3. *Infection of Pregnant Cows*

Each of the seven cows inoculated with Bungowannah virus were found to shed Bungowannah virus RNA in nasal secretions. Shedding began in most cows at day three, with peak virus shedding between days seven and 10. One cow was still shedding viral RNA in nasal secretions on day 14 (Table 5). Bungowannah virus RNA was not detected in conjunctival swabs but was detected in the serum of four of the seven cows for a period of one to three days. No signs of respiratory disease or pyrexia were observed in the cows.

Table 5. qRT-PCR results for pregnant cows infected with Bungowannah virus.

		Days Post Inoculation														
Cow		**0**	**1**	**2**	**3**	**4**	**5**	**6**	**7**	**8**	**9**	**10**	**11**	**12**	**13**	**14**
1	Nasal	-	-	-	36.6	-	31.8	33.9	33.6	30.2	31.0	31.6	32.7	-	-	
	Serum	-	-	-	-	-	-	-	-	-	33.4	35.5	35.0	-	-	-
2	Nasal	-	-	-	30.5	35.6	30.9	36.2	28.7	29.3	27.8	31.8	34.0	34.3	-	-
	Serum	-	-	-	-	-	-	-	-	-	-	-	-	-	-	-
3	Nasal	-	-	-	35.6	36.0	-	37.4	35.4	33.2	33.6	31.8	34.6	35.4	-	-
	Serum	-	-	-	-	-	-	-	-	36.8	-	-	-	-	-	-
4	Nasal	-	-	-	-	34.3	34.0	29.8	28.5	28.5	33.1	-	37.4	-	-	-
	Serum	-	-	-	-	-	-	37.9	27.2	34.0	37.5	-	-	-	-	-
5	Nasal	-	-	-	-	-	-	37.4	36.3	-	-	-	37.8	36.4	34.7	32.2
	Serum	-	-	-	-	-	-	-	-	-	-	-	-	-	-	-
6	Nasal	-	-	-	37.3	31.8	36.7	35.8	31.1	32.1	-	-	-	-	-	-
	Serum	-	-	-	-	-	-	-	-	-	-	-	-	-	-	36.5
7	Nasal	-	37.4	35.4	36.2	-	33.4	32.2	29.7	30.5	27.2	31.8	35.6	-	-	-
	Serum	-	-	-	-	-	-	-	-	-	-	-	-	-	-	-

A dash indicates that Bungowannah virus RNA was not detected.

All cows developed neutralising antibodies against Bungowannah virus (Table 6). One cow had developed antibodies by day 11, and the remainder by day 14. Titres peaked on day 44 post infection. Neutralising antibody was detected for the duration of the pregnancy for all cows. A biphasic antibody titre developed in Cow 3 with a peak of 2048 on days 37 and 44, a decline to 512 on days 66 and 100, and then a rise to 1024 on day 128 and 2048 on day 163.

Table 6. Bungowannah VN titres from pregnant cows intranasally inoculated.

	Days Post Inoculation												
Cow	**0**	**7**	**11**	**14**	**21**	**31**	**37**	**44**	**52**	**66**	**100**	**128**	**163**
1	-	-	-	32	1024	2048	4096	4096	2048	1024	1024	512	1024
2	-	-	-	32	128	512	512	1024	256	512	512	256	128
3	-	-	-	8	256	512	2048	2048	1024	512	512	1024	2048
4	-	-	-	8	128	512	1024	2048	512	512	512	512	256
5	-	-	8	8	128	512	256	256	128	512	128	256	64
6	-	-	-	8	128	512	512	1024	512	NT	2048	512	1024
7	-	-	-	16	16	64	64	64	64	64	64	64	32

A dash indicates that neutralising anti-Bungowannah virus antibodies were not detected.

Pregnancies in the cows were unremarkable with no abnormalities were detected by ultrasound. At birth and before suckling three of the seven calves were found to have antibodies against Bungowannah virus. Bungowannah virus RNA was also detected in ear skin biopsy samples from these same three calves, with Ct values of 26.0, 32.7 and 33.6. The calf with the highest RNA concentration (lowest Ct value) was the progeny of Cow 3. Bungowannah virus RNA was also detected in the serum of this calf. Viral RNA was not detected in conjunctival, oral, nasal or rectal swabs. All calves appeared normal at birth and displayed normal behaviour. The virus was detected in the vaginal swabs from dams of the three positive calves, but not from the other cows. Cow 3 had a Ct value of 30.4, while the other two were 35.8 and 36.6.

Post-mortem examination of all calves was unremarkable, with no gross abnormalities detected. The three calves found to have Bungowannah virus RNA in the skin at birth were euthanased and necropsied at four weeks of age. Viral RNA was again detected in the skin of these calves, and also detected in the testicle of the single male calf among these three (Table 7). Viral RNA was not detected in any other tissues (oesophagus, stomach, duodenum, jejunum, ileum, caecum, colon, liver, bile, mesenteric lymph node, tracheobronchial lymph node, prescapular lymph node, tonsil, thymus, spleen, trachea, lung, thyroid, adrenal gland, kidney, urine, epididymis, ovary, uterus, skeletal muscle, bone marrow, myocardium, cortex, cerebellum or medulla). Virus isolation was attempted on all qRT-PCR positive samples but was not successful.

Table 7. Summary of laboratory results for progeny of cows infected with Bungowannah virus.

Calf	Gestational Age at Infection (Days)	Gestational Age at Birth (Days)	Skin (Copies/mL)	Serum (Copies/mL)	Testicle (Copies/mL)	VNT
1	63	274	1.5×10^6	-	Female	>512
2	65	276	8.1×10^4	-	Female	>512
3	65	276	2.4×10^7	5.0×10^2	6.4×10^5	>512
4	65	274	-	-	-	-
5	65	276	-	-	-	-
6	63	274	-	-	-	-
7	53	255	-	-	Female	-

A dash indicates that a result was undetected. Calf numbers correspond to cow numbers.

4. Discussion

In this study we have demonstrated that Bungowannah virus can infect a proportion of both cattle and sheep after intranasal or subcutaneous inoculation, although infection was less efficient by the intra-nasal route when compared to pigs [5]. Additionally, viral RNA was detected sporadically at low levels in the serum of a proportion of infected sheep and cattle. These results suggest that following intra-nasal or subcutaneous exposure, the infection is predominantly localized. Systemic infection, as determined by the detection of viraemia, only occurred in a proportion of infections.

No significant disease was observed in the infected sheep and cattle, despite all the intranasally infected cattle shedding Bungowannah virus RNA in nasal swabs for up to two weeks after infection. The qRT-PCR results indicated that only low levels of virus were shed. The pattern of detection of RNA was consistent with virus replication rather than residual inoculum as there was a short period after inoculation when no RNA was detected. Similarly, viral RNA was detected in the nasal swabs of five of the six sheep infected intranasally, yet no disease was observed. It is interesting to note that the development of the humoral immune response in the two subcutaneously inoculated sheep occurred over a similar period as the intranasally inoculated sheep. We speculate the infection was localised at the injection site in these two animals and, in the absence of a viraemia, shedding via the upper respiratory tract did not occur. The low level of viral RNA detection in nasal secretions of sheep and cattle would suggest that ruminants are unlikely to play a significant role in transmission of the virus.

It was also possible to infect a calf through co-habitation with a chronically infected pig for 48 h. Interestingly, this calf was the most curious of the four calves in the trial and interacted the most with the pigs. It is possible that if given longer exposure to the infected pigs the other calves may have become infected.

Although all cows became infected, transplacental infection of the calf only occurred in less than half of the cases. Viraemia was associated with in utero infection in two of the four cows in which it was detected. It is noteworthy that Cow 3 developed a rising anti-Bungowannah virus antibody titre during the last trimester of pregnancy. The progeny of this cow was the calf with the highest levels of Bungowannah RNA at birth. We speculate that the dam's immune system may have been stimulated by transplacental transfer of Bungowannah virus antigen during the last trimester of pregnancy.

Each of the three calves that became infected had seroconverted by the time of birth. Persistent infections were not established, despite the cow being infected at a stage of gestation where immunotolerance and persistent infections would be expected, as usually occurs with BVDV infection. These three calves had significant levels of Bungowannah virus RNA present in skin samples at birth, over 210 days after their dams were infected. We hypothesise that these calves did develop a generalised infection in utero, but subsequently cleared the infection some time prior to birth. The skin and testes were the only sites where the viral RNA was not cleared. The testes are an immunoprivileged site, and so it may be that the infection continued at this site for a longer period of time, perhaps much closer to birth. It has also been suggested that skin is also immunoprivileged [14]. It has been shown that a live BVDV vaccine is able to cross the bovine placenta, cause a prolonged transient infection in the foetus and result in detection of viral RNA in the skin for many weeks after birth [15]. It appears a similar mechanism of infection has occurred in these calves with prolonged detection in the skin. The detection of viral RNA in the post-partum vaginal swabs of the three dams provides further evidence for a prolonged infection in these calves and is comparable to what was observed following infection in pregnant pigs [9].

The timing of infection of pregnant cows was designed to maximise the likelihood of producing persistently infected calves. Infection of cattle with BVDV during the first 90 days of gestation will generally produce calves that are immunotolerant to BVDV [16,17]. The results of this trial indicate that Bungowannah virus and BVDV infections in cattle behave differently, with Bungowannah failing to establish immunotolerance and persistence in the bovine foetus.

Understanding why Bungowannah virus resulted in a humoral response in these calves may lead to a deeper understanding of the functional differences between specific proteins in BVDV and Bungowannah virus and how they do or do not affect evasion of the host innate immune response in the bovine or porcine foetus [18].

In conclusion, while cattle and sheep can be infected with Bungowannah virus, they appear to be less susceptible to infection compared to pigs when challenged with a similar dose of virus. When compared to the infection of pigs [9], ruminant species appear to shed less virus and for a shorter period of time. As a result, transmission of Bungowannah virus is likely to be inefficient in ruminants and, without further host adaptation, it will probably not be able to be sustained in these species.

Author Contributions: Conceptualization, A.J.R., D.S.F. and P.D.K.; methodology, A.J.R., D.S.F. and P.D.K.; investigation, A.J.R.; data curation, A.J.R.; writing—original draft preparation, A.J.R.; writing—review and editing, A.J.R., D.S.F. and P.D.K. All authors have read and agreed to the published version of the manuscript.

Funding: This research received no external funding.

Acknowledgments: We wish to acknowledge Bob Rheinberger for his generosity and expertise while conducting the ultrasound examinations of the cattle fetuses. We are indebted to the staff of the Virology Laboratory at EMAI (in particular Katherine King and Melinda Frost) for their invaluable assistance during the testing of the samples described in this study, and to Xingnian Gu for his veterinary assistance during sample collection.

Conflicts of Interest: The authors declare no conflict of interest.

References

1. McOrist, S.; Thornton, E.; Peake, A.; Walker, R.; Robson, S.; Finlaison, D.; Kirkland, P.; Reece, R.; Ross, A.; Walker, K.; et al. An infectious myocarditis syndrome affecting late-term and neonatal piglets. *Aust. Vet. J.* **2004**, *82*, 509–511. [CrossRef]
2. Kirkland, P.D.; Frost, M.; Finlaison, D.S.; King, K.R.; Ridpath, J.F.; Gu, X. Identification of a novel virus in pigs-Bungowannah virus: A possible new species of pestivirus. *Virus Res.* **2007**, *129*, 26–34. [CrossRef] [PubMed]
3. Finlaison, D.S.; King, K.R.; Frost, M.J.; Kirkland, P.D. Field and laboratory evidence that Bungowannah virus, a recently recognised pestivirus, is the causative agent of the porcine myocarditis syndrome (PMC). *Vet. Microbiol.* **2009**, *136*, 259–265. [CrossRef] [PubMed]
4. Kirkland, P.D.; Read, A.J.; Frost, M.J.; Finlaison, D.S. Bungowannah virus—A probable new species of pestivirus—What have we found in the last 10 years? *Anim. Health Res. Rev.* **2015**, *16*, 60–63. [CrossRef] [PubMed]

5. Finlaison, D.S.; King, K.R.; Gabor, M.; Kirkland, P.D. An experimental study of Bungowannah virus infection in weaner aged pigs. *Vet. Microbiol.* **2012**, *160*, 245–250. [CrossRef] [PubMed]
6. Finlaison, D.S.; Cook, R.W.; Srivastava, M.; Frost, M.J.; King, K.R.; Kirkland, P.D. Experimental infections of the porcine foetus with Bungowannah virus, a novel pestivirus. *Vet. Microbiol.* **2010**, *144*, 32–40. [CrossRef] [PubMed]
7. Finlaison, D.S.; Kirkland, P.D. The outcome of porcine foetal infection with Bungowannah virus is dependent on the stage of gestation at which infection occurs. Part 1: Serology and virology. *Viruses* **2020**, *12*, 691.
8. Finlaison, D.S.; Kirkland, P.D. The outcome of porcine foetal infection with Bungowannah virus is dependent on the stage of gestation at which infection occurs. Part 2: Clinical signs and gross pathology. *Viruses*, under review.
9. Kirkland, P.D.; Frost, M.J.; King, K.R.; Finlaison, D.S.; Hornitzky, C.L.; Gu, X.; Richter, M.; Reimann, I.; Dauber, M.; Schirrmeier, H.; et al. Genetic and antigenic characterization of Bungowannah virus, a novel pestivirus. *Vet. Microbiol.* **2015**, *178*, 252–259. [CrossRef] [PubMed]
10. Smith, D.B.; Meyers, G.; Bukh, J.; Gould, E.A.; Monath, T.; Scott Muerhoff, A.; Pletnev, A.; Rico-Hesse, R.; Stapleton, J.T.; Simmonds, P.; et al. Proposed revision to the taxonomy of the genus Pestivirus, family Flaviviridae. *J. Gen. Virol.* **2017**, *98*, 2106–2112. [CrossRef] [PubMed]
11. Chakraborty, A.K.; Karam, A.; Mukherjee, P.; Barkalita, L.; Borah, P.; Das, S.; Sanjukta, R.; Puro, K.; Ghatak, S.; Shakuntala, I.; et al. Detection of classical swine fever virus E2 gene in cattle serum samples from cattle herds of Meghalaya. *Virus Dis.* **2018**, *29*, 89–95. [CrossRef] [PubMed]
12. Richter, M.; Reimann, I.; Wegelt, A.; Kirkland, P.D.; Beer, M. Complementation studies with the novel "Bungowannah" virus provide new insights in the compatibility of pestivirus proteins. *Virology* **2011**, *418*, 113–122. [CrossRef] [PubMed]
13. Kirkland, P.D.; MacKintosh, S.G. Ruminant Pestivirus Infections. In *Australia and New Zealand Standard Diagnostic Procedures*; Elizabeth Macarthur Agricultural Institute: Camden, Australia, 2006; pp. 1–30.
14. Frazer, I.H.; Thomas, R.; Zhou, J.; Leggatt, G.R.; Dunn, L.; McMillan, N.; Tindle, R.W.; Filgueira, L.; Manders, P.; Barnard, P.; et al. Potential strategies utilised by papillomavirus to evade host immunity. *Immunol. Rev.* **1999**, *168*, 131–142. [CrossRef] [PubMed]
15. Wernike, K.; Michelitsch, A.; Aebischer, A.; Schaarschmidt, U.; Konrath, A.; Nieper, H.; Sehl, J.; Teifke, J.P.; Beer, M. The Occurrence of a Commercial N(pro) and E(rns) Double Mutant BVDV-1 Live-Vaccine Strain in Newborn Calves. *Viruses* **2018**, *10*, 274. [CrossRef] [PubMed]
16. Liess, B.; Orban, S.; Frey, H.R.; Trautwein, G.; Wiefel, W.; Blindow, H. Studies on transplacental transmissibility of a bovine virus diarrhoea (BVD) vaccine virus in cattle. II. Inoculation of pregnant cows without detectable neutralizing antibodies to BVD virus 90-229 days before parturition (51st to 190th day of gestation). *Zent. Vet. B* **1984**, *31*, 669–681. [CrossRef]
17. Duffell, S.J.; Harkness, J.W. Bovine virus diarrhoea-mucosal disease infection in cattle. *Vet. Rec.* **1985**, *117*, 240–245. [CrossRef] [PubMed]
18. Peterhans, E.; Schweizer, M. BVDV: A pestivirus inducing tolerance of the innate immune response. *Biologicals* **2013**, *41*, 39–51. [CrossRef] [PubMed]

viruses

MDPI

Brief Report

Single-Round Infectious Particle Production by DNA-Launched Infectious Clones of Bungowannah Pestivirus

Anja Dalmann [1], Kerstin Wernike [1], Eric J. Snijder [2], Nadia Oreshkova [2], Ilona Reimann [1] and Martin Beer [1,*]

1 Institute of Diagnostic Virology, Friedrich-Loeffler-Institut, 17493 Greifswald-Insel Riems, Germany; anja.dalmann@fli.de (A.D.); kerstin.wernike@fli.de (K.W.)

2 Molecular Virology Laboratory, Department of Medical Microbiology, Leiden University Medical Center, 2333 ZA Leiden, The Netherlands; E.J.Snijder@lumc.nl (E.J.S.); nadia.oreshkova@wur.nl (N.O.)

* Correspondence: martin.beer@fli.de

Received: 15 July 2020; Accepted: 31 July 2020; Published: 4 August 2020

Abstract: Reverse genetics systems are powerful tools for functional studies of viral genes or for vaccine development. Here, we established DNA-launched reverse genetics for the pestivirus Bungowannah virus (BuPV), where cDNA flanked by a hammerhead ribozyme sequence at the 5′ end and the hepatitis delta ribozyme at the 3′ end was placed under the control of the CMV RNA polymerase II promoter. Infectious recombinant BuPV could be rescued from pBuPV-DNA-transfected SK-6 cells and it had very similar growth characteristics to BuPV generated by conventional RNA-based reverse genetics and wild type BuPV. Subsequently, DNA-based E^{RNS} deleted BuPV split genomes (pBuPV$\Delta E^{RNS}/E^{RNS}$)—co-expressing the E^{RNS} protein from a separate synthetic CAG promoter—were constructed and characterized in vitro. Overall, DNA-launched BuPV genomes enable a rapid and cost-effective generation of recombinant BuPV and virus mutants, however, the protein expression efficiency of the DNA-launched systems after transfection is very low and needs further optimization in the future to allow the use e.g., as vaccine platform.

Keywords: Bungowannah virus; flavivirus; reverse genetics; single round infectious particle

1. Introduction

Bungowannah virus (BuPV) is an atypical pestivirus (species *Pestivirus F*) within the genus *Pestivirus* of the *Flavivirdae* family [1]. The virus was isolated for the first time in 2003 from a large Australian integrated pig farm during an outbreak of sudden death in young pigs, followed by an increase in stillbirth [2,3]. Although BuPV represents a potential threat to commercial pig farming, it has not yet been reported from any other region or country [4,5].

BuPV has a positive sense RNA genome that is approximately 12.6 kb in length. A single open reading frame, flanked by 5′ and 3′ non-translated regions, encodes a polyprotein, which is co- and post-translationally processed into structural proteins (C, E^{RNS}, E1, E2) and non-structural proteins (N^{PRO}, p7, NS2/NS3 (NS2, NS3), NS4A, NS4B, NS5A, NS5B) [2]. The envelope protein E^{RNS} as well as the non-structural protein N^{PRO} are unique to pestiviruses.

Genomic and antigenic properties of BuPV, as well as its broad in vitro host cell tropism, indicate remarkable distance to previously described pestiviruses [6–9]. In general, pestiviruses infect host-specific cells of ruminant, porcine, or sheep origin. Classical swine fever virus (CSFV) can only infect porcine cells efficiently, while bovine viral diarrhea virus (BVDV) and border disease virus (BDV) have the potential to infect broader host spectra [10–14]. However, only BuPV could infect cell lines of African green monkey, bat, human, and mouse origin [9].

To study the special characteristics of BuPV in detail, a robust reverse genetics system (RGS) is essential. For other pestiviruses, such as CSFV or BVDV, many RGSs based on cDNA copies of the viral genome cloned into plasmid or bacterial artificial chromosome (BAC) vectors have already been reported [15–20]. Furthermore, recombinant pestiviruses were generated by full-length genome RT-PCR-based amplification and direct RNA generation from the amplicons without cloning steps [21]. All these techniques have in common the use of a bacteriophage T7 or SP6 RNA polymerase promoter for in vitro transcription to synthesize infectious positive strand RNA, which is subsequently transfected into cells to produce infectious virus progeny.

Here, we report a first dual promoter DNA-launched BuPV RGS, which is based on a cDNA plasmid (pBuPV), with a cytomegalovirus (CMV) immediate-early promoter as well as the bacteriophage RNA polymerase T7 (T7) promoter upstream of the BuPV genome in a mammalian expression vector. For the generation of correct 5′ and 3′ ends, self-cleaving ribozyme sequences were inserted. This construct enables the transcription of the BuPV DNA by the CMV promoter in the nucleus and by the T7 promoter in the cytoplasm of polymerase expressing BSR cells (BSR-T7/5), where the latter serves as proof of principle to demonstrate that the modifications of the genome do not affect the transcription. This plasmid allows the rescue of infectious virus without in vitro RNA synthesis, and the virus can be passaged efficiently. We also established a split genome construct (pBuPVΔE^{RNS}/E^{RNS}) with a large deletion in the E^{RNS} gene, preventing the efficient generation of virus progeny, and a synthetic CAG promoter followed by the genomic region encoding the BuPV-E^{RNS} protein downstream the T7 termination signal (T7$_{term}$) for the expression of BuPV-E^{RNS} and the production of single-round infectious particles (SRIPs) via *trans*-complementation. The particles generated in this way should be able to pass through an additional replication cycle to infect surrounding cells, but are not expected to be capable of further propagation.

2. Materials and Methods

2.1. Cells and Viruses

SK-6 cells (RIE262, Collection of Cell Lines in Veterinary Medicine (CCLV), Friedrich-Loeffler-Institut, Insel Riems, Germany) and BSR-T7/5 cells, constitutively expressing the T7 polymerase (RIE583, CCLV) [22], were grown in Dulbecco's Modified Eagle Medium (DMEM) supplemented with 10% fetal calf serum (FCS) at 37 °C and 5% CO_2. rBuPV^{RNA} was generated after RNA transfection of SK-6 cells with in vitro transcribed RNA from the previously described synthetic cDNA clone pA/BV [9]. The virus was propagated in SK-6 cells and virus stocks were generated after three cell culture passages.

2.2. Plasmid Construction

All plasmids were prepared by standard molecular biological methods and plasmid DNA was purified by the Qiagen Plasmid Midi kit (Qiagen, Hilden, Germany). The identity of the constructs was confirmed by Sanger sequencing using the Big Dye® Terminator v1.1 Cycle sequencing kit (Applied Biosystems, Foster City, CA, USA) and appropriate primers. Nucleotide sequences were read with an automatic sequencer (3130 Genetic Analyzer; Applied Biosystems, Foster City, CA, USA) and analyzed using Geneious software (version 10.2.3.). Primers used for cloning procedures are shown in Table 1.

Table 1. Primer sequences for plasmid construction.

Construct	Primer	Sequence 5′-3′
pBV_opt	Ph_Donor_C-Erns_F	GCCTGCCTATTGGTCGTGCCCGTGGGCTCCACCAACGTGACACAATG
	Ph_Donor_Erns_R	TCACCCCTAAGTCTGCATCGTATCTGCATGTGACTGCGCACTC
	Ph_Donor_Erns_F	GTTGACGGTTACACCGAGGTGGTGGAGAAGGCCAGGTCAAGTGG
	Bungo_2534R	CGCTAATGCGTACATGAATTC
	Bungo_Npro_Mut_Donor_F	CTTTGTACAAACCAAGAGAGATGTGAGGGATCCAAGTGTGTA
	Bungo_1079R	GTGGCATCTGGTGGTCTAG
	Bungo_Donor_IV_F	GGCACTTGTATTGACAAAGAGGGTAGCGTGCAATGCTACATAGGGGA
	Bungo_Donor_IV_R_new	TCTTTAGTTCCCTCTTCGGCGCGTACTAAACCGACGAAGTAGACCAC
	Bungo_Donor_V_F	GCCTACACACACCCTGGAGGTGTAAGCAGTGTGATGCATGTCACCGC
	Bungo_6186_R	CACCGAACCTATGTATTTTTGACATCACTGCCAACTGTTC
	Bungo_Donor_VI_F	ATCACCAAATCCAACAAATTCTCGAGGGTGGGAAAGAATATGTCGGCCAAGCCTA
	Bungo_Donor_VI_R_new	GGACCCCCCATAGACCGTATTTCTTGATGTCACCGGCATGCTCTTGCAAGTATTC
	Bungo_Donor_VII_F	GGCCAGAAAAATTGCCAGTAGTAAGGGCCCAGACCAGTACCAAAG
	Bungo_Donor_VII_R	CTGGTTGACCACTTCCCCTTTGTCCTTCTCTTATGTAGACGTTTC
	Bungo_Donor_VIII_F	GTAGATGATTGGATGGAAGGAGATTATGTAGAAGAAAAAAGACC
	Bungo_Donor_VIII_R	GGCCCCTTGATCGCAAAGGCTTCGCCAAAACTTTTCTCAGTTATC
	Bungo_Donor_IX_F	GGTCAACCAGACACTAGCGCTGGAAATAGTATGTTGAATGTACT
	Bungo_Donor_IX_R	GACAAGCAGGCATATTCTTCGTACGAGGGGGTTCCAAGAATAC
pBuPV	Bungo_LLHR_F	CGTCGTTATACCTGATGAGTCCGTGAGGACGAAACCCGGAGTCCCGGGTCGTATAACGACAGTAGTTCAA
	Bungo_LLHR_R	TTCGGATGCCCAGGTCGGACCGCGAGGAGGTGGAGATGCCATGCCGACCCAGGGCTTTTTGGAACTGTGC
	pHaHd_F	TGGGTCGGCATGGCATCTCC
	pHaHd_R	GACCCGGGACTCCGGGTTTCGTCCTCACGGACTCATCAGGTATAACGACGACTAGCCAGCTTG
pBuPVΔE^{RNS}	Bungo_dERNS_F	CATCTAGCAGCAGACTATGAAAGTAAGATTGAAAACACCAAGA
	Bungo_2164R	CATCACGAAGTCCCTGTTGTC

The infectious clone pBuPV, the basis for DNA-launched BuPV, was constructed by multiple cloning steps. In a first step, donor splice sites, that were detected in the BuPV genome with a confidence >0.65 using the NetGene2 Server (http://www.cbs.dtu.dk/services/NetGene2 [23,24]; Table S1), were mutated in the full-length clone "pA/BV." For this, fusion PCR was applied, which is a restriction-free cloning method [7], using the Phusion® High-Fidelity PCR kit (New England Biolabs, Ipswich, MA, USA) and the primers Ph_Donor_C-Erns_F, Ph_Donor_Erns_R, Ph_Donor_Erns_F, Bungo_2534R, Bungo_Npro_Mut_Donor_F, Bungo_1079R, Bungo_Donor_IV_F, Bungo_Donor_IV_R_new, Bungo_Donor_V_F, Bungo_6186_R, Bungo_Donor_VI_F, Bungo_Donor_VI_R_new, Bungo_Donor_VII_F, Bungo_Donor_VII_R, Bungo_Donor_VIII_F, Bungo_Donor_VIII_R, Bungo_Donor_IX_F, and Bungo_Donor_IX_R (biomers.net GmbH, Ulm Germany), resulting in plasmid "pBV_opt."

For pBuPV generation, the well characterized plasmid pHaHd was used, which contains in addition to the CMV and T7 promoters, ribozyme sequences for the generation of correct genome sequences. In order to insert the optimized BuPV-specific cDNA into plasmid pHaHd [25] by linear-to-linear homologous recombination (LLHR) [26], plasmid pHaHd was linearized with *BglII* and used as template for PCR amplification of a linear vector fragment with the Phusion® High-Fidelity PCR kit and primers pHaHd_F and pHaHd_R. A full-length BuPV-specific PCR fragment was amplified by using plasmid DNA pBV_opt as template and primers Bungo_LLHR_F and Bungo_LLHR_R. To allow homologous recombination of the virus- and vector-specific fragments, 50 nucleotide-long vector specific homology arms were included in the primer sequences. Furthermore, for correct cleavage by the synthetic hammerhead ribozyme (HHr), five nucleotides complementary to the 5′ end of the BuPV-genome had to be inserted into primer Bungo_LLHR_F (Table 1). Thereafter, the PCR fragments were digested with *DpnI* to remove residual template DNA and gel purified (QIAquick Gel Extraction kit; Qiagen, Hilden, Germany). Subsequently, both PCR fragments were subjected to LLHR [26]. Recombination was performed by electroporation of both DNA fragments in the *E. coli* strain GB05-dir (Gene Bridges, Heidelberg, Germany). In brief, fresh overnight cultures in lysogeny broth (LB) medium were incubated at 37 °C for 1.5 h and RecE/RecT recombination was induced by L-Arabinose. After an additional incubation period at 37 °C for 30 min, the cells were washed two times with ice-cold water. A total of 100 ng of the amplified DNA fragments were added to the pelleted bacteria and electroporation was done at 1350 V, 50 μF, and 600 Ω by using the Gene pulser Xcell Electroporation System (Bio-Rad, Hercules, CA, USA).

Plasmid pBuPVΔE^{RNS} with a deletion of 448 bases within the E^{RNS} protein (aa 328–483) was generated by fusion PCR using pBuPV as DNA template and primer pair Bungo_dERNS_F and Bungo_2164R. For construction of the split genome plasmid pBuPV$\Delta E^{RNS}/E^{RNS}$, plasmid pCAGGS_BuPV-E^{RNS} [27] was digested with *SmaI* and *NotI* and the E^{RNS} comprising fragment was ligated into plasmid pBuPVΔE^{RNS}, digested with *PmeI* and *NotI*. Further details of the plasmid constructions are available on request.

2.3. cDNA Stability

The plasmid pBuPV was propagated for 10 passages in *E. coli*. Subsequently, the DNA was purified using the Qiagen Plasmid Mini kit (Qiagen, Hilden, Germany) and analyzed by *HindIII* digestion. DNA preparations of passages 5 and 10 were also used to transfect SK-6 or BSR-T7/5 cells and investigate for virus rescue.

2.4. Transfection and Virus Rescue

DNA transfections of plasmids pBuPV, pBuPVΔE^{RNS} and pBuPV$\Delta E^{RNS}/E^{RNS}$ (2 μg DNA each) into SK-6 (plasmid pBuPV) or BSR-T7/5 (all plasmids) cells were performed by using Lipofectamine™ 2000 Transfection Reagent (Invitrogen, Carlsbad, CA, USA) according to the manufacturer's protocol. The transfected cells were seeded and incubated for three days at 37 °C and 5% CO_2. For recovery of infectious viruses, supernatants of the transfected cells were harvested three days post transfection (p.t.) and passaged on SK-6 cells. At the day of collection, replication of BuPV was monitored by immunofluorescence (IF) staining using monoclonal antibodies. Virus stocks of pBuPV were prepared after four cell culture passages. The identity of the recombinant viruses was confirmed by RT-PCR and sequence analysis using appropriate primers. For RT-PCR, total RNA of virus-infected cells was extracted using the QIAamp Viral RNA Mini kit (Qiagen, Hilden, Germany) according to manufacturer's instructions, and the cDNA was amplified using the OneStep RT-PCR kit (Qiagen, Hilden, Germany).

2.5. *Immunofluorescence Assay*

Transfected or infected cells were fixed and permeabilized with 80% acetone on ice for 15 min. After 30 to 45 min incubation with monoclonal antibodies specific for BuPV-E^{RNS} (682/43C3, diluted 1:20; M. Dauber, Friedrich-Loeffler-Institut, Insel Riems, Germany), for BuPV-E2 (682/45F12, diluted 1:20; M. Dauber, Friedrich-Loeffler-Institut, Insel Riems, Germany) or the pan-*Pestivirus* NS3 antibody WB112 (diluted 1:500; CVL, Weybridge, UK), the cells were washed twice with phosphate buffered saline (PBS). Thereafter, cell cultures were incubated with a goat anti-mouse Ig Alexa-488 conjugate (1:1000; Thermo Fischer scientific Inc., Waltham, MA, USA) for 30 min and analyzed by using a fluorescence microscope (Nikon Eclipse; Nikon GmbH, Düsseldorf, Germany).

2.6. *Virus Titration and Growth Kinetics*

SK-6 cells were infected with recombinant BuPV recovered from pBuPV ($rBuPV^{DNA}$) and recombinant BuPV recovered from pA/BV ($rBuPV^{RNA}$) at a multiplicity of infection (M.O.I.) of 1. Supernatants were collected at 0, 8, 24, 48, and 72 h post infection (p.i.), and virus titers were calculated as a 50% tissue culture infective dose per ml ($TCID_{50}/_{mL}$) after IF staining.

3. Results and Discussion

Reverse genetics systems are important tools that enable the investigation of viral genes, viral replication cycles, or pathogenesis, and allow for the development of safe and efficacious vaccines. Infectious virus production from DNA plasmid transfections into mammalian cells using RNA polymerase I (pol I) or RNA pol II systems had been described for several RNA viruses [28–32]. These systems allow a simple and stable virus rescue, faster and less costly than conventional RNA-based reverse genetics. RGS using a pol I-promoter [33] or a pol II-promoter [34] were also described previously for the pestivirus CSFV. Both systems allowed the rescue of infectious CSFV with high virus titers. In our study, we established a dual promoter BuPV infectious clone "pBuPV" with both the CMV pol II promoter and the bacteriophage T7 promoter, which allows virus generation via DNA through the nucleus (CMV pol II promoter) or cytoplasmatic T7-based generation. The T7-based generation served as a proof of principle to demonstrate that the modifications of the genome do not affect the transcription.

In order to prevent the viral RNA from being spliced in the nucleus, donor splice sites with confidence >0.65 detected in N^{PRO}, C, E^{RNS}, E2, p7, and the other non-structural proteins were mutated in pA/BV (Table S1). Subsequently, the optimized BuPV-specific cDNA was used for construction resulting in plasmid pBuPV. In this construct, downstream of the CMV and bacteriophage T7 promoters, the BuPV genome termini were flanked by sequences coding for a synthetic HHr and hepatitis delta virus ribozyme (HDVr) to generate precise 5′- and 3′-terminal sequences. Downstream of the HDVr sequence, a T7 termination ($T7_{term}$) signal allows transcription termination (Figure 1A).

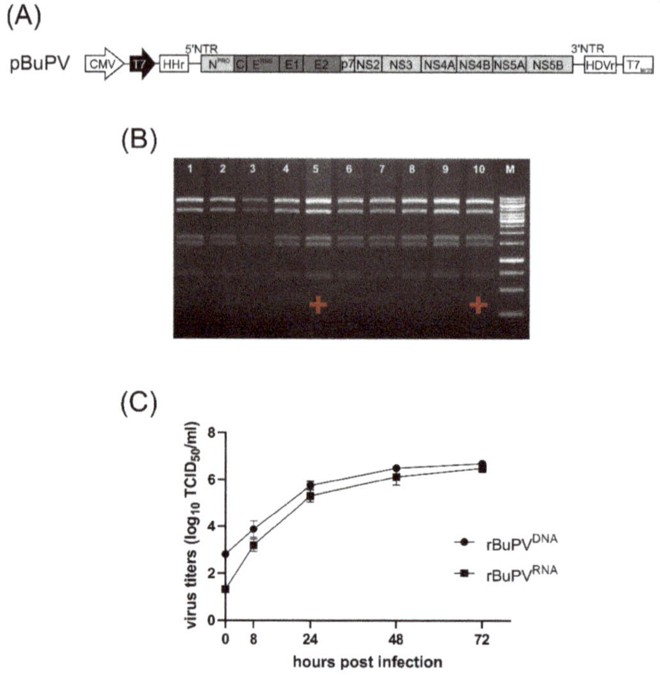

Figure 1. Plasmid pBuPV, genetic stability and rescue of recombinant rBuPV^{DNA}. (**A**) Schematic representation of the RNA Polymerase II-based plasmid pBuPV encoding full-length BuPV cDNA. Indicated are cytomegalovirus immediate-early (CMV) RNA Polymerase II promoter (open arrow), bacteriophage RNA polymerase T7 (T7) promoter (shaded arrow), hammerhead ribozyme (HHr), 3′ hepatitis delta virus ribozyme (HDVr), and T7 terminator sequence (T7$_{term}$). (**B**) Stability of the full-length cDNA clone pBuPV. The primary plasmid (P0) was passaged 10 times in *E. coli* DH10B (P1–P10) and investigated by restriction analysis using *Hind*III. + indicates generation of infectious rBuPV^{DNA} in rescue experiments. (**C**) Multi-step growth curves determined after infection of SK-6 cells with rBuPV^{DNA} (rescued by DNA transfection of pBuPV) or rBuPV^{RNA} (rescued by transfection of in vitro transcribed RNA of the infectious cDNA clone pA/BV) at an M.O.I. of 1 showed similar growth characteristics for both viruses.

The stability of plasmid pBuPV in bacteria was investigated by 10 serial cloning and passaging cycles in *E. coli* DH10B cells. Restriction enzyme analysis using *Hind*III provided some indication about the genetic stability of the construct, and the same restriction pattern for P0 (primary construct) and passages P1 to P10 were observed (Figure 1B). In transfection experiments, the DNA-launched recombinant rBuPV^{DNA} was analyzed for RNA replication, expression of BuPV proteins, and virus growth in both SK-6 and BSR-T7/5 cells. At 72 h p.t., CMV-driven expression of the BuPV proteins NS3 (anti-NS3), E2 (anti-E2), and E^{RNS} (anti-E^{RNS}) was detected by IF staining of the pBuPV-transfected SK-6 cells.

Bacteriophage T7 RNA polymerase (T7-RNA-Polymerase)-driven cytoplasmatic expression of all proteins could be observed as well after transfection in BSR-T7/5 cells; DNA-transfection of SK-6 cells and transfection in BSR-T7/5 cells resulted in single NS3, E2, and E^{RNS} expressing cells (Figure 2). Cell culture supernatants collected from both transfected cell lines were inoculated into fresh SK-6 cells and the presence of recombinant BuPV (rBuPV^{DNA}) particles in supernatants of both cell lines could be confirmed by IF analysis. The rescued rBuPV^{DNA} could be efficiently passaged in SK-6 cells (Figure 2). Virus rescue was possible, regardless of the bacterial passage number of pBuPV in *E. coli*, indicating once more the general stability of the plasmid (data not shown). Virus stocks produced from pBuPV-transfected SK-6 cells after four passages in SK-6 cells with a titer of $10^{6.25}$ $TCID_{50}/_{mL}$ were used for growth kinetics analyses in comparison to rBuPV generated by the previously established T7-RNA-Polymerase based RGS [9]. Analysis of the multi-step growth curves in SK-6 cells revealed similar growth characteristics for both viruses, and final virus titers of $10^{6.6}$ $TCID_{50}/_{mL}$ ($rBuPV^{DNA}$) and $10^{6.5}$ $TCID_{50}/_{mL}$ ($rBuPV^{RNA}$) could be determined at 72 h p.i. (Figure 1C).

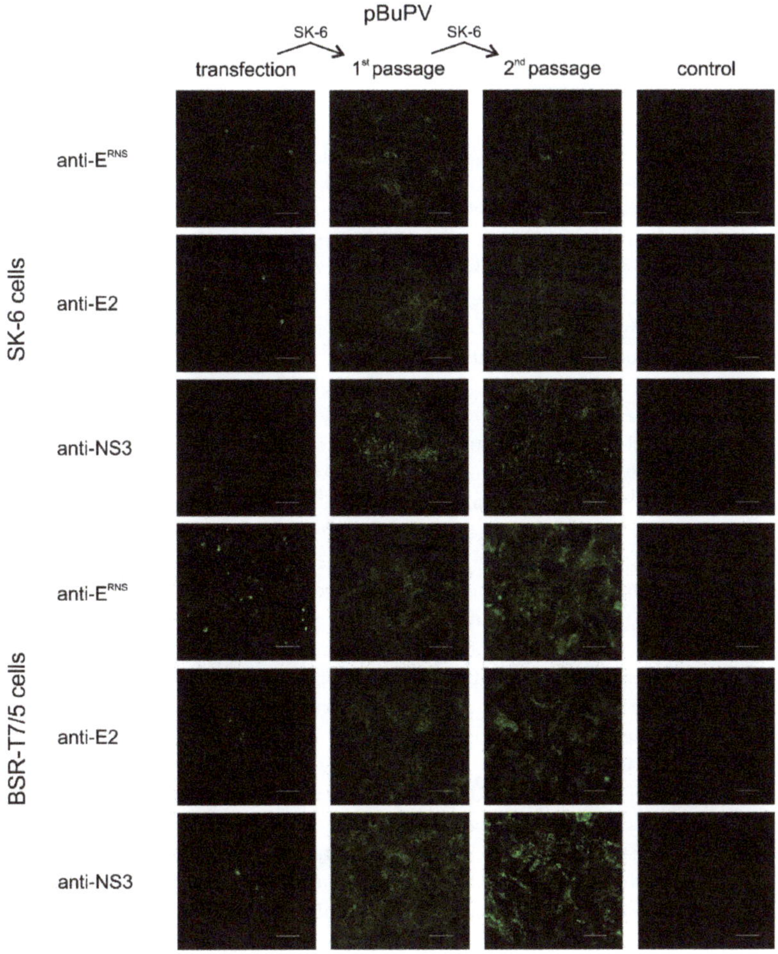

Figure 2. Rescue of $rBuPV^{DNA}$ in BSR-T7/5 and SK-6 cells. Cells were transfected with plasmid pBuPV. At 72 h p.t, IF staining with pan-pesti NS3-specific mab WB112 (anti-NS3), and E2-specific and E^{RNS}-specific mabs verified expression of NS3, E2 and E^{RNS} in transfected cells. At this time, recombinant virus in the supernatants was transferred to SK-6 cells (1st passage), and later on transferred for a 2nd passage. E2, E^{RNS} and NS3-positive cells indicated the generation of infectious progeny virus at 72 h p.i. in both transfected SK-6 and BSR-T7/5 cells. Scale bars indicate 100 μm.

In a next step, we were interested in the production of BuPV single-round infectious particles (SRIPs), since RNA-based SRIPs generating systems have already been described for several other flaviviruses [35–38]. SRIP production relies on the transfection of in vitro transcribed replicon

RNA with deletions within the genomic region encoding for one of the structural proteins C, E^{RNS}, E1, or E2 in cells stably expressing either the protein missing in the replicon or all structural proteins [39–43]. The packaged replicon particles are infectious, but progeny virus cannot spread from the infected cells, because the packaged replicon genome lacks the respective structural protein genes. In experimental animal studies, packaged replicon particles were proven to be appropriate for the development of non-transmissible, life attenuated pestivirus marker vaccine candidates [44–46]. However, the production of the replicon particles is time-consuming and needs the establishment of *trans*-complementing cell lines, which in many cases do not allow further passaging of the packaged pestivirus replicon particles.

Here, DNA-based SRIPs were produced as packaged BuPV replicon particles. Other DNA-based flavivirus SRIPs generating systems are mostly based on co-transfection of two expressing plasmids directly in eukaryotic cells, a subgenomic replicon plasmid, which lacks the structural protein-coding region, and a structural protein-expressing plasmid [35–38]. In addition, split genomes with two CMV promoters in back-to-back orientation, directing either the transcription of a capsid-deleted replicon RNA or the transcription of capsid-encoding mRNA had been described [47]. Since this strategy was described for the flavivirus West Nile virus, but had not been applied to pestiviruses up to now, we first constructed pBuPVΔE^{RNS}, which is a DNA-based replicon plasmid with a deletion of a large portion (codons 328–483) of the genome region encoding E^{RNS} (Figure 3A, upper panel). Subsequently, this plasmid was used for the establishment of the split genome construct pBuPV$\Delta E^{RNS}/E^{RNS}$ (Figure 3A, lower panel). In transfection experiments using pBuPVΔE^{RNS} and BSR-T7/5 cells, transient expression of NS3 could be detected at 24 h p.t. by IF staining, while expression of BuPV-E^{RNS} could not be observed (Figure 3C, panels c–d). No infectious recombinant BuPV could be recovered, even after serial passages in SK-6 cells (Figure 3C, panels k–l and s–t).

By insertion of a synthetic CAG promoter and the genomic region encoding the N-terminal signal sequence and the BuPV-E^{RNS} protein in plasmid pBuPVΔE^{RNS} downstream the $T7_{term}$ signal, the split genome plasmid pBuPV$\Delta E^{RNS}/E^{RNS}$ was generated (Figure 3A, lower panel). This construct is capable of transcribing two separate RNA species from two different promoters.

The CMV promoter directs the transcription of BuPV replicon RNA BuPVΔE^{RNS}, which expresses all non-structural protein genes and the structural protein genes *C*, *E1*, *E2*, and a truncated E^{RNS} gene (ΔE^{RNS}), whereas a synthetic CAG promoter [48] downstream of the BuPV replicon genome directs transcription of mRNA encoding full-length E^{RNS} for complementation. Together, the two promoters allow the expression of the complete BuPV genome including all structural proteins. The E^{RNS}-deleted replicon genome is amplified by the BuPV non-structural proteins NS3, NS4A, NS4B, NS5A, and NS5B and can be packaged by the structural proteins C, E^{RNS}, E1, and E2, essential for virus assembly, to generate SRIPs (Figure 3B). Secreted SRIPs are able to infect surrounding cells, where the replicon RNA can be replicated. As the structural proteins C, E1, and E2, but not BuPV-E^{RNS} can be expressed from this RNA by the non-structural proteins, the RNA cannot be packaged again into new particles and no further spread of the SRIPs is possible (Figure 3B) resulting in a self-restricted system.

To examine the ability of the split genome plasmid pBuPV$\Delta E^{RNS}/E^{RNS}$ to produce SRIPs, BSR-T7/5 cells were transfected with pBuPV$\Delta E^{RNS}/E^{RNS}$ and compared with cells transfected with the replicon plasmid pBuPVΔE^{RNS} and the full-length plasmid pBuPV. Autonomous replication of the newly synthesized BuPV RNA was shown by the expression of BuPV-NS3 in cells transfected with plasmids pBuPVΔE^{RNS} and full-length pBuPV. However, unexpectedly no NS3 expression was observed for the SRIPs, which might be due to the low transfection efficiency, especially since also only single positive cells could be shown for the other constructs. Expression of E^{RNS} could only be detected in cells transfected with pBuPV$\Delta E^{RNS}/E^{RNS}$ and the full-length pBuPV, but not in pBuPVΔE^{RNS} DNA transfected SK-6 cells (Figure 3C, panels a, c, and e). The observation that the SRIPs showed an increased E^{RNS} expression (Figure 3C, panel a) could be related to the fact that the inserted sequence was optimized for the expression system applied in this study.

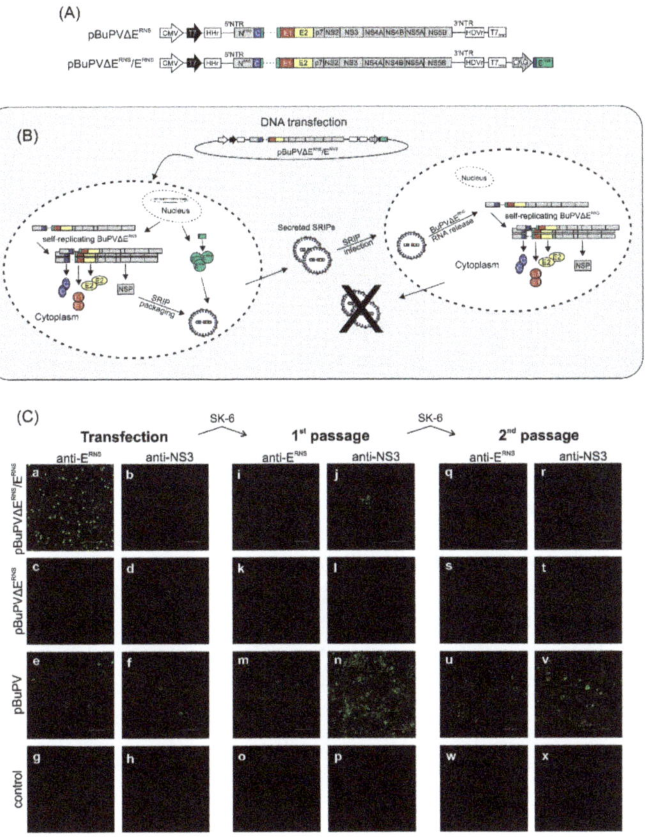

Figure 3. Schematic representation of the plasmids pBuPVΔE^{RNS} and pBuPVΔE^{RNS}/E^{RNS} and production of rBuPV and BuPV-SRIPs. (**A**) The replicon construct pBuPVΔE^{RNS} was generated on the basis of the CMV immediate early promoter containing plasmid pBuPV by partial deletion of the E^{RNS} encoding genomic region (aa 328–483); the split genome plasmid pBuPVΔE^{RNS}/E^{RNS} contains two eukaryotic promoters. The CMV promoter (open arrow) controls transcription of BuPV replicon RNA, BuPVΔE^{RNS}, which expresses the non-structural protein genes and the structural protein genes C, E1, and E2. The CAG promoter (shaded arrow) downstream T7$_{term}$ directs the expression of BuPV-E^{RNS}. Indicated is also the T7 promoter (black arrow), and the hammerhead ribozyme (HHr), and the 3′ hepatitis delta virus ribozyme (HDVr), which are important for the generation of the correct termini of the transcribed replicon RNAs. (**B**) Generation and operation mode of BuPV single round infectious particles (SRIPs) [47]. When cDNA of pBuPVΔE^{RNS}/E^{RNS} is transfected into susceptible cells, RNA-transcription starts in the nucleus under the control of the CMV promoter. The structural proteins C, E1, E2, and the non-structural proteins (NSP) are then expressed in the cytoplasm, while E^{RNS} is expressed by the CAG promoter. The self-replicating, truncated RNAs can be packaged in SRIPs by the four essential structural proteins. The secreted SRIPs are able to infect new cells. The released RNA replicates autonomously in the cytoplasm and allows the expression of the structural proteins C, E1, and E2 but not of E^{RNS}. Therefore, no further SRIPs can be produced and spread again (self-restriction). (**C**) IF analysis of BSR-T7/5 cells transfected with pBuPVΔE^{RNS}, pBuPVΔE^{RNS}/E^{RNS} or pBuPV (a–f) or SK-6 cells infected with supernatants of DNA-transfected cells (1st passage, i–n) or infected with supernatants collected from cells after the first infection (2nd passage, q–v). IF staining using anti-NS3 or anti-E^{RNS} monoclonal antibodies was performed at 72 h p.t. and 72 h p.i., respectively. Non-transfected or uninfected cells were used as control (g–h, o–p, and w–x). Scale bars indicate 100 μm.

When in vitro-transcribed RNA produced from an infectious cDNA clone of BuPV was transfected, the deletion of E^{RNS} still allowed the generation of infectious particles [27]. However, in this study there were no indications of cell-to-cell spread in pBuPVΔE^{RNS} DNA-transfected cells or production of SRIPs in transfection supernatants (Figure 3C, panels c, d, k, and l), while transfection with pBuPV resulted in single NS3 and E^{RNS} positive cells. In cells transfected with pBuPVΔE^{RNS}/E^{RNS}, amplified BuPVΔE^{RNS} replicon RNA was packaged into BuPV-SRIPs (Figure 3C, panel a). The infectivity of BuPV-SRIPs and rBuPV was also demonstrated after inoculation of the supernatants of the pBuPVΔE^{RNS}/E^{RNS} DNA-transfected cells to fresh SK-6 cells (1st passage) as indicated by the IF-detection of NS3 at 72 h p.i. Here, replicon RNA amplified itself since no additional SRIPs were produced; only single cells infected with BuPV-SRIPs were observed by IF staining using the anti-NS3 mab and no positive signals could be detected by E^{RNS} staining (Figure 3C, panels i and j).

In contrast, after one passage, the cells infected with rBuPV produced large foci of BuPV-NS3 and -E^{RNS} protein-positive cells resulting from virus replication and spread (Figure 3C, panel m and n). The supernatants from all investigated clones obtained after infection were again transferred to fresh cells (2nd passage), but infectious progeny virus was only detected in rBuPV infected cells (Figure 3C, panels u, v).

While the newly established DNA-based system is well suited to produce infectious viruses, it unfortunately shows only a very low transfection efficiency, as demonstrated by the IF staining of relatively few positive cells at 72 h after transfection (Figures 2 and 3). A possible explanation for the reduced efficiency might be the still insufficient removal of donor splice sites. As BuPV does not naturally replicate in the cell nucleus, we have modified several donor splice sites, although additional splice sites may have an impact on efficiency [49,50]. In addition, further optimization attempts such as the modification of the CMV promoter or the insertion of a polyA tail were made following other flavivirus systems, but they did not lead to an increase in efficiency after transfection for the newly established BuPV system (data not shown).

However, also SRIP systems for the flavivirus West Nile virus showed reduced efficiency after transfection, but still were able to induce neutralizing antibodies and confer protection in immunized mice and horses [47]. Whether this is the case for the BuPV system as well, needs to be evaluated in vaccination experiments. Nevertheless, since a high transfection efficiency might be important for SRIP systems, further optimization of the BuPV system will be necessary and useful. Roby et al. could increase the SRIP production of their beta-galactosidase expression system by using an elongations factor EF1α promoter for the expression and optimized the codon of the capsid protein [51]. Thus, the replacement of the CMV promoter by another, more efficient promoter in SK-6 cells, e.g., the CAG promoter, might be an option to improve the BuPV system in the future. In addition, it could be explored whether other structures of the system, such as the ribozyme sequences, influence the efficiency as was shown previously [31,52,53].

4. Conclusions

In summary, we established a cDNA clone for the rescue of infectious BuPV using an RNA polymerase II-driven system. The full-length clone pBuPV enables further investigation of the atypical pestivirus BuPV by rapid and cost-effective generation of BuPV mutants. Even though the split-genome strategy has to be optimized due to its very low transfection efficiency in cell culture, it could allow the establishment of a platform for *trans*-complementation of viral proteins with a single plasmid. A more efficient version could be an attractive alternative to conventional complementation approaches and will therefore be a major focus for the future research in this field.

Supplementary Materials: Supplementary materials can be found at http://www.mdpi.com/1999-4915/12/8/847/s1. Table S1: Mutated sequences for donor splicing site deletion.

Author Contributions: Conceptualization, E.J.S., N.O., I.R., and M.B.; methodology, A.D. and I.R.; formal analysis, I.R.; investigation, A.D., K.W., and I.R.; writing—original draft preparation, A.D. and I.R.; writing—review and editing, K.W., E.J.S., N.O., and M.B.; visualization, A.D. and K.W.; supervision, K.W., I.R., and M.B. All authors have read and agreed to the published version of the manuscript.

Funding: This research was funded by the Zoonoses Anticipation and Preparedness Initiative (ZAPI, Grant Agreement No. 115760) within the Innovative Medicines Initiative (IMI Call 11—IMI-JU-11-2013-04).

Acknowledgments: We thank Doreen Schulz and Gabriela Adam for excellent technical assistance and Stefan Finke for kindly providing the plasmid pHaHd.

Conflicts of Interest: The authors declare no conflict of interest. The funders had no role in the design of the study; in the collection, analyses, or interpretation of data; in the writing of the manuscript, or in the decision to publish the results.

References

1. Simmonds, P.; Becher, P.; Bukh, J.; Gould, E.A.; Meyers, G.; Monath, T.; Muerhoff, S.; Pletnev, A.; Rico-Hesse, R.; Smith, D.B.; et al. ICTV Virus Taxonomy Profile: Flaviviridae. *J. Gen. Virol.* **2017**, *98*, 2–3. [CrossRef] [PubMed]
2. Kirkland, P.D.; Frost, M.J.; Finlaison, D.S.; King, K.R.; Ridpath, J.F.; Gu, X. Identification of a novel virus in pigs—Bungowannah virus: A possible new species of pestivirus. *Virus Res.* **2007**, *129*, 26–34. [CrossRef] [PubMed]
3. McOrist, S.; Thornton, E.; Peake, A.; Walker, R.; Robson, S.; Finlaison, D.; Kirkland, P.; Reece, R.; Ross, A.; Walker, K.; et al. An infectious myocarditis syndrome affecting late-term and neonatal piglets. *Aust. Vet. J.* **2004**, *82*, 509–511. [CrossRef] [PubMed]
4. Abrahante, J.E.; Zhang, J.W.; Rossow, K.; Zimmerman, J.J.; Murtaugh, M.P. Surveillance of Bungowannah pestivirus in the upper Midwestern USA. *Transbound. Emerg. Dis.* **2014**, *61*, 375–377. [CrossRef]
5. Michelitsch, A.; Dalmann, A.; Wernike, K.; Reimann, I.; Beer, M. Seroprevalences of newly discovered porcine pestiviruses in German pig farms. *Vet. Sci.* **2019**, *6*, 86. [CrossRef]
6. Kirkland, P.D.; Frost, M.J.; King, K.R.; Finlaison, D.S.; Hornitzky, C.L.; Gu, X.; Richter, M.; Reimann, I.; Dauber, M.; Schirrmeier, H.; et al. Genetic and antigenic characterization of Bungowannah virus, a novel pestivirus. *Vet. Microbiol.* **2015**, *178*, 252–259. [CrossRef]
7. Richter, M.; König, P.; Reimann, I.; Beer, M. N^{pro} of Bungowannah virus exhibits the same antagonistic function in the IFN induction pathway than that of other classical pestiviruses. *Vet. Microbiol.* **2014**, *168*, 340–347. [CrossRef]
8. Richter, M.; Reimann, I.; Wegelt, A.; Kirkland, P.D.; Beer, M. Complementation studies with the novel "Bungowannah" virus provide new insights in the compatibility of pestivirus proteins. *Virology* **2011**, *418*, 113–122. [CrossRef]
9. Richter, M.; Reimann, I.; Schirrmeier, H.; Kirkland, P.D.; Beer, M. The viral envelope is not sufficient to transfer the unique broad cell tropism of Bungowannah virus to a related pestivirus. *J. Gen. Virol.* **2014**, *95 Pt 10*, 2216–2222. [CrossRef]
10. Liess, B.; Moennig, V. Ruminant pestivirus infection in pigs. *Rev. Sci. Tech.* **1990**, *9*, 151–161. [CrossRef]
11. Fernelius, A.L.; Lambert, G.; Hemness, G.J. Bovine viral diarrhea virus-host cell interactions: Adaptation and growth of virus in cell lines. *Am. J. Vet. Res.* **1969**, *30*, 1561–1572. [PubMed]
12. Ames, T.R. Hosts. In *Bovine Viral Diarrhea Virus: Diagnosis, Management, and Control*; Goyal, S.M., Ridpath, J.F., Eds.; Blackwell Publishing: Oxford, UK, 2005; p. 175.
13. Nettleton, P.F. Pestivirus infections in ruminants other than cattle. *Rev. Sci. Tech.* **1990**, *9*, 131–150. [CrossRef] [PubMed]
14. Løken, T. Ruminant pestivirus infections in animals other than cattle and sheep. *Vet. Clin. N. Am. Food Anim. Pract.* **1995**, *11*, 597–614. [CrossRef]
15. Rasmussen, T.B.; Risager, P.C.; Fahnøe, U.; Friis, M.B.; Belsham, G.J.; Höper, D.; Reimann, I.; Beer, M. Efficient generation of recombinant RNA viruses using targeted recombination-mediated mutagenesis of bacterial artificial chromosomes containing full-length cDNA. *BMC Genom.* **2013**, *14*, 819. [CrossRef] [PubMed]

16. Mischkale, K.; Reimann, I.; Zemke, J.; König, P.; Beer, M. Characterisation of a new infectious full-length cDNA clone of BVDV genotype 2 and generation of virus mutants. *Vet. Microbiol.* **2010**, *142*, 3–12. [CrossRef] [PubMed]
17. Moormann, R.J.; van Gennip, H.G.; Miedema, G.K.; Hulst, M.M.; van Rijn, P.A. Infectious RNA transcribed from an engineered full-length cDNA template of the genome of a pestivirus. *J. Virol.* **1996**, *70*, 763–770. [CrossRef] [PubMed]
18. Ruggli, N.; Tratschin, J.D.; Mittelholzer, C.; Hofmann, M.A. Nucleotide sequence of classical swine fever virus strain Alfort/187 and transcription of infectious RNA from stably cloned full-length cDNA. *J. Virol.* **1996**, *70*, 3478–3487. [CrossRef]
19. Vassilev, V.B.; Collett, M.S.; Donis, R.O. Authentic and chimeric full-length genomic cDNA clones of bovine viral diarrhea virus that yield infectious transcripts. *J. Virol.* **1997**, *71*, 471–478. [CrossRef]
20. Kümmerer, B.M.; Meyers, G. Correlation between point mutations in NS2 and the viability and cytopathogenicity of Bovine viral diarrhea virus strain Oregon analyzed with an infectious cDNA clone. *J. Virol.* **2000**, *74*, 390–400. [CrossRef]
21. Rasmussen, T.B.; Reimann, I.; Uttenthal, A.; Leifer, I.; Depner, K.; Schirrmeier, H.; Beer, M. Generation of recombinant pestiviruses using a full-genome amplification strategy. *Vet. Microbiol.* **2010**, *142*, 13–17. [CrossRef]
22. Buchholz, U.J.; Finke, S.; Conzelmann, K.K. Generation of bovine respiratory syncytial virus (BRSV) from cDNA: BRSV NS2 is not essential for virus replication in tissue culture, and the human RSV leader region acts as a functional BRSV genome promoter. *J. Virol.* **1999**, *73*, 251–259. [CrossRef] [PubMed]
23. Hebsgaard, S.M.; Korning, P.G.; Tolstrup, N.; Engelbrecht, J.; Rouzé, P.; Brunak, S. Splice site prediction in Arabidopsis thaliana pre-mRNA by combining local and global sequence information. *Nucleic Acids Res.* **1996**, *24*, 3439–3452. [CrossRef] [PubMed]
24. Brunak, S.; Engelbrecht, J.; Knudsen, S. Prediction of human mRNA donor and acceptor sites from the DNA sequence. *J. Mol. Biol.* **1991**, *220*, 49–65. [CrossRef]
25. Orbanz, J.; Finke, S. Generation of recombinant European bat lyssavirus type 1 and inter-genotypic compatibility of lyssavirus genotype 1 and 5 antigenome promoters. *Arch. Virol.* **2010**, *155*, 1631–1641. [CrossRef] [PubMed]
26. Nolden, T.; Pfaff, F.; Nemitz, S.; Freuling, C.M.; Höper, D.; Müller, T.; Finke, S. Reverse genetics in high throughput: Rapid generation of complete negative strand RNA virus cDNA clones and recombinant viruses thereof. *Sci. Rep.* **2016**, *6*, 23887. [CrossRef] [PubMed]
27. Dalmann, A.; Reimann, I.; Wernike, K.; Beer, M. Autonomously replicating RNAs of Bungowannah pestivirus: E^{RNS} is not essential for the generation of infectious particles. *J. Virol.* **2020**. [CrossRef] [PubMed]
28. Hoffmann, E.; Neumann, G.; Kawaoka, Y.; Hobom, G.; Webster, R.G. A DNA transfection system for generation of influenza a virus from eight plasmids. *Proc. Natl. Acad. Sci. USA* **2000**, *97*, 6108–6113. [CrossRef]
29. Steel, J.J.; Henderson, B.R.; Lama, S.B.; Olson, K.E.; Geiss, B.J. Infectious alphavirus production from a simple plasmid transfection. *Virol. J.* **2011**, *8*, 356. [CrossRef]
30. Tretyakova, I.; Nickols, B.; Hidajat, R.; Jokinen, J.; Lukashevich, I.S.; Pushko, P. Plasmid DNA initiates replication of yellow fever vaccine in vitro and elicits virus-specific immune response in mice. *Virology* **2014**, *468*, 28–35. [CrossRef]
31. Tan, C.W.; Tee, H.K.; Lee, M.H.; Sam, I.C.; Chan, Y.F. Enterovirus A71 DNA-launched infectious clone as a robust reverse genetic tool. *PLoS ONE* **2016**, *11*, e0162771. [CrossRef]
32. Almazán, F.; González, J.M.; Pénzes, Z.; Izeta, A.; Calvo, E.; Plana-Durán, J.; Enjuanes, L. Engineering the largest RNA virus genome as an infectious bacterial artificial chromosome. *Proc. Natl. Acad. Sci. USA* **2000**, *97*, 5516–5521. [CrossRef] [PubMed]
33. Li, L.; Pang, H.; Wu, R.; Zhang, Y.; Tan, Y.; Pan, Z. Development of a novel single-step reverse genetics system for the generation of classical swine fever virus. *Arch. Virol.* **2016**, *161*, 1831–1838. [CrossRef] [PubMed]

34. Li, C.; Huang, J.; Li, Y.; He, F.; Li, D.; Sun, Y.; Han, W.; Li, S.; Qiu, H.J. Efficient and stable rescue of classical swine fever virus from cloned cDNA using an RNA polymerase II system. *Arch. Virol.* **2013**, *158*, 901–907. [CrossRef] [PubMed]
35. Yamanaka, A.; Moi, M.L.; Takasaki, T.; Kurane, I.; Matsuda, M.; Suzuki, R.; Konishi, E. Utility of Japanese encephalitis virus subgenomic replicon-based single-round infectious particles as antigens in neutralization tests for Zika virus and three other flaviviruses. *J. Virol. Methods* **2017**, *243*, 164–171. [CrossRef] [PubMed]
36. Suzuki, R.; Ishikawa, T.; Konishi, E.; Matsuda, M.; Watashi, K.; Aizaki, H.; Takasaki, T.; Wakita, T. Production of single-round infectious chimeric flaviviruses with DNA-based Japanese encephalitis virus replicon. *J. Gen. Virol.* **2014**, *95 Pt 1*, 60–65. [CrossRef]
37. Yamanaka, A.; Suzuki, R.; Konishi, E. Evaluation of single-round infectious, chimeric dengue type 1 virus as an antigen for dengue functional antibody assays. *Vaccine* **2014**, *32*, 4289–4295. [CrossRef]
38. Li, W.; Ma, L.; Guo, L.P.; Wang, X.L.; Zhang, J.W.; Bu, Z.G.; Hua, R.H. West Nile virus infectious replicon particles generated using a packaging-restricted cell line is a safe reporter system. *Sci. Rep.* **2017**, *7*, 3286. [CrossRef]
39. Khromykh, A.A.; Varnavski, A.N.; Westaway, E.G. Encapsidation of the flavivirus kunjin replicon RNA by using a complementation system providing Kunjin virus structural proteins in trans. *J. Virol.* **1998**, *72*, 5967–5977. [CrossRef]
40. van Gennip, H.G.; Bouma, A.; van Rijn, P.A.; Widjojoatmodjo, M.N.; Moormann, R.J. Experimental non-transmissible marker vaccines for classical swine fever (CSF) by trans-complementation of E^{rns} or E2 of CSFV. *Vaccine* **2002**, *20*, 1544–1556. [CrossRef]
41. Gehrke, R.; Ecker, M.; Aberle, S.W.; Allison, S.L.; Heinz, F.X.; Mandl, C.W. Incorporation of tick-borne encephalitis virus replicons into virus-like particles by a packaging cell line. *J. Virol.* **2003**, *77*, 8924–8933. [CrossRef]
42. Reimann, I.; Meyers, G.; Beer, M. Trans-complementation of autonomously replicating Bovine viral diarrhea virus replicons with deletions in the E2 coding region. *Virology* **2003**, *307*, 213–227. [CrossRef]
43. Scholle, F.; Girard, Y.A.; Zhao, Q.; Higgs, S.; Mason, P.W. trans-Packaged West Nile virus-like particles: Infectious properties in vitro and in infected mosquito vectors. *J. Virol.* **2004**, *78*, 11605–11614. [CrossRef] [PubMed]
44. Widjojoatmodjo, M.N.; van Gennip, H.G.; Bouma, A.; van Rijn, P.A.; Moormann, R.J. Classical swine fever virus E^{rns} deletion mutants: Trans-complementation and potential use as nontransmissible, modified, live-attenuated marker vaccines. *J. Virol.* **2000**, *74*, 2973–2980. [CrossRef] [PubMed]
45. Maurer, R.; Stettler, P.; Ruggli, N.; Hofmann, M.A.; Tratschin, J.D. Oronasal vaccination with classical swine fever virus (CSFV) replicon particles with either partial or complete deletion of the E2 gene induces partial protection against lethal challenge with highly virulent CSFV. *Vaccine* **2005**, *23*, 3318–3328. [CrossRef]
46. Reimann, I.; Semmler, I.; Beer, M. Packaged replicons of bovine viral diarrhea virus are capable of inducing a protective immune response. *Virology* **2007**, *366*, 377–386. [CrossRef]
47. Chang, D.C.; Liu, W.J.; Anraku, I.; Clark, D.C.; Pollitt, C.C.; Suhrbier, A.; Hall, R.A.; Khromykh, A.A. Single-round infectious particles enhance immunogenicity of a DNA vaccine against West Nile virus. *Nat. Biotechnol.* **2008**, *26*, 571–577. [CrossRef]
48. Miyazaki, J.; Takaki, S.; Araki, K.; Tashiro, F.; Tominaga, A.; Takatsu, K.; Yamamura, K. Expression vector system based on the chicken beta-actin promoter directs efficient production of interleukin-5. *Gene* **1989**, *79*, 269–277.
49. Yamshchikov, V.; Mishin, V.; Cominelli, F. A new strategy in design of +RNA virus infectious clones enabling their stable propagation in *E. coli*. *Virology* **2001**, *281*, 272–280. [CrossRef]
50. Shiu, J.S.; Liu, S.T.; Chang, T.J.; Ho, W.C.; Lai, S.S.; Chang, Y.S. The presence of RNA splicing signals in the cDNA construct of the E2 gene of classical swine fever virus affected its expression. *J. Virol. Methods* **1997**, *69*, 223–230. [CrossRef]
51. Roby, J.A.; Bielefeldt-Ohmann, H.; Prow, N.A.; Chang, D.C.; Hall, R.A.; Khromykh, A.A. Increased expression of capsid protein in trans enhances production of single-round infectious particles by West Nile virus DNA vaccine candidate. *J. Gen. Virol.* **2014**, *95 Pt 10*, 2176–2191. [CrossRef]

52. Martin, A.; Staeheli, P.; Schneider, U. RNA polymerase II-controlled expression of antigenomic RNA enhances the rescue efficacies of two different members of the Mononegavirales independently of the site of viral genome replication. *J. Virol.* **2006**, *80*, 5708–5715. [CrossRef] [PubMed]
53. Kanai, Y.; Kawagishi, T.; Nouda, R.; Onishi, M.; Pannacha, P.; Nurdin, J.A.; Nomura, K.; Matsuura, Y.; Kobayashi, T. Development of stable rotavirus reporter expression systems. *J. Virol.* **2019**, *93*. [CrossRef] [PubMed]

MDPI
St. Alban-Anlage 66
4052 Basel
Switzerland
Tel. +41 61 683 77 34
Fax +41 61 302 89 18
www.mdpi.com

Viruses Editorial Office
E-mail: viruses@mdpi.com
www.mdpi.com/journal/viruses

www.ingramcontent.com/pod-product-compliance
Lightning Source LLC
LaVergne TN
LVHW070145100526
838202LV00015B/1895
* 9 7 8 3 0 3 9 4 3 5 7 7 7 *